Allosterism in Drug Discovery

RSC Drug Discovery Series

Titles in the Series:

1: Metabolism, Pharmacokinetics and Toxicity of Functional Groups
2: Emerging Drugs and Targets for Alzheimer's Disease; Volume 1
3: Emerging Drugs and Targets for Alzheimer's Disease; Volume 2
4: Accounts in Drug Discovery
5: New Frontiers in Chemical Biology
6: Animal Models for Neurodegenerative Disease
7: Neurodegeneration
8: G Protein-Coupled Receptors
9: Pharmaceutical Process Development
10: Extracellular and Intracellular Signaling
11: New Synthetic Technologies in Medicinal Chemistry
12: New Horizons in Predictive Toxicology
13: Drug Design Strategies: Quantitative Approaches
14: Neglected Diseases and Drug Discovery
15: Biomedical Imaging
16: Pharmaceutical Salts and Cocrystals
17: Polyamine Drug Discovery
18: Proteinases as Drug Targets
19: Kinase Drug Discovery
20: Drug Design Strategies: Computational Techniques and Applications
21: Designing Multi-Target Drugs
22: Nanostructured Biomaterials for Overcoming Biological Barriers
23: Physico-Chemical and Computational Approaches to Drug Discovery
24: Biomarkers for Traumatic Brain Injury
25: Drug Discovery from Natural Products
26: Anti-Inflammatory Drug Discovery
27: New Therapeutic Strategies for Type 2 Diabetes: Small Molecules
28: Drug Discovery for Psychiatric Disorders
29: Organic Chemistry of Drug Degradation
30: Computational Approaches to Nuclear Receptors
31: Traditional Chinese Medicine

32: Successful Strategies for the Discovery of Antiviral Drugs
33: Comprehensive Biomarker Discovery and Validation for Clinical Application
34: Emerging Drugs and Targets for Parkinson's Disease
35: Pain Therapeutics; Current and Future Treatment Paradigms
36: Biotherapeutics: Recent Developments using Chemical and Molecular Biology
37: Inhibitors of Molecular Chaperones as Therapeutic Agents
38: Orphan Drugs and Rare Diseases
39: Ion Channel Drug Discovery
40: Macrocycles in Drug Discovery
41: Human-based Systems for Translational Research
42: Venoms to Drugs: Venom as a Source for the Development of Human Therapeutics
43: Carbohydrates in Drug Design and Discovery
44:Drug Discovery for Schizophrenia
45: Cardiovascular and Metabolic Disease: Scientific Discoveries and New Therapies
46: Green Chemistry Strategies for Drug Discovery
47: Fragment-Based Drug Discovery
48: Epigenetics for Drug Discovery
49: New Horizons in Predictive Drug Metabolism and Pharmacokinetics
50: Privileged Scaffolds in Medicinal Chemistry: Design, Synthesis, Evaluation
51: Nanomedicines: Design, Delivery and Detection
52: Synthetic Methods in Drug Discovery: Volume 1
53: Synthetic Methods in Drug Discovery: Volume 2
54: Drug Transporters: Role and Importance in ADME and Drug Development
55: Drug Transporters: Recent Advances and Emerging Technologies
56: Allosterism in Drug Discovery

How to obtain future titles on publication:
A standing order plan is available for this series. A standing order will bring delivery of each new volume immediately on publication.

For further information please contact:
Book Sales Department, Royal Society of Chemistry, Thomas Graham House, Science Park, Milton Road, Cambridge, CB4 0WF, UK
Telephone: +44 (0)1223 420066, Fax: +44 (0)1223 420247,
Email: booksales@rsc.org
Visit our website at www.rsc.org/books

Allosterism in Drug Discovery

Edited by

Dario Doller
Alcyoneus/ScienceWorks, Sparta, New Jersey, USA
Email: dario_doller@yahoo.com

ROYAL SOCIETY OF CHEMISTRY

THE QUEEN'S AWARDS FOR ENTERPRISE: INTERNATIONAL TRADE 2013

RSC Drug Discovery Series No. 56

Print ISBN: 978-1-78262-459-2
PDF eISBN: 978-1-78262-927-6
EPUB eISBN: 978-1-78262-969-6
ISSN: 2041-3203

A catalogue record for this book is available from the British Library

Published by The Royal Society of Chemistry,
Thomas Graham House, Science Park, Milton Road,
Cambridge CB4 0WF, UK

Registered Charity Number 207890

For further information see our web site at www.rsc.org

Printed in the United Kingdom by CPI Group (UK) Ltd, Croydon, CR0 4YY, UK

Preface

I went to sleep one evening, very tired after a long, cold winter day in the hills of northwestern New Jersey. I felt like I could sleep for a thousand years... and then, like thunder, the thought struck my mind: if I did just that and woke up in the 30th century, what would drug discovery be like? How would it have changed? What diseases would still be incurable, and which ones will not exist anymore? And how would our current efforts be seen, in retrospect? I took the precaution of making a note of that idea, so that I would not forget (as it has happened in the past) and I could work on it the next morning.

Drug discovery is an interesting human endeavor in that nobody knows exactly how to succeed. We try different strategies hoping that the next time we will get it right. We swung from the purported, process-centered "industrialization" of the late 1990's to the explosion of innovation of the 2010's. When I began my pharmaceutical research career we worked in ultra-secretive laboratories in buildings without windows to preclude anyone walking outside from getting the slightest idea of what we were doing indoors. In contrast, openness is now flourishing and precompetitive collaborations are a necessity today, as alliances are forming at a high pace to integrate emerging knowledge generated in academia into industrial new drug projects. This interdependence among different sectors in the "Life Science Ecosystem" is key to the future of drug discovery.

As a testament to that spirit, this book is co-authored by scientists from academia and industry, roughly in equal proportion. And we, the co-authors, come from many different scientific fields, such as molecular pharmacology, cell biology, medicinal chemistry, computational chemistry, and chemical biology. Working together we are trying to make the future come faster. This spirit of collaboration should not be taken for granted, and I sincerely hope it is nurtured and developed for the next thousand years by future generations of scientists.

RSC Drug Discovery Series No. 56
Allosterism in Drug Discovery
Edited by Dario Doller

Published by the Royal Society of Chemistry, www.rsc.org

This book aims to be a time capsule documenting the initial stages of applying allosteric modulation principles to drug discovery. Allosterism has been known for some 50 years. However, it is relatively recently that we started trying to incorporate it systematically into our arsenal of drug modalities. As allosterism is broadly seen in nature, we are presenting examples from the work of experts in areas such as enzymes, transporters, ion channels, and G-protein-coupled receptors. I suspect allosterism will become a stepping stone in our journey to understanding human life at the molecular level.

Diversity of thought is an important component in the creation of new scientific knowledge. All authors have been challenged to present their most advanced views in their field of expertise. They were also given complete freedom to express their perspectives in areas that are constantly expanding. Every single one of them took the risk, and all deserve credit for doing so. With that in mind, I hope the reader will gain diverse, and perhaps not always converging, insight coming from many different perspectives gained throughout our efforts to understand allosterism. In editing this work, I have aimed to preserve a feeling that these are our current best approaches to study allosterism in drug discovery, but knowing that we still have a lot to learn. Indeed, I personally have greatly broadened my own views on the topic and learned a lot from editing the input to this book and related scientific reports.

So, whether you are reading this in 2016 or in 3016, you should know that we, the co-authors, came together to get to the future as soon as possible. As long as the time to deliver a new drug treatment is measured in decades, every difference than we can make along the way could have tremendous impact on the lives of patients and their families. That is the sacrament of those working in drug discovery, always.

I am grateful to my fellow co-authors for joining me in this endeavor, which I hope they found rewarding and worthy of their time and efforts. I thank Harriet Manning, Rowan Frame, and Antonia Pass, from the Royal Society of Chemistry, for their guidance. Special thanks to Dr Dave Rotella – an excellent medicinal chemist and colleague – for the opportunity he gave me to lead this effort and interact together.

Finally, and humbly, I would like to acknowledge five mentors who had major impact in my professional and personal life over a number of years. In chronological order, the late Professor Eduardo G. Gros, the late Professor Derek H. R. Barton, Dr Bill Greenlee, Dr Frank Larsen, and Dr Al Robichaud supported, challenged, and inspired me. Without them, I would not be where I am today. Wherever you are, I hope you see a part of yourselves reflected in this work.

Dario Doller

With admiration, to all drug hunters who pursue Nature's truths

To my wife Cecilia, our daughter Miranda, and our dogs, Willie Nelson, and Ricky, who make our home my place in the World

Contents

RSC Drug Discovery Series No. 56
Allosterism in Drug Discovery
Edited by Dario Doller

Published by the Royal Society of Chemistry, www.rsc.org

CHAPTER 1

Modulation of Biological Targets Using Allosteric Ligands: Food for Thought

DARIO DOLLER*[a] AND XINYAN HUANG[b]

[a]Alcyoneus/ScienceWorks, Sparta, NJ 07871, USA; [b]Office of Therapeutics Alliances, New York University Langone Medical Center, One Park Avenue, 6th Floor, New York, NY 10016, USA
*E-mail: dario_doller@yahoo.com

1.1 Drug Discovery in the Early 21st Century

The dawn of the new century found several major pharmaceutical companies facing significant challenges in keeping their pipelines populated with new drugs. For nearly two decades, robust increases in research and development investments had not produced the hoped for number of new medical entities (NMEs) approvals per year. In some therapeutic areas such as diseases of the central nervous system (CNS), and in spite of a growing patient population and major medical needs, only a handful of new drugs were approved during this period, and mostly based on mechanisms already known ("me too" drugs). As a response to this crisis of innovation, and propelled by major discoveries in molecular and cell biology research, in particular in the areas of oncology and immunology, a chemically broad range of new therapeutic modalities emerged, such as antibodies, proteins, nucleic acids, vaccines,

RSC Drug Discovery Series No. 56
Allosterism in Drug Discovery
Edited by Dario Doller

Published by the Royal Society of Chemistry, www.rsc.org

cell, or gene therapies. These are generally grouped under the denomination "biologics", to differentiate them from "small molecules". Drugs in the latter group have historically been mostly thought of as acting competitively with endogenous ligands, binding at their target receptors in a model that became known as "lock and key".[1] These two types of drugs differ in terms of a number of attributes, such as molecular weight, preparation, characterization of the active pharmaceutical ingredient (API), physicochemical properties, possible route of administration, pharmacokinetic distribution, metabolism, clearance mechanisms, drug–drug interactions, dosing regimen, safety and toxicology, antigenicity and hypersensitivity, and pharmacology – including side effects.[2,3] A comparison of these attributes highlights some of the challenges associated with the development of biologics as drugs. (Table 1.1) It also begs the question whether new strategies could be discovered to modulate drug targets with a *different* type of small molecules. This is, in essence, the promise of what allosteric modulators might do.

1.2 Allostery: A 50-Year Old Concept

Broadly, allostery has been defined as an indirect interaction between topographically distinct binding sites in a protein, mediated by a conformational change. Monod and Jacob used the term for the first time in the printed version of the Proceedings of the 26th Cold Spring Harbor Symposium on

Table 1.1 Comparison of key attributes for small molecules and biologics as drugs.[2,3]

	Small molecule drugs	Biological drugs
Size	- Small (single molecule)	- Large (mixture of related molecules)
	- Low molecular weight (<1000 amu)	- High molecular weight (>1000 amu)
Structure	Simple, well defined, independent of manufacturing process	Complex (heterogeneous), defined by the exact manufacturing process
Modification	Well defined	Many options
Manufacturing	- Produced by chemical synthesis	- Produced in living cell culture
	- Predictable chemical process	- Difficult to control from starting material to final API
	- Identical copies can be made	- Currently very difficult to ensure identical copies
	- Low cost of goods	- High cost to produce
Characterization	Easy to characterize completely at chemical and physical levels	Not completely characterized in terms of chemical composition and heterogeneity
Stability	Stable	Unstable, sensitive to external conditions
Immunogenicity	Mostly non-immunogenic	Immunogenic

Quantitative Biology entitled "Cellular Regulatory Mechanisms" in 1961, in the general discussion written as a conclusion of the conference. The earliest appearance of the term "allostery" in a publication listed in PubMed dates to more than half a century ago.[4,5] Since then, the concept has evolved and its use expanded notably. Today, allosterism is considered an inherent property of all dynamic proteins and other macromolecules of biological relevance.[6] Examples of important drugs exist that are allosteric modulators of diverse targets such as ion channels, G protein-coupled receptors (GPCRs), and enzymes. Benzodiazepines, one of the earlier drug classes with major impact on society, act as negative (flumazenil, **1**) or positive (diazepam, **2**; flunitrazepam, **3**) allosteric modulators of gamma aminobutyric acid A ($GABA_A$) ion channels. The first inhibitors of human immunodeficiency virus reverse transcriptase enzyme (HIVRT) useful in the treatment of AIDS were nucleoside analogs. These compounds act competitively and inhibit the binding of the appropriate deoxynucleotide triphosphate (ATP, TTP, GTP, or CTP) to the reverse transcriptase (RT) enzyme. Nevirapine (**4**) and efavirenz (**5**) are non-nucleoside RT inhibitors (NNRTIs). Both nucleoside and non-nucleoside RTIs inhibit the same target, the reverse transcriptase enzyme, an essential viral enzyme which transcribes viral RNA into DNA. Unlike nucleoside RTIs, which bind at the enzyme's active site, NNRTIs bind allosterically at a distinct site away from the active site called the NNRTI pocket. Maraviroc (**6**) is another antiretroviral drug used in the treatment of HIV infection, which works through a negative allosteric modulation (NAM) mechanism inhibiting the cell-surface chemokine CCR5 receptor. The chemokine receptor CCR5 is a co-receptor for most HIV strains, necessary for the entrance of the virus into the host cell. The drug binds to CCR5 and changes its conformation, thus blocking the HIV protein gp120 from associating with the receptor. HIV is then unable to enter human macrophages and T-cells. Cinacalcet (**7**), a calcimimetic, is a positive allosteric modulator (PAM) of the calcium-sensing receptor (CaSR), which is activated according to the natural rise and fall of endogenous calcium ion levels. This drug acts by lowering the threshold for activation of feedback on the parathyroid chief cells. Trametinib (**8**) is a highly selective reversible allosteric inhibitor of mitogen-activated protein kinase (MAPK)–extracellular-signal-regulated kinase enzymes MEK1 and MEK2 activity. Compound **8** is an ATP non-competitive inhibitor that binds MEK at a location adjacent to the ATP binding site. Approved in 2013, it is a first-in-class oral treatment for metastatic melanoma with BRAF V600 mutations (Figure 1.1).[7,8]

In addition to these marketed drugs, a number of compounds working through allosteric mechanisms are undergoing advanced clinical trials at the time of writing. The serine–threonine kinase AK thyoma oncoprotein homolog (Akt), also known as protein kinase B (PKB), plays a key role in the phosphoinositide 3-kinase–Akt–mammalian target of rapamycin (PI3K–Akt–mTOR) signaling cascade, tightly associated with cell proliferation, survival, migration, angiogenesis and other activities directly linked to tumor genesis and progression. The allosteric Akt inhibitor MK-2206 (**9**) is currently being

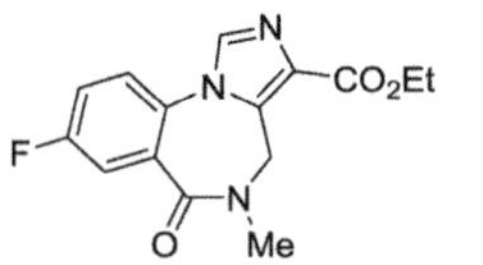

Flumazenil (**1**)
Anexate®

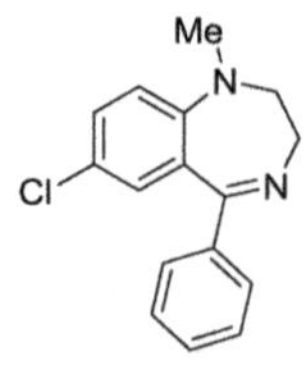

Diazepam (**2**)
Valium®

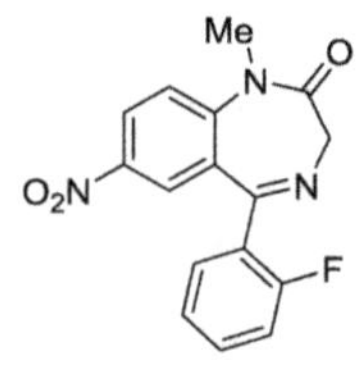

Flunitrazepam (**3**)
Rohypnol®

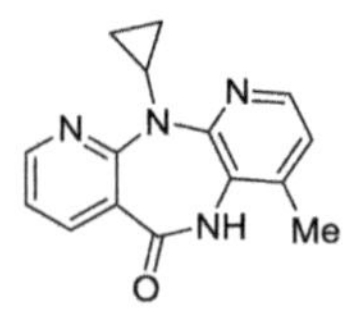

Nevirapine (**4**)
Viramune®

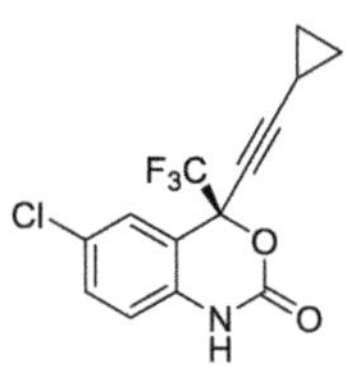

Efavirenz (**5**)
Sustiva®

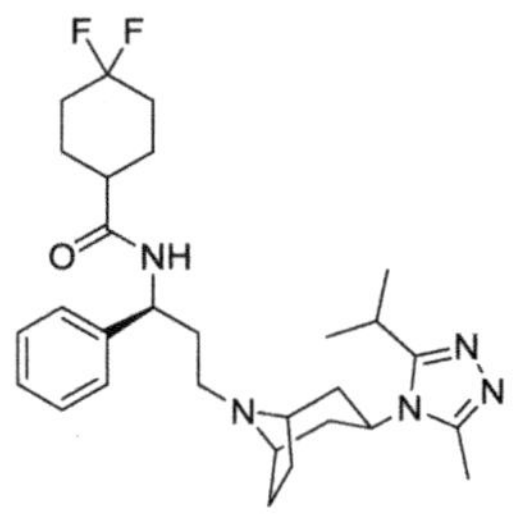

Maraviroc (**6**)
Selzentry®

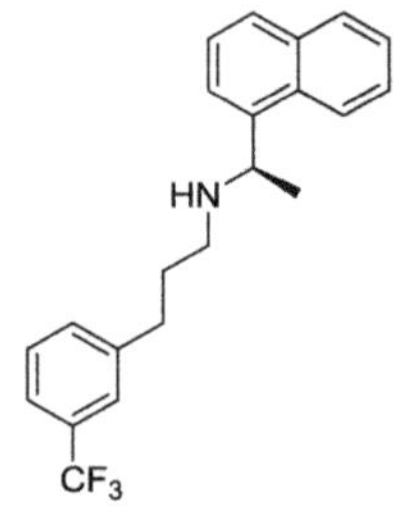

Cinacalcet (**7**)
Sensipar®

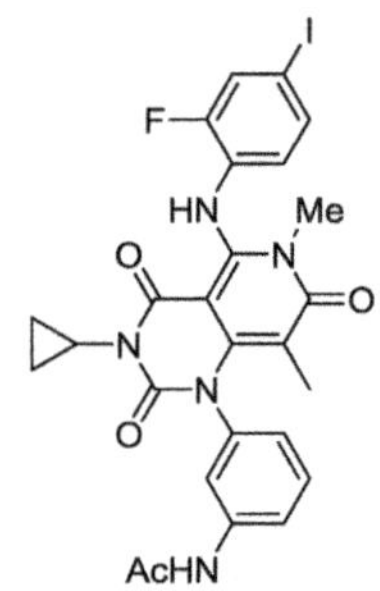

Trametinib (**8**)
Mekinist®

Figure 1.1 Some examples of marketed drugs that work by allosteric mechanisms at diverse biological targets such as GPCRs, ion channels, and enzymes.

evaluated in clinical Phase 1 and 2 studies on breast, colon/rectal and other cancers.[9] The muscarinic M1 receptor positive allosteric modulator (PAM) 1-(4-*cyano*-4-(pyridine-2-yl)piperidin-1-yl)methyl-4-oxo-4*H*-quinolizine-3-carboxylic acid (PQCA) (**10**) has shown *in vivo* efficacy in a number of preclinical models in both rodents and non-human primates, suggesting this mechanism may have therapeutic potential for the treatment of Alzheimer's disease and cognitive deficits.[10] The metabotropic glutamate receptor 4 (mGluR4) PAM Lu AF21934 (**11**) has been extensively studied in a number of *in vitro* and *in vivo* mechanistic tests. It differentiates from congeners by its remarkable ability to enhance *in vitro* agonist concentration–response curves for both, the mGluR4 homodimer as well as the mGlu2/4 heterodimer, as opposed to the earlier tool compounds, which are functionally selective for the mGluR4 homodimer (Figure 1.2).[11]

In some cases, strong support was obtained using X-ray crystallography to enable the discovery of allosteric drugs. Phosphodiesterase 4 (PDE4) is an enzyme that hydrolyzes cAMP in cells, and a promising drug target for the treatment of Alzheimer's disease, Huntington's disease, schizophrenia, and depression. Drugs treating these diseases by aiming at central biological targets must cross the blood–brain barrier (BBB) and penetrate into the brain to exert therapeutic benefit, which restricts the available chemical space for potential competitive inhibitors. PDE4 inhibitors exist which are cAMP-competitive marketed drugs, but the lack of subtype selectivity is thought to lead

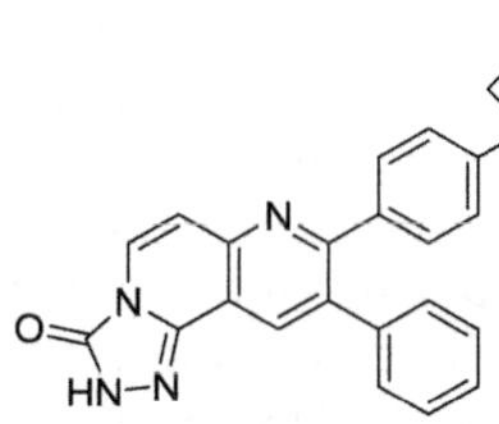

MK-2206 (**9**)
PKB inhibitor

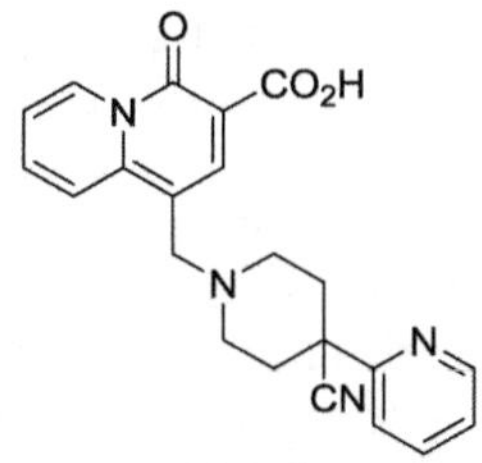

PQCA (**10**)
M1 PAM

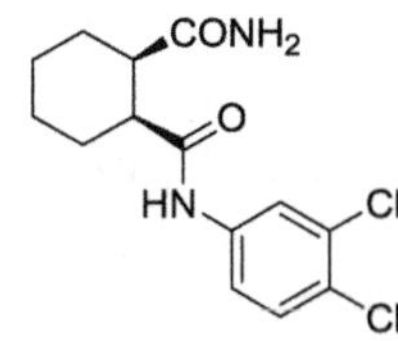

Lu AF21934 (**11**)
mGluR4 PAM

Figure 1.2 Examples of allosteric modulators used as chemical tools to decipher the fundamental biology of novel targets and compounds presently undergoing clinical investigation.

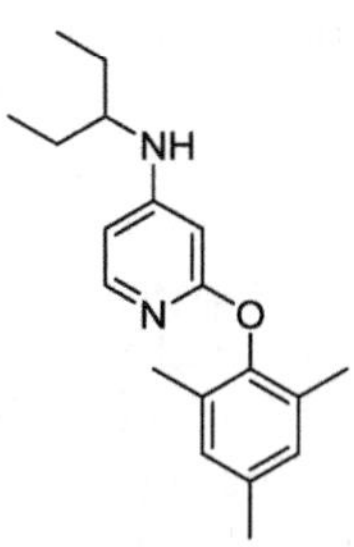

CP-376395 (**12**)

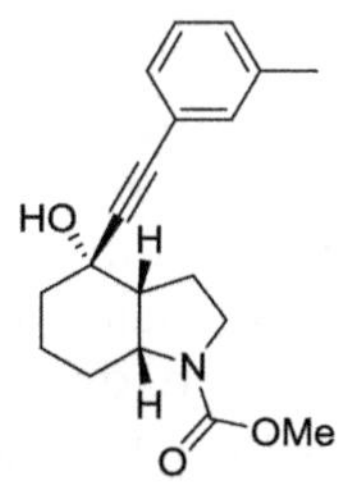

Mavoglurant (**13**)

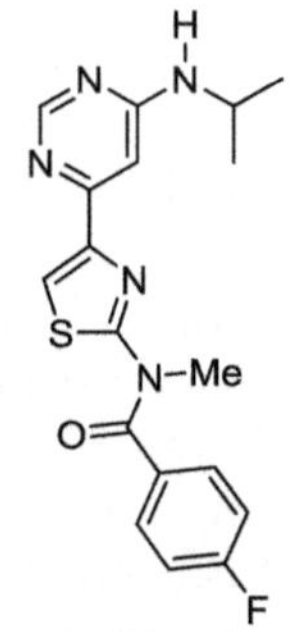

FITM (**14**)

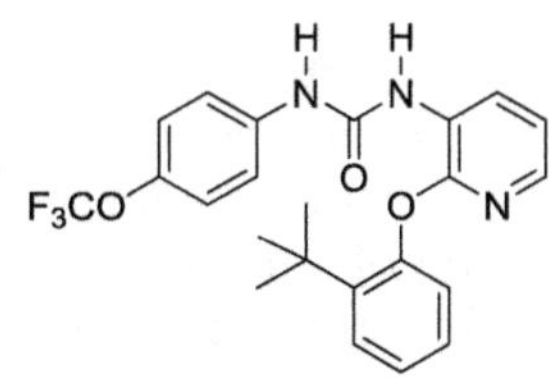

BPTU (**15**)

Figure 1.3 Some of the allosteric modulators for which X-ray crystal structures bound to their biological targets have been reported.

to side effects, including nausea, depression, and weight loss. Studying crystal structures of PDE4 with bound inhibitors exposed the structural basis of this enzyme's regulation and inspired, together with mutagenesis and kinetic studies, the design of subtype-selective negative allosteric modulators of PDE4D with reduced potential to cause the dose-limiting side effects observed with existing active-site-directed PDE4 inhibitors.[12] BPN14770, a first in class PDE4D NAM (structure not disclosed at present time) is being developed with support from the NIH Blueprint Neurotherapeutics Network and is the first compound funded by the program to reach a Phase 1 clinical trial.[13] Likewise, the recent disclosure of crystallographic studies with GPCR ligands confirmed their allosteric nature. Notable examples are the corticotropin-releasing factor-1 (CRF-1) functional inhibitor CP-376395 (**12**),[14] the mGluR5 NAM mavoglurant (**13**),[15] the mGluR1 NAM FITM (4-fluoro-*N*-(4-(6-(isopropylamino)pyrimidin-4yl)thiazol-2-yl)-*N*-methylbenzamide) (**14**),[16] and the $P2Y_1$ receptor non-nucleotide antagonist BPTU (1-(2-(2-(*tert*-butyl)phenoxy)pyridin-3-yl)-3-(4-(trifluoromethoxy)phenyl)urea) (**15**). The latter has the peculiarity of being the first structurally characterized selective GPCR ligand located entirely outside of the helical bundle (Figure 1.3).[17]

These last several examples strongly suggest that the use of allosteric ligands has the potential to help solve druggability issues in areas where ligands binding at the orthosteric site present challenges, thus providing a new chemistry modality to deliver therapeutic treatments using small molecules, with which chemists are highly familiar working. Yet, the purposeful targeting of allosteric sites in proteins that are biological targets of drug discovery efforts is a relatively recent undertaking in drug discovery projects.

1.3 Allosteric Drugs: The Right Tool at the Right Time

The pharmaceutical industry has witnessed a major change over the last decade, as it went from working on relatively well-understood biological targets with clinical validation, where efficacy in the patient population was expected, to novel biological targets with unprecedented track records in clinical settings. The reasons for this change are not only scientific, and include social, political, and financial factors, which are beyond the scope of this book. The bottom-line is that today, unmet medical needs and financial incentives dictate the discovery of drugs with enhanced efficacy over existing treatments. A way to achieve this goal is by using novel mechanisms of action. However, the novelty entails a risky task: demonstrating clinical efficacy in a broad patient population, usually requiring Phase 3 clinical studies. Late failures are very expensive, and while they might be explicated by our imperfect understanding of disease etiology and flawed replication of human (patho)physiology in the preclinical disease models used, that is no consolation after a billion-dollar drug development exercise yields negative clinical data. The current premise is that translational sciences will, at least in part, mitigate this risk and enhance success rates.

Allosteric drugs were envisioned as having the potential to provide two ways to mitigate the risk of late clinical failure:

1. By providing an alternative molecular mechanism to attack biological targets (clinically validated or not) that require a different, friendlier drug-like chemistry space than originally attempted with orthosteric ligands (*e.g.*, GABA, glutamate or acetylcholine receptors); and
2. Giving novel ways to attack previously intractable targets (*e.g.*, the calcium sensing receptors or sweet taste receptors; voltage gated receptors).

Therefore, it seemed reasonable that some of the same biological targets previously aimed at, and which had been proven recalcitrant to drug discovery efforts using orthosteric ligands, yet not technically invalidated,[18] were given a second life and addressed using allosteric ligands.

1.4 Potential Advantages of Allosteric Modulators Over Orthosteric Ligands… or are They?

A number of excellent publications have discussed in great detail the potential advantages of using allosteric ligands in drug discovery, including comparisons with their orthosteric counterparts.[19–23] A review of this literature consistently points to a variety of factors as drivers to encourage the exploration of allosteric ligands as drugs and chemical probes. These include factors such as: being easier to achieve selectivity, lack of intrinsic activity, ability to "dim" endogenous agonist function while preserving spatial and temporal effects, *etc.* (Figure 1.4). Some of these initial views have evolved as they were challenged by experimental data, suggesting a reconsideration of these idealized properties.

Target selectivity is generally thought to be easier to obtain with allosteric ligands than with orthosteric ligands. This belief would be a consequence of a highly conserved amino acid sequence in the orthosteric binding site. Remote allosteric sites tend to be less well conserved, and sufficiently different that selectivity for allosteric ligands may be expected. For example, when comparing different mGluR subtypes binding the same endogenous agonist (glutamate), the amino acids forming the binding site are indeed highly conserved.[24,25] Furthermore, these allosteric sites are thought to be an evolutionary artifact and less sensitive to mutations in terms of the receptor functional output, as there would be no endogenous ligands binding at such sites. This would make the design of selective orthosteric ligands somewhat challenging.

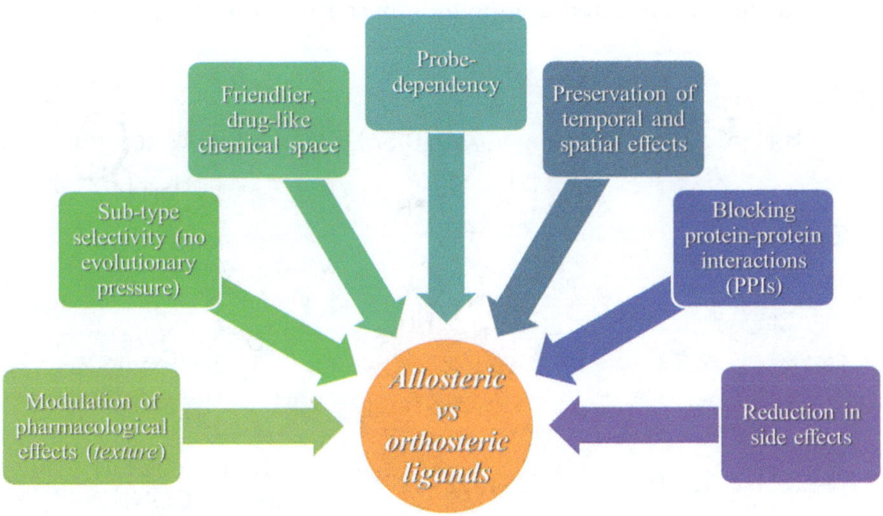

Figure 1.4 Potential favorable attributes supporting the use of allosteric ligands compared with their orthosteric counterparts.

However, this thinking may in part be a consequence of how orthosteric analogs were conceived in the past. Historically, these ligands were often designed by constraining the conformation of the endogenous natural ligand and keeping a similar ligand size. Thus, the same amino acids are involved in the binding process of both the endogenous ligand and the putative drug. Indeed, recent work has shown that orthosteric ligand selectivity may be achieved by making analogs with added substituents, so that additional amino acids that may no longer be conserved among receptor subtypes are involved in the binding process. Examples of such selective ligands are LSP4-2022 (**16**,[26] mGluR4-selective) and LY2812223 (**17**,[27] mGluR2-selective) (Figure 1.5; glutamate backbone shown in color). Furthermore, evidence suggesting that allosteric sites might play a role in determining receptor function came from the study of the $GABA_A$ receptor, and the discovery of endogenous ligands binding at the benzodiazepine site, including oleamides, nonpeptidic endozepines and the protein diazepam-binding inhibitor (DBI).[28]

Potent and selective adenosine A_1R receptor orthosteric agonists and antagonists have also been discovered and tested in clinical settings for a number of indications.[29] The use of bulky N^6-adamantyl substituents on adenosine has led to agonists previously shown to be A_1R-selective with respect to binding affinity at rat receptors.[30] Recently, the discovery of a series of subtype-selective N^6-bicyclic and N^6-(2-hydroxy)cyclopentyl derivatives of adenosine as agonists was reported and their A_1R/A_2R selectivity assessed based on a modified *Saccharomyces cerevisiae* strains screening platform. In addition, the activity at A_1R and A_3R of a few select analogs which were active against the A_1R in the yeast screen was studied in mammalian CHOK1 cells; compounds such as **18** and **19** (Figure 1.5) were shown to be highly selective A_1R agonists. Preferred compounds based on potency and selectivity

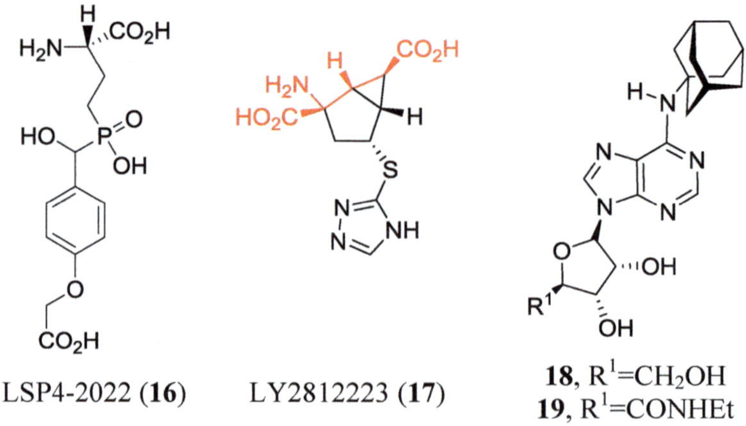

LSP4-2022 (**16**) LY2812223 (**17**) **18**, R^1=CH_2OH **19**, R^1=CONHEt

Figure 1.5 Examples of selective orthosteric agonists for GPCRs of pharmaceutical relevance.

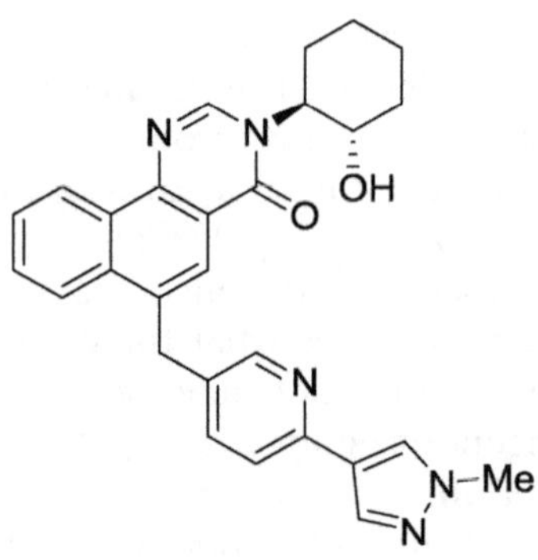

20

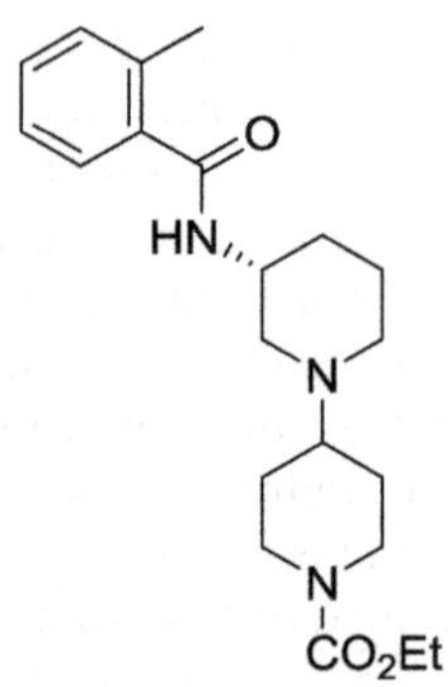

21

Figure 1.6 Structures of M1-selective compounds with mixed ago-PAM profiles.

contained N^6-adamantyl substitution in combination with 5′-*N*-ethylcarboxamido or 5′-hydroxymethyl groups.[31]

In some cases, allosteric ligands have provided a path forward to selectively activate CNS receptors when orthosteric ligands failed to do so. One such example is the use of compounds **10**, **20** and **21** (Figure 1.6), characterized as having a mixed agonist–positive allosteric modulator profile highly functionally selective for the M1 receptor. M1 agonists such as xanomeline produce beneficial cognitive effects in Alzheimer's disease and schizophrenia patients. Unfortunately, its therapeutic use is limited due to cholinergic side effects (sweating, salivation, gastrointestinal distress). These are suggested to result from nonselective activation of other muscarinic receptor subtypes such as M2 and M3. However, this theory was refuted as the M1-selective activators **10**, **20** and **21** tested in rats, dogs, and cynomologous monkeys still elicited cholinergic side effects such as salivation, diarrhea, and emesis.[32] This suggests that activation of M1 receptors alone is sufficient to produce unwanted cholinergic side effects such as those seen with xanomeline.

Another example of an idealized property of allosteric modulators is that they frequently exhibit little or no intrinsic activity, since their mode of action is to enhance or inhibit the action of the endogenous agonist.[33] The observation of agonist-independent activity (allosteric agonism) by allosteric modulators is a rather complex phenomenon. Studies have shown that the specific nature of the biological system under interrogation may play a role in the functional response observed to the same PAM. For example, higher levels of receptor expression in a cell line under investigation may facilitate the observation of intrinsic activity, as was reported for the case of the glucagon-like peptide 1 (GLP-1) receptor (*vide infra*).[34] Small modifications to the chemical structure of an allosteric potentiator may provide compounds in the same chemotype with varying degrees of agonist-independent activity, including none. Thus, distinct drug target product profiles may be defined. Allosteric agonists or "ago-PAMs" are compounds with intrinsic activity independent of the endogenous ligand. On the other hand, "pure PAMs" show activity only in the presence of the orthosteric agonist and no activity in its absence. In any case, medicinal chemistry optimization to a preferred target product profile is feasible.[35]

Allosteric ligands are also thought to allow fine tuning of functional responses. While this may be the case for properly designed and characterized PAMs, most notably for mGluR5 PAMs,[36,37] it is also possible to achieve super-physiological functional responses, above what the endogenous natural system may deliver.[38] This design attribute may be an advantage in diseases where the tone of the endogenous agonist is high but the receptor functional response needs augmentation.[39]

As acknowledged by Dougall and Unitt, the "potential advantages of allosteric modulators remain largely theoretical as very few such agents have to date reached the market".[40] About a decade into conducting research with the intent to engineer certain attributes into allosteric drugs using structure–activity relationships (SAR) as has been done with orthosteric ligands, it seems prudent to maintain some level of skepticism regarding the extent to which this potential has been realized.

1.5 Looking Under the Hood

Once the idea of modulating the functional activity of a biological target using allosteric drugs found support in reliable early experimental data, it also became clear than the concept was more complex than initially thought. Empirically, differences could be found between developing SAR for allosteric compounds that are inhibitors (NAMs) or enhancers (PAMs) of biological target function. Furthermore, subtler aspects further differentiate different types of allosteric enhancers. For enzymes, K-type modulators affect the affinity of the substrate, while V-type modulators affect the catalytic rate. Distinct cases (Type II, Types Ia and Ib) generate unique kinetic signatures which may be of diagnostic value.[41] For membrane-bound targets (*e.g.*, GPCRs, transporters), some compounds may act by enhancing the affinity of the orthosteric ligand (K_d), others by increasing the efficacy at the receptor (E_{max}), or by acting as allosteric agonists activating the receptor in the absence of orthosteric agonist (*e.g.*, Figure 1.7). In practice, compounds with a combination of these modes of action are often encountered.[42] Indeed, examples of allosteric modulators exist which enhance the binding affinity of the orthosteric ligand (attribute of a PAM) while reducing the receptor efficacy (attribute of a NAM).[43]

One of the models used to describe the effect of an allosteric ligand on the functional response elicited by the action of an orthosteric ligand at a receptor is the Operational Model.[44] Mathematically, this model can be represented by eqn (1.1), where pK_A and pK_B are the affinity of orthosteric and allosteric ligand for the free receptor, respectively; α is the affinity cooperativity factor, β is the efficacy cooperativity factor, τ_A and τ_B represent the agonist and modulator efficacy, and n is a scaling factor.[45] Indeed, the simulations shown in Figure 1.7 were generated using this equation. Recommendations for its proper application have been discussed.[46]

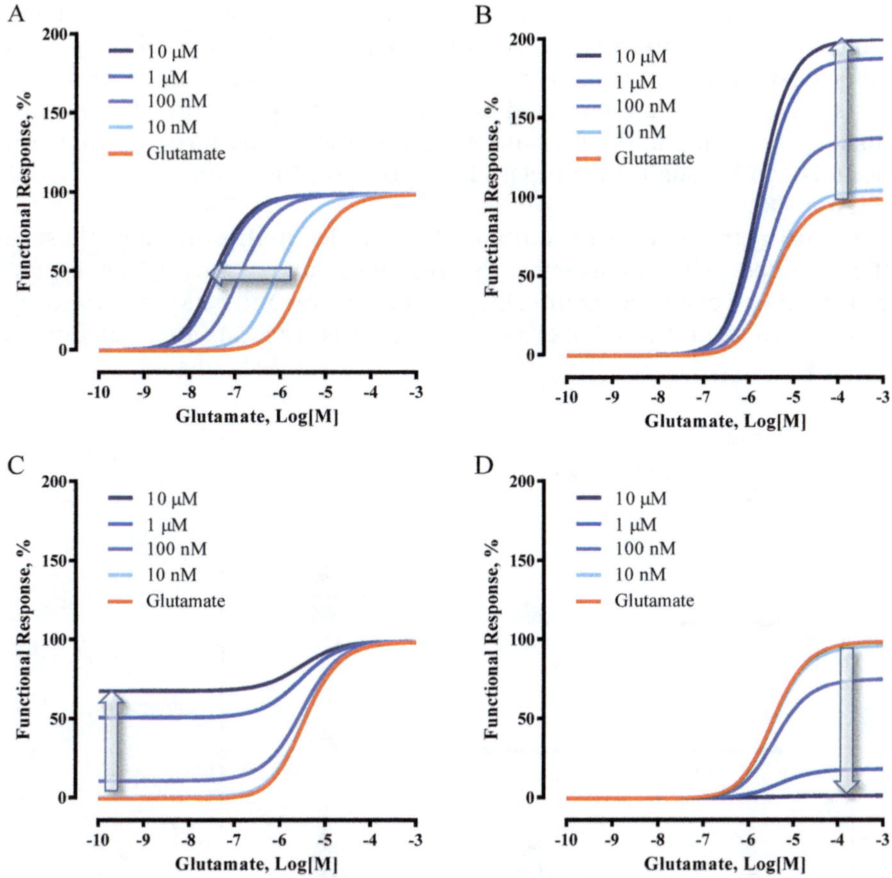

Figure 1.7 Graphical depiction of concentration–response curves for different types of modulators of GPCR function, exemplified for the metabotropic glutamate receptor 4 (mGluR4). Simulations computed using the following parameters: $E_m = 268$; $pK_A = 5.5$; $pK_B = 6.5$; $\tau_A = 0.7$; $n = 1.5$. Ligand concentrations are as indicated in the legends. A: $\alpha = 100$; $\beta = 1$; $\tau_B = 0$. B: $\alpha = 1$; $\beta = 3$; $\tau_B = 0$. C: $\alpha = \beta = 1$; $\tau_B = 0.5$. D: $\alpha = 1$; $\beta = 0.01$; $\tau_B = 0$.

$$E = \frac{E_m\left(\tau_A[A]\left(K_B + \alpha\beta[B]\right) + \tau_B[B]K_A\right)^n}{\left([A]K_B + K_AK_B + K_A[B] + \alpha[A][B]\right)^n + \left(\tau_A[A]\left(K_B + \alpha\beta[B]\right) + \tau_B[B]K_A\right)^n} \tag{1.1}$$

A protocol known as "triple addition" has become broadly used in allosteric drug research.[47] This methodology enables the fast and efficient screening of compound libraries in projects seeking to identify novel modulators of receptor function by exploring in a single assay the effects of compound alone (which identifies new agonists, orthosteric or allosteric), followed by orthosteric agonist at two concentrations. A 20% maximal effective concentration (EC_{20}) identifies PAMs, and this is followed by an 80% maximal effective concentration (EC_{80}) to identify NAMs or antagonists. This screen is conducted at a single test compound concentration (usually 10 μM), and delivers different potentiation patterns, so that compounds can be qualitatively classified based on their functional response (Figure 1.8).

To qualify the potency of compounds identified in this first step, these are then studied using a concentration–response curve (CRC). PAMs are typically screened at a fixed agonist EC_{20} concentration and NAMs at a fixed agonist EC_{80} concentration. For example, for the case of PAMs, these data-points

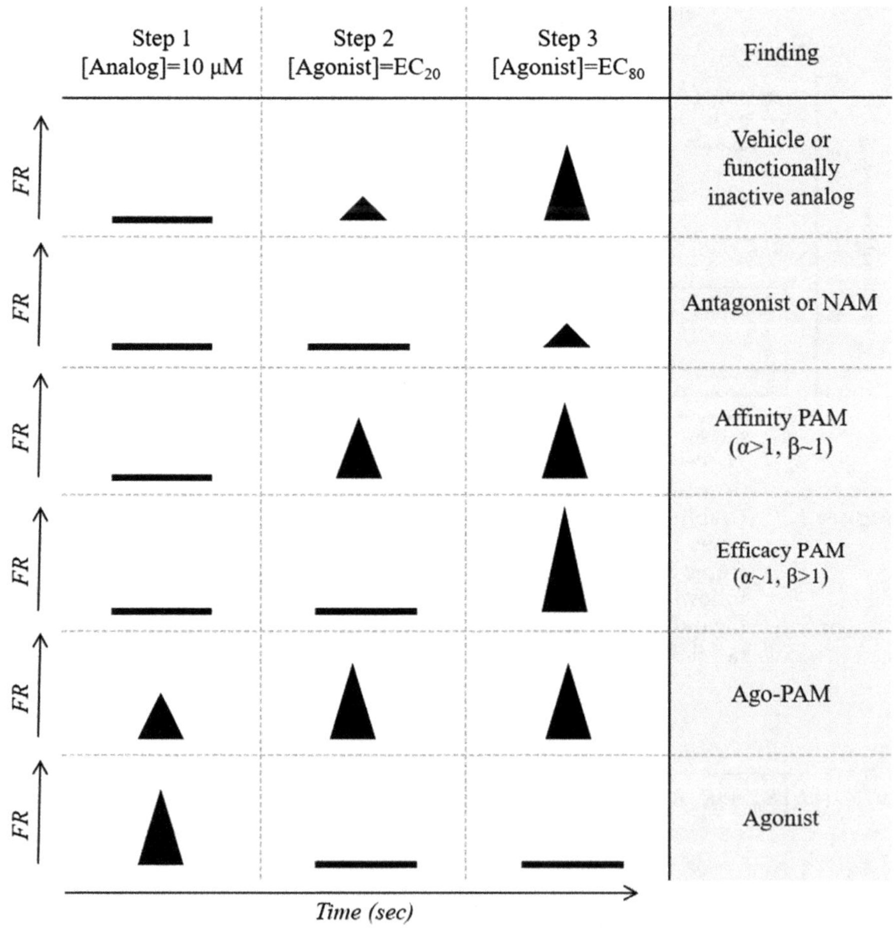

Figure 1.8 Schematic with examples of possible outcomes in a "triple addition" protocol and possible interpretation of the results. FR = functional response in an *in vitro* screen. It is assumed that orthosteric agonists will lead to receptor desensitization, whereas ago-PAMs generally won't.

correspond to those inside the black boxes in Figure 1.9A–C, and are usually plotted as in Figure 1.9D, to provide half maximal effective concentration (EC_{50}) and E_{max} values. These two values are then used by medicinal chemists during the lead optimization stage of allosteric drug discovery projects. All three compounds have low nanomolar EC_{50} values, but their E_{max} values, and therefore their functional effects at the target, are different.

Analyzing PAM EC_{50} data in this paradigm may lead to observations that could be interpreted as non-tractable SAR. While the approach is fast and efficient, using these data mechanically may lead to comparisons of compounds of fundamentally different nature, as the simulation in Figure 1.10 shows. In this simulation, three compounds were chosen to show similar

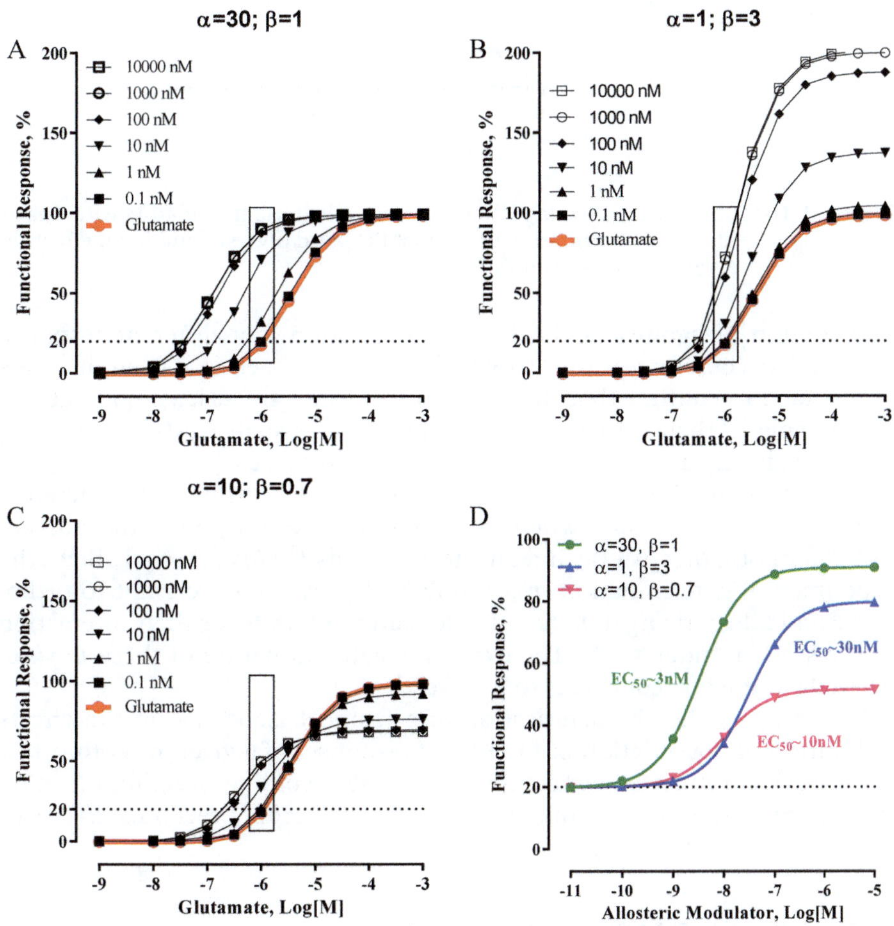

Figure 1.9 Simulations of full concentration–response curves (CRC) for an affinity PAM (**A**), and efficacy PAM (**B**) and a mixed allosteric ligand (enhanced affinity, reduced efficacy; **C**). Inside the black rectangles are shown the data points for the agonist EC_{20} used in the CRC shown on **D**.

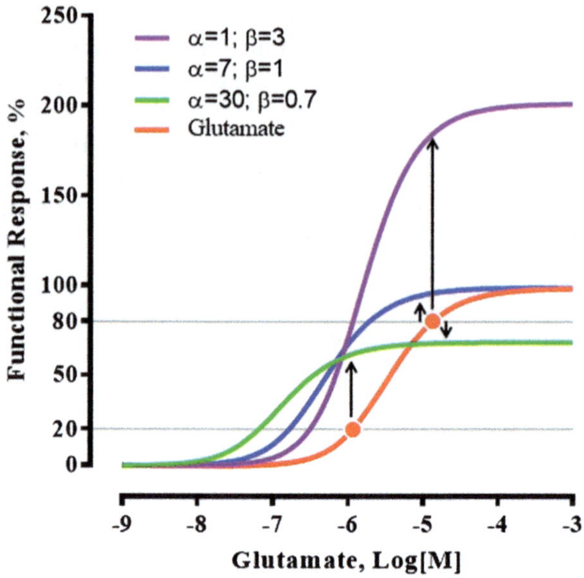

Figure 1.10 Simulation showing three different allosteric modulators with nearly identical potentiation of agonist EC_{20} and diverse functional effects at agonist EC_{80} concentration.

potentiation of agonist at the EC_{20} concentration, but different response at the EC_{80}. Therefore, if the "triple addition" approach is chosen, it is recommended to confirm that the compounds tested are at least qualitatively comparable in their functional profile (*e.g.*, they are all PAMs maintaining maximal efficacy at 100% E_{max}, they are all full NAMs, *etc.*).

In contrast to this methodology, albeit more laborious, full functional CRCs can be determined where the range of relevant concentrations are tested for both, orthosteric and allosteric ligands.[48] This protocol allows the experimental determination of a number of parameters relevant to orthosteric and allosteric ligands and the derivation of SARs for their interaction in the system under study. The exact translational nature of these *in vitro*-derived parameters continues to be investigated.

In summary, the characterization of a ligand based only on the effects (inhibition or potentiation of functional response) of a given receptor in an *in vitro* system must be conducted judiciously, avoiding oversimplification that might lead to translational disconnects with other preclinical assays or even in the clinical setting.

1.6 "Pure" PAMs and Ago-PAMs

The extent to which effects of allosteric ligands are interrogated leads to an important, and often overlooked consequence: compounds are nowadays tested more in-depth than in the past. In so doing, compounds that were considered as having essentially the same pharmacology based on a reduced number of screens may now be differentiated by assays based on a new technology. This suggests that concepts such as "class effects" may have been misleading. This newly found differentiation also provides an opportunity to re-purpose older compounds.

Given the diverse and complex effects that an allosteric modulator may produce on the functional response of a biological target, understanding the exact nature of these effects is key in drug discovery. Ideally, an appropriate preclinical characterization of allosteric drug candidates (or chemical probes) using a number of *in vitro* test systems and *in vivo* models will enable the correct translation to human pharmacology. This is generally far from trivial, and efforts at an early stage will go a long way to avoid costly mistakes and minimize the risk of preventable clinical attrition.

Some of these precautions are common to all ligands, whether they act allosterically or not. For example, when testing compounds with receptors cloned on a certain cell line, even when receptor homology across species is high, using the correct receptor sequence for the appropriate species (*e.g.*, rat for *in vivo* studies in rat, human for clinical drug candidates) is consistent with a robust translational approach.

As mentioned above, understanding the effects of receptor expression levels on functional response is particularly important when testing allosteric ligands using receptors overexpressed in a cell line. This is critical when studying compounds with allosteric agonist effects. For example, a tendency to an increase in ligand-independent functional effects has been correlated with an increased level of heterologous GPCR expression for the dopamine D_{1B} receptor subtype.[49]

The GLP-1 receptor is highly conserved in mammals. An interesting report described cloning of the homologous GLP-1 receptors from mouse, rabbit, pig, cynomolgous monkey, and chimpanzee. Receptors were expressed by stable transfection of BHK cells at both high expression levels and lower, more physiologically relevant, expression levels. In the system with high expression levels of cloned GLP-1 receptors, an increased potency of GLP-1 peptide and other high-affinity ligands was observed, while K_d values stayed the same. When studying a low-affinity compound thought to be an allosteric agonist (known as "compound 2", see Chapter 12), expression levels of the human GLP-1 receptor were found to be important for maximal efficacy and potency. The two natural peptidic metabolites of GLP-1, GLP-1(9–37) and GLP-1(9–36) amide, behaved as agonists in the cell line with high expression of human GLP-1 receptor, but they were low-potency antagonists when the receptor was expressed at lower levels. Thus, it was concluded that for the proper functional characterization of endogenous or synthetic ligands of the GLP-1 receptor, the level of receptor expression is a key parameter.[34]

1.7 Flat SAR

The early years of research into the discovery of GPCR PAMs saw a number of statements in the primary literature such as:

"...*both series highlight a major challenge in the development of mGluR4 allosteric ligands—the common theme of HTS hits confirming upon re-synthesis, but displaying little to no SAR... upon the slightest structural modifications. Clearly, ligands that potentiate mGluR4, and their allosteric binding sites, represent significant challenges in the hit-to-lead stage of drug discovery*",[50]

or

"*The ability to predict structure–activity relationships in allosteric modulators is in its infancy... Several of these series...demonstrate a surprisingly sensitive structure–activity relationship, because even a slight modification of the parent compound almost invariably results in inactive derivatives*",[51]

or, referring to mGluR4 PAMs,

"... *medicinal chemistry publications have revealed that obtaining good chemical leads is difficult but in particular transforming these leads into clinically relevant molecules is very demanding*".[52]

Very quickly, allosteric modulators became notorious as research accounts were surfacing reporting "flat SAR" with a higher propensity than expected based on classical medicinal chemistry knowledge. Why is this so? Or is it?

To begin answering this question, we can consider an interesting example from the area of mGluR4 PAMs. First, 4-(1-phenyl-1*H*-pyrazol-4-yl)quinoline (**22**, Figure 1.11) was reported as a relatively potent, selective, and brain-penetrant mGluR4 PAM. This compound was a confirmed high-throughput screening (HTS) hit, quite attractive based on its low molecular weight (271.3 amu) and appropriate physicochemical properties as a CNS lead compound (*e.g.*, $\text{Log}D_{7.4}$ = 3.9 and PSA = 31 Å^2). However, after making several close analogs across the molecule, most were inactive and only a couple of them had activity at the target.[53] Clearly, this was an HTS hit with poorly tractable SAR. Second, an early lead compound, (±)-VU0155041 (**23**, peripherally restricted distribution) and the closely related amide (+)-Lu AF21934 (**11**, brain-penetrant)

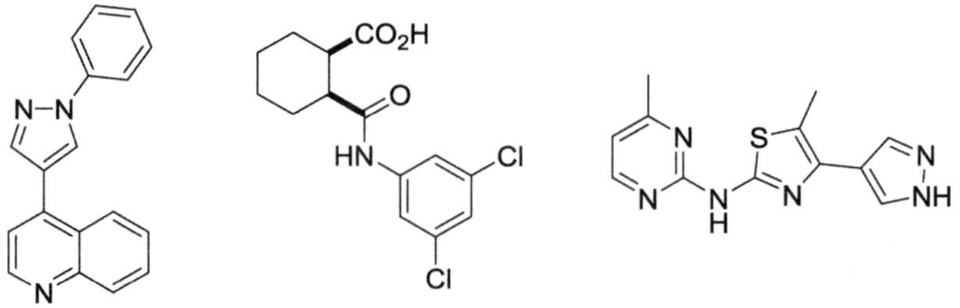

22 **23** **24**

Figure 1.11 Structures of 4-(1-phenyl-1*H*-pyrazol-4-yl)quinoline **22**, (±)-VU0155041 (**23**) and ADX88178 (**24**), three mGluR4 PAM tool compounds with varying degrees of chemical tractability.

showed some SAR trends as mGluR4 PAMs. For example, replacing Cl with F was tolerated, as well as moving around the halogens in the 3- and 4- positions of the aniline ring. Still, even with this somewhat narrow SAR, these compounds became useful probe molecules to help develop the target's preclinical validation package. Third, the *N*-heteroaryl-4-(1*H*-pyrazol-4-yl)thiazol-2-amine chemotype showed nicely tractable and additive SAR, and also led to an excellent tool compound that enabled interrogation of the biology in the field (ADX88178, **24**),[54,55] as well as publications,[56] and a number of patent applications all based on the same scaffold,[57–59] indicating the highly tractable nature of this chemical lead. Thus, all three hits derived from HTSs presented different attributes in terms of SAR tractability as mGluR4 PAMs. It is inferred that SAR tractability (or the lack thereof, a.k.a. flat SAR) is an attribute of a chemotype interrogating a biological system, and not necessarily a reflection of the mechanism of action of the compounds (*e.g.*, PAMs or NAMs).

Could SAR tractability be predicted? It is important to note that following an HTS campaign, confirmed validated hits are often times rank-ordered based on their physicochemical attributes. However, to our best knowledge, there is no established link between tractability of SAR and physicochemical descriptors or other easily calculated molecular parameters that can be estimated prior to making a few analogs. Thus, a validated way to establish whether a series has tractable SAR is to make a number of compounds purposely designed to explore SAR and/or surface plasmon resonance (SPR) trends, rather than to enhance potency or stability.

Also, tractable SAR should not be considered as an entitlement of all confirmed hits derived from an HTS campaign. As written by Robert Rydzewski,

> "*many compounds with weak activity will only lead to a plethora of analogs that are similarly anemic, so it's best to rule out this kind of flat SAR as quickly as possible*".[60]

A number of *bona fide* reasons may account for flat SAR. For example, a molecular fragment may be partially solvent-exposed and therefore, a significant portion is unable to interact with amino acids on the receptor and contribute to constructive binding interactions.[61] More importantly for the case of allosteric modulators, when establishing SAR trends for PAMs, some researchers compare EC_{50} and E_{max} values for the analogs synthesized. As explained in Section 1.5, these values are determined at a single agonist concentration. EC_{50} values so obtained reflect a combination of compound contributions to affinity and efficacy. This may be the reason why SAR trends are hard to interpret (leading to the conclusion of flat SAR). Instead, when the same compounds are analyzed using the operational model methodology, trends are found for α, β, τ_B, and pK_B, in that specific system. pK_A and τ_A may be used as a control, since they are independent of the PAM used.[62–64]

In summary, finding novel, good quality chemical lead series with tractable SAR or SPR may not be an exclusive attribute of allosteric ligands, but rather one of the many challenges faced in drug discovery.

1.8 Functional Switches

During SAR studies of allosteric modulators, it is not uncommon to encounter chemotypes where relatively minor structural differences have resulted in dramatically altered, even opposite functional effects—as in going from agonist to antagonist. The structural differences involved may be as small as replacing an –H for $-CH_3$, –OH, or –F, and may derive from medicinal chemistry design or be the consequence of *in vivo* metabolism.[65] Thus, when establishing pharmacokinetic–pharmacodynamic relationships during *in vivo* efficacy studies, ignoring the presence of circulating metabolites may lead to the wrong conclusions. For example, a metabolite with opposite functionality to the parent compound may lead to a diminished observed effect.

It is important to recognize that these functional switches are not an exclusive attribute of allosteric modulators. Indeed, interconversion of GPCR agonists and antagonists in the same chemical scaffold has been shown to be a general phenomenon, and considered to be a viable tactical approach to generate new leads.[66] Allosteric substrate-enabled switching was recently reported in a voltage-sensing lipid phosphatase.[67] Chemical switches have been exploited to enable the transformation of an mGluR2 silent allosteric modulator (SAM) into dual metabotropic glutamate receptor 2/3 negative/positive allosteric modulators.[68]

1.9 Concluding Remarks

The 20th century has been called the "century of physics", as technologies were developed to take advantage of energy; namely combustion, electricity and nuclear power. Likewise, the following hundred-year period has been envisioned as the "century of biology", as it was anticipated that decoding the complete genetic blueprint of the human species would define scientific advance, and play a major impact on our understanding and treatment of disease.[69]

We began this chapter discussing the crisis of productivity of the pharmaceutical industry and the potential role that allosterism could play in helping to establish a more successful drug discovery paradigm. Towards this end, there seems to be a consensus about the importance of understanding in enhanced detail the biology of life, and the pathophysiology of disease. The key question is: what is the best strategy to move forward?

We believe that the ideal path forward to understand human disease is to engage and empower all branches of the life science ecosystem in the research. Some organizations are following this strategy. Others prefer to focus on understanding new aspects of biology and leave the discovery of new drugs to third parties, which are not fully integrated into the major research path.[70] Time will tell how this story progresses, but just one-and-a-half decades into the century, the interface between biology and chemistry, where allosterism lies, is developing at a robust pace and as a result, exciting new knowledge is emerging that we think will be vital to arrive at some of the disease therapies of the future. Allosterism is a broad attribute of molecular components of live systems, and learning to design allosteric drugs mandates that we understand the fundamental nature of these interactions. We have a lot of work ahead of us. So, shall we start?

Acknowledgements

The authors are grateful to team members and management at Lundbeck Research for supporting the scientific research conducted on allosteric modulation CNS drug discovery projects. We also thank many colleagues in academic and contract research organizations for so generously sharing their ideas and making this area of research lots of fun. In particular, the contributions of Dr A. Pilc and Dr J. Wieronska (Polish Academy of Sciences), Dr F. Acher (Université Paris Descartes, France), Dr P. Gubellini (Aix-Marseille Université, France), Dr S. Duty, Dr S. Salvage and colleagues (King's College London, UK), Dr T. Johnstone and colleagues (Atuka Ltd, Canada), Dr J. Sprouce and colleagues (Neuroservice, France), and Dr T. Wolinsky (Porsolt & Associates, France) are warmly acknowledged.

References

1. E. Fischer, *Ber. Dtsch. Chem. Ges.*, 1894, **27**, 2985–2993.
2. J. Samanen, Similarities and differences in the discovery and use of biopharmaceuticals and small-molecule chemotherapeutics, in *Introduction to Biological and Small Molecule Drug Research and Development: Theory and Case Studies*, ed. R. Ganellin, S. M. Roberts and R. Jefferies, 2013, ch. 5, ISBN: 978-0-12-397176-0, pp. 161–203.
3. P. J. Declerck, *GaBIJ.*, 2012, **1**(1), 13–16, DOI: 10.5639/gabij.2012.0101.005.
4. J. Monod, J. P. Changeux and F. Jacob, *J. Mol. Biol.*, 1963, **6**, 306–329.
5. J. P. Changeux, *Protein Sci.*, 2011, **20**(7), 1119–1124.
6. K. Gunasekaran, B. Ma and R. Nussinov, *Proteins*, 2004, **57**(3), 433–443.
7. A. G. Gilmartin, M. R. Bleam, A. Groy, K. G. Moss, E. A. Minthorn, S. G. Kulkarni, C. M. Rominger, S. Erskine, K. E. Fisher, J. Yang, F. Zappacosta, R. Annan, D. Sutton and S. G. Laquerre, *Clin. Cancer Res.*, 2011, **17**(5), 989–1000.
8. J. F. Ohren, H. Chen, A. Pavlovsky, C. Whitehead, E. Zhang, P. Kuffa, C. Yan, P. McConnell, C. Spessard, C. Banotai, W. T. Mueller, A. Delaney, C. Omer, J. Sebolt-Leopold, D. T. Dudley, I. K. Leung, C. Flamme, J. Warmus, M. Kaufman, S. Barrett, H. Tecle and C. A. Hasemann, *Nat. Struct. Mol. Biol.*, 2004, **11**(12), 1192–1197.
9. H. Hirai, H. Sootome, Y. Nakatsuru, K. Miyama, S. Taguchi, K. Tsujioka, Y. Ueno, H. Hatch, P. K. Majumder, B. S. Pan and H. Kotani, *Mol. Cancer Ther.*, 2010, **9**(7), 1956–1967.
10. H. S. Lange, C. E. Cannon, J. T. Drott, S. D. Kuduk and J. M. Uslaner, *J. Pharmacol. Exp. Ther.*, 2015, **355**(3), 442–450.

11. S. Yin, M. J. Noetzel, K. A. Johnson, R. Zamorano, N. Jalan-Sakrikar, K. J. Gregory, P. J. Conn and C. M. Niswender, *J. Neurosci.*, 2014, **34**(1), 79–94.
12. A. B. Burgin, O. T. Magnusson, J. Singh, P. Witte, B. L. Staker, J. M. Bjornsson, M. Thorsteinsdottir, S. Hrafnsdottir, T. Hagen, A. S. Kiselyov, L. J. Stewart and M. E. Gurney, *Nat. Biotechnol.*, 2010, **28**(1), 63–70.
13. http://www.nih.gov/news-events/news-releases/new-nih-funded-memory-drug-moves-into-phase-1-clinical-study, accessed on January 7, 2016.
14. K. Hollenstein, J. Kean, A. Bortolato, R. K. Cheng, A. S. Doré, A. Jazayeri, R. M. Cooke, M. Weir and F. H. Marshall, *Nature*, 2013, **499**(7459), 438–443.
15. A. S. Doré, K. Okrasa, J. C. Patel, M. Serrano-Vega, K. Bennett, R. M. Cooke, J. C. Errey, A. Jazayeri, S. Khan, B. Tehan, M. Weir, G. R. Wiggin and F. H. Marshall, *Nature*, 2014, **511**(7511), 557–562.
16. H. Wu, C. Wang, K. J. Gregory, G. W. Han, H. P. Cho, Y. Xia, C. M. Niswender, V. Katritch, J. Meiler, V. Cherezov, P. J. Conn and R. C. Stevens, *Science*, 2014, **344**(6179), 58–64.
17. D. Zhang, Z. G. Gao, K. Zhang, E. Kiselev, S. Crane, J. Wang, S. Paoletta, C. Yi, L. Ma, W. Zhang, G. W. Han, H. Liu, V. Cherezov, V. Katritch, H. Jiang, R. C. Stevens, K. A. Jacobson, Q. Zhao and B. Wu, *Nature*, 2015, **520**(7547), 317–321.
18. A. Bespalov, T. Steckler, B. Altevogt, E. Koustova, P. Skolnick, D. Deaver, M. J. Millan, J. F. Bastlund, D. Doller, J. Witkin, P. Moser, P. O'Donnell, U. Ebert, M. A. Geyer, E. Prinssen, T. Ballard and M. Macleod, *Nat. Rev. Drug Discovery*, 2016, **15**, 516.
19. C. W. Lindsley, K. A. Emmitte, C. R. Hopkins, T. M. Bridges, K. J. Gregory, C. M. Niswender and P. J. Conn, *Chem. Rev.*, 2016, **116**(11), 6707–6041.
20. T. Kenakin, *Br. J. Clin. Pharmacol.*, 2016, **81**(1), 41–51.
21. D. Wootten, A. Christopoulos and P. M. Sexton, *Nat. Rev. Drug Discovery*, 2013, **12**(8), 630–644.
22. L. T. May, K. Leach, P. M. Sexton and A. Christopoulos, *Annu. Rev. Pharmacol. Toxicol.*, 2007, **47**, 1–51.
23. P. J. Flor and F. C. Acher, *Biochem. Pharmacol.*, 2012, **84**(4), 414–424.
24. P. D. Suzdak, C. Thomsen, E. Mulvihill and P. Kristensen, Molecular Cloning, Expression, and Characterization of Metabotropic Glutamate Receptor Subtypes, in *The Metabotropic Glutamate Receptors*, ed. P. Jeffrey Conn and J. Patel, Springer Science+Business Media, New York, 1994, ch. 2.
25. S. Topiol, M. Sabio and M. Uberti, *Neuropharmacology*, 2011, **60**(1), 93–101.
26. C. Goudet, B. Vilar, T. Courtiol, T. Deltheil, T. Bessiron, I. Brabet, N. Oueslati, D. Rigault, H. O. Bertrand, H. McLean, H. Daniel, M. Amalric, F. Acher and J.-P. Pin, *FASEB J.*, 2012, **26**(4), 1682–1693.
27. J. A. Monn, L. Prieto, L. Taboada, J. Hao, M. R. Reinhard, S. S. Henry, C. D. Beadle, L. Walton, T. Man, H. Rudyk, B. Clark, D. Tupper, S. R. Baker, C. Lamas, C. Montero, A. Marcos, J. Blanco, M. Bures, D. K. Clawson, S. Atwell, F. Lu, J. Wang, M. Russell, B. A. Heinz, X. Wang, J. H. Carter,

B. G. Getman, J. T. Catlow, S. Swanson, B. G. Johnson, D. B. Shaw and D. L. McKinzie, *J. Med. Chem.*, 2015, **58**, 7526–7548.
28. Z. Farzampour, R. J. Reimer and J. Huguenard, *Adv. Pharmacol.*, 2015, **72**, 147–164.
29. S. Schenone, C. Brullo, F. Musumeci, O. Bruno and M. Botta, *Curr. Top. Med. Chem.*, 2010, **10**(9), 878–901.
30. Z.-G. Gao, J. B. Blaustein, A. S. Gross, N. Melman and K. A. Jacobson, *Biochem. Pharmacol.*, 2003, **65**, 1675–1684.
31. A. Knight, J. L. Hemmings, I. Winfield, M. Leuenberger, E. Frattini, B. G. Frenguelli, S. J. Dowell, M. Lochner and G. Ladds, *J. Med. Chem.*, 2016, **59**(3), 947–964.
32. A. Alt, A. Pendri, R. L. Bertekap, G. Li, Y. Benitex, M. Nophsker, K. L. Rockwell, N. T. Burford, C. S. Sum, J. Chen, J. J. Herbst, M. Ferrante, A. Hendricson, M. E. Cvijic, R. Westphal, J. O'Connell, M. Banks, L. Zhang, R. Gentles, S. Jenkins, J. Loy and J. E. Macor, *J. Pharmacol. Exp. Ther.*, 2016, **356**(2), 293–304.
33. S. Daniels, Allosteric modulation of receptor function, in *Allosteric Receptor Modulation in Drug Targeting*, ed. N. G. Bowery, Taylor & Francis, New York, 2006, ch. 1, pp. 1–18.
34. L. B. Knudsen, S. Hastrup, C. R. Underwood, B. S. Wulff and J. Fleckner, *Regul. Pept.*, 2012, **175**, 21–29.
35. M. Turlington, M. J. Noetzel, A. Chun, Y. Zhou, R. D. Gogliotti, E. D. Nguyen, K. J. Gregory, P. N. Vinson, J. M. Rook, K. K. Gogi, Z. Xiang, T. M. Bridges, J. S. Daniels, C. Jones, C. M. Niswender, J. Meiler, P. J. Conn, C. W. Lindsley and S. R. Stauffer, *J. Med. Chem.*, 2013, **56**, 7976–7996.
36. M. Turlington, C. Malosh, J. Jacobs, J. T. Manka, M. J. Noetzel, P. N. Vinson, S. Jadhav, E. J. Herman, H. Lavreysen, C. Mackie, J. M. Bartolomé-Nebreda, S. Conde-Ceide, M. L. Martín-Martín, H. M. Tong, S. López, G. J. MacDonald, T. Steckler, J. S. Daniels, C. D. Weaver, C. M. Niswender, C. K. Jones, P. J. Conn, C. W. Lindsley and S. R. Stauffer, *J. Med. Chem.*, 2014, **57**, 5620–5637.
37. F. Yang, L. B. Snyder, A. Balakrishnan, J. M. Brown, D. V. Sivarao, A. Easton, A. Fernandes, M. Gulianello, U. M. Hanumegowda, H. Huang, Y. Huang, K. M. Jones, Y.-W. Li, M. Matchett, G. Mattson, R. Miller, K. S. Santone, A. Senapati, E. E. Shields, F. J. Simutis, R. Westphal, V. J. Whiterock, J. J. Bronson, J. E. Macor and A. P. Degnan, *ACS Med. Chem. Lett.*, 2016, 7, 289–293.
38. C. J. Langmead and A. Christopoulos, *Br. J. Pharmacol.*, 2013, **169**(2), 353–356.
39. J. M. Wierońska, S. H. Zorn, D. Doller and A. Pilc, *Pharmacol. Ther.*, 2016, **157**, 10–27.
40. I. G. Dougall and J. Unitt, Evaluation of the Biological Activity of Compounds: Techniques and Mechanism of Action Studies, in *The Practice of Medicinal Chemistry*, ed. C. G. Wermuth, D. Aldous, P. Raboisson, and D. Rognan, Academic Press, London, 2015, ch. 2, pp. 15–44.
41. E. Di Cera, *Methods Enzymol.*, 2009, **466**, 259–271.

42. D. R. Groebe, *Drug Discovery Today*, 2006, **11**, 632–639.
43. T. P. Kenakin, *Br. J. Pharmacol.*, 2012, **165**(6), 1659–1669.
44. J. W. Black and P. Leff, *Proc. R. Soc. London, Ser. B*, 1983, **220**(1219), 141–162.
45. K. J. Gregory, P. M. Sexton and A. Christopoulos, *Curr. Protoc. Pharmacol.*, 2010, **1**, 1–21.
46. J. Giraldo, *Trends Pharmacol. Sci.*, 2015, **36**(1), 1–2.
47. B. J. Melancon, C. R. Hopkins, M. R. Wood, K. A. Emmitte, C. M. Niswender, A. Christopoulos, P. J. Conn and C. W. Lindsley, *J. Med. Chem.*, 2012, **55**, 1445–1464.
48. N. T. Burford, K. E. Livingston, M. Canals, M. R. Ryan, L. M. Budenholzer, Y. Han, Y. Shang, J. J. Herbst, J. O'Connell, M. Banks, L. Zhang, M. Filizola, D. L. Bassoni, T. S. Wehrman, A. Christopoulos, J. R. Traynor, S. W. Gerritz and A. Alt, *J. Med. Chem.*, 2015, **58**(10), 4220–4229.
49. M. Tiberi and M. G. Caron, *J. Biol. Chem.*, 1994, **269**(45), 27925–27931.
50. R. Williams, C. M. Niswender, Q. Luo, U. Le, P. J. Conn and C. W. Lindsley, *Bioorg. Med. Chem. Lett.*, 2009, **19**(3), 962–966.
51. L. Wang, B. Martin, R. Brenneman, L. M. Luttrell and S. Maudsley, *J. Pharmacol. Exp. Ther.*, 2009, **331**(2), 340–348.
52. S. P. East and K. Gerlach, *Expert Opin. Ther. Pat.*, 2010, **20**(3), 441–445.
53. H. N. Jimenez, K. G. Liu, S. P. Hong, M. S. Reitman, M. A. Uberti, M. D. Bacolod, M. Cajina, M. Nattini, M. Sabio and D. Doller, *Bioorg. Med. Chem. Lett.*, 2012, **22**(9), 3235–3239.
54. E. Le Poul, C. Boléa, F. Girard, S. Poli, D. Charvin, B. Campo, J. Bortoli, A. Bessif, B. Luo, A. J. Koser, L. M. Hodge, K. M. Smith, A. G. DiLella, N. Liverton, F. Hess, S. E. Browne and I. J. Reynolds, *J. Pharmacol. Exp. Ther.*, 2012, **343**(1), 167–177.
55. M. Kalinichev, E. Le Poul, C. Boléa, F. Girard, B. Campo, M. Fonsi, I. Royer-Urios, S. E. Browne, J. M. Uslaner, M. J. Davis, J. Raber, R. Duvoisin, S. T. Bate, I. J. Reynolds, S. Poli and S. Celanire, *J. Pharmacol. Exp. Ther.*, 2014, **350**(3), 495–505.
56. S. P. Hong, K. G. Liu, G. Ma, M. Sabio, M. A. Uberti, M. D. Bacolod, J. Peterson, Z. Z. Zou, A. J. Robichaud and D. Doller, *J. Med. Chem.*, 2011, **54**(14), 5070–5081.
57. C. Bolea, WO 2010/079239, 2010.
58. C. Bolea and S. Celanire, WO 2010/079238, 2010.
59. N. J. Liverton, C. Bolea, S. Celanire and Y. Luo, WO 2012/006760, 2012.
60. R. M. Rydzewski, Real World Drug Discovery, in *Hit Generation*, Elsevier, Oxford, UK, 1st edn, 2008, ch. 6, pp. 220–221.
61. P. J. Ala, L. Gonneville, M. Hillman, M. Becker-Pasha, E. W. Yue, B. Douty, B. Wayland, P. Polam, M. L. Crawley, E. McLaughlin, R. B. Sparks, B. Glass, A. Takvorian, A. P. Combs, T. C. Burn, G. F. Hollis and R. Wynn, *J. Biol. Chem.*, 2006, **281**(49), 38013–38021.
62. P. J. Conn, C. W. Lindsley, J. Meiler and C. M. Niswender, *Nat. Rev. Drug Discovery*, 2014, **13**(9), 692–708.

63. S. N. Mistry, C. Valant, P. M. Sexton, B. Capuano, A. Christopoulos and P. J. Scammells, *J. Med. Chem.*, 2013, **56**(12), 5151–5172.
64. D. Doller, *Not All mGluR PAMs Are Created Equal: Designing the Right Allosteric Ligand for Your Clinical Indication*, World Preclinical Congress, Boston, MA, June 10–12, 2015.
65. C. W. Lindsley, *J. Med. Chem.*, 2014, **57**(18), 7485–7498.
66. P. I. Dosa and E. A. Amin, *J. Med. Chem.*, 2016, **59**(3), 810–840.
67. S. S. Grimm and E. Y. Isacoff, *Nat. Chem. Biol.*, 2016, **12**(4), 261–267.
68. S. Schann, S. Mayer, C. Franchet, M. Frauli, E. Steinberg, M. Thomas, L. Baron and P. Neuville, *J. Med. Chem.*, 2010, **53**(24), 8775–8779.
69. C. Venter and D. Cohen, The Century of Biology, *New Perspect. Q.*, 2004, **21**(4), 73–77.
70. E. E. Schadt, S. Buchanan, K. J. Brennand and K. M. Merchant, *Front. Pharmacol.*, 2014, **5**(252), 1–15.

CHAPTER 2

Identifying and Quantifying Allosteric Drug Function

TERRY KENAKIN[a]

[a]Department of Pharmacology, University of North Carolina School of Medicine, 120 Mason Farm Road, Room 4042 Genetic Medicine Building, CB# 7365, Chapel Hill, NC 27599, USA
*E-mail: kenakin@email.unc.edu

2.1 Introduction: Receptor Allosterism

Allosteric molecules are fundamentally different from molecules that bind to the natural receptor binding pocket for endogenous hormones and neurotransmitters, and this gives them a unique repertoire of dose-response behaviors. The underlying principle that differentiates allosteric from orthosteric molecules is the concept of permissiveness. Specifically, allosteric ligands may permit some aspects of the receptor system to function; in contrast, orthosteric ligands do not. Orthosteric ligands are pre-emptive in that if an orthosteric ligand is bound to the natural receptor binding site, then no other molecule can bind to the receptor, *i.e.* steric hindrance prevents further interaction. This being the case, the ligand-bound species are somewhat uniform, with respect to different orthosteric ligands, in that the observed properties are related only to the bound new molecule and not to the endogenous natural ligand. For instance, a β-blocker such as propranolol pre-empts the binding of natural catchecholamines in the same way as does the orthosteric β-blocker pindolol; the differences in this regard only relate to potency.

RSC Drug Discovery Series No. 56
Allosterism in Drug Discovery
Edited by Dario Doller

Published by the Royal Society of Chemistry, www.rsc.org

The important difference between orthosteric and allosteric systems is that in orthosteric systems there never is a pharmacologically relevant receptor species with both the orthosteric ligand and natural ligand bound simultaneously. In contrast, thc pharmacologically relevant receptor species in allosteric systems has both the allosteric and the natural ligand bound simultaneously. This permissive nature makes allosteric receptor systems extremely flexible with respect to the fine tuning of receptor function. Within the repertoire of allosteric receptor behaviors lie a number of behaviors that can appear to be identical to the behaviors seen with orthosteric molecules. Failure to recognize which of these modes of action are operative with respect to drug action can lead to dissimulations in the interpretation of drug behavior. Therefore, it is important to determine whether a ligand interaction is orthosteric or allosteric in nature. Clearly if potentiation of effect is observed, this can only occur allosterically. However, allosteric antagonism can appear identical to simple competitive antagonism if it is surmountable (no depression of maximal response), and orthosteric non-competitive antagonism if it is insurmountable (depressed maximal response). As a preface to a discussion of the detection of allosteric antagonism, it is useful to outline the different properties of allosteric antagonists as used in drug therapy.

2.2 Unique Effects of Allosteric Antagonists

It is worth considering the behaviors of allosteric antagonists that differentiate them from standard orthosteric antagonists:

a. Allosteric modulators can alter the interaction of very large proteins. For instance, the allosteric HIV-1 entry inhibitor aplaviroc blocks the interaction of the chemokine CCR5 receptor and the HIV-1 virus coat protein gp120 (aplaviroc is 1/200 the size of these proteins) to prevent HIV-1 infection.[1]
b. Allosteric antagonists can modulate but need not completely inactivate receptor function. This is because, unlike orthosteric ligands which occlude the orthosteric binding site, allosteric antagonists change the conformation of the receptor to reset the affinity of that receptor to other molecules.
c. Allosteric effects are saturable because the maximal occupation of the allosteric binding site results in a limit to the maximal effect. This fact can dissociate the length of time the effect is operative (*i.e.* target coverage) from the intensity of effect. Therefore, high doses of a limited allosteric antagonist effectively become a pool of drug to give long-lasting response without overdose.
d. Allosteric antagonists may modify the quality of endogenous agonist efficacy. While orthosteric antagonists block all signaling of the receptor to the cell, allosteric antagonists may allow some of these to continue as in the case of indole1, an allosteric antagonist of prostaglandin

D2 (PDG2) response at the chemoattractant receptor homologous molecule expressed on Th2 cells (CRTH) receptor which blocks G protein signaling but not β-arrestin effects of the endogenous agonist (*vide infra*).[2]

e. Allosteric antagonism can be textured in that allosteric antagonists produce antagonism through alteration of protein target conformation, therefore the resulting protein species need not be identical for all agonists. This has been shown for the chemokine receptor CCR5 where the allosteric antagonists TAK779, SCH–C (Schering-C) and aplaviroc produce different antibody binding profiles for Ab45531 and Ab45523; this suggests that different tertiary conformations are stabilized by the modulators.[3] In the case of HIV-1 this may be relevant to the therapeutic use of entry inhibitors since it is expected that viral mutation eventually will lead to tolerance to an HIV-1 entry inhibitor as the virus mutates in response to selective pressure of viral expression.[4] The different conformations stabilized by these modulators could allow treatment of viral resistance due to mutation.
f. Allosteric modulators can have differential effects on agonist affinity and efficacy. The stabilization of a new receptor conformation can produce changes in natural ligand affinity and/or efficacy and there is no rule to cause those changes to be in the same direction, *i.e.* a modulator may increase affinity and decrease efficacy as in the case of positive allosteric modulation (PAM) antagonists (*vide infra*). For example, considering the effects of the HIV-1 entry inhibitor aplaviroc, this antagonist blocks the function of the chemokine CCL3 and CCL5 and the binding of CCL3 but does not affect the binding of CCL5.[1,5] A similar effect is seen with the ester CPCCOEt (7-hydroxyiminocyclopropan[*b*] chromen-1*a*-carboxylic acid ethyl ester), which blocks the responses to glutamate in Chinese Hamster Ovary (CHO) cells naturally expressing human GluR1b glutamate receptors while not affecting glutamate binding.[6] A combination of increased affinity and decreased efficacy (PAM antagonists) may be particularly useful therapeutically due to the reciprocal flow of allosteric energy between the allosteric and orthosteric sites. Specifically, as the modulator increases the affinity of the agonist the agonist also will increase the affinity of the modulator (shown experimentally).[7] Thus, antagonism will increase with increased agonism, *i.e.* the modulator will adjust its activity according to the activity of the system.
g. Allosteric antagonists have the potential for extraordinary selectivity for receptor subtypes. While orthosteric ligands must bind to the natural agonist binding site (and these may be conserved across receptor subtypes since they must recognize identical natural agonists), such conservation is not necessary for allosteric binding sites.[8–12]
h. Allosteric antagonists, by virtue of the property of probe dependence, may block some stimuli to the receptor more than others. This also can be seen for HIV-1 entry inhibitors where the receptor mediating HIV-1 infection (CCR5) also mediates the internalization of the receptor

in response to the binding of chemokines. A selection of allosteric antagonists for CCR5 have up to a 500-fold differential potency for the blockade of these two effects,[13] a property that may provide a useful therapeutic advantage. Specifically, studies with AIDS patients have shown that a naturally functional chemokine response system offers favorable protection in progression to AIDS after HIV-1 infection.[14] Therefore, if this function could be spared through probe-dependent allosteric blockade of HIV-1 infection, it would provide an increased therapeutic value.

2.3 Detecting Allosteric Effect

There are two main properties of allosteric systems that make them uniquely different from orthosteric systems and these can be used to experimentally identify allosteric antagonism, namely (1) saturation of effect and (2) probe dependence. It is worth considering how these concepts can be applied to pharmacological experiments to identify allosteric function.

2.3.1 Saturation of Effect

In orthosteric simple competitive antagonism, agonist concentration–response curves are shifted to the right in the presence of antagonist in accordance with the Schild relationship:[15]

$$\mathrm{DR} = [\mathrm{B}]/K_{\mathrm{B}} + 1 \tag{2.1}$$

where the concentration of antagonist is [B], K_B is the equilibrium dissociation constant of the antagonist–response complex and DR is the dose ratio for the agonist (defined as the equiactive ratio of half maximal effective concentration (EC_{50}) values in the presence of the antagonist divided by the EC_{50} in the absence of the antagonist). It can be seen from eqn (2.1) that DR is limitless, *i.e.* as more antagonist is added, more shift of the curve will occur. In contrasts, antagonism by an allosteric antagonist occurs through the binding to the ligand to a separate site on the receptor and once this site is fully occupied, no further antagonism will occur. This leads to a saturation of effect, *i.e.* a limit to the degree of antagonism produced. The corresponding equation to eqn (1) for an allosteric system is:

$$\mathrm{DR} = \frac{[\mathrm{B}](1-\alpha)}{\alpha[\mathrm{B}] + K_{\mathrm{B}}} + 1 \tag{2.2}$$

where α refers to the effect of the allosteric modulator on the affinity of the agonist. From eqn (2.2) it can be seen that as $[\mathrm{B}] \rightarrow \infty$, $\mathrm{DR} \rightarrow (\alpha^{-1} + 1)$. Therefore, the first hallmark of allosteric action is the observation of a limit to the effect. It will be seen from a comparison of data to the functional allosteric model (*vide infra*) that a caveat to the general application of this rule is that the maximal effect of an allosteric ligand can be such that it may be difficult to observe the saturation level, thus this criterion is a one-way experiment, *i.e.* if saturation is observed then allosterism is confirmed. If saturation is not observed, it may still be an allosteric effect with a maximal limit beyond the bounds of the particular experiment (*i.e.* α may be so small as to cause very large values of DR).

2.3.2 Probe Dependence

The second hallmark of allosteric function is probe dependence. This occurs because allosteric effects on the interaction of the natural agonist and the receptor occur through the induction of a conformational change in the receptor protein. There are no specific rules for the effects of a conformational change in the receptor protein on the interaction of the protein with a ligand, and it is often observed that a conformational change producing a large effect on the interaction of one ligand with the receptor may produce a very different or even minimal change in the interaction of the same receptor with a different ligand (probe). For example, the allosteric ligand alcuronium produces inhibition of the binding of [^{3}H] quinuclidinyl benzilate (QNB) to muscarinic M2 receptors but a *stimulation* of the binding of [^{3}H] atropine to the same receptor.[16] In contrast, orthosteric ligands will have uniform effects on all receptor probes (including agonists and radioligands). Therefore, if probe-dependent effects are observed experimentally, then allosterism is confirmed. However, this too is a one-way experimental situation in that if no probe dependence is observed it may be that the wrong probes were used to detect it. As a preface to the application of these principles towards the identification and quantification of allosteric function, it is useful to define the functional model of receptor allosterism.

2.4 The Functional Allosteric Receptor Model

Seven transmembrane receptors (7TMRs) form a communication link between extracellular molecules (*i.e.* neurotransmitters, hormones) and intracellular signaling proteins (G proteins, β-arrestin, *etc.*) through changes in protein conformation. The primary unit in the allosteric functional model is the allosteric binding equation which itself is a combination of mass action binding reactions in the form:[17,18]

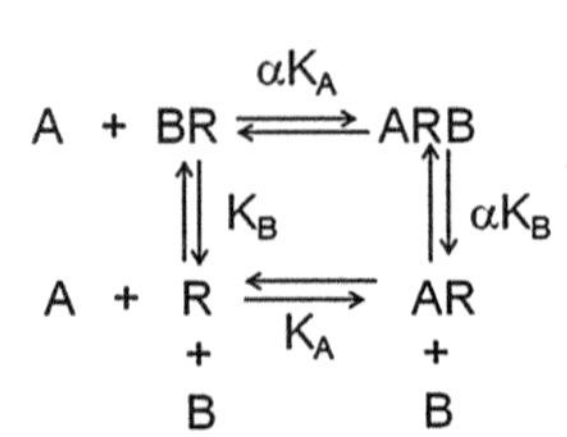

(2.3)

where the modulator is B, the receptor R and the guest molecule is A. The parameter α defines the change in the affinity of A for the receptor in the presence of bound allosteric ligand B. Interestingly, the allosteric binding model formally predicts allosteric saturability of effect. The relationship describing the affinity of the probe [A] for the receptor, in the presence of varying concentrations of antagonist is:

$$K_{obs} = K_A([B]/K_B)/(1 + \alpha[B]/K_B) \tag{2.4}$$

Thus as $[B] \rightarrow \infty$, K_{obs} (the new value for the location parameter for the probe curve) $\rightarrow K_A/\alpha$. The various protein species are then linked to response mechanisms in the cell through the Black–Leff operational model for agonism:[19]

$$\text{Response} = [A]\tau E_{max}/([A](1 + \tau) + K_A) \tag{2.5}$$

where K_A is the equilibrium dissociation constant of the agonist–receptor complex and τ is the efficacy of the agonist; this represents the intrinsic power of the molecule to induce response and also the efficiency of the system to transduce that stimulation into cell response. This extends the binding model to one where the ligand bound species [AR], [BR] and [ABR] produce response through the a mechanism in the cell:

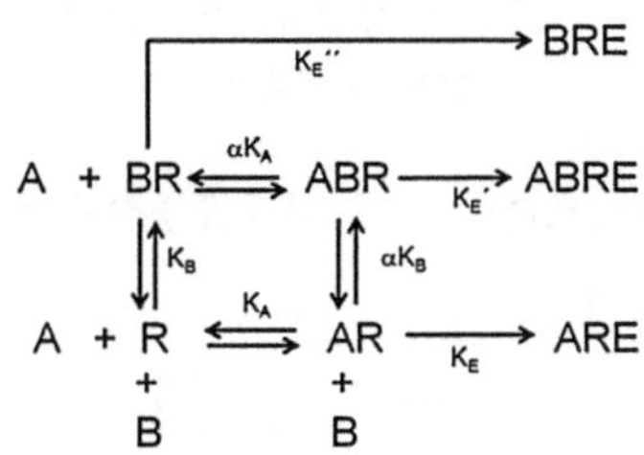

(2.6)

The response to an agonist [A], as modified by an allosteric modulator [B] is given as:[20–22]

$$\text{Response} = \frac{\tau_A[A]/K_A\left(1 + \alpha\beta[B]/K_B\right) + \tau_B[B]/K_B}{[A]/K_A\ (1 + \alpha[B]/K_B\ + \tau_A(1 + \alpha\beta[B]/K_B)) + [B]/K_B\ (1 + \tau_B) + 1)} \tag{2.7}$$

This equation can model all possible effects of an allosteric modulator on a receptor system. Direct production of receptor response can occur through direct efficacy of the modulator (τ_B), alteration of the affinity of the primary agonist occurs through α and alteration of the efficacy of the primary agonist through β. This leads to three phenotypic effects of allosteric ligands:

(1) negative allosteric modulation (modulators referred to as NAMs) where response is blocked (a variant of this is NAMs with direct partial agonist activity through $\tau_B > 0$). The effect occurs through the product $\alpha\beta < 1$,
(2) PAM-antagonists where the affinity of the agonist is increased ($\alpha > 1$) by the modulator but the efficacy is decreased ($\beta < 1$) with total $\alpha\beta < 1$,
(3) positive allosteric modulation where response is potentiated through $\alpha\beta > 1$ (with a variant of PAM agonism for direct activation by the modulator).

Since this chapter relates to the detection of masked allosterism and PAM effects are clearly allosteric, discussion will focus only on NAMs and PAM-antagonists.

2.5 Negative Allosteric Modulators (NAMs)

Figure 2.1 shows an array of possible patterns of antagonism through allosteric mechanisms. In general, depression of the maximal response occurs through $\beta < 1$ while dextral displacement of concentration–response curves occurs through $\alpha < 1$. The relative degree in which each of these effects is observed depends upon the receptor reserve of the system, *i.e.* with high receptor reserve, $\beta < 1$ may produce dextral displacement as well until the receptor reserve is blocked. It should be noted that the phenotypes shown in Figure 2.1 are also common to slowly dissociating orthosteric antagonists; therefore, the principle of saturation of effect and/or probe dependence may need to be applied to the dataset to identify an allosteric mechanism. This is demonstrated in Figure 2.2 where two ligands produce surmountable antagonism. The orthosteric ligand produces a linear Schild regression whereas the allosteric antagonist yields a curvilinear Schild

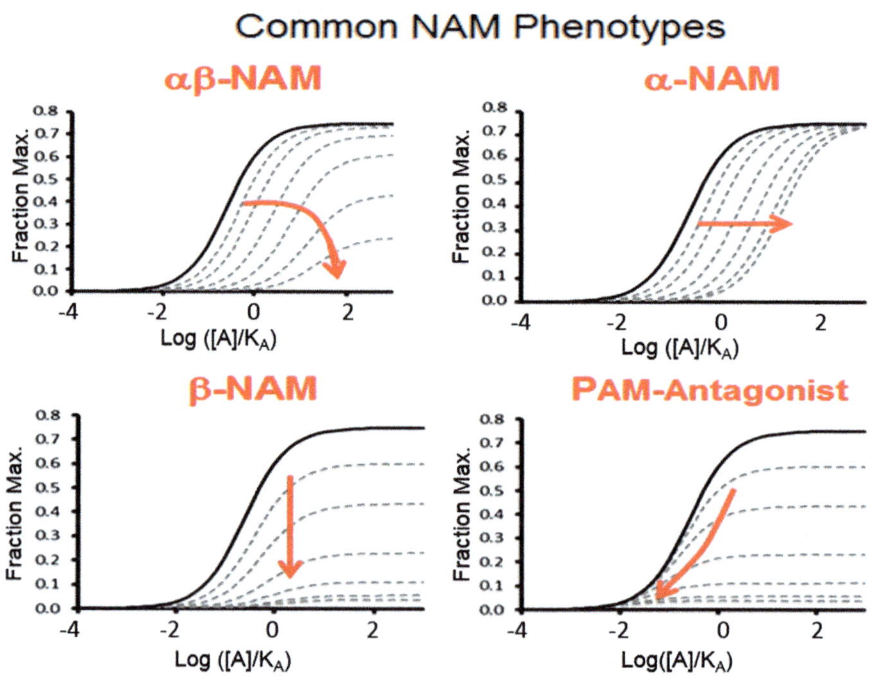

Figure 2.1 Antagonism of agonist response by NAMs with differing effects on affinity (α effect) and efficacy (β effect). Various combinations of dextral displacement of concentration–response curves and depressions of maximal response can be observed.

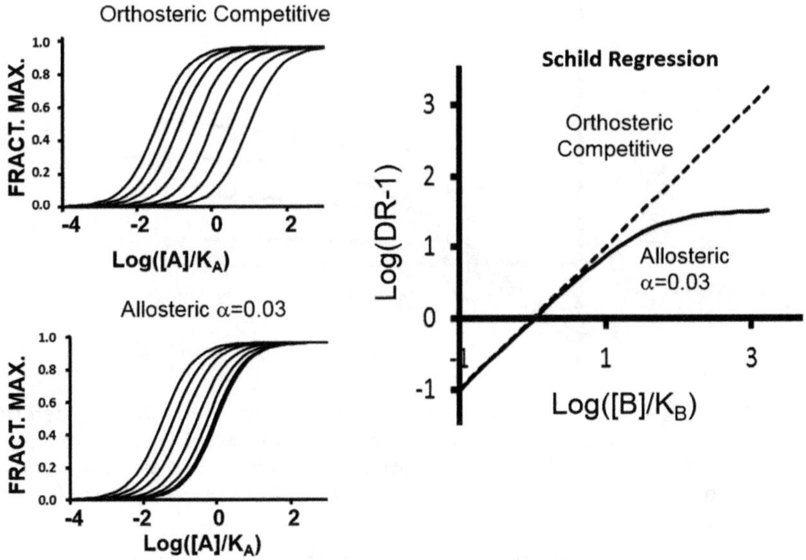

Figure 2.2 Effects of surmountable antagonists. The top left panel shows the effects of an orthosteric simple competitive antagonist—this yields a linear Schild regression (right panel). In contrast (bottom left panel), a NAM with an α value of 0.03 produces a maximal 33.3-fold shift to the right of the agonist concentration–response curves; this is reflected in a curvilinear Schild regression.

regression reflecting saturation of effect. Therefore, obtaining non-linear Schild regressions is an effective method of determining allosteric mechanisms for surmountable antagonists. The caveat to this approach is that very large dextral displacement values may be observed for NAMs with very small α values. Therefore, as large a concentration as possible should be applied to these analyses.

Figure 2.3 illustrates another possible strategy. If one agonist does not obviously yield a non-linear Schild regression, then it is possible that another will. This is because probe dependence dictates that α and β values are unique to the receptor probe (agonist). Therefore, an agonist with a larger value of α may demonstrate non linearity. Figure 2.3 also shows the dependence of non-linear Schild regressions on the value of α. As seen in Figure 2.3, the intercept of the Schild regression (estimated pA_2) is quite insensitive to the use of different agonist probes. This can also be seen from eqn (2.2) for the pA_2 (*i.e.*, for DR = 2, pA_2 = log(K_B/(1 − 2α)); for low values of α (NAMs) differences in α values, as would be seen with probe dependence, would cause little change in the pA_2).

Saturation of allosteric effect also can be utilized for insurmountable antagonists. However, if β is sufficiently low, then a given NAM may produce depression of the maximal response to basal levels with no obvious

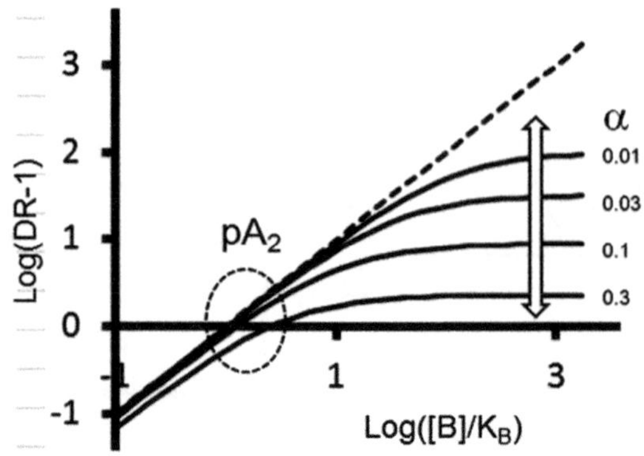

Figure 2.3 Values of α, which vary with agonist probe dependence, determine where along the NAM concentration axis of a Schild regression curvilinearity occurs. However, the intercept of Schild regression is insensitive to probe dependence and varies little with different agonists for NAMs.

saturation of effect (see Figure 2.4). One strategy that can be employed to unveil saturation of effect is to increase the receptor reserve of the system, *i.e.* increase the receptor density of the tissue and/or increase the coupling efficiency of each receptor to the stimulus–response mechanisms of the cell. Under these circumstances, the saturation of allosteric effect may better be observed (Figure 2.4).

Another strategy to unveil NAM activity is to detect possible differentiation of blockade of various signaling pathways – *i.e.*, observation of induced bias by NAMs on agonism. Orthosteric antagonists will block all signaling of a given agonist equally; in contrast, a NAM may not do so. Specifically, there are NAMs that block one signaling pathway to an agonist but not others. For example, Figure 2.5 shows the effects of the modulator indole1 on PDG2 activation of CRTH2 receptors. While PDG2 normally activates Gq protein and β-arrestin, after binding of indole1, the β-arrestin activation is blocked but the Gq signaling remains.[2] Induction of signaling bias also has been reported for NAMs for neurokinin receptor 2 (NK2) receptors[23] and calcium-sensing receptors.[24,25] Therefore, if maximal antagonism of different pathway signals can be demonstrated for a given antagonist, this would be presumptive evidence that the antagonism is allosteric.

2.5.1 Pharmacologic Resultant Analysis

Another potential strategy to detect allosteric mechanisms is to take advantage of the additive nature of orthosteric competitive antagonists. Specifically, it can be shown that for a defined magnitude of competitive blockade (eqn (2.1)), the dose-ratio to a mixture of competitive antagonists $[B_1]$ and $[B_2]$ is given by:

$$DR' = [B1]/K_{B1} + [B2]/K_{B2} + 1 \tag{2.8}$$

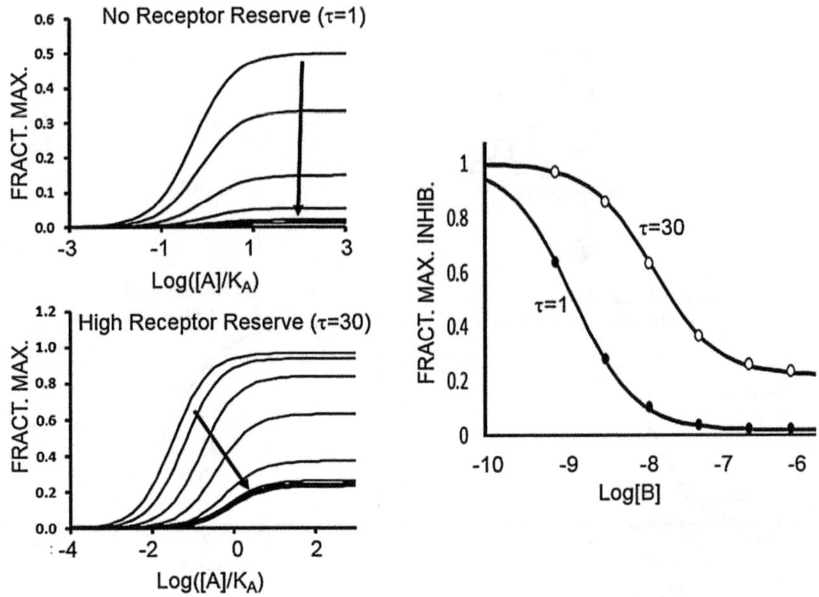

Figure 2.4 The effect of receptor reserve on the observed effects of insurmountable antagonists. In systems with little agonist receptor reserve, insurmountable antagonists, such as NAMs with low β values produce immediate depression of the agonist maximal response (top left panel). In contrast, in systems with a high receptor reserve for the agonist, the same NAM will produce dextral displacement of agonist concentration–response curves before depression of the maximal response (bottom left panel). The magnitude of the receptor reserve (different values of τ for the agonist) also can affect the value of the half maximal inhibitory concentration (IC_{50}) of the NAM (right panel).

It can further be shown that the Schild plot to a competitive antagonist will be shifted to the right along the antagonist concentration axis in the presence of an added competitive antagonist and that the magnitude of this shift depends upon the concentration and equilibrium dissociation constant of the added antagonist. Figure 2.6 shows Schild plots to the muscarinic competitive antagonist scopolamine in the absence and presence of added concentrations of the muscarinic competitive antagonist atropine; it can be seen that the scopolamine Schild regressions are shifted to the right by the increasing concentrations of atropine. The shifts to the right of the Schild regressions can then be utilized in the resultant analysis equation[26] of the form:

$$\mathrm{Log}(Y-1) = \mathrm{Log}(\mathrm{B}_{\mathrm{test}}) - \mathrm{Log}\,K_{\mathrm{B\,-\,test}} \tag{2.9}$$

where Y is the ratio of K_B values for scopolamine in the presence and absence of atropine at the various concentrations of atropine. Eqn (2.9) describes a straight line referred to as a resultant plot where the abscissal intercept of

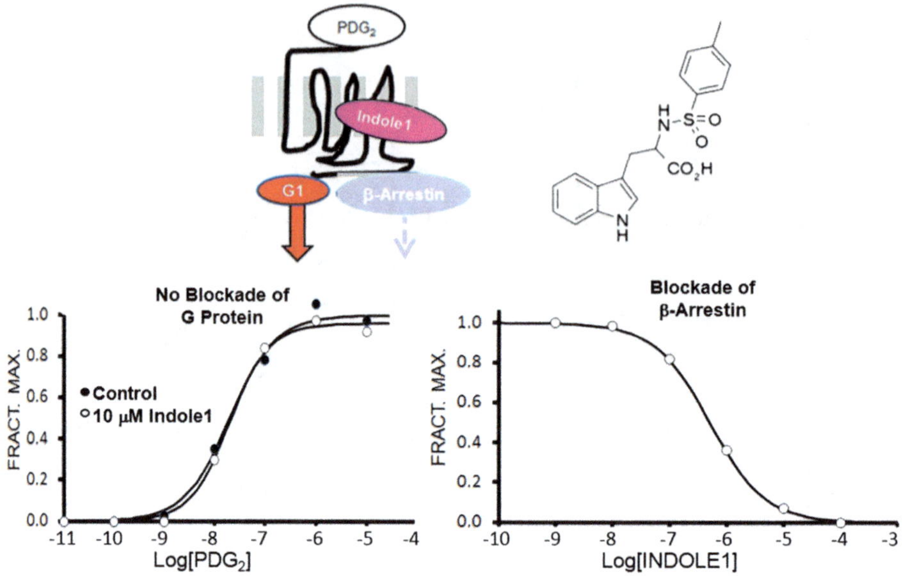

Figure 2.5 Blockade of CRTH2-mediated agonist effects of PDG2 by the NAM indole1; whereas β-arrestin signaling is blocked, G protein effects are not affected. Data from ref. 2.

which is the pK_B for the added test antagonist (Atropine). The same procedure done with the NAM alcuronium shows how dose-ratio additivity does not hold for ligands that do not compete for the same binding site. As seen in Figure 2.6, the resultant plot for alcuronium (according to eqn (2.9)) is curvilinear and does not adhere to a system of co-binding orthosteric competitive antagonists.[27]

2.5.2 PAM-Antagonists

Since allosteric ligands can have separate effects on the affinity and efficacy of other ligands for the receptor, there is always a possibility that an allosteric molecule can have opposing effects for these two receptor properties; For example as discussed previously the ester CPCCOEt (7-hydroxyiminocyclopropan[*b*]chromen-1*a*-carboxylic acid ethyl ester) is a non-competitive antagonist of responses to glutamate in CHO cells naturally expressing human GluR1b receptors but has no effect on glutamate binding.[6] Similarly, the CCR5 receptor allosteric modulator aplaviroc has no effect on the binding of CCL5 to the receptor but blocks the physiological response to the chemokine CCL5.[1,5] A particularly interesting and therapeutically relevant type of modulator in this regard is the PAM-antagonist. These modulators increase the affinity of the receptor for the agonist but decrease the efficacy. Thus, the receptor forms more ternary complex of agonist–receptor–modulator but the complex is sterile with respect to signaling, *i.e.* no agonism is

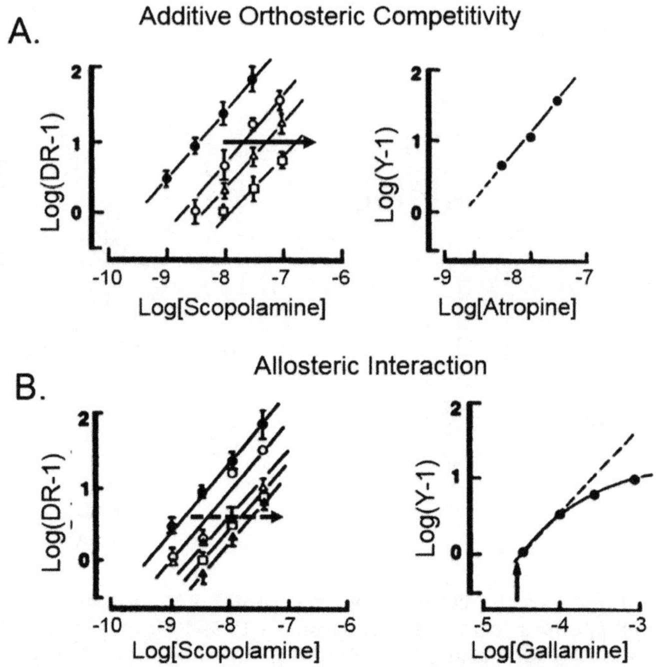

Figure 2.6 Resultant analysis. Panel A shows the effect of increasing concentrations of atropine on the Schild regression for scopolamine as a blocker of muscarinic response. The additive effect of atropine on scopolamine receptor occupancy results in a linear resultant plot. Panel B shows the different effects of the allosteric antagonist gallamine on the Schild regressions to scopolamine. The curvilinearity of the resultant plot indicates the scopolamine and gallamine do not bind to the same site on the receptor. Redrawn from ref. 27.

produced. This effect can be seen with the cannabinoid receptor PAM-antagonist Org27569 (Figure 2.7).[21] The important feature of such an antagonist is that, just as the modulator increases the affinity for the receptor for the agonist, so too does the agonist increase the affinity of the receptor for the modulator. Under these circumstances, the antagonist becomes more potent as the concentrations of agonist increase. This is because the energetic effects of modulators are reciprocal through the conduit of the receptor protein, *i.e.* an allosteric effect from the modulator through the conduit to the guest travels equally well in the opposite direction of guest through conduit to the modulator; this has been shown in dual radioligand binding studies.[7] Currently, PAM-antagonists have been reported for *N*-methyl-D-aspartate (NMDA) receptors,[28] free fatty acid (FFA)-3 receptors[29] and cannabinoid CB-1 receptors.[21,30] In terms of detection, a hallmark of PAM-antagonists is that their negative log of the half maximal inhibitory concentration (pIC_{50}) values are the opposite of what would be expected of a conventional competitive antagonist, *i.e.* the half maximal inhibitory

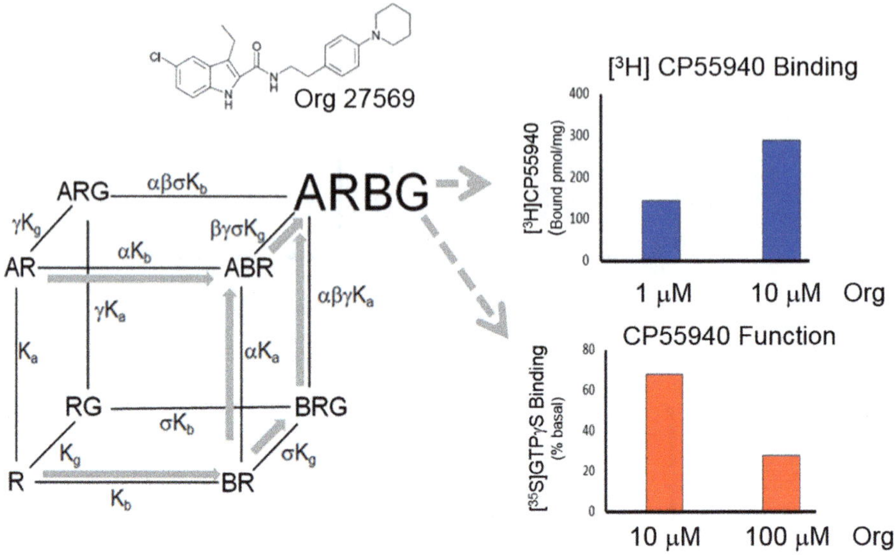

Figure 2.7 Effects of the cannabinoid receptor 1 PAM-antagonist Org 27569 on responses to the opioid agonist CP55940 (antagonism shown in red bars) and binding of [^{3}H]-CP55940 (blue bars). The PAM-antagonist promotes formation of a ternary complex between CP55940 and the receptor that cannot signal. Redrawn from ref. 28.

concentration (IC_{50}) values for the blockade of high concentrations of agonist are actually lower than the IC_{50} values required to block lower concentrations of agonist.

2.6 Kinetics

Finally, the positive proof of allosteric effect is through demonstration of changes in ligand binding kinetics after treatment with the putative allosteric molecule. Specifically, the indicators of receptor conformation are k_1, the rate of ligand association with the receptor and k_2 the rate of ligand dissociation from the receptor; it is assumed that a change in the conformation of the receptor will cause a change in one or both of these. Therefore, the offset (k_2) or onset (k_1) of a test ligand is measured in the absence and presence of a putative allosteric modulator and if a change is seen, then allosterism is confirmed.[31] Orthosteric ligands cannot modify these values since they obstruct the binding to preclude measurement, *i.e.* only an allosteric modulator will change the kinetics of a test ligand. Figure 2.8 shows the effect of the allosteric modulator ρ-TIA on the offset of [^{3}H]-prazosin; it can be seen that the rate of dissociation increases, indicating NAM activity for ρ-TIA.[32]

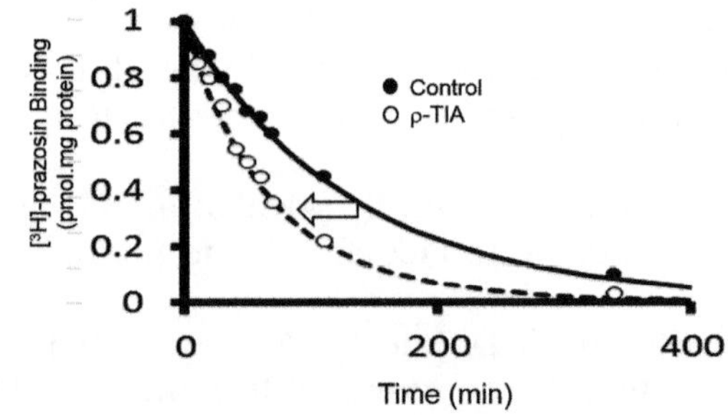

Figure 2.8 Allosteric increase in the rate of dissociation of [^{3}H]-prazosin from α adrenoceptors by the NAM ρ-TIA (ρ-conopeptide).[32]

2.7 Conclusions

In view of the vastly different effects of allosteric and orthosteric ligands on agonist response, it is advantageous to identify and differentiate allosteric modes of action. In cases where ligands increase the response to agonists, there is no question that allosteric mechanisms are responsible, However, for antagonists, there are cases where it may not be clear, as many negative allosteric antagonists (NAMs) can produce patterns of antagonism very similar to those produced by orthosteric antagonists. In these cases, special experiments may need to be done to uncover differences. The first major principle that can be employed is the observation of saturation of effect. NAMs produce limited degrees of antagonism determined by the values of the allosteric cooperativity constants α and β. Thus, the antagonism produced by a NAM will reach a maximal asymptote when the allosteric binding site is fully occupied. Observance of such an asymptote is evidence of an allosteric mode of action. The second operative principle is that of probe dependence in that the maximal antagonism produced to agonists may depend on the agonist and not be constant. In this regard observance of different maximal blockade of different signaling pathways by an antagonist (induced-bias) is presumptive evidence of an allosteric mechanism. Finally, detailed quantitative assessment of additive effects of antagonists (resultant analysis) can unveil allosteric mechanisms. Once an allosteric mechanism has been identified, demonstration of the ability of the ligand to modify the kinetics of binding of other ligands for the receptor is confirmatory evidence of the mechanism.

References

1. C. Watson, S. Jenkinson, W. Kazmierski and T. Kenakin, *Mol. Pharmacol.*, 2005, **67**, 1268–1282.
2. J. M. Mathiesen, T. Ulven, L. Martini, L. O. Gerlach, A. Heinemann and E. Kostenis, *Mol. Pharmacol.*, 2005, **68**, 393–402.

3. T. Kenakin, *Trends Pharmacol. Sci.*, 2007, **28**, 407–415.
4. A. Trkola, S. E. Kuhmann, J. M. Strizki, E. Maxwell, T. Ketas, T. Morgan, P. Pugach, S. Xu, L. Wojcik and J. Tagat, *Proc. Natl. Acad. Sci. U. S. A.*, 2002, **99**, 395–400.
5. K. Maeda, H. Nakata, Y. Koh, T. Miyakawa, H. Ogata, Y. Takaoka, S. Shibayama, K. Sagawa, D. Fukushima and J. Moravek, *J. Virol.*, 2004, **78**, 8654–8662.
6. S. Litschig, F. Gasparini, D. Rueegg, N. Stoehr, P. J. Flor, I. Vranesic, L. Pre′zeau, J. P. Pin, C. Thomsen and R. Kuhn, *Mol. Pharmacol.*, 1999, **55**, 453–461.
7. C. Tränkle, O. Weyand, A. Schröter and K. Mohr, *Mol. Pharmacol.*, 1999, **56**, 962–965.
8. C. Melchiorre, A. Minarini, P. Angeli, D. Giardina, U. Gulini and W. Quaglia, *Trends Pharmacol. Sci.*, 1989, **10**, 55–59.
9. J. Ellis, J. Huyler and M. R. Brann, *Biochem. Pharmacol.*, 1991, **42**, 1927–1932.
10. J. S. Liang, J. Carsi-Gabrenas, J. L. Krajesewski, J. M. McCafferty, S. L. Parkerson and P. M. Santiago, *Toxicon*, 1996, **34**, 1257–1267.
11. A. L. Gnagey, M. Seidenberg and J. Ellis, *Mol. Pharmacol.*, 1999, **56**, 1245–1253.
12. M. P. Johnson, E. S. Nisenbaum, T. H. Large, R. Emkey, M. Baez and A. E. Kingston, *Biochem. Soc. Trans.*, 2004, **32**, 881–887.
13. V. M. Muniz-Medina, S. Jones, J. M. Maglich, C. Galardi, R. E. Hollingsworth, W. M. Kazmierski, R. G. Ferris, M. P. Edelstein, K. E. Chiswell and T. P. Kenakin, *Mol. Pharmacol.*, 2009, **75**, 490–501.
14. E. Gonzalez, H. Kulkarni, H. Bolivar, A. Mangano, R. Sanchez, G. Catano, R. J. Nibbs, B. I. Freedman, M. P. Quinones, M. J. Bamshad, K. K. Murthy, B. H. Rovin, W. Bradley, R. A. Clark, S. A. Anderson, R. J. O'connell, B. K. Agan, S. S. Ahuja, R. Bologna, L. Sen, M. J. Dolan and S. K. Ahuja, *Science*, 2005, **307**, 1434–1440.
15. O. Arunkakshana and H. O. Schild, *Br. J. Pharmacol.*, 1959, **14**, 48.
16. L. Hejnová, S. Tuce and E. E. el-Fakahany, *Eur. J. Pharmacol.*, 1995, **291**, 427–430.
17. J. M. Stockton, N. J. Birdsall, A. S. Burgen and E. C. Hulme, *Pharmacol.*, 1983, **23**, 551.
18. F. J. Ehlert, *Mol. Pharmacol.*, 1988, **33**, 187.
19. J. W. Black and P. Leff, *Proc. R. Soc. London, Ser. B*, 1983, **220**, 141.
20. T. Kenakin, *Nat. Rev. Drug Discovery*, 2005, **4**, 919.
21. M. R. Price, G. L. Baillie, A. Thomas, L. A. Stevenson, M. Easson, R. Goodwin, A. McLean, L. McIntosh, G. Goodwin and G. Walker, *Mol. Pharmacol.*, 2005, **68**, 1484.
22. F. J. Ehlert, *J. Pharmacol. Exp. Ther.*, 2005, **315**, 740–754.
23. E. L. Maillet, N. Pellegrini, C. Valant, B. Bucher, M. Hibert, J. J. Bourguignon and J. L. Galzi, *FASEB J.*, 2007, **21**, 2124–2134.
24. A. E. Davey, K. Leach, C. Valant, A. D. Conigrave, P. M. Sexton and A. Christopoulos, *Endocrinology*, 2012, **153**, 1232–1241.

25. A. E. Cook, S. N. Mistry, K. J. Gregory, S. G. B. Furness, P. M. Sexton and P. J. Scammells, *Br. J. Pharmacol.*, 2015, **172**, 185–200.
26. J. W. Black, V. P. Gerskowich and P. Leff, *Br. J. Pharmacol.*, 1986, **89**, 547–555.
27. T. P. Kenakin and C. Boselli, *J. Pharmacol. Exp. Ther.*, 1989, **250**, 944–952.
28. J. N. C. Kew, G. Trube and J. A. Kemp, *J. Physiol.*, 1996, **497**(3), 761–772.
29. B. D. Hudson, E. Christiansen, H. Murdoch, L. Jenkins, A. H. Hansen and O. Madsen, *Mol. Pharmacol.*, 2014, **86**, 200–210.
30. G. L. Baillie, J. G. I. Horswil, S. Anavi-Goffer, P. H. Reggio, D. Bolognini, M. E. Abood, S. McAllister, P. G. Strange, G. J. Stephens, R. G. Pertwee and R. A. Ross, *Mol. Pharmacol.*, 2013, **83**, 322.
31. K. Leach, P. M. Sexton and A. Christopoulos, *Curr. Protoc. Pharmacol.*, 2011, **1**(22), 1–41.
32. I. A. Sharpe, L. Thomas, M. Loughnan, L. Motin, E. Palant, D. E. Croker, D. Alewood, S. Chen, R. M. Graham, P. F. Alewood, D. J. Adams and R. J. Lewis, *J. Biol. Chem.*, 2003, **278**, 34451–34457.

CHAPTER 3

Targeting Catalytic and Non-Catalytic Functions of Protein Kinases

SUSANNE MÜLLER[a,b,c] AND STEFAN KNAPP*[a,b,c]

[a]Nuffield Department of Clinical Medicine, University of Oxford, Structural Genomics Consortium, Old Road Campus Research Building, Roosevelt Drive, Oxford OX3 7DQ, UK; [b]Nuffield Department of Clinical Medicine, University of Oxford, Target Discovery Institute (TDI), Roosevelt Drive, Oxford OX3 7BN, UK; [c]Johann Wolfgang Goethe-University, Institute for Pharmaceutical Chemistry and Buchmann Institute for Life Sciences, Max-von-Laue-Str. 9, D-60438 Frankfurt am Main, Germany
*E-mail: knapp@pharmchem.uni-frankfurt.de

3.1 Introduction

Protein kinases constitute a family of more than 500 proteins in human with central functions in regulating cell fate.[1] Due to their key roles in cellular signalling, kinases are frequently deregulated in disease and have therefore been extensively targeted for the development of new drugs. To date, more than 30 kinase inhibitors have been approved for treatment. Most of these drugs have been developed for applications in the oncology area. Applications outside oncology have been limited, which is mainly a consequence of a lack of design strategies for highly selective inhibitors and certain

RSC Drug Discovery Series No. 56
Allosterism in Drug Discovery
Edited by Dario Doller

Published by the Royal Society of Chemistry, www.rsc.org

limitations of classical ATP-competitive kinase inhibitors that are often flat and hydrophobic compounds with limited tissue distribution *in vivo*. For instance, development of brain-penetrant kinase inhibitors has remained challenging. The main focus of kinase drug discovery has therefore been on a few well-studied kinases in cancer that are often over-expressed, activated by mutations or kinase oncogenic fusions with little change in target selection.[2] Currently only approximately 40 kinase targets have entered clinical testing.[3] Thus, the remaining target space provides ample opportunities for future inhibitor development projects.[4]

Most small-molecule kinase inhibitors bind to the nucleotide binding site, thereby preventing ATP from binding. The kinase family has been structurally well explored. Kinase catalytic domain X-ray crystal structures are now available for more than 200 kinase targets and for many kinases descriptions of multiple inhibitor complexes have been published. This large repository of kinase structural information revealed an enormous diversity of conformations that can be targeted. A comprehensive analysis of kinase inhibitor binding mode has recently been made available by van Linden *et al.* This effort resulted also in the development of the Kinase–Ligand Interaction Fingerprints and Structure (KLIFS) database.[5] Despite the large structural diversity of kinase inhibitor complexes, the main binding modes have been classified as type I (inhibitors binding the active state), type II (inhibitors binding to an inactive conformation where the tripeptide motif Asp–Phe–Gly (DFG) is in an "out" position), type III (allosteric inhibitors that don't occupy the ATP binding site), and type IV (inhibitors binding distal to the ATP site, for instance occupying the substrate binding site). To date the majority of kinase inhibitor complexes revealed a canonical ATP-competitive type-I binding mode whereas structures of type-II binding modes are underrepresented, in particular in the Ser–Thr kinase family.[6] However, development of multitargeting type-II kinase libraries demonstrated that most, if not all kinases, can adopt a DFG-out conformation offering *per se* no advantage in targeting the DFG-out pocket for the development of selective kinase inhibitors.[6] Only few examples of allosteric (type-III) inhibitors have been described so far. Interestingly most type-III inhibitors are highly selective. Thus, a broader development effort of type-III inhibitors is likely to lead to new drug development opportunities for applications where exquisite selectivity would be desired. In addition, allosteric kinase inhibitors are chemically highly diverse, enabling the exploitation of a much wider chemical space. Considering the high degree of domain plasticity of kinases and the often drastic structural rearrangements during kinase activation, the comparably small contribution of allosteric inhibitors to our kinase drug and tool compound repertoire is surprising. The main reason for this is that allosteric kinase inhibitors are difficult to identify by general screening methods or to rationally develop and there are to date no promiscuous type-III inhibitors or inhibitor libraries that target allosteric binding modes available. In fact, most allosteric kinase inhibitors known to date have been identified serendipitously by structural biology studies on kinase–ligand complexes.

The importance of domain dynamics and plasticity in kinase signalling and the influence of kinase non-catalytic functions is well known.[7] Kinase mutations that drive disease development and progression also significantly modulate non-catalytic kinase function. It is therefore not surprising that the binding mode of kinase inhibitors and the effects of inhibitors on catalytic domain plasticity has a major effect on inhibitor response. RAF (RAS associated factor) kinases for instance depend on allosteric activation by heterodimerization of BRAF or CRAF following RAS activation.[8] In contrast, the oncogenic driver mutation BRAF V600E is active independent of RAS and stimulates the growth-promoting mitogen-activated protein kinase (MAPK) pathway through the active monomeric V600 BRAF mutant or constitutively dimeric mutant BRAF.[9–11] Thus, kinase inhibitors that stabilize dimerization lead to activation of wild type BRAF signalling and to growth-promoting effects in tissues that express wild-type BRAF. Similarly, allosteric AKT inhibitors but not ATP competitive type-I inhibitors alter AKT scaffolding and non-catalytic functions resulting in differential cellular activity and *in vivo* efficacy and different effects on the AKT signalling network.[12]

These clinically highly relevant observations prompt the question of what is the structural mechanism mediating non-catalytic functions in kinases and the relevance of these structural changes to kinase signalling networks and inhibitor response. Thus, allosteric kinase inhibitors will be essential chemical tools not only for our understanding of the role of the catalytic, but also kinase non-catalytic function in disease development, signalling network response and therapeutic drug efficacy. Such studies would also include catalytically inactive, so call pseudokinases that serve as essential scaffold and activators in many signalling pathways.[13,14] Here we review the main binding modes of kinase inhibitors with a particular focus on allosteric kinase inhibitors.

3.2 The Kinase Active State

Despite large sequence variations, the overall folding of the kinase catalytic domain is highly conserved. The canonical kinase fold consists of two domains, also called lobes (Figure 3.1A). The core structure of the smaller N-terminal lobe (N-lobe) comprises five β-sheets (β1–β5) and helix αC whereas the larger C-terminal lobe (C-lobe) is mainly alpha helical. Both lobes are linked by a flexible hinge. The ATP binding site is located between the two kinase lobes and the co-factor is coordinated by structural elements located in both kinase lobes as well as by conserved hydrogen bonds formed between the adenosine ring and the hinge backbone. The loop region between sheets β1 and β2 contains a flexible glycine-rich sequence motif (GXGXφG, where φ is a tyrosine or phenylalanine residue) called the phosphate binding loop or P-loop. The flexibility of the glycine residues in the P-loop facilitates coordination of the ATP phosphate moieties *via* backbone interactions. A second highly conserved motif is also essential for ATP binding: the tripeptide motif Asp–Phe–Gly (DFG). In addition, a conserved lysine residue present in the

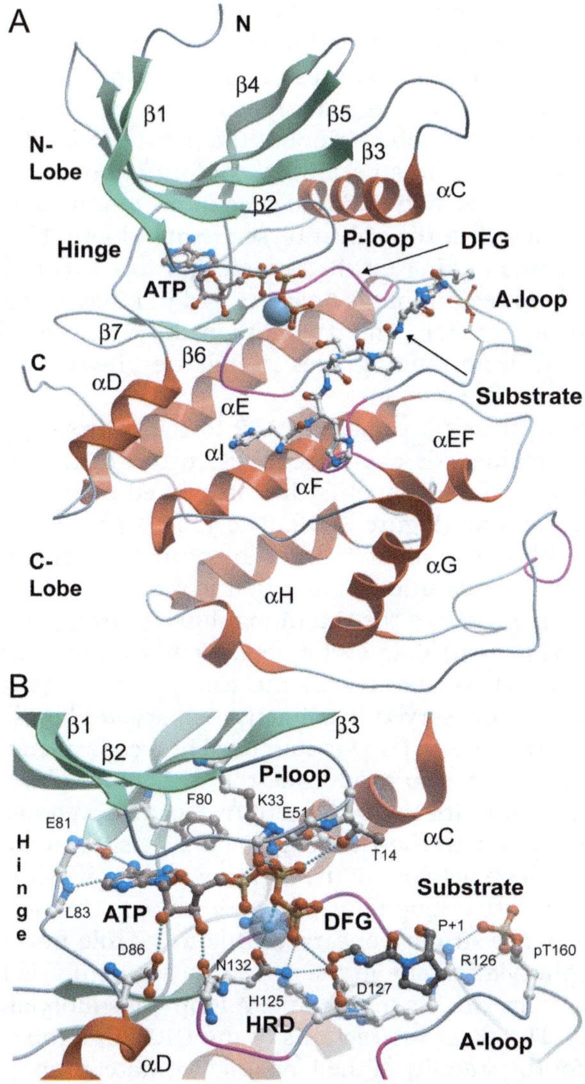

Figure 3.1 Structural elements of the kinase catalytic domain. A: Shown is the structure of a cyclin-dependent kinase 2 (CDK2) ATP and peptide complex (pdb entry: 1qmz).[93] The main secondary structure elements are labelled and conserved kinase catalytic domain structural elements as well as the N and C termini are highlighted in bold type. ATP and the substrate peptide are shown as ball and stick representations and the Mg^{2+} ion is shown as a sphere. The termini are labelled N and C respectively. B: Kinase active site and coordination of ATP. Shown is a CDK2 substrate–ATP complex (pdb entry: 1qmz) that is phosphorylated at the activation segment (pT160). Some side chains and part of the substrate (grey peptide carbon atoms) have been truncated for clarity. Hydrogen bonds are shown as dotted lines.

VIAK (Val–Ile–Ala–Lys) motif located in the β3 sheet stabilizes the ATP phosphate groups by salt bridges. In active kinases, this lysine also forms a salt bridge to a conserved glutamate residue located in αC, indicating correct positioning of this important regulatory helix.

The C-terminal kinase lobe harbours the substrate binding site. Kinase substrates typically bind in an extended conformation to a shallow surface groove that is formed by side chains located in the four main C-terminal lobe helices (αD–αG) as well as the activation segment (A-loop). The activation segment has a key role in kinase substrate recognition. It comprises usually a 20–30 residue loop region that is unstructured in kinases that are regulated by A-loop phosphorylation. The flexibility of the A-loop results in partially unstructured substrate binding sites in inactive kinases, at least partially impeding substrate interactions.

The transfer of the γ-phosphate of ATP to the hydroxyl group of serine, threonine, or tyrosine side chains requires correct spacial arrangement of conserved catalytic domain motifs that are linked by a network of largely conserved interactions (Figure 3.1B). The kinase active state is therefore structurally highly constrained in order to optimally align structural motifs for catalysis. A key structural element that is required for kinase activity is the catalytic loop located in the C-terminal lobe following helix αE. The catalytic loop (Asp127–Asn132 in cyclin-dependent kinase 2 (CDK2)) harbours the His–Arg–Asp (HRD) motif with the catalytic base Asp127. The HRD aspartate is strictly conserved in all active eukaryotic kinases. The proper side chain conformation of D127 is stabilized by a conserved hydrogen bond with Asn132. The other two residues of the HRD motif play a key role linking the catalytic loop with the DFG motif (*via* a conserved interaction of the histidine side chain with the DGF backbone) as well as with the activation loop (*via* a salt bridge network of the arginine side chain with the phosphate moiety of the activation loop phosphorylation site (pThr160)). However, the HRD arginine is not strictly conserved. Due to its role in coordinating the primary phosphorylation site in the activation segment it is mostly but not exclusively found in kinases that require A-loop phosphorylation for activity (RD kinases).[15] The polar interactions of the HRD arginine with the A-loop are essential for the stability of the kinase active state, proper activation segment conformation, and for some kinases A-loop phosphorylation provides also specificity of substrate binding.[16]

Catalytic activity also requires proper orientation of the N-terminal and C-terminal kinase lobes. Recently, Taylor and co-workers discovered that in active kinases two spatially highly conserved but non-contiguous sets of hydrophobic residue align to form hydrophobic "spines" that span both kinase lobes in active kinases.[17] The two spines interconnect and position all structural elements important for kinase catalytic activity and link them to the central C-terminal-lobe helix αF. The "Catalytic spine" (C-spine) integrates also the ATP co-factor. For example, in the active conformation of the kinase PKA (Protein Kinase A) the C-spine is initiated by two bulky hydrophobic residues (Met231 and Leu227) located in helix αF (Figure 3.2). Helix αD contributes Met128 which in turn interacts with two

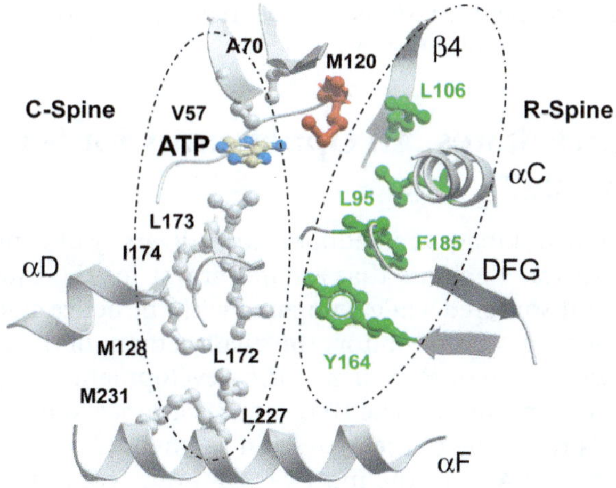

Figure 3.2 Hydrophobic spines in active kinases. The two hydrophobic spines are shown using the kinase PKA as an example (pdb-code: 1atp). Residues participating in the C- (white) and R-spines (green) as well as the linking residue M120 (red) and secondary structure elements that build a platform for these residues are highlighted.

residues located in sheet β7 (Leu172, Ile174). The third hydrophobic residue in the β7 sheet (Leu173) extends the C-spine to the ATP binding side and creates a hydrophobic docking site for the adenine ring of ATP which bridges the C-spine to the N-terminal kinase lobe in the ATP-bound state. In agreement with the important structural role of the C-spine, mutation of Leu173 to alanine in PKA and the analogous residue in CDK2 results in kinase inactivation.[18] Val57 and Ala70 located in the N-terminal lobe complete the catalytic spine.

The "Regulatory spine" (R-spine) in PKA is initiated by Tyr164 in the C-terminal kinase lobe. The R-spine continues *via* the DFG phenylalanine (Phe185) to the N-terminal kinase lobe by forming tight interactions with the αC helix Leu95 and Leu106 located in the β4 strand. Since the regulatory spine undergoes considerable structural rearrangements in inactive kinases the intact R-spine can be used as a structural hallmark of the kinase active state and the correct alignment of regulatory structural elements in active kinases. The gatekeeper residue located on the N-terminal hinge region occupies an interesting position linking both hydrophobic spine residues. Gatekeeper residues are variable in kinases and determine the shape and size of the ATP pocket, but they are never a glycine in human kinases. Introducing glycine at the gatekeeper position is an elegant way to exclusively target mutated kinases in cells with nucleotide analogues binding in the generated pocket.[19,20] It is interesting to note that introduction of a large hydrophobic residue at the gatekeeper position in the tyrosine kinases SRC (the product of the Rous sarcoma oncogene homolog) and ABL (encoded by the Abelson murine leukaemia viral oncogene homolog) instead of the small threonine present in the wild-type protein stabilizes the active state by locking the

R-spine, and gatekeeper mutations have been detected in oncogenic mutants resulting in constitutively active kinases.[21,22]

3.3 Inactive States: An Opportunity for Selective Targeting?

The inactive state of kinases can adopt a large diversity of conformations in which key structural elements that are important for the catalytic domain are displaced or disordered when compared with the active state.[23–25] For the development of drugs and inhibitors the displacement of the DFG motif has been extensively explored, resulting in the development of a large number of so called type-II inhibitors. The DFG motif is a conserved tripeptide motif located at the N-terminus of the activation segment (A-loop) important for binding of Mg^{2+} and ATP. In the inactive "DFG-out" conformation the DFG phenylalanine flips around the peptide backbone dislodging a hydrophobic pocket at the C-terminus of helix αC. This structural rearrangement opens a large allosteric pocket (Figure 3.3A and B). The large extension of the ATP binding site has been targeted by a large diversity of inhibitors including the first Food and Drug Administration (FDA) approved ABL inhibitor gleevec.[26,27] Selectivity screening data and molecular dynamics simulations showed that

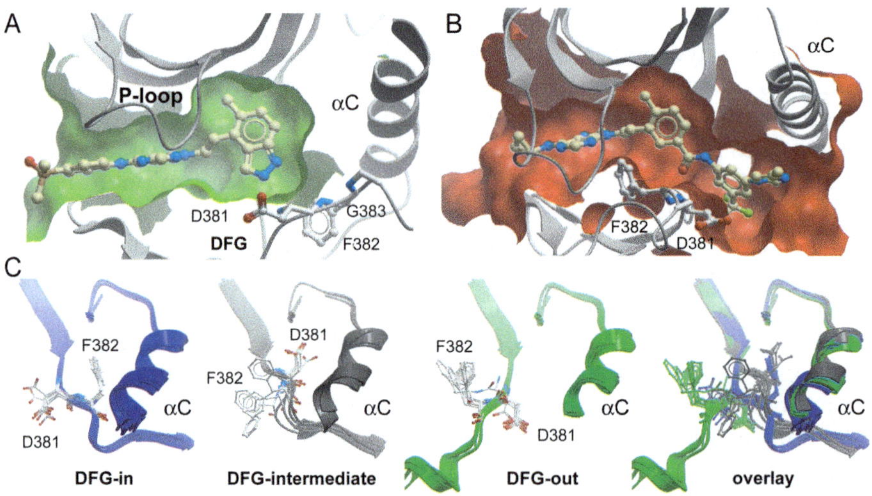

Figure 3.3 Comparison of the ATP binding site in the DFG-in and DFG-out conformation of ABL kinase. Shown are the active state (A) (pdb-code: 3kf4) and the DFG-out conformation (B) (pdb-code: 3kfa).[94] The ATP pocket is shown as a solid surface to demonstrate the structural differences within the pocket. The DFG phenylalanine is labelled in both panels. Inhibitors binding to the active state and inactive DFG out state are shown in ball and stick representation. C: Comparison of canonical DFG-in (blue) and DFG-out (green) conformations as well as intermediate states (grey) observed in ABL inhibitor complexes. An overlay of the DFG motifs in 19 ABL inhibitor complexes is shown in the right panel.

differences in domain dynamics and in energy barriers associated with the DFG-out movement may result in selectivity to closely related kinases such as SRC.[28] In some kinases such as the insulin receptor, the DFG-out conformation has even been observed in apo-structures, indicating that it resembles a low-energy state.[24] However, results from large comparative studies showed that DFG-out inhibitors are not *per se* more selective, indicating that more kinases seem to be able to assume this conformation than originally anticipated.[6,29] Type-II inhibitors typically resemble lipophilic interactions within the DFG-out pocket formed by the DFG phenylalanine, thus stabilizing αC in an active "in" conformation. Interestingly, DFG-out conformations can be highly diverse. Depending on interactions formed with the type-II inhibitor the DFG motif can assume a diversity of intermediate conformations (Figure 3.3C). In ABL kinase for instance, in some of the DFG intermediate conformations the aspartate may replace the phenylalanine closing, at least partially, the DFG-out pocket. The outward-flipped sidechain of the DFG phenylalanine often engages in aromatic π-stacking interaction in DFG-out conformations, providing a rationale for some of the unusual conformations that have been observed in crystal structures.

The phosphate-binding loop (P-loop), also called the glycine-rich loop, is another dynamic element of the kinase catalytic domain. In its ATP bound "active conformation" the P-loop engages in a network of backbone hydrogen bonds with the phosphate moieties in ATP. The tip of the P-loop contains a conserved aromatic residue (either tyrosine or phenylalanine) that lines the ATP binding pocket. In kinase apo-structures, the P-loop is often disordered due to the lack of stabilizing interactions with ATP or inhibitors. Interestingly, a number of inactive conformations have been described in which the P-loop occupies part of the ATP binding site engaging in aromatic interactions mediated by the conserved aromatic residues at the P-loop tip (Figure 3.4A and B). These folded P-loop conformations have been associated with favourable inhibitor selectivity and ligand efficacy.[30] However, the consequence of folded P-loop conformation is also that they open an allosteric binding site between the P-loop and helix αC. If this binding site is accessible it can be targeted by ATP-competitive inhibitors as demonstrated by the highly selective extracellular signal-regulated kinase (ERK)1/2 inhibitor SCH772984[31] (Figure 3.4C). In the ERK1/2 complex with SCH772984, the P-loop residue Tyr36 interacts with the inhibitor resulting in a highly distorted and folded P-loop conformation and opening of the P-loop pocket. However, to date only few inhibitors have been developed targeting this binding pocket.

3.4 Highly Selective Kinase Inhibitors Target Unique Binding Pockets

Kinome-wide selectivity screening together with structural studies revealed a number of binding modes of highly selective inhibitors. These studies showed that usually more than one unique active site property has been explored in order to develop target selective inhibitors. For instance, the highly selective

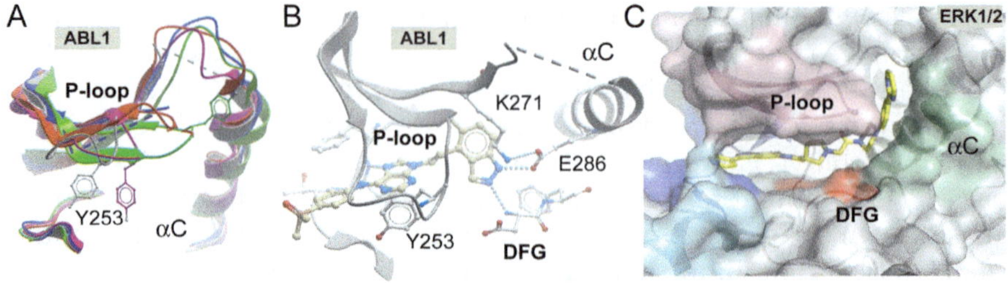

Figure 3.4 Plasticity of the P-loop and its implication for drug design. A: Overview of different P-loop conformations in ABL ranging from the "active" P-loop conformation (green; pdb-code: 2hz0), structures with disordered and partially disordered P-loop (blue and red; pdb-code: 2hz4; 3ue4), folded P-loop conformation (magenta; pdb-code: 2gqg) and distorted folded P-loop conformation (gray; pdb-code: 3kf4). B: Example of inhibitor interaction of a folded distorted P-loop conformation (pdb-code: 3kf4). C: Targeting of the P-loop pocket generated by folded P-loop conformations in the kinases ERK1/2.

MET (the hepatocyte growth factor receptor) inhibitor SGX523 binds to a DFG-in conformation with excellent shape complementarity and targets an additional binding pocket created by an unusual conformation of the activation segment.[32] Also the highly selective DFG-in inhibitor GSK2606414 targets a unique binding pocket created by an inactive activation segment conformation in the protein kinase R (PKR)-like endoplasmic reticulum kinase (PERK).[33] The approved anti-cancer drug lapatinib targets a DFG-in conformation inducing large, inactivating conformational rearrangements unique to the kinase domain in epidermal growth factor receptor (EGF1R)[34] and CGI1746, an inhibitor with exquisite selectivity for the unphosphorylated form of Bruton agammaglobulinemia tyrosine kinase (BTK), induces large conformational rearrangements resulting in formation of an allosteric binding site termed the H3 pocket, located between the P-loop and the activation segment. The development of these inhibitors demonstrates that target selectivity can be achieved for some kinases. However, the high selectivity of many of these inhibitors has often been an accidental finding, revealed by selectivity screening and subsequent determination of crystal structures rather than by rational design strategies. Some active site features however might be utilized to improve selectivity of several targets. For instance, only about 5% of all kinases contain a small gatekeeper residue. An exquisitely selective inhibitor for the kinase p38α/β has been developed, combining the enlarged back pocket created by the presence of a threonine gatekeeper with a hinge backbone rearrangement facilitated by an unusual glycine residue located in the kinase hinge region.[35]

Another approach improving inhibitor selectivity is the development of irreversible inhibitors, an approach that received a lot of attention by the kinase drug development community after the first irreversible inhibitors had been approved.[36,37] Cysteine residues are frequently found in key positions in kinase active sites and a number of highly selective inhibitors have been developed by improving the selectively of reversible inhibitors targeting a specific cysteine residue.[3,38,39]

3.5 Allosteric Inhibitors

A large diversity of allosteric inhibitor binding modes has been described in crystal structures. A kinase, which has been extensively targeted by type-III inhibitors is p38α (Figure 3.5). Early studies focussed on the allosteric back pocket binders that interact with p38α in a DFG-out conformation. Examples for this type of allosteric inhibitors are phenyl-pyrazole-substituted ureas (pdb-code: 1KV1). Extension of these inhibitors into the ATP binding site led to the DFG-out inhibitor BIRB796.[40] Other DFG-out back pocket binders include disubstituted pyrazoles[41] as well as fragments such as 2-amino-3-benzyloxypyridine and 3-(2-(4-pyridyl)ethyl)indole.[42] A number of inhibitors have been described that target the p38 MAPK insert region located in the lower kinase lobe,[43,44] a structural element that has also been targeted in the MAPK c-Jun N-terminal protein kinase 1 (JNK1).[43] Inhibitors of the DNA damage signalling kinase checkpoint kinase 1 (CHK1) have also been reported to bind to a surface pocket positioned at the lower lobe in a position where usually αD is located (pdb-code: 3F9N, 3JVR, 3JVS).[45,46]

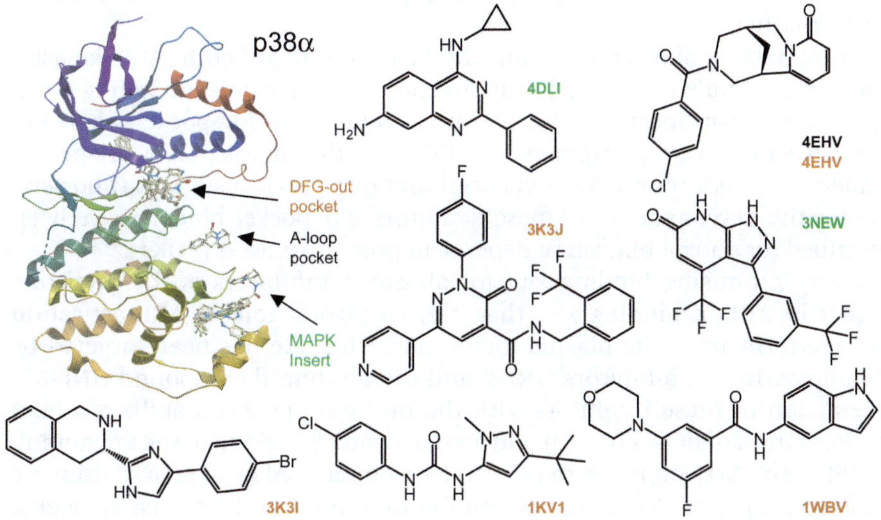

Figure 3.5 Examples of allosteric p38α inhibitors. Shown is a ribbon diagram in the left panel exemplifying the location of different co-crystallized inhibitors. Three different allosteric pockets (DFG-out, A-loop and MAPK insert) are labelled and are highlighted in red, black and green, respectively. The pdb-codes are coloured according to their corresponding binding modes.

Interestingly, fragments derived from natural products have been shown to bind to the p38α activation segment, providing potential starting points for the development of substrate-competitive inhibitors.[47] Allosteric p38α binders have been either identified by fragment screening approaches using crystallographic methods,[40,42] using fluorescent probes that target the allosteric site in a displacement assay format[41] or fluorescent labels in kinases (FLiK) that exclusively detect ligands which bind to inactive (DFG-out) kinase conformations.[44]

A recent review reported allosteric back pocket binders for 10 diverse kinases, making this binding pocket the most frequently explored site for the development of type-III inhibitors.[48] Two different types of back pocket binders have been developed: inhibitors that bind to a DFG-in conformation, which are exemplified by allosteric MEK1/2 and ITK inhibitors, and compounds that recognize a DFG-out conformation, including the p38 inhibitors described above, as well as inhibitors targeting for instance CDK2, AKT, focal adhesion kinase (FAK), insulin-like growth factor 1 receptor (IGF1R), LIM domain kinase 2 (LIMK2), p21-activated kinase (PAK1) and receptor-interacting protein kinase 1 (RIPK1). Both types of back pocket binders displace the N-terminal lobe helix αC. Dilation of this helix creates space for the allosteric binding site and inactivates the catalytic domain. Interestingly, kinases that are targeted by DFG-in back pocket binders have stable αC-out conformations also in the presence of ATP. Binding of DFG-out allosteric back pocket inhibitors usually results in occlusion of the ATP site while αC is similar to type-II binding modes, positioned in an active conformation.

An interesting allosteric binding site has been identified in AGC kinases. This kinase family is activated by interaction of a C-terminal activating motif with a binding site located at the N-terminal kinase lobe (PIF-pocket; where PIF is the PDK1 Interacting Fragment). Inhibition of this interaction by PIF-pocket binders inhibits kinase activation in cells but may also be used as an allosteric site for the development of kinase activators. PIF-pocket binders have been described for phosphoinositide-dependent protein kinase-1 (PDK1).[49–51]

A very promising binding site for allosteric inhibitors is the myristate pocket in ABL, a binding site that plays a pivotal role in ABL activation and recruitment to the plasma membrane. This site has been targeted by the allosteric ABL inhibitors GNF-2 and the optimized compound GNF-5.[52] Interaction of these inhibitors with the myristate pocket stabilizes a bent conformation of helix I resulting in an inactivating docking of the Src homology SH3 and SH2 domains to the catalytic domain. Also some ATP mimetic inhibitors such as gleevec have been found to bind to the myristate pocket with low micromolar activity[53] and some myristate pocket binders may also induce ABL activation.[54] Interestingly, when used in combination with clinical ATP-competitive inhibitors, GNF-5 reduced emergence of resistance mutations, a property that is now being explored clinically with the drug candidate ABL001.

3.6 Examples: Back Pocket Binders Recognizing a Stable DFG-In Conformation

MEK1/2 type-III inhibitors are characterized by a DFG-in conformation allowing simultaneous ATP binding as well as an αC-out conformation. The first structural information on this binding mode was provided by diaryl-amine complexes that revealed a binding pocket adjacent to the ATP binding site.[55] In these structures the arginine–glutamate ion bridge characteristic of active kinases is disrupted with K97 forming an intricate network of polar interactions with the inhibitor, the ATP phosphate backbone, the DFG aspartate (D208) as well as Mg^{2+} (Figure 3.6). In contrast, the αC glutamate (E115) was turned away from the binding site. Subsequently, a number of other highly potent diaryl-amine MEK1/2 inhibitors have been developed.[56,57] MEK1/2 have not been crystallized to date in an active conformation suggesting that the DFG-in/αC-out conformation is stable also in the absence of MEK back pocket binders. Development of MEK inhibitors has been clinically very successful with a large number of type-III inhibitors in clinical trials and the recently approved inhibitors trametinib and cobimetinib.[58–60]

For a long time MEK1/2 inhibitors were the only examples of DFG-in back pocket allosteric inhibitors. However, recently also a DFG-in back pocket inhibitor has been disclosed targeting interleukin-2-inducible T-cell kinase (ITK). Similar to MEK1/2 inhibitors, this naphthalenylpyrazole binds simultaneously with ATP. The compound has however a dual binding mode targeting also the ATP binding site. The developed inhibitor showed strong inhibition of the ITK kinase domain (20 nM) but inhibited the full-length kinase much more weakly (5 μM).[61] A shared feature of MEK1/2 and ITK DFG-in back pocket inhibitors is their high selectivity for the targeted kinase.

An interesting binding mode has been reported for inhibitors targeting the serine–threonine kinase inositol requiring enzyme 1 (IRE1), a kinase that mediates signalling pathways relevant to cellular stress induced by unfolded regulator of the unfolded protein response cascade.[62] The disclosed inhibitors are typical hinge binders containing a well-explored amino-pyrimidine ATP-mimetic group. However, similarly to allosteric MEK inhibitors, the αC helix is shifted outwards while the DFG is in an active "in" conformation, suggesting that development of allosteric back pocket binders with a DFG-in binding mode might also be possible for this kinase.[63]

3.7 Examples: Back Pocket Binders Recognizing a Stable DFG-Out Conformation

To date most allosteric kinase inhibitors target the back pocket, inducing a DFG-out conformation. Inhibitors targeting the p38α back pocket have already been discussed above. Recently, a highly selective back pocket DFG-out inhibitor has been disclosed by Novartis. This dibenzodiazepine

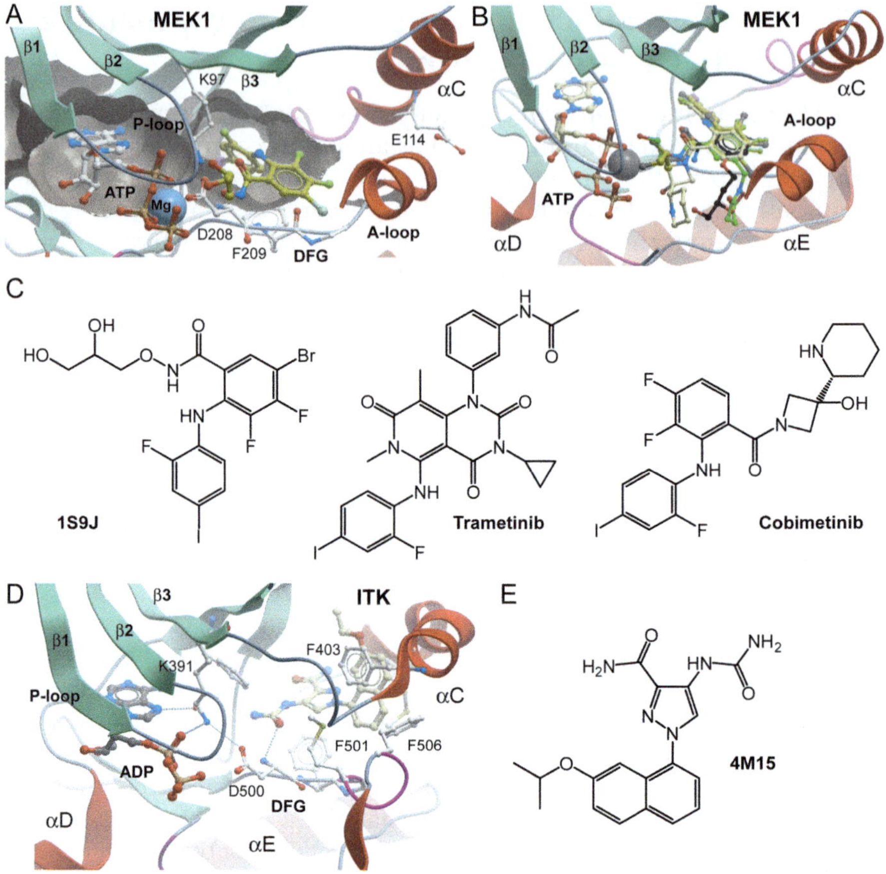

Figure 3.6 Examples of MEK and ITK DFG-in back pocket inhibitors. A: Shown is a structure of MEK1 in complex with ATP/Mg^{2+} (white carbon atoms) and a DFG-in back pocket binder (yellow carbon atoms) (pdb-code: 1S9J). B: Overlay of a selection of allosteric MEK1 inhibitors (pdb-code: 4AN2 (light yellow), 3ORN (green), 4ARK (black)). C: Chemical structures of the crystallized inhibitors shown in panel A and the clinical inhibitors Trametinib and Cobimetinib. D: Binding mode of a type-III ITK inhibitor. Shown is ADP (grey atoms) as well as the naphthalenylpyrazole inhibitor (yellow carbon atoms). E: Chemical structure of the ITK inhibitor.

(DBD) inhibitor was initially identified by nuclear magnetic resonance-fragment-based screening. The inhibitor class showed exceptional selectivity for PAK1 even against closely related isoforms such as PAK2, which shares 93% sequence conservation with the PAK1 catalytic domain.[64] Comparison of crystal structures of active PAK1 in complex with ATP (pdb-ID:3q53), an initial dibenzodiazepine fragment and the developed highly potent inhibitors revealed a progressive outward shift of αC as well as an N-terminal helix that docks onto αC, and an outward shift of the DFG motif (Figure 3.7).

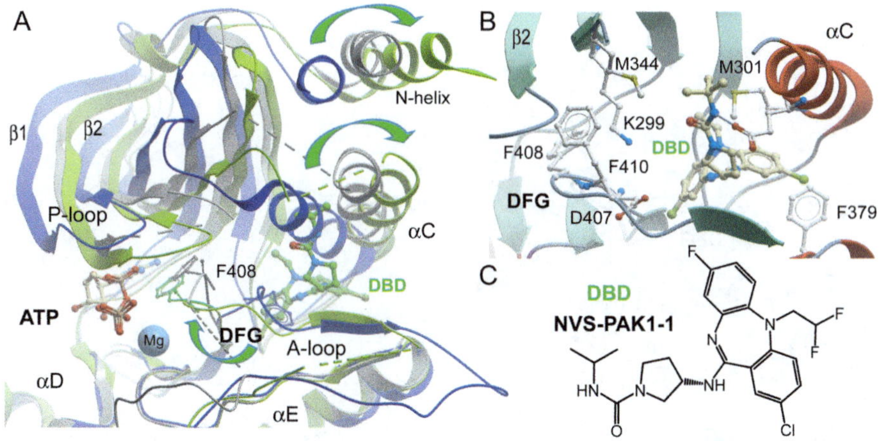

Figure 3.7 Binding mode of an allosteric PAK1-specific DFG-out back pocket inhibitor. A: Shown is a superimposition of active PAK1 in complex with ATP/Mg^{2+} (blue; pdb-code: 3Q53), the dibenzodiazepine (DBD) lead compound (grey; pdb-code: 4ZLO) and the developed inhibitor (green; pdb-code: 4ZJJ). The main structural elements are labelled and conformational changes are indicated by arrows. B: Details of the interactions made by the DBD PAK1 inhibitor. Inhibitor is highlighted by green carbon atoms. C: Chemical structure of the PAK1 chemical probe NVS-PAK1-1 released by Novartis.

A similar binding mode has been explored for the development of LIMK2 inhibitors. A series of *N*-phenylsulfonamide compounds showed surprising selectivity and potency for this tyrosine-like kinase even compared with its closely related isoform LIMK1.[65] To date no crystal structure of active LIMK2 has been published but comparison with the structure of the closely related LIMK1 (pdb-code: 3S95) revealed a large displacement of αC as indicated by disruption of the salt bridge between E376 and K360 and a DFG-out binding mode of the inhibitor. In addition, the P-loop is distorted, showing a folded conformation with the two aromatic residues F341 and F342 oriented towards the ATP pocket (Figure 3.8A and B).

An interesting variant of a back pocket DFG-out binding mode is a recently disclosed IGF1R inhibitor developed by Merck Serono.[66] This inhibitor binds to the DFG-out pocket, but without significantly displacing αC as indicated by the presence of the αC (E1050)–K1033 salt bridge (PDB-code: 3LW0; Figure 3.8C and D). In addition, binding of this inhibitor induced significant conformational changes in the activation loop, creating a binding pocket for the distal 3-cyano indole moiety.

DFG-out pocket binders based on the 3-tertbutyl pyrazole moiety led to discovery of allosteric inhibitors for diverse kinases. This DFG-pocket binding motif was first discovered for p38, resulting in the development of BIRB796.[40] Often coupled to urea 3-*tert*-butyl pyrazoles, led to inhibitors for c-SRC (pdb-code: 3F3U, 3F3T),[67] ABL1 (pdb-code: 3QRK)[68] and CDK8/CycC[69] (pdb-code:

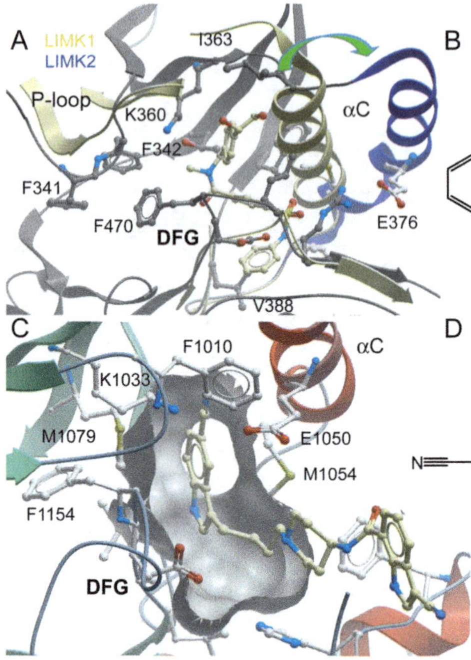

Figure 3.8 Binding mode of allosteric LIMK2 and IGF1R DFG-out back pocket inhibitors. A: Shown are details of the interaction with LIMK2. A superimposition with active LIMK1 (only αC and P-loop) is shown in yellow. B: Chemical structure of the LIMK2 inhibitor. C: Details of the interaction of an allosteric IGF1R inhibitor with the kinase back pocket. D: Chemical structure of the IGF1R inhibitor.

4F6S, 4F70). Similarly, isoxazole inhibitors were reported targeting vascular endothelial growth factor receptor (VEGFR) in a similar binding mode (pdb-code: 3VHK)[70] and a number of 1,5-dihydropyrazolo[4,3-*c*][2,1]benzothiazines demonstrate a DFG-out allosteric binding mode binding to focal adhesion kinase (pdb-code: 4EBV, 4EBW, 4I4F).[71] Interestingly, most of these inhibitors showed slow dissociation rates. Finally, RIPK1 binds inhibitors of the necrostatin class in an allosteric DFG-out type binding mode (pdb-code: 4ITH, 4ITI, 4ITJ) with a binding preference for the inactive kinase.[72]

3.8 Differential Effects of Type I and Type II Inhibitors in Signalling

From a therapeutic standpoint and for our understanding of the consequences of diverse kinase inhibitor binding modes on cellular signalling, it is of fundamental importance to assess how current allosteric and ATP-competitive inhibitors affect kinase non-catalytic functions. Certainly, the growing number of highly specific kinase allosteric and ATP-competitive inhibitors will be key reagents for unravelling the effects of inhibitor binding modes on signalling networks in normal physiology and in disease, as well as the possible advantages of also targeting kinases non-catalytic function for disease treatment.

A well-established example of the modulation of non-catalytic function are inhibitors targeting RAF kinases. This family of kinases comprises key regulators of MAPK signalling, and activating mutations in the family member BRAF are a strong driver of tumour development. In healthy tissue the activation of MAPK signalling depends on homo- or hetero-dimerization of BRAF and CRAF which is induced upon activation of the small GTPase RAS. In contrast, mutated BRAF bypasses RAS dependency by either signalling through monomeric BRAF or through constitutively dimeric BRAF.[9,10]

Wild type BRAF is predominantly monomeric in the absence of RAS activation but binding of type I inhibitors, as well as inhibitors of the sulfonamide type, induces dimer formation and hence activates the wild type enzyme.[73] The strong allosteric control of RAF through activation of dimerization is evident in crystal structures of BRAF type-I inhibitor complexes that usually show a dimeric arrangement with only one binding site occupied by the inhibitor with the associated apo-structure being in an active conformation.[11] The clinical consequences of this unexpected behaviour is that MAPK signalling is activated in tissues harbouring wild-type BRAF, resulting in formation of secondary tumours,[74] while in mutated cancer tissue bearing the monomeric BRAF V600E mutation inhibitor drug resistance may occur through a mechanisms leading to constitutive dimers.[8] Type-II BRAF inhibitors such as LY3009120 do not induce dimerization and are active on both monomeric and dimeric BRAF.[75] Thus, even though most RAF inhibitors show limited selectivity for mutant or wild type RAF family members, the investigated different binding modes showed pronounced differences in pharmacology in cancer and normal tissue due to the modulation of kinase scaffolding function. The complex pharmacology of the clinical BRAF inhibitor vemurafenib, that does induce MAPK activation in BRAF wild-type cells despite stabilizing an inactive BRAF conformation, not promoting dimerization indicates that models explaining allosteric mechanisms may also include other signalling partners such as the downstream RAF effector MEK as well as interaction with phosphatases.[10]

A compelling example of the role of kinase non-catalytic function was recently reported in a study that compared the effects of the allosteric AKT inhibitor MK2206[76] and the ATP-competitive AKT inhibitors GSK690693[77,78] and GDC0068[79] on signalling and cancer cell survival.[12] Inactive AKT is locked in a closed conformation stabilized by the interaction of the lipid binding pleckstrin homology (PH) domain with the kinase domain.[80] As a result of growth factor stimulation, AKT is recruited to the plasma membrane through PH-domain-mediated interaction with the phospholipids phosphatidyl-inositol-3,4,5-trisphosphate (PtdIns (3,4,5)P_3) and phosphatidyl-inositol-3,4-bisphosphate (PtdIns(3,4)P_2). The binding of lipids to the PH domain induced a structural change that renders the activating phosphorylation sites threonine 308 and serine 473 accessible, resulting in full activation of the kinase. Similarly to BRAF inhibitors, type-I AKT inhibitors stimulate the kinase active state, resulting paradoxically in hyper-phosphorylation of AKT at the activating phosphorylation sites threonine 308 and serine 473.[81] Comparing the three AKT inhibitors Vivanco *et al.*[12] showed that the allosteric inhibitor MK2206 that targets a unique binding site at the interface of the catalytic and PH domain in "closed" inactive AKT[82] is more efficient impairing cancer cell survival and in *in vivo* models than the conventional ATP-competitive inhibitors, despite reduced phosphorylation levels measured on AKT substrates. This intriguing observation highlights the pivotal role targeting non-catalytic, allosteric mechanisms for inhibitor efficacy. Indeed, analysis of lipid interaction with AKT in the presence and absence of MK2206 and the two ATP competitive inhibitors GSK690693 and GDC006 revealed impaired phosphoinositide binding in MK2206-treated cells and hence reduced localization of AKT to the plasma membrane. The non-catalytic role of AKT mediating cell survival was further demonstrated by over-expression of catalytically inactive AKT which protected cells from cell death induced by doxycycline treatment.[12]

Another recent example of changes in scaffolding function by kinase inhibitors are inhibitors that have been developed targeting the interaction of Aurora A with N-MYC and the ubiquitin ligase Skp, Cullin, F-box containing complex protein F-box and WD repeat-containing protein 7 (SCF^{Fbxw7}). Aurora kinases are key regulators of mitosis and significant over-expression of these kinases in cancer triggered a large number of drug development projects in oncology.[83,84] Aurora A interacts with N-MYC preventing degradation of this growth-promoting transcription factor. N-MYC interaction is independent of kinase catalytic activity as demonstrated by the ability of inactivating Aurora A mutations to protect N-MYC degradation.[85] While typical ATP-competitive Aurora A inhibitors do not affect N-MYC interaction, inhibitors that induce an unusual inactive conformation characterized by extreme displacement of αC, break the N-MYC interaction, resulting in N-MYC degradation.[86,87]

3.9 Pseudokinases as Drug Targets?

Pseudokinases represent a large group of kinases that have little or no catalytic activity but play essential roles in cellular signalling.[88] Interestingly, many pseudokinases are dysfunctional in disease, further highlighting the important role of the non-catalytic functions of these kinases. As outlined above, allosteric inhibitors developed for catalytically active kinases demonstrated that non-catalytic functions in kinases can be successfully targeted. These results indicate that also catalytically inert pseudokinases may present attractive drug targets. One prominent example is the tyrosine kinase Janus kinase 2 (JAK2) that contains an active catalytic domain (JAK homology domain 1 (JH1)) as well as a pseudokinase domain (JH2), an architecture that is shared among all JAK family kinases (JAK1 and 2 and tyrosine kinase 2(TYK2)). Mutations in the pseudokinase domain cause myoproliferative disorder due to aberrant activation of JAK2.[89,90] Recently, the first inhibitors targeting the pseudokinase domain of TYK2 have been disclosed.[91] The developed inhibitors showed good target selectivity for the pseudokinase domain of TYK2 and JAK1 and were shown to modulate IL-23-stimulated reporter gene assays most probably by stabilizing an auto-inhibitory conformation.

A second example targeting a catalytically inactive kinase has recently been reported for the pseudokinase mixed lineage kinase domain-like protein (MLKL). Screening a collection of typical ATP mimetic inhibitors identified MLKL binders that enhanced phosphorylation of MLKL by RIPK3, probably due to an induced structural change of the activation segment. The enhanced phosphorylation levels prevented localization of MLKL to the plasma membrane and impaired necroptosis.[92] These two recent examples demonstrate that catalytically inactive pseudokinases can be successfully targeted by conformation-selective inhibitors, resulting in significant phenotypic changes due to interference with kinase scaffolding functions.

3.10 Conclusions

The large number of developed kinase inhibitors, as well as the good coverage of the kinase family with high resolution crystal structures has identified a significant diversity of inhibitor binding modes and structural changes induced by the inhibitors. An increasing number of studies reported target-specific allosteric pockets, resulting in the development of unprecedented inhibitor specificity, sometimes differentiating between even closely related kinase isoforms, family members and mutants.[25,48] However, only few studies investigated the consequences of conformation-sensitive targeting and their implications for the development of more efficacious and specific drugs. For kinases with profound translational interest such as the RAF, AKT and the Aurora family, the implications of targeting kinase non-catalytic function have now been realized in clinical as well as mechanistic studies. Interestingly, the results of these studies indicate that the active kinase conformation is also essential for many non-catalytic functionalities, such as interaction with signalling partners, phosphatases and kinase regulatory proteins. It is likely that this is a logical consequence of signalling network regulation, which simultaneously activates catalytic as well as non-catalytic mechanisms in a concerted response to environmental cues. These considerations make the development of allosteric inhibitors even more exciting as catalytic and non-catalytic functions are simultaneously targeted. However, even though many non-catalytic functions of kinases are deregulated in disease, for the majority of kinases, in particular for catalytically inactive pseudokinases, the implications of targeting catalytic and non-catalytic kinase function remain largely unknown.[4] We believe that this large gap in knowledge is best addressed by developing highly selective conformation-specific tool compounds together with a comprehensive evaluation of these tools in cellular model systems. The development of high-quality chemical tools is, however, challenging and will require the development of new target-specific screening technologies and new chemical libraries that target kinase allosteric sites.

We predict that the availability of chemical tools for unexplored kinase targets and kinase binding sites will lead to new drug targets and medicines in the kinase area, as well as to a significant expansion of the current kinase chemistry space, which will be diversified from typical ATP-mimetic scaffolds, and to a more comprehensive understanding of cellular signalling networks and their deregulation in disease.

References

1. G. Manning, D. B. Whyte, R. Martinez, T. Hunter and S. Sudarsanam, *Science*, 2002, **298**, 1912–1934.
2. S. Knapp and M. Sundstrom, *Curr. Opin. Pharmacol.*, 2014, **17**, 58–63.
3. Q. Liu, Y. Sabnis, Z. Zhao, T. Zhang, S. J. Buhrlage, L. H. Jones and N. S. Gray, *Chem. Biol.*, 2013, **20**, 146–159.
4. O. Fedorov, S. Muller and S. Knapp, *Nat. Chem. Biol.*, 2010, **6**, 166–169.
5. O. P. van Linden, A. J. Kooistra, R. Leurs, I. J. de Esch and C. de Graaf, *J. Med. Chem.*, 2014, **57**, 249–277.
6. Z. Zhao, H. Wu, L. Wang, Y. Liu, S. Knapp, Q. Liu and N. S. Gray, *ACS Chem. Biol.*, 2014, **9**, 1230–1241.
7. J. Rauch, N. Volinsky, D. Romano and W. Kolch, *Cell Commun. Signaling*, 2011, **9**, 23.
8. P. I. Poulikakos, Y. Persaud, M. Janakiraman, X. Kong, C. Ng, G. Moriceau, H. Shi, M. Atefi, B. Titz, M. T. Gabay, M. Salton, K. B. Dahlman, M. Tadi, J. A. Wargo, K. T. Flaherty, M. C. Kelley, T. Misteli, P. B. Chapman, J. A. Sosman, T. G. Graeber, A. Ribas, R. S. Lo, N. Rosen and D. B. Solit, *Nature*, 2011, **480**, 387–390.
9. Z. Yao, N. M. Torres, A. Tao, Y. Gao, L. Luo, Q. Li, E. de Stanchina, O. Abdel-Wahab, D. B. Solit, P. I. Poulikakos and N. Rosen, *Cancer Cell*, 2015, **28**, 370–383.
10. J. R. Haling, J. Sudhamsu, I. Yen, S. Sideris, W. Sandoval, W. Phung, B. J. Bravo, A. M. Giannetti, A. Peck, A. Masselot, T. Morales, D. Smith, B. J. Brandhuber, S. G. Hymowitz and S. Malek, *Cancer Cell*, 2014, **26**, 402–413.
11. N. Thevakumaran, H. Lavoie, D. A. Critton, A. Tebben, A. Marinier, F. Sicheri and M. Therrien, *Nat. Struct. Mol. Biol.*, 2015, **22**, 37–43.
12. I. Vivanco, Z. C. Chen, B. Tanos, B. Oldrini, W. Y. Hsieh, N. Yannuzzi, C. Campos and I. K. Mellinghoff, *Elife*, 2014, **3**, e03751.
13. H. Zhang, A. Photiou, A. Grothey, J. Stebbing and G. Giamas, *Cell. Signalling*, 2012, **24**, 1173–1184.
14. M. Salazar, M. Lorente, A. Orea-Soufi, D. Davila, T. Erazo, J. Lizcano, A. Carracedo, E. Kiss-Toth and G. Velasco, *Biochem. Soc. Trans.*, 2015, **43**, 1122–1126.
15. L. N. Johnson and R. J. Lewis, *Chem. Rev.*, 2001, **101**, 2209–2242.
16. J. A. Adams, *Biochemistry*, 2003, **42**, 601–607.

17. A. P. Kornev and S. S. Taylor, *Trends Biochem. Sci.*, 2015, **40**, 628–647.
18. J. Yang, E. J. Kennedy, J. Wu, M. S. Deal, J. Pennypacker, G. Ghosh and S. S. Taylor, *J. Biol. Chem.*, 2009, **284**, 6241–6248.
19. C. Zhang, D. M. Kenski, J. L. Paulson, A. Bonshtien, G. Sessa, J. V. Cross, D. J. Templeton and K. M. Shokat, *Nat. Methods*, 2005, **2**, 435–441.
20. A. C. Bishop, J. A. Ubersax, D. T. Petsch, D. P. Matheos, N. S. Gray, J. Blethrow, E. Shimizu, J. Z. Tsien, P. G. Schultz, M. D. Rose, J. L. Wood, D. O. Morgan and K. M. Shokat, *Nature*, 2000, **407**, 395–401.
21. C. H. Yun, K. E. Mengwasser, A. V. Toms, M. S. Woo, H. Greulich, K. K. Wong, M. Meyerson and M. J. Eck, *Proc. Natl. Acad. Sci. U. S. A.*, 2008, **105**, 2070–2075.
22. M. Azam, M. A. Seeliger, N. S. Gray, J. Kuriyan and G. Q. Daley, *Nat. Struct. Mol. Biol.*, 2008, **15**, 1109–1118.
23. H. L. De Bondt, J. Rosenblatt, J. Jancarik, H. D. Jones, D. O. Morgan and S. H. Kim, *Nature*, 1993, **363**, 595–602.
24. S. R. Hubbard, L. Wei, L. Ellis and W. A. Hendrickson, *Nature*, 1994, **372**, 746–754.
25. S. Muller, A. Chaikuad, N. S. Gray and S. Knapp, *Nat. Chem. Biol.*, 2015, **11**, 818–821.
26. T. Schindler, W. Bornmann, P. Pellicena, W. T. Miller, B. Clarkson and J. Kuriyan, *Science*, 2000, **289**, 1938–1942.
27. J. Zhang, P. L. Yang and N. S. Gray, *Nat. Rev. Cancer*, 2009, **9**, 28–39.
28. M. Karplus and J. Kuriyan, *Proc. Natl. Acad. Sci. U. S. A.*, 2005, **102**, 6679–6685.
29. L. T. Alexander, H. Mobitz, P. Drueckes, P. Savitsky, O. Fedorov, J. M. Elkins, C. M. Deane, S. W. Cowan-Jacob and S. Knapp, *ACS Chem. Biol.*, 2015, **10**, 2116–2125.
30. C. R. Guimaraes, B. K. Rai, M. J. Munchhof, S. Liu, J. Wang, S. K. Bhattacharya and L. Buckbinder, *J. Chem. Inf. Model.*, 2011, **51**, 1199–1204.
31. A. Chaikuad, E. M. Tacconi, J. Zimmer, Y. Liang, N. S. Gray, M. Tarsounas and S. Knapp, *Nat. Chem. Biol.*, 2014, **10**, 853–860.
32. S. G. Buchanan, J. Hendle, P. S. Lee, C. R. Smith, P. Y. Bounaud, K. A. Jessen, C. M. Tang, N. H. Huser, J. D. Felce, K. J. Froning, M. C. Peterman, B. E. Aubol, S. F. Gessert, J. M. Sauder, K. D. Schwinn, M. Russell, I. A. Rooney, J. Adams, B. C. Leon, T. H. Do, J. M. Blaney, P. A. Sprengeler, D. A. Thompson, L. Smyth, L. A. Pelletier, S. Atwell, K. Holme, S. R. Wasserman, S. Emtage, S. K. Burley and S. H. Reich, *Mol. Cancer Ther.*, 2009, **8**, 3181–3190.
33. J. M. Axten, J. R. Medina, Y. Feng, A. Shu, S. P. Romeril, S. W. Grant, W. H. Li, D. A. Heerding, E. Minthorn, T. Mencken, C. Atkins, Q. Liu, S. Rabindran, R. Kumar, X. Hong, A. Goetz, T. Stanley, J. D. Taylor, S. D. Sigethy, G. H. Tomberlin, A. M. Hassell, K. M. Kahler, L. M. Shewchuk and R. T. Gampe, *J. Med. Chem.*, 2012, **55**, 7193–7207.
34. E. R. Wood, A. T. Truesdale, O. B. McDonald, D. Yuan, A. Hassell, S. H. Dickerson, B. Ellis, C. Pennisi, E. Horne, K. Lackey, K. J. Alligood, D. W. Rusnak, T. M. Gilmer and L. Shewchuk, *Cancer Res.*, 2004, **64**, 6652–6659.

35. S. C. Koeberle, J. Romir, S. Fischer, A. Koeberle, V. Schattel, W. Albrecht, C. Grutter, O. Werz, D. Rauh, T. Stehle and S. A. Laufer, *Nat. Chem. Biol.*, 2012, **8**, 141–143.
36. R. A. de Claro, K. M. McGinn, N. Verdun, S. L. Lee, H. J. Chiu, H. Saber, M. E. Brower, C. J. Chang, E. Pfuma, B. Habtemariam, J. Bullock, Y. Wang, L. Nie, X. H. Chen, D. R. Lu, A. Al-Hakim, R. C. Kane, E. Kaminskas, R. Justice, A. T. Farrell and R. Pazdur, *Clin. Cancer Res.*, 2015, **21**, 3586–3590.
37. M. Joshi, S. M. Rizvi and C. P. Belani, *Cancer Manage. Res.*, 2015, **7**, 75–82.
38. T. Zhang, F. Inesta-Vaquera, M. Niepel, J. Zhang, S. B. Ficarro, T. Machleidt, T. Xie, J. A. Marto, N. Kim, T. Sim, J. D. Laughlin, H. Park, P. V. LoGrasso, M. Patricelli, T. K. Nomanbhoy, P. K. Sorger, D. R. Alessi and N. S. Gray, *Chem. Biol.*, 2012, **19**, 140–154.
39. L. Tan, K. Akahane, R. McNally, K. M. Reyskens, S. B. Ficarro, S. Liu, G. S. Herter-Sprie, S. Koyama, M. J. Pattison, K. Labella, L. Johannessen, E. A. Akbay, K. K. Wong, D. A. Frank, J. A. Marto, T. A. Look, J. S. Arthur, M. J. Eck and N. S. Gray, *J. Med. Chem.*, 2015, **58**, 6589–6606.
40. C. Pargellis, L. Tong, L. Churchill, P. F. Cirillo, T. Gilmore, A. G. Graham, P. M. Grob, E. R. Hickey, N. Moss, S. Pav and J. Regan, *Nat. Struct. Biol.*, 2002, **9**, 268–272.
41. H. Tecle, F. Feru, H. Liu, C. Kuhn, G. Rennie, M. Morris, J. Shao, A. C. Cheng, D. Gikunju, J. Miret, R. Coli, S. H. Xi, S. L. Clugston, S. Low, S. Kazmirski, Y. H. Ding, Q. Cao, T. L. Johnson, G. D. Deshmukh, J. P. DiNitto, J. C. Wu and J. M. English, *Chem. Biol. Drug Des.*, 2009, **74**, 547–559.
42. A. L. Gill, M. Frederickson, A. Cleasby, S. J. Woodhead, M. G. Carr, A. J. Woodhead, M. T. Walker, M. S. Congreve, L. A. Devine, D. Tisi, M. O'Reilly, L. C. Seavers, D. J. Davis, J. Curry, R. Anthony, A. Padova, C. W. Murray, R. A. Carr and H. Jhoti, *J. Med. Chem.*, 2005, **48**, 414–426.
43. K. M. Comess, C. Sun, C. Abad-Zapatero, E. R. Goedken, R. J. Gum, D. W. Borhani, M. Argiriadi, D. R. Groebe, Y. Jia, J. E. Clampit, D. L. Haasch, H. T. Smith, S. Wang, D. Song, M. L. Coen, T. E. Cloutier, H. Tang, X. Cheng, C. Quinn, B. Liu, Z. Xin, G. Liu, E. H. Fry, V. Stoll, T. I. Ng, D. Banach, D. Marcotte, D. J. Burns, D. J. Calderwood and P. J. Hajduk, *ACS Chem. Biol.*, 2011, **6**, 234–244.
44. M. Getlik, J. R. Simard, M. Termathe, C. Grutter, M. Rabiller, W. A. van Otterlo and D. Rauh, *PLoS One*, 2012, **7**, e39713.
45. D. Vanderpool, T. O. Johnson, C. Ping, S. Bergqvist, G. Alton, S. Phonephaly, E. Rui, C. Luo, Y. L. Deng, S. Grant, T. Quenzer, S. Margosiak, J. Register, E. Brown and J. Ermolieff, *Biochemistry*, 2009, **48**, 9823–9830.
46. A. Converso, T. Hartingh, R. M. Garbaccio, E. Tasber, K. Rickert, M. E. Fraley, Y. Yan, C. Kreatsoulas, S. Stirdivant, B. Drakas, E. S. Walsh, K. Hamilton, C. A. Buser, X. Mao, M. T. Abrams, S. C. Beck, W. Tao, R. Lobell, L. Sepp-Lorenzino, J. Zugay-Murphy, V. Sardana, S. K. Munshi, S. M. Jezequel-Sur, P. D. Zuck and G. D. Hartman, *Bioorg. Med. Chem. Lett.*, 2009, **19**, 1240–1244.
47. B. Over, S. Wetzel, C. Grutter, Y. Nakai, S. Renner, D. Rauh and H. Waldmann, *Nat. Chem.*, 2013, **5**, 21–28.

48. S. W. Cowan-Jacob, W. Jahnke and S. Knapp, *Future Med. Chem.*, 2014, **6**, 541–561.
49. V. Hindie, A. Stroba, H. Zhang, L. A. Lopez-Garcia, L. Idrissova, S. Zeuzem, D. Hirschberg, F. Schaeffer, T. J. Jorgensen, M. Engel, P. M. Alzari and R. M. Biondi, *Nat. Chem. Biol.*, 2009, **5**, 758–764.
50. K. Busschots, L. A. Lopez-Garcia, C. Lammi, A. Stroba, S. Zeuzem, A. Piiper, P. M. Alzari, S. Neimanis, J. M. Arencibia, M. Engel, J. O. Schulze and R. M. Biondi, *Chem. Biol.*, 2012, **19**, 1152–1163.
51. E. V. Bobkova, M. J. Weber, Z. Xu, Y. L. Zhang, J. Jung, P. Blume-Jensen, A. Northrup, P. Kunapuli, J. N. Andersen and I. Kariv, *J. Biol. Chem.*, 2010, **285**, 18838–18846.
52. J. Zhang, F. J. Adrian, W. Jahnke, S. W. Cowan-Jacob, A. G. Li, R. E. Iacob, T. Sim, J. Powers, C. Dierks, F. Sun, G. R. Guo, Q. Ding, B. Okram, Y. Choi, A. Wojciechowski, X. Deng, G. Liu, G. Fendrich, A. Strauss, N. Vajpai, S. Grzesiek, T. Tuntland, Y. Liu, B. Bursulaya, M. Azam, P. W. Manley, J. R. Engen, G. Q. Daley, M. Warmuth and N. S. Gray, *Nature*, 2010, **463**, 501–506.
53. E. Salah, E. Ugochukwu, A. J. Barr, F. von Delft, S. Knapp and J. M. Elkins, *J. Med. Chem.*, 2011, **54**, 2359–2367.
54. J. Yang, N. Campobasso, M. P. Biju, K. Fisher, X. Q. Pan, J. Cottom, S. Galbraith, T. Ho, H. Zhang, X. Hong, P. Ward, G. Hofmann, B. Siegfried, F. Zappacosta, Y. Washio, P. Cao, J. Qu, S. Bertrand, D. Y. Wang, M. S. Head, H. Li, S. Moores, Z. Lai, K. Johanson, G. Burton, C. Erickson-Miller, G. Simpson, P. Tummino, R. A. Copeland and A. Oliff, *Chem. Biol.*, 2011, **18**, 177–186.
55. J. F. Ohren, H. Chen, A. Pavlovsky, C. Whitehead, E. Zhang, P. Kuffa, C. Yan, P. McConnell, C. Spessard, C. Banotai, W. T. Mueller, A. Delaney, C. Omer, J. Sebolt-Leopold, D. T. Dudley, I. K. Leung, C. Flamme, J. Warmus, M. Kaufman, S. Barrett, H. Tecle and C. A. Hasemann, *Nat. Struct. Mol. Biol.*, 2004, **11**, 1192–1197.
56. R. A. Heald, P. Jackson, P. Savy, M. Jones, E. Gancia, B. Burton, R. Newman, J. Boggs, E. Chan, J. Chan, E. Choo, M. Merchant, P. Rudewicz, M. Ultsch, C. Wiesmann, Q. Yue, M. Belvin and S. Price, *J. Med. Chem.*, 2012, **55**, 4594–4604.
57. K. D. Rice, N. Aay, N. K. Anand, C. M. Blazey, O. J. Bowles, J. Bussenius, S. Costanzo, J. K. Curtis, S. C. Defina, L. Dubenko, S. Engst, A. A. Joshi, A. R. Kennedy, A. I. Kim, E. S. Koltun, J. C. Lougheed, J. C. Manalo, J. F. Martini, J. M. Nuss, C. J. Peto, T. H. Tsang, P. Yu and S. Johnston, *ACS Med. Chem. Lett.*, 2012, **3**, 416–421.
58. G. V. Long, J. S. Weber, J. R. Infante, K. B. Kim, A. Daud, R. Gonzalez, J. A. Sosman, O. Hamid, L. Schuchter, J. Cebon, R. F. Kefford, D. Lawrence, R. Kudchadkar, H. A. Burris 3rd, G. S. Falchook, A. Algazi, K. Lewis, I. Puzanov, N. Ibrahim, P. Sun, E. Cunningham, A. S. Kline, H. Del Buono, D. O. McDowell, K. Patel and K. T. Flaherty, *J. Clin. Oncol.*, 2016, **34**, 871–878.
59. S. Inman, *Am. J. Manag. Care*, 2014, **20**, E10.
60. K. P. Garnock-Jones, *Drugs*, 2015, **75**, 1823–1830.

61. S. Han, R. M. Czerwinski, N. L. Caspers, D. C. Limburg, W. Ding, H. Wang, J. F. Ohren, F. Rajamohan, T. J. McLellan, R. Unwalla, C. Choi, M. D. Parikh, N. Seth, J. Edmonds, C. Phillips, S. Shakya, X. Li, V. Spaulding, S. Hughes, A. Cook, C. Robinson, J. P. Mathias, I. Navratilova, Q. G. Medley, D. R. Anderson, R. G. Kurumbail and A. Aulabaugh, *Biochem. J.*, 2014, **460**, 211–222.
62. Z. Liu, Y. Lv, N. Zhao, G. Guan and J. Wang, *Cell Death Dis.*, 2015, **6**, e1822.
63. P. E. Harrington, K. Biswas, D. Malwitz, A. S. Tasker, C. Mohr, K. L. Andrews, K. Dellamaggiore, R. Kendall, H. Beckmann, P. Jaeckel, S. Materna-Reichelt, J. R. Allen and J. R. Lipford, *ACS Med. Chem. Lett.*, 2015, **6**, 68–72.
64. A. S. Karpov, P. Amiri, C. Bellamacina, M. H. Bellance, W. Breitenstein, D. Daniel, R. Denay, D. Fabbro, C. Fernandez, I. Galuba, S. Guerro-Lagasse, S. Gutmann, L. Hinh, W. Jahnke, J. Klopp, A. Lai, M. K. Lindvall, S. Ma, H. Mobitz, S. Pecchi, G. Rummel, K. Shoemaker, J. Trappe, C. Voliva, S. W. Cowan-Jacob and A. L. Marzinzik, *ACS Med. Chem. Lett.*, 2015, **6**, 776–781.
65. N. C. Goodwin, G. Cianchetta, H. A. Burgoon, J. Healy, R. Mabon, E. D. Strobel, J. Allen, S. Wang, B. D. Hamman and D. B. Rawlins, *ACS Med. Chem. Lett.*, 2015, **6**, 53–57.
66. T. Heinrich, U. Gradler, H. Bottcher, A. Blaukat and A. Shutes, *ACS Med. Chem. Lett.*, 2010, **1**, 199–203.
67. J. R. Simard, S. Kluter, C. Grutter, M. Getlik, M. Rabiller, H. B. Rode and D. Rauh, *Nat. Chem. Biol.*, 2009, **5**, 394–396.
68. W. W. Chan, S. C. Wise, M. D. Kaufman, Y. M. Ahn, C. L. Ensinger, T. Haack, M. M. Hood, J. Jones, J. W. Lord, W. P. Lu, D. Miller, W. C. Patt, B. D. Smith, P. A. Petillo, T. J. Rutkoski, H. Telikepalli, L. Vogeti, T. Yao, L. Chun, R. Clark, P. Evangelista, L. C. Gavrilescu, K. Lazarides, V. M. Zaleskas, L. J. Stewart, R. A. Van Etten and D. L. Flynn, *Cancer Cell*, 2011, **19**, 556–568.
69. E. V. Schneider, J. Bottcher, R. Huber, K. Maskos and L. Neumann, *Proc. Natl. Acad. Sci. U. S. A.*, 2013, **110**, 8081–8086.
70. H. Iwata, H. Oki, K. Okada, T. Takagi, M. Tawada, Y. Miyazaki, S. Imamura, A. Hori, J. D. Lawson, M. S. Hixon, H. Kimura and H. Miki, *ACS Med. Chem. Lett.*, 2012, **3**, 342–346.
71. M. Iwatani, H. Iwata, A. Okabe, R. J. Skene, N. Tomita, Y. Hayashi, Y. Aramaki, D. J. Hosfield, A. Hori, A. Baba and H. Miki, *Eur. J. Med. Chem.*, 2013, **61**, 49–60.
72. T. Xie, W. Peng, Y. Liu, C. Yan, J. Maki, A. Degterev, J. Yuan and Y. Shi, *Structure*, 2013, **21**, 493–499.
73. G. Hatzivassiliou, K. Song, I. Yen, B. J. Brandhuber, D. J. Anderson, R. Alvarado, M. J. Ludlam, D. Stokoe, S. L. Gloor, G. Vigers, T. Morales, I. Aliagas, B. Liu, S. Sideris, K. P. Hoeflich, B. S. Jaiswal, S. Seshagiri, H. Koeppen, M. Belvin, L. S. Friedman and S. Malek, *Nature*, 2010, **464**, 431–435.
74. L. Boussemart, I. Girault, H. Malka-Mahieu, C. Mateus, E. Routier, M. Rubington, N. Kamsu-Kom, M. Thomas, G. Tomasic, S. Agoussi, M. Breckler,

M. Laporte, L. Lacroix, A. M. Eggermont, A. Cavalcanti, F. Grange, J. Adam, S. Vagner and C. Robert, *Cancer Res.*, 2016, **76**, 1476–1484.
75. S. B. Peng, J. R. Henry, M. D. Kaufman, W. P. Lu, B. D. Smith, S. Vogeti, T. J. Rutkoski, S. Wise, L. Chun, Y. Zhang, R. D. Van Horn, T. Yin, X. Zhang, V. Yadav, S. H. Chen, X. Gong, X. Ma, Y. Webster, S. Buchanan, I. Mochalkin, L. Huber, L. Kays, G. P. Donoho, J. Walgren, D. McCann, P. Patel, I. Conti, G. D. Plowman, J. J. Starling and D. L. Flynn, *Cancer Cell*, 2015, **28**, 384–398.
76. H. Hirai, H. Sootome, Y. Nakatsuru, K. Miyama, S. Taguchi, K. Tsujioka, Y. Ueno, H. Hatch, P. K. Majumder, B. S. Pan and H. Kotani, *Mol. Cancer Ther.*, 2010, **9**, 1956–1967.
77. D. A. Heerding, N. Rhodes, J. D. Leber, T. J. Clark, R. M. Keenan, L. V. Lafrance, M. Li, I. G. Safonov, D. T. Takata, J. W. Venslavsky, D. S. Yamashita, A. E. Choudhry, R. A. Copeland, Z. Lai, M. D. Schaber, P. J. Tummino, S. L. Strum, E. R. Wood, D. R. Duckett, D. Eberwein, V. B. Knick, T. J. Lansing, R. T. McConnell, S. Zhang, E. A. Minthorn, N. O. Concha, G. L. Warren and R. Kumar, *J. Med. Chem.*, 2008, **51**, 5663–5679.
78. N. Rhodes, D. A. Heerding, D. R. Duckett, D. J. Eberwein, V. B. Knick, T. J. Lansing, R. T. McConnell, T. M. Gilmer, S. Y. Zhang, K. Robell, J. A. Kahana, R. S. Geske, E. V. Kleymenova, A. E. Choudhry, Z. Lai, J. D. Leber, E. A. Minthorn, S. L. Strum, E. R. Wood, P. S. Huang, R. A. Copeland and R. Kumar, *Cancer Res.*, 2008, **68**, 2366–2374.
79. J. Lin, D. Sampath, M. A. Nannini, B. B. Lee, M. Degtyarev, J. Oeh, H. Savage, Z. Guan, R. Hong, R. Kassees, L. B. Lee, T. Risom, S. Gross, B. M. Liederer, H. Koeppen, N. J. Skelton, J. J. Wallin, M. Belvin, E. Punnoose, L. S. Friedman and K. Lin, *Clin. Cancer Res.*, 2013, **19**, 1760–1772.
80. V. Calleja, M. Laguerre, P. J. Parker and B. Larijani, *PLoS Biol.*, 2009, 7, e17.
81. T. Okuzumi, D. Fiedler, C. Zhang, D. C. Gray, B. Aizenstein, R. Hoffman and K. M. Shokat, *Nat. Chem. Biol.*, 2009, **5**, 484–493.
82. W. I. Wu, W. C. Voegtli, H. L. Sturgis, F. P. Dizon, G. P. Vigers and B. J. Brandhuber, *PLoS One*, 2010, **5**, e12913.
83. A. B. D'Assoro, T. Haddad and E. Galanis, *Front. Oncol.*, 2015, **5**, 295.
84. G. S. Falchook, C. C. Bastida and R. Kurzrock, *Semin. Oncol.*, 2015, **42**, 832–848.
85. T. Otto, S. Horn, M. Brockmann, U. Eilers, L. Schuttrumpf, N. Popov, A. M. Kenney, J. H. Schulte, R. Beijersbergen, H. Christiansen, B. Berwanger and M. Eilers, *Cancer Cell*, 2009, **15**, 67–78.
86. M. Brockmann, E. Poon, T. Berry, A. Carstensen, H. E. Deubzer, L. Rycak, Y. Jamin, K. Thway, S. P. Robinson, F. Roels, O. Witt, M. Fischer, L. Chesler and M. Eilers, *Cancer Cell*, 2013, **24**, 75–89.
87. W. C. Gustafson, J. G. Meyerowitz, E. A. Nekritz, J. Chen, C. Benes, E. Charron, E. F. Simonds, R. Seeger, K. K. Matthay, N. T. Hertz, M. Eilers, K. M. Shokat and W. A. Weiss, *Cancer Cell*, 2014, **26**, 414–427.
88. J. Boudeau, D. Miranda-Saavedra, G. J. Barton and D. R. Alessi, *Trends Cell Biol.*, 2006, **16**, 443–452.

89. D. P. Steensma, *J Mol Diagn*, 2006, **8**, 397–411, quiz 526.
90. O. Silvennoinen and S. R. Hubbard, *Blood*, 2015, **125**, 3388–3392.
91. J. S. Tokarski, A. Zupa-Fernandez, J. A. Tredup, K. Pike, C. Chang, D. Xie, L. Cheng, D. Pedicord, J. Muckelbauer, S. R. Johnson, S. Wu, S. C. Edavettal, Y. Hong, M. R. Witmer, L. L. Elkin, Y. Blat, W. J. Pitts, D. S. Weinstein and J. R. Burke, *J. Biol. Chem.*, 2015, **290**, 11061–11074.
92. J. M. Hildebrand, M. C. Tanzer, I. S. Lucet, S. N. Young, S. K. Spall, P. Sharma, C. Pierotti, J. M. Garnier, R. C. Dobson, A. I. Webb, A. Tripaydonis, J. J. Babon, M. D. Mulcair, M. J. Scanlon, W. S. Alexander, A. F. Wilks, P. E. Czabotar, G. Lessene, J. M. Murphy and J. Silke, *Proc. Natl. Acad. Sci. U. S. A.*, 2014, **111**, 15072–15077.
93. N. R. Brown, M. E. Noble, J. A. Endicott and L. N. Johnson, *Nat. Cell Biol.*, 1999, **1**, 438–443.
94. T. Zhou, L. Commodore, W. S. Huang, Y. Wang, T. K. Sawyer, W. C. Shakespeare, T. Clackson, X. Zhu and D. C. Dalgarno, *Chem. Biol. Drug Des.*, 2010, **75**, 18–28.

CHAPTER 4

Molecular Biology Techniques Applied to GPCR Allosteric and Biased Ligands

MÉLANIE FRAULI[a], CHRISTEL FRANCHET[a], ISMET DORANGE[a], ARTURO MANCINI[b], BILLY BRETON[b] AND STEPHAN SCHANN*[a]

[a]Domain Therapeutics SA, Bioparc, 850 Boulevard Sébastien Brant, Illkirch, F-67400, France; [b]Domain Therapeutics NA Inc., 7171 rue Fréderick-Banting, H4S 1Z9 St-Laurent (Montréal), Canada
*E-mail: sschann@domaintherapeutics.com

4.1 Introduction

Involved in all physiological pathways, G-protein coupled receptors (GPCRs) have proven to be highly successful therapeutic targets. Indeed, marketed GPCR compounds represent more than 40% of the top 200 worldwide best-selling drugs, and are projected to represent 120 billion US dollars of revenues per year by 2017.[1] Interestingly, these marketed compounds modulate the activity of a restricted set of 100 GPCRs out of the 350 that are considered to be potential drug targets, indicating that in spite of the powerful drug discovery arsenal that is now available, it remains hard to "crack" new GPCRs.[2] This is probably due to the inherent complexity of GPCR signaling, that arises from the multiplicity of intracellular pathways that can be elicited

RSC Drug Discovery Series No. 56
Allosterism in Drug Discovery
Edited by Dario Doller

Published by the Royal Society of Chemistry, www.rsc.org

by a single receptor, according to the downstream effectors it is interacting with. Moreover, the development of selective and druggable compounds remains challenging due to the high evolutionary conservation of orthosteric binding sites.

To overcome these obstacles, an interesting strategy consisting of targeting allosteric modulator (AM) sites has emerged. AMs are a novel class of small-molecule ligands that can be used to therapeutically modulate GPCRs.[3,4] Mostly inactive in absence of agonist, they can either potentiate the agonist action (Positive Allosteric Modulator; PAM) or decrease it (Negative Allosteric Modulator; NAM). They offer significant advantages in terms of drug discovery and development. First, subtype selectivity is easier to achieve with AMs due to a lower evolutionary pressure on their binding sites;[5] consequently, PAMs and NAMs exert better specificity of action as they are acting solely where and when the endogenous ligand is present, preserving physiological GPCR signaling. Similarly, AMs offer the possibility of "fine tuning" the endogenous physiological response by selectively affecting signaling pathway(s),[6,7] by exerting partial NAM properties or by modulating the activity of inactive metabolites.[8–10] Finally, allosteric modulation turns challenging GPCRs such as peptide, lipid, or family C receptors into more druggable targets.

Another emerging concept, referred to as functional selectivity, is based on the now well-acknowledged ability of a specific GPCR to elicit multiple cellular responses in response to an extracellular stimulus, such as the ones mediated by the primary effectors in direct contact with the GPCR, namely heterotrimeric G-proteins or β-arrestins. Heterotrimeric G-proteins are classified into four main families. Stimulatory G protein (Gs) family proteins couple to adenylyl cyclase (AC) and thereby induce production of cyclic AMP (cAMP). Members of the inhibitory G protein (Gi) family (namely Gi1, Gi2, Gi3, GoA (G other A), GoB, and pertussis toxin-insensitive Gz) are negatively coupled to AC, their activation leading to a decrease in intracellular cAMP levels. Gq, G11, G14, G15 and G16 form the so-called Gq/11 family. They activate phospholipase C (PLC) which produces diacylglycerol (DAG) and inositol triphosphate (IP3), leading to an increase of intracellular calcium. G12 and G13 activate phospholipases and can lead to activation of the mitogen-activated protein kinases (MAP kinases) phosphorylation cascade or Ras homolog gene family engagement. It is noteworthy that induction of the pathways described above is mostly mediated by the effector interaction with the alpha subunit of the heterotrimeric G-protein; however, downstream effectors like ion channels, phospholipases, ACs, and extracellular-signal-regulated kinases (ERK) or phosphoinositide 3-kinases (PI3-kinases) can also be activated by the β–γ subunits of the heterotrimeric G-protein.[11] Initially thought to be involved only in termination of receptor interaction with the G-protein and receptor desensitization, β-arrestins are more and more documented as promoting a full range of G-protein-independent signaling functions. Serving as multiprotein scaffolds, they bring into close proximity of the receptor the elements of specific signaling pathways, especially kinases such as MAP kinases, (*e.g.* ERK), PI3 kinase and Ras homolog gene family, member A (RhoA).[12] Since the latter effectors are also regulated by G-proteins, both G-protein-dependent and -independent signaling act in parallel. Exerting its effect after receptor endocytosis, β-arrestin-mediated signaling probably develops more "deeply" inside the cells while G-protein-dependent signaling remains closer to the cell surface.[13] Therefore, compared with G-protein-mediated signaling, signaling *via* β-arrestins is spatially and temporally distinct and results in different biological outcomes. The complexity and potential cross-talk of the pathways described above represents a challenge but also an opportunity in drug discovery. Indeed, the full understanding of a GPCR's "signaling signature" can lead to the development of ligands, including AMs, which are biased for a specific signaling pathway and thus tuned for the desired beneficial effects.

Current technologies used for AM high-throughput screening (HTS) and signaling characterization are reviewed in this article (Figure 4.1), as well as the utility of combining novel technological approaches to efficiently discover allosteric modulators and thoroughly characterize them for their ability to activate or inhibit specific signaling pathways.

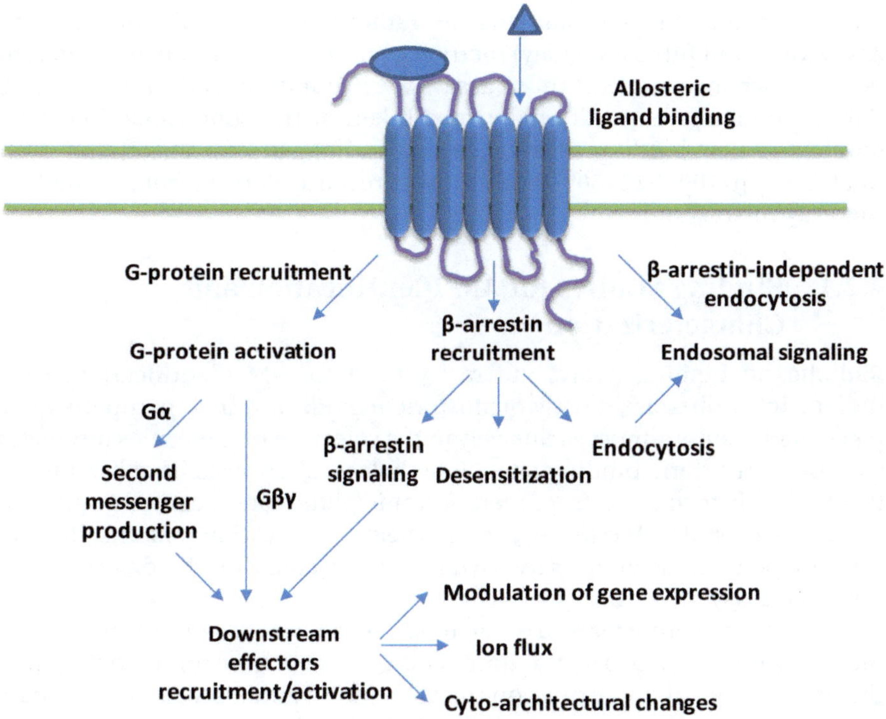

Figure 4.1 GPCR signaling events to be monitored by molecular biology techniques.

4.2 Primary HTS Assays for Allosteric Modulators of GPCRs

During the last decade, functional assays were mainly used as the primary screening assay to identify new GPCR ligands. Thoroughly designed, these assays can discriminate between the different ligand classes and can be used to define the structure–activity relationship (SAR) of synthetic ligands in a drug discovery program. Most of the time, cell-based assays are run with cells transfected with a given receptor. One of the advantages of such assays is the direct appreciation of a ligand's impact on the function of the receptor. A major limitation is that the data interpretation depends heavily on the assay conditions such as the probe used, the cell system and the receptor expression level.[14] A complementary approach that is the direct binding assay plays an important role in the GPCR drug discovery process and especially in pharmaceutical companies. Recently, Balderud and colleagues presented a comparison of HTS tests used across several AstraZeneca sites with external assays published in the PubChem database.[15,16] The authors highlighted that while cell-based functional tests (using gene reporter, second messengers) are mostly used in HTS by researchers, binding assays are complementary and offer the advantage of using different detection technologies at AstraZeneca. Indeed, within AstraZeneca, several binding assays were set up with different detection methods such as radiometric (Scintillation Proximity Assay (SPA) and filtration assay) methods, "fluorescence intensity" and luminescence whereas none of those assays were found for GPCR binding in the PubChem database. Additionally, as a token of the continuous interest in developing new binding assays, the first radioligand for an orphan receptor, namely [^{3}H]PSB-12150 for GPR17, was recently derived from a small synthetic agonist ligand.[17]

4.2.1 Binding Studies for AM Identification and Characterization

Radioligand binding assays are rarely used for AM identification. GPCR radioactive probes are mostly orthosteric ligands and their competitive displacement results almost exclusively in the selection of molecules interacting with the orthosteric binding site. Nonetheless, a few AMs have been identified with this technique (*e.g.*, metabotropic glutamate receptor 2 (mGluR2) NAMs discovered by Woltering *et al.*);[18] their identification remains, however, series-dependent as alternative mGluR2 NAMs could not be detected using the same assay.[19]

Another interesting feature of the binding assay is the possibility of following the binding kinetics in real time. This gives the opportunity to determine the impact of small molecules on the association and dissociation constants of a given orthosteric ligand and thus to better characterize allosteric interaction.[20] The residence time (*i.e.*, the time that a drug remains bound to its target) is also an important parameter that can be obtained from the binding kinetics. Dedicated review articles dealing with this concept in the field of GPCRs were recently published.[21–23] Tuning this parameter early on can lead to, among other things, longer duration of action or/and a better compound profile. In recognition of the importance of binding kinetics, a consortium financially supported by the European Innovative Medicines Initiative program and major pharma companies named Kinetics for Drug Discovery (K4DD) was created in 2012 to analyze the kinetics of drug interaction.

The next generation of GPCR binding assays may be label-free. Recent technical advancements in surface plasmon resonance (SPR) rendered this approach amenable to investigation of GPCR ligands and detecting both orthosteric and allosteric ligands with the appropriate experimental design, as very recently reviewed by Aristotelous *et al.*[24] However, this approach is still in its infancy since it requires purification of the target protein. Therefore, the most popular GPCR binding techniques still remain those based on a radiolabeled or fluorescent ligand as reporter (Table 4.1).

4.2.1.1 Radiolabeled Ligands

The radioligand assay is often used to identify new GPCR ligands at the early stages of the drug discovery process. Most of the time, the assay is not homogenous (except for the SPA which uses beads) and requires separation of the bound fraction from the free one using a sequence of washing and filtration steps. Technological advances allow for the miniaturization of specific parts of the assay, facilitating its implementation in the drug screening process. The probe used in the assay is often a selective orthosteric ligand that displays high affinity for a given target (dissociation constant (K_d) in the low nM range) and bears a radionuclide (such as ^3H or ^{125}I). The radioligand binding technology is applicable to different preparation types such as whole cells, tissue sections, or cellular membrane preparations. It is of interest that this assay is applicable to the identification of allosteric ligands. In 2007, Knudsen and colleagues showed that a small ligand (Compound 1 from Novo Nordisk) described as a glucagon-like peptide-1 receptor (GLP-1R) ago-allosteric ligand changed the binding of the [^{125}I] GLP-1 (7–36) amide probe in saturation-binding experiments.[25] Another approach to study allosteric interaction is based on the study of dissociation kinetics as presented by Goupil *et al.*[26] The main concept here is the study of the probe dissociation (labeled orthosteric ligand) in the presence of an allosteric molecule with a saturating concentration of the orthosteric ligand, which competes with the probe to occupy the orthosteric site. It is recognized as the common approach for monitoring an allosteric interaction.[27] In this type of experiment, the allosteric agent will modulate the dissociation rate of the radioligand by affecting its k_{on} and/or k_{off}, *i.e.*, association and dissociation rate constants, respectively. Unfortunately, technical constraints and radioactive waste management issues make the radioligand binding approach poorly amenable to HTS. It should be noted that for allosteric ligand characterization, the choice of the orthosteric ligand that will compete with the probe to occupy the site

Table 4.1 Binding assays.

	Assay type	Reporter type	Readout type	*In vitro*	*Ex vivo/in vivo*
Radiolabeled ligands	Saturation binding assay	Radioligand	Radioactivity	Membrane preparation or whole-cell	Tissue slices
	Scintillation proximity assay (SPA)	Radioligand	Radioactivity	Membrane preparation or purified protein immobilized on beads	No
Fluorescent ligands	Fluorescence intensity	Labeled ligand	Fluorescence	Membrane preparation or whole-cell	Tissue slices
	Fluorescence polarization-anisotropy	Labeled ligand	Fluorescence	Membrane preparation	No
	Dissociation-enhanced lanthanide fluorescence immunoassay–time-resolved fluorescence (DELFIA®–TRF)	Lanthanide chelates	Time-resolved fluorescence	Transfected cell	No
	Fluorescence correlation spectroscopy (FCS)	Labeled ligand	Fluorescence	Membrane preparation or whole-cell	No
	DTect-All™	Green fluorescent protein (GFP)–GPCR / labeled ligand	Fluorescence	Transfected cell	No
	Tag-Lite®	Tag–GPCR / lanthanide chelates	HTRF®	Transfected cell	No
	NanoLuc®	NanoLuc-GPCR / labeled ligand	BRET	Transfected cell	No
Label-free	Surface plasmon resonance	Label-free	Biacore	Purified protein or whole-cell	No

is important. Indeed, the observed effect induced by the allosteric molecule is directly dependent on the orthosteric ligand that occupies the orthosteric site. This effect is well exemplified by muscarinic receptors.[28] Thus, for receptors having several endogenous orthosteric ligands such as the GLP-1R, the task at hand can become quite complex.

4.2.1.2 Fluorescent Ligands

Introduced into pharmacological studies about 40 years ago, the initial purpose of fluorescent ligands was the visualization of GPCR localization in the cell by microscopy. Today, taking advantage of the large fluorophore spectrum and the wide fluorescent ligand diversity, they are commonly used in drug discovery. In addition to reducing the costs due to use of radioactive materials and associated waste management, the use of fluorescent ligands in binding assays offers the significant advantage of being homogenous (with no need to separate the bound fluorescent ligands from the free ones) and having a real-time readout (permitting the possibility of performing kinetic studies). However, fluorescent ligands also have drawbacks. Their synthesis requires the existence of a small ligand with a high affinity (nM) for the receptor since the addition of the fluorophore, which is generally large, can alter its affinity. In addition, in the screening mode, any optical interference of the screened molecules at the wavelength of the probe and/or any quenching effect of compounds can interfere with the readout. Thus, different counter screens should be performed.

Not surprisingly, multiple technologies based upon fluorescent ligands were developed, such as time-resolved fluorescence (TRF) intensity, fluorescence polarization-anisotropy, fluorescence correlation spectroscopy (FCS), fluorescence resonance energy transfer (FRET), time-resolved FRET assay and bioluminescence resonance energy transfer (BRET). A complete review that illustrates well the use of these technologies was recently published.[29]

Developed by Perkin Elmer, DELFIA® (dissociation-enhanced lanthanide fluorescence immunoassay) is a TRF intensity technology. Assays are designed to detect binding using lanthanide-chelate labeled reagents, separating unbound reagent using wash steps.[30] The fluorescence decay time of these lanthanide chelate labels is much longer than those of traditional fluorophores, allowing efficient use of temporal resolution for reduction of auto-fluorescent background. Methods based on fluorescence polarization anisotropy measure the rotational time of fluorescent molecules, which changes when the fluorescent molecule is bound to the receptor.[31] FCS enables the distinction between free and bound fluorescent ligand based on the differences in their diffusion coefficients. This method has a resolution that allows the study of interactions in the membrane microdomains.[32] Methods based on resonance energy transfer (RET) involve non-radiative transfer of energy from an excited donor to a suitable acceptor. The development of these methods has had a great impact on the current understanding of the functionality of GPCRs. Multiple studies were run *in vitro* and *in vivo* to follow ligand binding, G-protein coupling, cellular responses, receptor desensitization and oligomerization. A case-study is given by Ciruela and colleagues for the adenosine and purinergic (P2Y) receptor fluorescent ligands that were used across different technologies.[33] Advances in technologies allow assay multiplexing, meaning the possibility of following at least two cellular events using two different fluorophores emitting at two different wavelengths.

Technologies based on energy transfer (RET-assay)[29,34] are approaches that are currently widely used in the drug screening context and will be further discussed in the following sections. In a FRET-based binding assay, an energy transfer occurs from the donor fluorophore to the acceptor fluorophore. Three main conditions are required for an efficient energy transfer: (1) the two fluorophores must be compatible for an energy transfer (emission spectra of the donor overlaps the excitation spectra of the acceptor); (2) the two fluorophores must be close to each other with a distance up to 10 nm; (3) the dipole moments of both fluorophores must be parallel.

4.2.1.3 Library Filtering with FRET-Based DTect-All™

The DTect-All™ FRET-based binding platform set up at Domain Therapeutics is used to identify small compounds, peptides, as well as antibodies interacting with the receptor 7 transmembrane (7TM) domain. Independent of the GPCR signaling pathways, this technology can serve as first filter to reduce the number of compounds to be functionally characterized. Running using 384-well microplates, the assay can be performed on intact cells, frozen cells, or cell membrane preparations. The energy donor, the Green Fluorescent Protein (GFP), is fused to the amino terminal (Nter) part of the target GPCR, and the energy acceptor is a fluorophore (or a chromophore) linked to a probe binding the receptor.[35,36] The intensity of GFP fluorescence variation is proportional to the occupancy of receptor sites. Counter-screening using another GPCR confirms the interaction specificity. This technology is specifically suited for the pre-selection of GPCR AMs for two reasons: (1) thanks to the truncation of the Nter domain, only ligands interacting with the 7TM domain, the privileged site for AMs, and extracellular loops can be detected, and (2) by acting as a "surrogate marker" of the conformational change of the GPCR, it will detect ligands binding at any site of the 7TM domain. This technology can therefore identify a large diversity of binders that can be further characterized as agonists and antagonists (for those GPCRs still possessing an active endogenous site after Nter truncation), as well as allosteric agonists (PAMs or NAMs). As such, DTect-All™ demonstrates unique features overcoming major limitations of ligand binding (no bias toward orthosteric ligand – no need for a known high-affinity ligand) and functional assays (binding is a very upstream phenomenon – no need for receptor activation – no influence of cellular environment or signaling pathway allowing the discovery of silent allosteric modulators (SAMs), a new source of ligands not detectable by a single functional assay).[37]

DTect-All™ was successfully used with GPCRs belonging to the three different GPCR classes and notably also with orphan GPCRs. For instance for class B GPCRs, the use of a GLP-1R assay enabled the first identification of quercetin,[38] a novel PAM displaying various functional outcomes according to the probes and signaling read-out used in the assays.[39] Similarly, the same binding assay detects the recently described GLP-1R ago-allosteric modulators whose functional characterizations were shown to be "context-dependent".[10] Another DTect-All™ assay led to the identification of the first mGluR2 SAM that was further converted into dual mGluR2/3 NAMs/PAMs.[37] The later example illustrates the potential of these silent binders that can serve as chemical starting points for discovery of novel series of active AMs *via* "molecular switches".[37,40]

4.2.1.4 Other RET-Based Binding Assays

With the development of time-resolved FRET assay, the performance of the energy transfer assay was improved. This technology uses fluorophores with longer fluorescence lifetimes (greater than 1 ms, compared with less than 20 ns for conventional fluorophores) such as lanthanide ions (Eu^{3+}, Tb^{3+}).[41] The key point here is the existence of a delay between the donor excitation and the measurement of the acceptor emission. During that time, the fluorescent background noise has decayed and this improves the ratio of signal to noise significantly. Combining this technology, that is homogenous time-resolved FRET (HTRF®) with a self-labeling protein tag system, Cisbio Bioassays developed a new method to study ligand–GPCR interactions.[41] In these assays, the ligands are labeled with lanthanide ions (energy acceptor) and the GPCR is fused to a suicide enzyme-tag (energy donor) at its N-terminal part. The suicide enzyme (SNAP-Tag, CLIP-Tag, Halo-Tag enzymes)[42] emits fluorescence in the presence of substrate. This substrate is non-permeable and this allows the labeling of the cell-surface receptor only, thereby reducing the background noise. Quantification of the ligand interaction is directly correlated to the ratio of signal acceptor to donor. A conceptually similar BRET-based approach was developed by Promega, and uses a fused nanoluciferase (NanoLuc) at the N-terminal part of the receptor.[43] The NanoLuc (the energy-donor) displays a high brightness with a relatively long bioluminescence half-life (>2 h) in the presence of its substrate (furimazine), conferring a good sensitivity to the assay.[44] One of the limiting factors of these methods is the necessity to have a high-affinity nanomolar ligand that interacts with the given GPCR to be able to fuse the fluorophore. Thus, these approaches are not directly applicable to orphan GPCRs.

To summarize, the RET binding assay strategy has gained momentum in recent years and several evolutions have emerged. The ligand can be an endogenous ligand, a synthetic small ligand, an antibody labeled with fluorophore, quencher, or lanthanide ions; on the opposite side, the receptor can be labeled using a fluorescent protein, an enzyme-tag (suicide or nanoluciferase), or an antibody directed against a Tag fused at the N-terminal region of the receptor.[34,45] A refinement of this strategy consists of the site-specific introduction of genetically encoded unnatural amino acids (uaa) into expressed GPCRs. In a recent review, Huber and Sakmar reported the use of bioorthogonal chemical reactions to introduce fluorescent labels onto a GPCR with the aim of studying the conformational change process dynamically.[46] Indeed, the introduction of one or two uaas enables the development of binding assays wherein the GPCR-intramolecular energy transfer is detected without the need of a ligand. This assay requires confirmation that the functional receptor activity is not modified by the uaa insertion.

4.2.2 Functional Tests

As demonstrated in the previous sections, binding assays can be extremely helpful in GPCR ligand characterization in the course of the drug discovery process; however, functional tests remain the current standard for GPCR AM HTS.[47] These functional assays may provide a direct measure of GPCR activation by monitoring stimulation and/or recruitment of receptor primary effectors such as G protein(s) or β-arrestins; other assays follow events that happen downstream in the signaling cascade such as second-messenger production (Tables 4.2 and 4.3). If accumulation or dynamic cAMP functional assays are described, most HTS campaigns use Ca^{2+}-mobilization assays, either by taking advantage of the natural coupling to the Gq–PLC pathway of the targeted receptor, or by redirecting the signaling pathway to a promiscuous or chimeric G-protein co-expressed with the receptor.

In screening campaigns, a protocol combining a single concentration of both the tested compound and an endogenous agonist is often used. Two successive additions are performed on the cell system expressing the receptor of interest, the first one to assess the potential agonist (or allosteric agonist) activity of the tested compound, and the second one its modulation effect, using the concentration giving 10–20% of the maximum effect (EC_{10-20}) or that giving 80% (EC_{80}) of the endogenous agonist for PAM and NAM, respectively.[48,49] A triple addition can also be is sometimes used to screen for agonists, PAMs and NAMs in a single campaign.[48,50] To further quantify AM effects, dose–response curves of the allosteric agent in presence of a fixed orthosteric agonist concentration can be performed so as to determine the AM concentration giving 50% of the maximum effect (EC_{50}) and maximal effect (E_{max}) values. Alternatively, a dose–response of orthosteric agonist can be tested in presence of a fixed AM concentration; shifts in agonist EC_{50} or E_{max} can then be used to quantify and compare AM effects. In the particular case of NAMs, a Schild regression plot should be performed to confirm the allosteric *vs.* competitive character of the antagonist effect.

The major limitation of functional assays in an HTS campaign is the risk of missing active AMs due to inadequate assay conditions. Indeed, subtle changes in the assay can result in conflicting results. For instance, the use of

Table 4.2 Functional assays for direct measurement of GPCR activation.

	Assay type	Reporter type	Readout type	Signaling pathways	*In vitro*	*Ex vivo/in vivo*
G-protein engagement assay	GTPγS	[^{35}S]GTPγS binding	Radioactive + Immunoprecipitation (IP)	Mainly Gi + all Gα with IP	Membrane preparation	Yes
		[α-^{32}P] GTP-γ-azidoanilide	Radioactive + IP	Mainly Gi + all Gα with IP	Membrane preparation	Yes
		Europium GTPγS binding	TRF + IP	Mainly Gi + all Gα with IP	Membrane preparation	Yes
		BODIPY FL GTPγS binding	Fluorescence + IP	Mainly Gi + all Gα with IP	Membrane preparation	Yes
	GTP specific antibody	Monoclonal IgG antibody	Enzyme-linked immunosorbent assay (ELISA) + IP + fluorescence	Gs, Go, Gi, Gz, G13	Lysed cell	Yes
		Nanobody	Fluorescence	Gs	Transfected cell	No
	Conformational change	Gα-tag/GPCR-tag	BRET, FRET, or Protein complementation assay (PCA)	Gs, GoA, GoB, Gi1, Gi2, Gi3, Gz, Gq, G12, G13	Transfected cell	No
		Gαβγ biosensor	BRET, FRET, or PCA	Gs, GoA, GoB, Gi1, Gi2, Gi3, Gz, Gq, G12, G13	Transfected cell	No
		Gα- and Gβγ-interacting protein biosensors	BRET, FRET, or PCA	Gs, GoA, GoB, Gi1, Gi2, Gi3, Gz, Gq, G11 G12, G13, G14, G15/16, Golf, Ggust, Gt	Transfected cell	No
β-Arrestin assay	β-Arrestin membrane localization	Translocation	Fluorescence microscopy	β-Arrestin1, β-Arrestin2	Transfected cell	No
		β-Arrestin-tag or membrane-tag	BRET, FRET, or PCA	β-Arrestin1, β-Arrestin2	Transfected cell	No
	Protein–protein interaction	β-Arrestin-tag or GPCR-tag	BRET, FRET, or PCA chemiluminescent/reporter gene	β-Arrestin1, β-Arrestin2	Transfected cell	No
	Double brilliance	Double tag β-Arrestin	BRET	β-Arrestin1, β-Arrestin2	Transfected cell	No

Table 4.3 Functional assays for indirect measurement of GPCR activation.

	Assay type	Reporter type	Readout type	Signaling pathways	*In vitro*	*Ex vivo/ in vivo*
Second messengers	cAMP	Cyclase assay	Radioactive	Gs, Gi, Gβγ	Membrane preparation	Yes
		IgG antibody	AlphaScreen™, HTRF®, or CatchPoint™	Gs, Gi, Gβγ	Lysed cell	Yes
		cAMP–protein interaction	BRET, FRET, or Glosensor™	Gs, Gi, Gβγ	Transfected cell	Yes
		cAMP response element (CRE)	Reporter assay	Gs, Gi, Gβγ	Transfected cell	Yes
	cGMP	IgG antibody	AlphaScreen™, HTRF®, ELISA, or CatchPoint™	Nitric oxide	Lysed cell	Yes
	Calcium	Fluorescent dye	Fluorescence	Gq, Gβγ, ion channel	Labeled cell	Yes
		Photoprotein (Aequorin, Obelin, Photina)	Bioluminescence	Gq, Gβγ, ion channel	Transfected cell	No
		Calmodulin-M13 conformational change	BRET, FRET, PCA, or Cameleon	Gq, Gβγ, ion channel	Transfected cell	No
		PKC substrate conformational change	BRET, or FRET	Gq, Gβγ, ion channel	Transfected cell	No
		Calcium signaling response element (NFAT)	Reporter assay	Gq, Gβγ, ion channel	Transfected cell	No
	IP3	[^{3}H]-IP3 separation	High-performance liquid chromatography (HPLC) isolation radioactivity	Gq, Gβγ	Lysed cell	Yes
		IgG antibody	ELISA	Gq, Gβγ	Lysed cell	Yes
		IP3 binding protein	Fluorescence polarization	Gq, Gβγ	Transfected cell	No

	Inositol phosphate (IP1)	IgG antibody	HTRF®, or ELISA	Gq, Gβγ	Lysed cell	Yes
	DAG	DAG purification	Thin-layer chromatography (TLC), or HPLC	Gq, Gβγ	Lysed cell	Yes
		IgG antibody	ELISA	Gq, Gβγ	Lysed cell	Yes
		PKCδ conformational change	Fluorescence	Gq, Gβγ	Transfected cell	No
		Translocation of PH domain of PLCδ	BRET, or FRET	Gq, Gβγ	Transfected cell	No
Label-free	Electric-based sensor	Impedance	xCELLigence™, or CellKey™	Gs, Gq, Gi, G12/13, Gβγ	No labeling	Yes
	Optical-based sensor	Resonant waveguide grating	BIND®	Gs, Gq, Gi, G12/13, Gβγ	No labeling	Yes
		Surface plasmon resonance	Epic™	Gs, Gq, Gi, G12/13, Gβγ	No labeling	Yes

different probes can give opposite activities for the same AM.[8] This phenomenon, known as "probe dependency", and which is of particular relevance for GPCRs having multiple endogenous ligands, was recently illustrated with the characterizations of GLP-1R and relaxin/insulin-like family peptide receptor 3 (RXFP3) PAMs.[39,51,52] Similarly, "cell-system dependency" and "receptor-level dependency" were reported for a mGluR7 NAM and a GLP-1R ago-PAM respectively.[53,54]

4.2.2.1 Calcium

Using Ca^{2+} as readout for GPCR drug discovery is a common practice given the development of cell-permeable, wash-free, calcium-sensitive fluorescent dyes and automated real-time fluorescence detection systems. In these assays, cells are loaded with a dye that fluoresces upon Ca^{2+} binding. Importantly, the fluorescence magnitude emission is directly correlated to calcium response magnitude. The commonly-used Molecular Devices FLIPR™ Calcium Assay (now in the sixth generation) contains a proprietary quenching agent that allows cellular loading of dye without the need for subsequent washing to remove excess dye.[55] This assay can be fully automated and amenable to ultra HTS when used with the FLIPR™ Tetra High-Throughput Cellular Screening System or to high-content drug screening (HCS) when used with the Opera™ High Content Screening System (PerkinElmer). However, fluorescent compounds can interfere with this assay. Other dye-based Ca^{2+} assays include BDTM Biosciences' Calcium Assay Kit and DiscoveRx's HitHunter™ Calcium No WashPLUS. Alternatively, Ca^{2+}-activated photoproteins are employed as biosensors for intracellular Ca^{2+} flux. Such proteins include the photoproteins aequorin[56] and Photina (a derivate of obelin)[57] from PerkinElmer, and a variant of clytin (Millipore).[58] Upon binding Ca^{2+}, and in the presence of coelenterazine, these proteins immediately release a quantity of light that is directly proportional to the magnitude of the evoked Ca^{2+} response. In addition to displaying an appreciable sensitivity and dynamic range, the photoprotein-based Ca^{2+} biosensors can be targeted to specific subcellular compartments and organelles and thus allow for site-specific measurements of Ca^{2+} flux. Finally, Invitrogen's Premo™ Cameleon Calcium Sensor is a FRET-based ratiometric sensor consisting of a calmodulin and a M13 domain (Ca^{2+}-sensitive moiety) tagged at either end with fluorophores cyan fluorescent protein (CFP) and yellow fluorescent protein (YFP). Binding of four Ca^{2+} ions to the calmodulin-M13 core causes a conformational change that brings the cyan fluorescent protein (CFP) and yellow fluorescent protein (YFP) domains closer, allowing a FRET to occur. Due to the requirement for external light excitation, this assay is subject to a high background signal compared with bioluminescent emitters.[59] As for the FLIPR™ assay, fluorescent interference from compounds remains an issue. One of the drawbacks linked to the use of Ca^{2+} as a measure of ligand activity relates to the rapid and transient nature (occurring within seconds) of Ca^{2+} efflux. Thus the use of Ca^{2+} assays in detecting slow-binding ligands is not recommended.[60]

Another drawback consists of the requirement for using specialized instrumentation and long acquisition times when done in microplate format.

4.2.2.2 *IP1, IP3, and DAG*

An alternative way to monitor Ca^{2+} flux involves the measurement of inositol phosphate (IP1) or IP3. A radiometric scintillation proximity assay for IP3 (based on the incorporation of tritiated inositol) amenable to HTS has been described.[61] Non-radioactive alternatives include PerkinElmer's IP3 AlphaScreen™, and DiscoveRx's HitHunter™ IP3 Fluorescence Polarization Assay. In the latter assay, cells are treated with a ligand and are immediately quenched and lysed with perchloric acid. Fluorescent IP3 tracer and a proprietary IP3-binding protein that is highly selective for the active isomer of IP3 are subsequently added. Labeled IP3 tracer that is bound rotates less than free fluorescent IP3 when excited with plane-polarized light. Consequently, the polarized light used for excitation remains "unperturbed" and highly polarized, ultimately generating a higher polarization signal. Since unlabeled IP3 from cell lysates competes for binding of the fluorescent IP3 tracer to the IP3-binding protein, the magnitude of the polarization signal is inversely proportional to the amount of IP3 in the cell lysates. Cisbio's HTRF-based IP-One™ assay enables the quantification of IP1. The choice of IP1 as a substitute for IP3 is based on the fact that IP3 displays a relatively short half-life (*i.e.*, ~10 s).[62] Pretreatment of cells with LiCl inhibits inositol monophosphatase, the final enzyme in the IP3 degradation pathway (*i.e.*, inositol monophosphatase hydrolyzes IP1 to inorganic phosphate and *myo*-inositol). As such, LiCl promotes the accumulation of IP1, which can be subsequently used in endpoint measures of Gq pathway activity. This method is especially relevant for the characterization of slow-binding ligands, whose activities are too slow to be quantified with assays measuring rapid and transient second messenger fluxes (as seen for Ca^{2+} and IP3).[60]

Finally, genetically encoded biosensors for DAG are available from Montana Molecular. These biosensors consist of a circularly permuted GFP positioned either between the pseudo substrate and C1 domains or in the hinge region of the PKC isoform PKC delta. Binding of DAG to the C1 domain results in conformational changes that alter the fluorescence intensity of GFP. The change in fluorescence, which is determined *via* fluorescence imaging, is subsequently used as a readout for fluxes in intracellular DAG levels.[63] As for other biosensor-based approaches, this assay allows for the monitoring of DAG levels in real-time in living cells.

4.2.2.3 *cAMP and cGMP*

Significant progress has been achieved over the past few decades with regards to the quantification of cAMP. Such progress was mainly driven by the development of cAMP-specific antibodies and the discovery of cAMP-binding G-proteins. The cAMP assays described below could also be adapted to the analysis of cGMP (with the sole difference being the specificity of the anti-analyte antibody). The DiscoverX HitHunter™ system is a competitive immunoassay that utilizes enzyme fragment complementation (EFC) technology where a fragment β-galactosidase (β-gal) enzyme donor (ED) is conjugated with cAMP.[64] This ED–cAMP conjugate and cellular cAMP compete for binding to an anti-cAMP antibody. When levels of cellular cAMP increase, the anti-cAMP antibody becomes saturated allowing the ED–cAMP complex to complement with the β-gal acceptor and form a functional β-galactosidase holoenzyme, which then hydrolyzes a substrate to produce a chemiluminescent signal that is directly proportional to the amount of intracellular cAMP.[65] PerkinElmer's scintillant proximity-centered FlashPlate™ assay consists of a competitive radioimmunoassay in 384-well format. Wells are coated with polystyrene-based scintillant. Cells are grown in the assay wells and following stimulation with the compound(s) of interest cells are lysed in order to release their cAMP contents. Intracellular cAMP competes with [^{125}I]-labeled cAMP to bind to the cAMP-specific antibody coated onto the wells. As the antibody-bound [^{125}I]-cAMP will excite scintillant on the plate surface, a decreased signal is observed with increasing intracellular cAMP levels. Similar competitive non-radioactive immunoassays were developed by CisBio (HTRF®-based),[66] Molecular Devices (CatchPoint™), Electrochemiluminescence (Meso™)[67] and PerkinElmer (AlphaScreen™).[68] In the CatchPoint™ assay, intracellular cAMP competes with a horseradish peroxidase (HRP)-labeled cAMP for binding to plate-adsorbed antibody. Addition of their proprietary Stoplight Red substrate generates a stable fluorescent readout that can be measured from 10 minutes to 24 hours after substrate addition. The AlphaScreen™ (Amplified Luminescent Proximity Homogeneous Assay) assay contains streptavidin-coated donor beads that, upon laser excitation at 680 nm, release an oxygen singlet (ΔO_2). When in close proximity (*i.e.*, ~200 nm and less) to the acceptor beads that are coated with an anti-analyte antibody, the ΔO_2 will react with a chemiluminescent component contained within the acceptor beads, which in turn emits light at ~600 nm. In the cAMP AlphaScreen™, biotinylated cAMP brings the two beads together to generate a signal.[68] Similarly, in the CisBio (HTRF®-based), FlashPlate™, CatchPoint™ and Electrochemiluminescence (Meso™) assays, the cell lysates exogenous cAMP competes with the tagged cAMP to bind to the antibody sites on the acceptor beads or the capture plate. This results in an inversely proportional relationship between signal and intracellular cAMP concentration.

Finally, Promega has developed a luciferase-based biosensor for use in living cells. Their Glosensor™ cAMP assay consists of a mutant *Photinus pyralis* luciferase-tagged PKA type II-b regulatory subunit (denoted RIIbB). Cells are transfected with the RIIbB-based biosensor which upon binding cAMP, undergoes a conformational change that increases luciferase activity. The Glosensor™ assay was used for HTS of ligands acting at Gs- or Gi-coupled GPCRs.[69,70] This biosensor-type assay exhibits a broad dynamic range, with up to 500-fold changes in light output. Importantly, it does not require cell lysis, thus enabling kinetic measurements of cAMP levels in real-time in living cells.[71]

4.2.2.4 Reporter Assays

An alternative strategy for the direct measurement of second-messenger molecules involves the analysis of ligand-induced GPCR downstream signal transduction. Such assays are based on the observation that GPCRs modulate gene expression *via* response elements that bind second-messenger-responsive transcription factors (*i.e.*, cAMP response element, the nuclear factor of activated T-cells response element, the serum response element and the serum response factor response element). In both SABiosciences' Cignal Finder™ Reporter Array and Promega's Rapid Response™ assays,[72] cells are transfected with a pathway-selective luciferase reporter gene containing one of the aforementioned response elements in its promoter region. Following cell stimulation, cells are lysed and reporter activity is used to decipher which pathways are activated downstream of a given GPCR. Besides, activation of non-second-messenger-mediated pathways known to be downstream of GPCRs (mitogen-activated protein kinase (MAPK), stress-activated protein kinase (SAPK), activating transcription factor (ATF), nuclear factor κb (NFκb)) can also be determined with the Cignal Finder™ assay. These reporter assays are scalable down to a 1536-well format and provide significant dynamic range and large signal-to-noise ratio, thus facilitating HTS and detection of weakly active agonists or allosteric modulators. However, given their dependence on gene transcription and translation, such assays (1) require long incubation times and thus cannot be used to study rapid and transient responses, and (2) are sensitive to compounds that influence gene expression. Furthermore, the requirement for cell lysis precludes their use in the monitoring of real-time signaling kinetics.

Overall, second-messenger-based assays provide a high-throughput and sensitive means to conduct drug screens. However, such assays based on receptor downstream readouts are likely to be more sensitive to experimental conditions (*i.e.*, cell type, culture medium composition, *etc.*). Such environmental influences on experimental outcomes contribute to variability in reported ligand and receptor properties. The ability to obtain receptor-proximal measures of ligand-induced G-protein or pathway-specific GPCR coupling (*e.g.*, by probing G-protein activation) is therefore significantly advantageous.

4.2.2.5 RET-Based Techniques

The development of protein interaction assays has opened novel avenues for the measurement of GPCR activation and G-protein coupling. Energy transfer assays like FRET and BRET are used to measure the recruitment of different G-protein subtypes to a given receptor. These assays are used to directly measure G-protein activation following receptor stimulation. In the BRET assay, the subunits of the G-protein heterotrimer are fused to the energy donor and acceptor moieties. Given that GPCR activation produces a conformational intermolecular and intramolecular shift in the G-protein heterotrimer, energy transfer efficiency between the tagged subunits can be used as measure of G-protein activation. Importantly, this strategy is applicable to the monitoring of G-protein engagement and activation in real time in living cells.[73] At Domain Therapeutics NA Inc., BRET-based biosensors developed on the BioSensAll™ platform are available not only for measurement of G-protein engagement, but also for β-arrestin recruitment, second messenger (such as cAMP, DAG, and phosphatidylinositol (3,4,5)-trisphosphate (PIP3)) quantification and monitoring of downstream effectors or events. Taken alone, each of these biosensors are amenable to HTS; when considered in parallel, this assay panel allows characterization of allosteric modulator behavior on every pathway potentially activated by the targeted GPCR.

4.2.2.6 β-Arrestin Monitoring

Once believed to solely regulate receptor desensitization, internalization and recycling, it is now clear that β-arrestins (mainly β-arrestin-1 and -2) also serve as GPCR-proximal signaling scaffolds that may modulate the activity of various signaling networks (including ERK1/2, NF-κB, and PI3-K).[74] Assays for ligand-induced engagement of β-arrestins are becoming increasingly important components of drug discovery programs. Such assays are based on microscopic imaging and (more commonly) non-imaging strategies. The first commercialized β-arrestin recruitment assay, TransFluor™ (Molecular Devices), is performed using cells overexpressing GFP-tagged β-arrestin. In this assay, ligand-induced translocation of β-arrestin-GFP from the cytosol to the activated GPCR-containing pits at the plasma membrane, endosomal vesicles and/or nucleus is monitored in a quantitative and temporal manner with epifluorescence microscopy (ArrayScan™), confocal microscopy (INCell™ Analyzer, Opera™, ImageXpress™) or microplate laser-scanning cytometry (Acumen™)-based imaging systems.[75–77] In addition to being well-suited for HCS, other advantages of the TransFluor™ system include the ability to use untagged receptors (either endogenous or exogenous) and the capacity for direct microscopic monitoring of compound-related cytotoxic effects. However, despite the various advantages of imaging-based assays, they fail to provide information related to the "quality" of β-arrestin recruitment. Indeed, it is now appreciated that β-arrestins can assume ligand-specific conformations once recruited to a GPCR and that each conformation may have distinct biological functions.[77,78]

Numerous non-imaging-based β-arrestin recruitment approaches are now available. The PathHunter™ system from DiscoveRx is an EFC-based assay built on the same enzyme acceptor and donor as for the HitHunter™ system described above.[79] Invitrogen's Tango™ Assay is an extension of the EFC strategy used in the PathHunter™ system. In this modified (or protease-mediated) reporter assay, β-arrestin is fused to the tobacco etch virus (TEV)

N1a protease and the GPCR's C-terminal tail is tethered to a protease cleavage site followed by the transcription factor Gal-VP16.[80] Upon recruitment of β-arrestin to the activated GPCR, Gal-VP16 is cleaved off by the TEV protease and subsequently translocates to the nucleus to induce transcription of β-lactamase. β-Lactamase then cleaves a modified substrate tagged with two fluorophores, which subsequently transitions from green fluorescence emission to blue fluorescence emission. The cleavage-induced change in the FRET signal can be monitored as a measure of the GPCR–β-arrestin interaction. β-Arrestin recruitment assays based on EFC and modified reporter strategies afford the capacity of monitoring GPCR–β-arrestin interactions without the necessity of an imaging instrument (as for TransFluor™). These assays also provide high signal amplification with large detection windows, which is especially important when working with weakly active agonists and allosteric modulators. Furthermore, the use of a chemiluminescent readout makes the PathHunter™ assay insensitive to potential fluorescent test compounds' interference.

However, these assays suffer from various disadvantages. In the Tango™ assay, the GPCR–β-arrestin interaction is measured following transcription and translation of a reporter gene. As such, measurement of real-time interaction kinetics is not possible. Moreover, this assay is subject to compound interference at the level of (1) the signaling events leading to reporter gene expression, and (2) gene transcription itself. Although measurements of GPCR–β-arrestin interaction kinetics are possible with the EFC-based systems, acquisition of a positive signal following the interaction is delayed due to the time required for the reconstitution of a functional holoenzyme. As such, true real-time kinetics are not possible with the PathHunter™ system. Besides, neither EFC nor modified reporter assays provide insight into potential ligand-directed effects on β-arrestin conformation. Most importantly, these approaches require the modification of both the receptor and β-arrestin. Although tagging the receptor enhances signal specificity (especially when the experimental system contains several receptors responding to the same ligand), such modification may alter and/or perturb normal receptor function. PathHunter™ and Tango™ assays were successfully applied to receptors such as the sphingosine 1-phosphate receptor and lipid-activated GPCRs.[79,81,82]

Another non-imaging-based approach to study the recruitment of β-arrestins to activated GPCRs involves the use of FRET- and BRET-based biosensors. Such assays are suitable for HTS and were successfully applied to the study of a broad range of receptors as has been reviewed.[83,84] The main drawback related to this approach involves the necessity to tag β-arrestin and the GPCR (on its C-terminal tail) of interest. RET-based assays are also incompatible with the use of fluorescent compounds. However, the sensitivity of such technologies allows for the detection of intramolecular (conformational) changes in proteins that is not possible with TransFluor™, PathHunter™ or Tango™ assays. A "double brilliance" BRET biosensor available from Domain Therapeutics NA Inc. allows for the assessment of ligand-induced conformational shifts for β-arrestins using untagged receptors.[78,85] Given the instantaneous and high efficiency of RET between donor and acceptor moieties, true real-time measurements of interaction kinetics is now possible. Use of Prolume Purple (a new coelenterazine from NanoLight™ Technologies that exhibits an up to 13-fold higher luminescence than DeepBlueC) allows for measurements to be recorded over a period of 1 h. One commonly cited disadvantage for BRET with regards to the study of protein–protein interactions relates to the inability to discern the subcellular localization of the interaction. This is particularly relevant to GPCR–β-arrestin interactions since the stability (*i.e.*, longevity) of β-arrestin–GPCR complexes, particularly in endosomes, correlates with the magnitude of ERK1/2 activity.[78,86] Using new biosensors (Domain Therapeutics NA Inc.), it is now possible to spatially and temporally resolve interactions occurring at the plasma membrane from those present in endosomal compartments. As such, currently-available BRET-based sensors provide both qualitative (conformational and spatial) and quantitative (kinetics) information about GPCR–β-arrestin interactions.

4.3 Complementary Assays for AM Characterization

4.3.1 GTP Gamma S

The first GPCR membrane-based [^{35}S]GTPγS binding assays were described for muscarinic acetylcholine receptors in cardiac (atrial) membranes.[87] The main disadvantage of this assay is the requirement for filtration steps to eliminate unbound [^{35}S]GTPγS. The need for filtration was elegantly bypassed with the conception of bead- or plate-based (FlashPlate) scintillation proximity assays (SPA). Moreover, to avoid the use of radioactive GTPγS, a europium fluorescence-dye-based method was developed.[88] However, this nonradioactive approach requires the isolation of GPCR-expressing membranes and the use of complex assay buffers.[89] The low-throughput nature of this technique limits its use for large-scale GPCR coupling analysis. Recently, transfectable single-domain antibodies (nanobodies) capable of recognizing an active (Gs-bound) state of the β2 adrenergic receptor and an active conformation of Gαs were described.[90] In addition to offering an innovative approach for the measurement of G-protein-subtype-specific coupling, such nanobody technology allows for temporal and spatial resolution of G-protein engagement following GPCR activation (*i.e.*, endosomal *vs.* plasma membrane coupling).

4.3.2 Label-Free Assays

Label-free technologies are based on the principle that upon stimulation, cells undergo complex cyto-architectural changes that ultimately influence cell adhesion, shape, intercellular interactions, proliferation, and/or viability. Signaling events downstream of GPCRs (*i.e.*, G α-dependent signaling and second messenger generation, G β–γ-dependent signaling, small GTPase activation) are frequently involved in such cytoskeletal changes. Label-free technologies exploit these ligand-induced cytoskeletal changes as integrated and cumulative (*i.e.*, not pathway-specific) readouts in the study of GPCR biology and ligand pharmacology.

Label-free biosensors used for whole-cell sensing are mostly classified into two categories optical, and electric (*i.e.* impedance) biosensors. Optical biosensors (a.k.a., resonant waveguide grating (RWG), or surface plasmon resonance (SPR) biosensors) utilize grating surfaces embedded in the bottom of microtiter plates. When illuminated with white or broadband light, these grating surfaces reflect a very narrow band of light. Optical biosensors detect shifts in the resonant wavelength and angle of reflected light, which is characteristic of local refractive index near the grating (*i.e.* cell adhesion) surface. The refractive index itself is mainly influenced by the physical properties of the cell layer in contact with the grating surface and is altered by GPCR-mediated signaling. Indeed, GPCR activation results in temporally- and spatially-regulated protein trafficking, cytoskeletal remodeling, cell adhesion and morphological alterations, all of which result in significant changes in mass density and distribution (*i.e.*, dynamic mass redistribution, DMR) of biomolecules within the cells.[91] Sensor-detected DMR changes occurring within approximately 200 nm from the contact surface (the adherent portion of cells) are used as qualitative (*i.e.*, discrimination of G-protein class coupling) and quantitative (*i.e.*, calculation of ligand potency and efficacy) readouts.[92,93] Commercial optical biosensor systems include Epic™ (Corning Inc.) and BIND™ (SRU BioSystems).[94]

Electrical sensors use a gold microelectrode array to measure changes in impedance of a cell layer under electric fields generated by the continuous sweeping of AC voltages through a range of frequencies.[95,96] Under assay conditions, the cellular plasma membrane serves as an insulating barrier and thus impedes current flow, with the measured impedance value being a composite derivation of the relative density of cells over the electrode surface (influenced by cell number, size and shape) and the relative adherence of these cells (*i.e.* the distance between the cell surface and the electrodes). Commercially-available impedance biosensor systems include ECIS™ (Applied BioPhysics), xCELLigence™ (Roche/Acea Biosciences), and Cellkey™ (Molecular Devices).[94]

Label-free assays provide fine temporal resolution, allowing for real-time monitoring of ligand-induced changes over a period of hours and up to days. They also permit for the use of endogenously expressed receptors[93,97,98] and eliminate the necessity for reporter systems or protein tagging. Consequently, label-free technologies enable screens to be conducted in native or primary cell settings that are closer to *in vivo* contexts.[99] Particularly interesting is the possibility of performing ligand screens in tissues isolated from disease-afflicted subjects.[100] Label-free evaluation of GPCR signaling has allowed for the characterization of agonists, antagonists, and inverse agonists, and more recently allosteric modulators.[101] Comparison of ligand pharmacology data derived using optical and impendence biosensor systems with those obtained *via* traditional assays (*i.e.*, Ca^{2+} and cAMP) revealed similar rank-order potency values but notable variability in absolute potency values.[94,102,103] Importantly, these assays were applied to the determination of ligand signaling signatures.[104–107] The ability of label-free assays to discriminate coupling to specific G-proteins was reported. Indeed, receptors coupled to different G-protein pathways were shown to produce distinguishable kinetics profiles.[108]

Despite holding significant promise for GPCR drug discovery, label-free assays present some disadvantages relative to traditional assays. The major pitfall associated with these assays is inherent to their holistic nature. Occasionally denoted "black box" assays, the label-free kinetic profiles generated following GPCR activation are not specific to one receptor or one signaling pathway. As such, deconvolution of such profiles into their principal molecular components can be challenging and necessitates the use of complementary assays. The current lack of a clear understanding of the molecular processes linking GPCR activation to label-free signals is another source of complexity when interpreting the kinetics profiles. Finally, label-free assays cannot distinguish the concomitant activation of separate pathways having opposing (and thus neutralizing) biological effects and would thus generate a false negative outcome despite the presence of pathway engagement. Indeed, this scenario was described for the corticotropin-releasing hormone receptor 1 (a dual Gi–Gs coupler) when tested using the CellKey™ system.[94]

4.4 GPCR Biased Ligands: Concepts and Promises

While initially considered as an artifact related to the high expression level of the receptors in recombinant systems, a multiplicity of G-protein couplings has now been demonstrated for the vast majority of GPCRs. This recent paradigm shift stems from the observation that rather than functioning as simple switches, turning pre-selected linear signaling cascades "on" or "off", GPCRs are signaling hubs that can regulate alternative subsets of signaling modes depending on the receptor conformations stabilized by specific ligands. In addition to "selecting" particular receptor–G protein combinations which differ in their ability to stimulate particular effectors, certain GPCR ligands can also promote G-protein-independent signaling. By promoting specific conformational rearrangements, ligands can selectively regulate the engagement of distinct subsets of signaling pathways.[109] For instance, stimulation of the parathyroid hormone type 1 receptor PTH1R with its endogenous agonist leads to activation of Gs, Gq, Gi/o, and G12/13 proteins.[110,111] In the case of the CXCR4 receptor, the endogenous agonist stromal cell-derived factor 1 (SDF-1) can promote β-arrestin engagement in addition to the Gi (subtypes 1, 2 and 3) and G13 proteins.[112] Coupling of the GPCR to these different pathways is cell-specific, depending on the receptor interactome (*i.e.* trafficking, scaffolding, and accessory signaling G-proteins). Of particular interest, association with receptor activity-modifying proteins (RAMPs) induces modulation of receptor conformation and thereby function,[113] while PDZ (post-synaptic density protein 95 (PSD-95) *Drosophila* disc large tumor suppressor (Dlg1) and zonula occludens 1 (ZO-1)) family proteins like Na(+)/H(+) exchange regulatory cofactor 1 (NHERF1) are scaffolding G-proteins mediating the formation of multimeric complexes involving a large variety of partners.[114]

This concept referred to as "functional selectivity" or "ligand-biased signaling" has obvious implications for drug discovery as it raises the possibility of developing drugs that more selectively target the therapeutically relevant pathways, resulting in pharmacological treatments with fewer side effects. The multiplicity of pathways elicited by a GPCR can therefore be of interest when developing new therapeutic agents, if one particular pathway is well acknowledged to be associated with adverse effects. For example, morphine triggers both Gi (linked to analgesia) and β-arrestin (responsible for respiratory depression, constipation, and analgesic tolerance) pathways. Trevena compound TRV130, which is currently undergoing clinical trials for postoperative pain, only activates the Gi pathway and is therefore expected to induce analgesia without the side effects described here above. So called "biased agonism" is more and more documented, and seems to be a really promising approach for developing safer drugs.[115–119] Recently the concept of functional selectivity was extended from orthosteric to allosteric ligands. These biased allosteric ligands can selectively potentiate or inhibit a subset of the signaling pathways triggered by an orthosteric ligand. For instance, an allosteric modulator of the FP receptor which has no intrinsic activity on its own was found to be a PAM for the prostaglandin 2Fα (PG2Fα)-stimulated Gq, PKC and MAPK pathway but a NAM on the G12/13, Rho, Rho-associated protein kinase (ROCK) activity stimulated by PG2Fα.[120]

With the multiplicity of G-protein couplings now well acknowledged, the GPCR is to be seen as a "microprocessor" forwarding the information it has compiled through numerous pathways. Furthermore, with the increasing data related to β-arrestin-mediated signaling, it is more and more important to take all the different pathways into account when analyzing events emerging downstream of a receptor. Ideally, the sensors chosen to monitor the different pathways elicited by a receptor, and the respective bias of a given ligand for those pathways relative to a reference agonist, should be developed using the same cellular background (to avoid system bias due to different receptor interactome) and the same readout technology (to avoid observational bias).[121]

4.4.1 Multiparametric Profiling with BRET-Based Biosensors

Despite the increasing recognition that identifying the appropriate signaling profile is essential to optimize beneficial properties of a given drug and minimize impact of unwanted side effects, current drug discovery programs mostly rely on a very limited number of "practical" generic assays for both screening and SAR campaigns. Yet, broad-spectrum efficacy profiling of selected lead compounds toward multiple pathways would generate global signaling signatures that would refine the concept of target selection to the level of specific signaling pathways and have predictive value for therapeutic efficacy.

Establishing a complete signaling signature for even a limited number of reference compounds and lead drugs represents a significant challenge as the available assays remain laborious, time-consuming and expensive. BRET-based biosensors address these challenges. Indeed, the strict dependence on molecular proximity between donor and acceptor for energy transfer makes it a system of choice for monitoring protein–protein interactions and structural rearrangements in living cells, providing ideal probes to detect signaling activity.

Genetically encoded BRET-based biosensors can monitor most signaling pathways regulated by GPCRs in a homogeneous high-throughput format in living cells.[85,122–125] In the next decade, these sensors are also expected to provide real-time *in vivo* GPCR imaging.[126] In addition to detecting engagement of proximal GPCR interactors (G-proteins, G-protein receptor kinases, and β-arrestins), BRET-based biosensors, such as the ones developed on the BioSens-All™ platform of Domain Therapeutics NA, Inc., can also detect the production of second messengers such as calcium, cAMP, and others as well as the activation of more distal effectors such as protein kinases, cytoskeletal elements, and ion channels. Interestingly, sensors can be used in a multiplexing mode for the simultaneous detection in the same cell system of the direct engagement of a G-protein by a receptor as well as the production of a second messenger allowing detection of multiple steps of a signaling cascade.[124]

Such profiling of hit and lead candidate series against large arrays of cells expressing BRET-based biosensors detecting multiple signaling pathways allows the clustering of compounds according to their signaling signatures. These global signatures can then be correlated to their biological and therapeutic activities so as to facilitate and accelerate the identification of more efficient and safer drug candidates.

4.5 Concluding Remarks

4.5.1 Combining Technologies to Discover Biased Allosteric Modulators

In the literature, there are now several demonstrations that context-dependent pharmacology can interfere with identification of GPCR allosteric modulator activity: in GLP1 receptor case studies, cell-system, receptor-level, agonist, and stimulus dependency were described.[10,51,54] The idea therefore emerged to combine a binding strategy as first filter then to perform larger functional characterization of a reduced set of compounds in assays measuring all pathways potentially triggered by the target GPCR in the same pharmacological context.[127]

In that context, DTect-All™ FRET-based binding assays & BioSens-All™ BRET-based biosensor technologies complement each other very well. In this integrative HTS approach, the first step could consist of screening compound libraries against a DTect-All™ assay to identify a list of binders. These binders would then be tested against a second FRET-based binding assay. This counter-screen, performed using a second assay developed for a distant GPCR, enables the identification and elimination of false positives interacting in a non-specific manner with the FRET-based readout. In the following step, specific binders would be characterized using a customized panel of biosensor assays from the BioSens-All™ platform, all performed in the same pharmacological context (agonist, cells, readout). Compounds could then be clustered according to their signaling signatures. For all screens, cells expressing either the biosensor alone, or alternatively a direct fusion between Rluc and GFP, would be used to eliminate compounds that could non-specifically alter the luminescence, fluorescence or energy transfer by directly binding to the Rluc or GFP and/or quenching the emission. Representative compounds of each specific cluster would then be tested in native system models to select those compounds with beneficial effects and no unwanted side effects.[128] These chemical series, validated with regards to their physiological responses, would constitute the hits and would be the starting points for further optimization (Figure 4.2). This combination of innovative technologies minimizes the biological and financial risks associated with selecting hits based on a single and biased functional test monitoring a signaling pathway potentially irrelevant in terms of physiology. This approach would therefore constitute a more comprehensive strategy for selecting the most promising chemical series.

4.5.2 Further Considerations

Additional trends that bring both challenges and promises to the field of GPCR drug discovery are not discussed in this paper but are subject of recent dedicated reviews. Since the first determination of the X-ray crystallographic structure of a non-rhodopsin GPCR namely beta-2 adrenergic receptor by Kobilka's team in 2007,[129,130] more than 20 solved GPCR structures have been made available.[131] The identification of GPCR ligands can therefore become structure-based and virtual screening of allosteric modulators using molecular docking should deliver next-generation hits.[132,133] As a token of this aspect, Topiol and Sabio recently proposed a structure-based explanation of the molecular switches between mGluR5 PAMs and their very close NAM analogs.[134]

Furthermore, oligomerization state (homodimerization, heterodimerization, multimerization) should be taken into account in future GPCR drug development efforts. Indeed, the GPCR oligomerization process, the complexity of which is only beginning to be unraveled,[135] increases the number of functional entities which may exhibit distinct pharmacological profiles;[136] this process opens as well the avenue to the identification of GPCR oligomerization modulators as a new class of therapeutic compounds.[137]

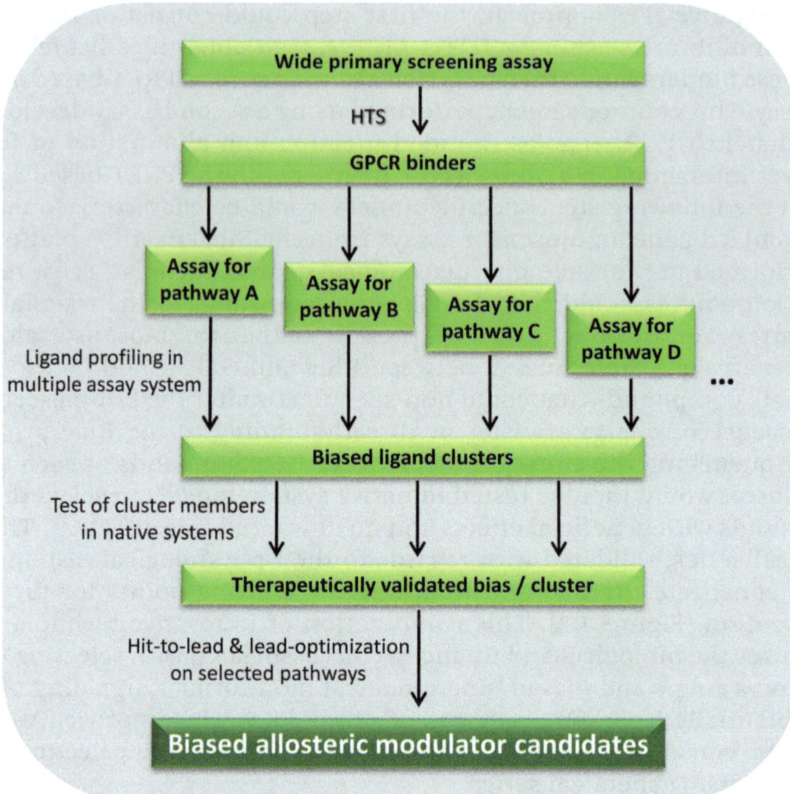

Figure 4.2 Screening cascade leading to validated biased allosteric drug candidates.

Conflict of Interest

Authors are employees of Domain Therapeutics SA (M. Frauli, C. Franchet, I. Dorange, S. Schann) or Domain Therapeutics NA Inc. (A. Mancini, B. Breton).

Acknowledgements

The authors acknowledge Dr Dario Doller for giving them the exciting opportunity to participate in this book chapter.

References

1. Global Industry Analysts, Inc. (GIA). G-Protein-Coupled Receptors (GPCRs): A Global Market Report. II Executive Summary, 5 Global Market Perspective, Chart II-38. (2012).
2. S. L. Garland, *J. Biomol. Screening*, 2013, **18**, 947–966.

3. K. Leach, P. M. Sexton and A. Christopoulos, *Trends Pharmacol. Sci.*, 2007, **28**, 382–389.
4. B. J. Melancon, C. R. Hopkins, M. R. Wood, K. A. Emmitte, C. M. Niswender, A. Christopoulos, P. J. Conn and C. W. Lindsley, *J. Med. Chem.*, 2012, **55**, 1445–1464.
5. A. Christopoulos, L. T. May, V. A. Avlani and P. M. Sexton, *Biochem. Soc. Trans.*, 2004, **32**, 873–877.
6. W. Stallaert, A. Christopoulos and M. Bouvier, *Expert Opin. Drug Discovery*, 2011, **6**, 811–825.
7. S. Galandrin, G. Oligny-Longpré and M. Bouvier, *Trends Pharmacol. Sci.*, 2007, **28**, 423–430.
8. T. Kenakin, *Mol. Interventions*, 2004, **4**, 222–229.
9. Z.-G. Gao, D. Verzijl, A. Zweemer, K. Ye, A. Göblyös, A. P. IJzerman and K. A. Jacobson, *Biochem. Pharmacol.*, 2011, **82**, 658–668.
10. D. Wootten, E. E. Savage, C. Valant, L. T. May, K. W. Sloop, J. Ficorilli, A. D. Showalter, F. S. Willard, A. Christopoulos and P. M. Sexton, *Mol. Pharmacol.*, 2012, **82**, 281–290.
11. G. Milligan and E. Kostenis, *Br. J. Pharmacol.*, 2006, **147**(1), S46–S55.
12. S. M. DeWire, S. Ahn, R. J. Lefkowitz and S. K. Shenoy, *Annu. Rev. Physiol.*, 2007, **69**, 483–510.
13. L. M. Luttrell and D. Gesty-Palmer, *Pharmacol. Rev.*, 2010, **62**, 305–330.
14. S. Schann, M. Bouvier and P. Neuville, *Drug Discovery Today: Technol.*, 2013, **10**, e261–267.
15. L. Zander Balderud, D. Murray, N. Larsson, U. Vempati, S. C. Schürer, M. Bjäreland and O. Engkvist, *J. Biomol. Screening*, 2015, **20**, 402–415.
16. Y. Wang, J. Xiao, T. O. Suzek, J. Zhang, J. Wang, Z. Zhou, L. Han, K. Karapetyan, S. Dracheva, B. A. Shoemaker, E. Bolton, A. Gindulyte and S. H. Bryant, *Nucleic Acids Res.*, 2012, **40**, D400–D412.
17. M. Köse, K. Ritter, K. Thiemke, M. Gillard, E. Kostenis and C. E. Müller, *ACS Med. Chem. Lett.*, 2014, **5**, 326–330.
18. T. J. Woltering, G. Adam, A. Alanine, J. Wichmann, F. Knoflach, V. Mutel and S. Gatti, *Bioorg. Med. Chem. Lett.*, 2007, **17**, 6811–6815.
19. M. Kalinichev, *40th Annual Meeting Society for Neuroscience*, San Diego, CA, USA (2010).
20. A. Christopoulos and T. Kenakin, *Pharmacol. Rev.*, 2002, **54**, 323–374.
21. D. C. Swinney, B. A. Haubrich, I. Van Liefde and G. Vauquelin, *Curr. Top. Med. Chem.*, 2015, **15**, 2504–2522.
22. C. Hoffmann, M. Castro, A. Rinken, R. Leurs, S. J. Hill and H. F. Vischer, *Mol. Pharmacol.*, 2015, **88**, 552–560.
23. J. D. Hothersall, A. J. Brown, I. Dale and P. Rawlins, *Drug Discovery Today*, 2016, **21**, 90–96.
24. T. Aristotelous, A. L. Hopkins and I. Navratilova, *Methods Enzymol.*, 2015, **556**, 499–525.
25. L. B. Knudsen, D. Kiel, M. Teng, C. Behrens, D. Bhumralkar, J. T. Kodra, J. J. Holst, C. B. Jeppesen, M. D. Johnson, J. C. de Jong, A. S. Jorgensen, T. Kercher, J. Kostrowicki, P. Madsen, P. H. Olesen, J. S. Petersen, F.

Poulsen, U. G. Sidelmann, J. Sturis, L. Truesdale, J. May and J. Lau, *Proc. Natl. Acad. Sci. U. S. A.*, 2007, **104**, 937–942.
26. E. Goupil, S. A. Laporte and T. E. Hébert, *Methods Cell Biol.*, 2013, **117**, 165–179.
27. A. Christopoulos, J. P. Changeux, W. A. Catterall, D. Fabbro, T. P. Burris, J. A. Cidlowski, R. W. Olsen, J. A. Peters, R. R. Neubig, J. P. Pin, P. M. Sexton, T. P. Kenakin, F. J. Ehlert, M. Spedding and C. J. Langmead, *Pharmacol. Rev.*, 2014, **66**, 918–947.
28. C. Valant, C. C. Felder, P. M. Sexton and A. Christopoulos, *Mol. Pharmacol.*, 2012, **81**, 41–52.
29. W. C. Jaeger, S. P. Armstrong, S. J. Hill and K. D. G. Pfleger, *Front. Endocrinol.*, 2014, **5**, 26.
30. S. Lee, G.-D. Kim, W.-K. Park, H. Cho, B. H. Lee, S.-E. Yoo and J. K. Kong, *J. Pharmacol. Toxicol. Methods*, 2006, **53**, 242–247.
31. W. A. Lea and A. Simeonov, *Expert Opin. Drug Discovery*, 2011, **6**, 17–32.
32. S. J. Briddon and S. J. Hill, *Trends Pharmacol. Sci.*, 2007, **28**, 637–645.
33. F. Ciruela, V. Fernández-Dueñas and K. A. Jacobson, *Neuropharmacology*, 2015, **98**, 58–67.
34. M. Cottet, O. Faklaris, D. Maurel, P. Scholler, E. Doumazane, E. Trinquet, J. P. Pin and T. Durroux, *Front. Endocrinol.*, 2012, **3**, 92.
35. B. Ilien, C. Franchet, P. Bernard, S. Morisset, C. O. Weill, J. J. Bourguignon, M. Hibert and J. L. Galzi, *J. Neurochem.*, 2003, **85**, 768–778.
36. C. Tahtaoui, F. Guillier, P. Klotz, J. L. Galzi, M. Hibert and B. Ilien, *J. Med. Chem.*, 2005, **48**, 7847–7859.
37. S. Schann, S. Mayer, C. Franchet, M. Frauli, E. Steinberg, M. Thomas, L. Baron and P. Neuville, *J. Med. Chem.*, 2010, **53**, 8775–8779.
38. S. Schann, *Proceedings of the 238th American Chemical Society National Meeting*, Washington, DC, USA, 2009.
39. D. Wootten, J. Simms, C. Koole, O. L. Woodman, R. J. Summers, A. Christopoulos and P. M. Sexton, *J. Pharmacol. Exp. Ther.*, 2011, **336**, 540–550.
40. M. R. Wood, C. R. Hopkins, J. T. Brogan, P. J. Conn and C. W. Lindsley, *Biochemistry (Moscow)*, 2011, **50**, 2403–2410.
41. N. Oueslati, C. Hounsou, A. Belhocine, T. Rodriguez, E. Dupuis, J. M. Zwier, E. Trinquet, J. P. Pin and T. Durroux, *Methods Mol. Biol.*, 2015, **1272**, 23–36.
42. Y. Yano and K. Matsuzaki, *Biochim. Biophys. Acta*, 2009, **1788**, 2124–2131.
43. L. A. Stoddart, E. K. M. Johnstone, A. J. Wheal, J. Goulding, M. B. Robers, T. Machleidt, K. V. Wood, S. J. Hill and K. D. G. Pfleger, *Nat. Methods*, 2015, **12**, 661–663.
44. M. P. Hall, J. Unch, B. F. Binkowski, M. P. Valley, B. L. Butler, M. G. Wood, P. Otto, K. Zimmerman, G. Vidugiris, T. Machleidt, M. B. Robers, H. A. Benink, C. T. Eggers, M. R. Slater, P. L. Meisenheimer, D. H. Klaubert, F. Fan, L. P. Encell and K. V. Wood, *ACS Chem. Biol.*, 2012, **7**, 1848–1857.
45. Z. Ma, L. Du and M. Li, *J. Med. Chem.*, 2014, **57**, 8187–8203.
46. T. Huber and T. P. Sakmar, *Chem. Biol.*, 2014, **21**, 1224–1237.
47. L. T. May, K. Leach, P. M. Sexton and A. Christopoulos, *Annu. Rev. Pharmacol. Toxicol.*, 2007, **47**, 1–51.

48. T. M. Bridges and C. W. Lindsley, *ACS Chem. Biol.*, 2008, **3**, 530–541.
49. N. T. Burford, J. Watson, R. Bertekap and A. Alt, *Biochem. Pharmacol.*, 2011, **81**, 691–702.
50. S. D. Pratt, M. Mezler, H. Geneste, M. H. Bakker, P. J. Hajduk and S. M. Gopalakrishnan, *Comb. Chem. High Throughput Screening*, 2011, **14**, 631–641.
51. C. Koole, D. Wootten, J. Simms, C. Valant, R. Sridhar, O. L. Woodman, L. J. Miller, R. J. Summers, A. Christopoulos and P. M. Sexton, *Mol. Pharmacol.*, 2010, **78**, 456–465.
52. L. Alvarez-Jaimes, S. W. Sutton, D. Nepomuceno, T. Motley, M. Cik, E. Stocking, J. Shoblock and P. Bonaventure, *PloS One*, 2012, **7**, e30792.
53. C. M. Niswender, K. A. Johnson, N. R. Miller, J. E. Ayala, Q. Luo, R. Williams, S. Saleh, D. Orton, C. D. Weaver and P. J. Conn, *Mol. Pharmacol.*, 2010, **77**, 459–468.
54. L. B. Knudsen, S. Hastrup, C. R. Underwood, B. S. Wulff and J. Fleckner, *Regul. Pept.*, 2012, **175**, 21–29.
55. Y. Zhang, D. Kowal, A. Kramer and J. Dunlop, *J. Biomol. Screening*, 2003, **8**, 571–577.
56. R. M. Eglen and T. Reisine, *Assay Drug Dev. Technol.*, 2008, **6**, 659–671.
57. S. Bovolenta, M. Foti, S. Lohmer and S. Corazza, *J. Biomol. Screening*, 2007, **12**, 694–704.
58. S. Inouye, *J. Biochem.*, 2008, **143**, 711–717.
59. S. P. Laptenok, I. H. M. van Stokkum, J. W. Borst, B. van Oort, A. J. W. G. Visser and H. van Amerongen, *J. Phys. Chem. B*, 2012, **116**, 3013–3020.
60. S. J. Charlton and G. Vauquelin, *Br. J. Pharmacol.*, 2010, **161**, 1250–1265.
61. P. E. Brandish, L. A. Hill, W. Zheng and E. M. Scolnick, *Anal. Biochem.*, 2003, **313**, 311–318.
62. S. S. Wang, A. A. Alousi and S. H. Thompson, *J. Gen. Physiol.*, 1995, **105**, 149–171.
63. P. Tewson, M. Westenberg, Y. Zhao, R. E. Campbell, A. M. Quinn and T. E. Hughes, *PLoS One*, 2012, **7**, e42791.
64. R. M. Eglen, *Assay Drug Dev. Technol.*, 2002, **1**, 97–104.
65. M. Weber, M. Ferrer, W. Zheng, J. Inglese, B. Strulovici and P. Kunapuli, *Assay Drug Dev. Technol.*, 2004, **2**, 39–49.
66. F. Degorce, *Expert Opin. Drug Discovery*, 2006, **1**, 753–764.
67. G. F. Blackburn, H. P. Shah, J. H. Kenten, J. Leland, R. A. Kamin, J. Link, J. Peterman, M. J. Powell, A. Shah, D. B. Talley, S. K. Tyagi, E. Wilkins, T.-G. Wu and R. J. Massey, *Clin. Chem.*, 1991, **37**, 1534–1539.
68. L. Beaudet, J. Bedard, B. Breton, R. J. Mercuri and M. L. Budarf, *Genome Res.*, 2001, **11**, 600–608.
69. S. S. Wigdal, J. L. Anderson, G. J. Vidugiris, J. Shultz, K. V. Wood and F. Fan, *Curr. Chem. Genomics*, 2008, **2**, 16–28.
70. M. Buccioni, G. Marucci, D. Dal Ben, D. Giacobbe, C. Lambertucci, L. Soverchia, A. Thomas, R. Volpini and G. Cristalli, *Purinergic Signalling*, 2011, **7**, 463–468.
71. F. Fan, B. F. Binkowski, B. L. Butler, P. F. Stecha, M. K. Lewis and K. V. Wood, *ACS Chem. Biol.*, 2008, **3**, 346–351.

72. Z. Cheng, A. Paguio, D. Garvin, P. Stecha, K. Wood and F. Fan, *Promega Corporation Web site*, 2009.
73. C. Janetopoulos, T. Jin and P. Devreotes, *Science*, 2001, **291**, 2408–2411.
74. E. J. Whalen, S. Rajagopal and R. J. Lefkowitz, *Trends Mol. Med.*, 2011, **17**, 126–139.
75. W. P. Bowen and P. G. Wylie, *Assay Drug Dev. Technol.*, 2006, **4**, 209–221.
76. C. Eggeling, L. Brand, D. Ullmann and S. Jager, *Drug Discovery Today*, 2003, **8**, 632–641.
77. D. Haasen, M. Wolff, M. J. Valler and R. Heilker, *Comb. Chem. High Throughput Screening*, 2006, **9**, 37–47.
78. B. Zimmerman, A. Beautrait, B. Aguila, R. Charles, E. Escher, A. Claing, M. Bouvier and S. A. Laporte, *Sci. Signaling*, 2012, **5**, ra33.
79. M. M. van Der Lee, M. Bras, C. J. van Koppen and G. J. Zaman, *J. Biomol. Screening*, 2008, **13**, 986–998.
80. G. Barnea, W. Strapps, G. Herrada, Y. Berman, J. Ong, B. Kloss, R. Axel and K. J. Lee, *Proc. Natl. Acad. Sci. U. S. A.*, 2008, **105**, 64–69.
81. H. Yin, A. Chu, W. Li, B. Wang, F. Shelton, F. Otero, D. G. Nguyen, J. S. Caldwell and Y. A. Chen, *J. Biol. Chem.*, 2009, **284**, 12328–12338.
82. J. A. Wetter, C. Revankar and B. J. Hanson, *J. Biomol. Screening*, 2009, **14**, 1134–1141.
83. A. Salahpour, S. Espinoza, B. Masri, V. Lam, L. S. Barak and R. R. Gainetdinov, *Front. Endocrinol. (Lausanne)*, 2012, **3**, 105.
84. M. J. Lohse, S. Nuber and C. Hoffmann, *Pharmacol. Rev.*, 2012, **64**, 299–336.
85. P. G. Charest, S. Terrillon and M. Bouvier, *EMBO Rep.*, 2005, **6**, 334–340.
86. J. E. Murphy, B. E. Padilla, B. Hasdemir, G. S. Cottrell and N. W. Bunnett, *Proc. Natl. Acad. Sci. U. S. A.*, 2009, **106**, 17615–17622.
87. G. Hilf, P. Gierschik and K. H. Jakobs, *Eur. J. Biochem.*, 1989, **186**, 725–731.
88. H. Frang, V.-M. Mukkala, R. Syystö, P. Ollikka, P. Hurskainen, M. Scheinin and I. Hemmilä, *Assay Drug Dev. Technol.*, 2003, **1**, 275–280.
89. P. G. Strange, *Br. J. Pharmacol.*, 2010, **161**, 1238–1249.
90. R. Irannejad, J. C. Tomshine, J. R. Tomshine, M. Chevalier, J. P. Mahoney, J. Steyaert, S. G. F. Rasmussen, R. K. Sunahara, H. El-Samad, B. Huang and M. von Zastrow, *Nature*, 2013, **495**, 534–538.
91. Y. Fang, A. M. Ferrie and G. Li, *FEBS Lett.*, 2005, **579**, 4175–4180.
92. Y. Fang, *Drug Discovery Today Technol.*, 2011, **7**, e5–e11.
93. Y. Fang, G. Li and A. M. Ferrie, *J. Pharmacol. Toxicol. Methods*, 2007, **55**, 314–322.
94. M. F. Peters, F. Vaillancourt, M. Heroux, M. Valiquette and C. W. Scott, *Assay Drug Dev. Technol.*, 2010, **8**, 219–227.
95. Y. Fang, *Expert. Opin. Drug Discovery*, 2010, **5**, 1237–1248.
96. R. McGuinness, *Curr. Opin. Pharmacol.*, 2007, **7**, 535–540.
97. N. Yu, J. M. Atienza, J. Bernard, S. Blanc, J. Zhu, X. Wang, X. Xu and Y. A. Abassi, *Anal. Chem.*, 2006, **78**, 35–43.

98. P. H. Lee, A. Gao, C. van Staden, J. Ly, J. Salon, A. Xu, Y. Fang and R. Verkleeren, *Assay Drug Dev. Technol.*, 2008, **6**, 83–94.
99. M. Rocheville and J. C. Jerman, *Curr. Opin. Pharmacol.*, 2009, **9**, 643–649.
100. S. Goldbard, *Curr. Opin. Drug Discovery Dev.*, 2006, **9**, 110–116.
101. A. N. Y. Chen, D. T. Malone, K. Pabreja, P. M. Sexton, A. Christopoulos and M. Canals, *J. Biomol. Screening*, 2015, **20**, 646–654.
102. P. Scandroglio, R. Brusa, G. Lozza, I. Mancini, R. Petro, A. Reggiani and M. Beltramo, *J. Biomol. Screening*, 2010, **15**, 1238–1247.
103. C. W. Scott and M. F. Peters, *Drug Discovery Today*, 2010, **15**, 704–716.
104. W. Stallaert, J. F. Dorn, E. van der Westhuizen, M. Audet and M. Bouvier, *PLoS One*, 2012, **7**, e29420.
105. Y. Fang and A. M. Ferrie, *FEBS Lett.*, 2008, **582**, 558–564.
106. M. Kammermann, A. Denelavas, A. Imbach, U. Grether, H. Dehmlow, C. M. Apfel and C. Hertel, *Biochem. Biophys. Res. Commun.*, 2011, **412**, 419–424.
107. M. Morse, E. Tran, H. Sun, R. Levenson and Y. Fang, *PLoS One*, 2011, **6**, e25643.
108. M. F. Peters and C. W. Scott, *J. Biomol. Screening*, 2009, **14**, 246–255.
109. B. K. Kobilka and X. Deupi, *Trends Pharmacol. Sci.*, 2007, **28**, 397–406.
110. W. F. Schwindinger, J. Fredericks, L. Watkins, H. Robinson, J. M. Bathon, M. Pines, L. J. Suva and M. A. Levine, *Endocrine*, 1998, **8**, 201–209.
111. A. T. Singh, A. Gilchrist, T. Voyno-Yasenetskaya, J. M. Radeff-Huang and P. H. Stern, *Endocrinology*, 2005, **146**, 2171–2175.
112. J. Quoyer, J. M. Janz, J. Luo, Y. Ren, S. Armando, V. Lukashova, J. L. Benovic, K. E. Carlson, S. W. Hunt III and M. Bouvier, *Proc. Natl. Acad. Sci. U. S. A.*, 2013, **110**, E5088–E5097.
113. M. Heroux, B. Breton, M. Hogue and M. Bouvier, *Biochemistry (Moscow)*, 2007, **46**, 7022–7033.
114. J. A. Ardura and P. A. Friedman, *Pharmacol. Rev.*, 2011, **63**, 882–900.
115. J. D. Violin, A. L. Crombie, D. G. Soergel and M. W. Lark, *Trends Pharmacol. Sci.*, 2014, **35**, 308–316.
116. J. D. Violin, S. M. DeWire, D. Yamashita, D. H. Rominger, L. Nguyen, K. Schiller, E. J. Whalen, M. Gowen and M. W. Lark, *J. Pharmacol. Exp. Ther.*, 2010, **335**, 572–579.
117. J. A. Allen, J. M. Yost, V. Setola, X. Chen, M. F. Sassano, M. Chen, S. Peterson, P. N. Yadav, X.-P. Huang, B. Feng, N. H. Jensen, X. Che, X. Bai, S. V. Frye, W. C. Wetsel, M. G. Caron, J. A. Javitch, B. L. Roth and J. Jin, *Proc. Natl. Acad. Sci. U. S. A.*, 2011, **108**, 18488–18493.
118. R. P. Xiao, S. J. Zhang, K. Chakir, P. Avdonin, W. Zhu, R. A. Bond, C. W. Balke, E. G. Lakatta and H. Cheng, *Circulation*, 2003, **108**, 1633–1639.
119. N. H. Moniri, D. Covington-Strachan and R. G. Booth, *J. Pharmacol. Exp. Ther.*, 2004, **311**, 274–281.
120. E. Goupil, D. Tassy, C. Bourguet, C. Quiniou, V. Wisehart, D. Pétrin, C. Le Gouill, D. Devost, H. H. Zingg, M. Bouvier, H. U. Saragovi, S. Chemtob, W. D. Lubell, A. Claing, T. E. Hébert and S. A. Laporte, *J. Biol. Chem.*, 2010, **285**, 25624–25636.

121. T. Kenakin and A. Christopoulos, *Nat. Rev. Drug Discovery*, 2013, **12**, 205–216.
122. C. Galés, R. V. Rebois, M. Hogue, P. Trieu, A. Breit, T. E. Hébert and M. Bouvier, *Nat. Methods*, 2005, **2**, 177–184.
123. C. Galés, J. J. Van Durm, S. Schaak, S. Pontier, Y. Percherancier, M. Audet, H. Paris and M. Bouvier, *Nat. Struct. Mol. Biol.*, 2006, **13**, 778–786.
124. B. Breton, E. Sauvageau, J. Zhou, H. Bonin, C. Le Gouill and M. Bouvier, *Biophys. J.*, 2010, **99**, 4037–4046.
125. B. Breton, M. Lagacé and M. Bouvier, *FASEB J.*, 2010, **24**, 4733–4743.
126. J. van Unen, J. Woolard, A. Rinken, C. Hoffmann, S. J. Hill, J. Goedhart, M. R. Bruchas, M. Bouvier and M. J. W. Adjobo-Hermans, *Mol. Pharmacol.*, 2015, **88**, 589–595.
127. S. Schann, *13th Annual Discovery on Target*, Boston, MA, USA, 2013.
128. T. Kenakin, *J. Pharmacol. Exp. Ther.*, 2011, **336**, 296–302.
129. V. Cherezov, D. M. Rosenbaum, M. A. Hanson, S. G. F. Rasmussen, F. S. Thian, T. S. Kobilka, H.-J. Choi, P. Kuhn, W. I. Weis, B. K. Kobilka and R. C. Stevens, *Science*, 2007, **318**, 1258–1265.
130. S. G. F. Rasmussen, H.-J. Choi, D. M. Rosenbaum, T. S. Kobilka, F. S. Thian, P. C. Edwards, M. Burghammer, V. R. P. Ratnala, R. Sanishvili, R. F. Fischetti, G. F. X. Schertler, W. I. Weis and B. K. Kobilka, *Nature*, 2007, **450**, 383–387.
131. J. Zhang, J. Yang, R. Jang and Y. Zhang, *Structure (London, United Kingdom)*, 2015, **1993**(23), 1538–1549.
132. S. Topiol, *Expert Opin. Drug Discovery*, 2013, **8**, 607–620.
133. D. Rodríguez, A. Ranganathan and J. Carlsson, *Curr. Top. Med. Chem.*, 2015, **15**, 2484–2503.
134. S. Topiol and M. Sabio, *Bioorg. Med. Chem. Lett.*, 2016, **26**, 484–494.
135. H. F. Vischer, M. Castro and J.-P. Pin, *Mol. Pharmacol.*, 2015, **88**, 561–571.
136. S. Yin, M. J. Noetzel, K. A. Johnson, R. Zamorano, N. Jalan-Sakrikar, K. J. Gregory, P. J. Conn and C. M. Niswender, *J. Neurosci.*, 2014, **34**, 79–94.
137. M. Hurevich, A. Talhami, D. E. Shalev and C. Gilon, *Curr. Top. Med. Chem.*, 2014, **14**, 1842–1863.

CHAPTER 5

Examining Allosterism in a Dimeric G-Protein-Coupled Receptor Context

JESÚS GIRALDO*[a], JORDI ORTIZ[b], JAMES DALTON[a] AND BIN ZHOU[a]

[a]Universitat Autònoma de Barcelona, Laboratory of Molecular Neuropharmacology and Bioinformatics, Institut de Neurociències and Unitat de Bioestadística, 08193 Bellaterra, Spain; [b]Universitat Autònoma de Barcelona, Laboratory of Molecular Neuropharmacology and Bioinformatics, Institut de Neurociències and Department of Biochemistry & Molecular Biology, 08193 Bellaterra, Spain
*E-mail: Jesus.Giraldo@uab.es

5.1 G-Protein-Coupled Receptors: Paradigms of Allosteric Machines

G-protein-coupled receptors (GPCRs) are integral membrane-bound receptors that transmit signals from outside to inside the cell. The signal can be either a light photon, in the case of rhodopsin, or chemical compounds of diverse size and structure, in the case of the remaining receptors. GPCRs, as proteins, are versatile entities able to adopt multiple conformations which may activate different signalling pathways depending on whether they bind G proteins (see, for example, the crystal structure of β_2-adrenergic receptor in

RSC Drug Discovery Series No. 56
Allosterism in Drug Discovery
Edited by Dario Doller

Published by the Royal Society of Chemistry, www.rsc.org

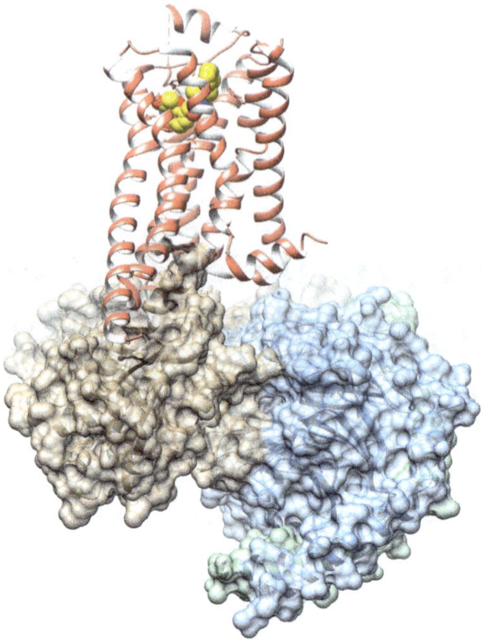

Figure 5.1 Side-view of crystal structure (PDB id: 3SN6) of β_2-adrenergic receptor (terracotta ribbon) with agonist (yellow) bound to its G-protein: alpha subunit (beige), beta-subunit (blue), gamma-subunit (green).

Figure 5.1) or β-arrestin molecules.[1] The possibility that ligands may induce or stabilize these signalling-dependent conformations in a differential manner led to the concept of functional selectivity,[2,3] also known as biased agonism,[4] pluridimensional efficacy,[5] and stimulus trafficking.[6]

It is worth noting that GPCRs are, in a general sense, allosteric in nature because the information communicated by an agonist at the orthosteric site is transmitted to another region of the protein, the site to which the signalling protein binds. However, the situation is much more complex than this because allosteric sites to which synthetic or endogenous allosteric compounds can bind have been found for a number of GPCRs.[7,8] Moreover, not only can there be interactions between GPCRs and signalling proteins and between ligands (orthosterically and allosterically bound) within a particular GPCR but also between GPCRs, because homo- and hetero-oligomeric GPCRs have been reported (for example, see the homodimer of adenosine A_{2A} receptor in Figure 5.2), and between GPCRs and modulating proteins such as Receptor Activity Modifying Proteins (RAMPs).[9,10] In an attempt to rationalize and generalize this complexity, Terry Kenakin defined GPCRs as conduits allowing the allosteric interaction between modulators and guests, with these entities being either ligands or proteins.[11] The flow of energy between modulators and guests is bidirectional, that is to say of the same magnitude from the modulator to the guest and *vice versa*, and can be either vertical or horizontal, depending on whether the interaction is across or along the

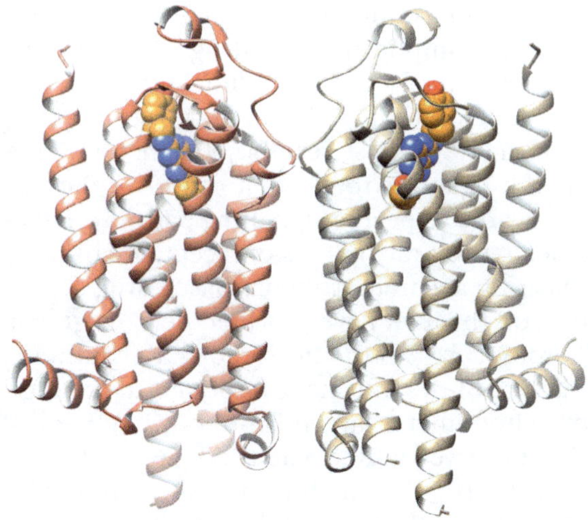

Figure 5.2 Side-view of the crystal structure of an adenosine A_{2A} receptor homodimer (PDB id: 4EIY, monomers shown as terracotta and beige ribbons, each with bound antagonist, predominantly in orange).

7-transmembrane (7TM) helical bundle.[11] This extremely versatile variety of molecular interactions at both structural and energetic levels makes GPCRs paradigms of allosteric machines.

5.2 A Decision to be Made by the Receptor: To Bind a G Protein or to Bind a β-Arrestin?

As mentioned above, GPCRs signal not only through G proteins, as was originally thought and was the reason for which they were given the name of G-protein-coupled receptors, but also through other accessory proteins such as β-arrestins.[12] Yet what can be said about the differential modes by which GPCRs recognize these proteins? Which are the structural features associated with this differential recognition? And moreover, what are the consequences and advantages of oligomerisation in functional selectivity? And finally, how can these structural and functional aspects be embodied into pharmacological parameters that can be estimated or simulated in mathematical modelling?

5.2.1 How GPCRs Recognize G Proteins: The μ-Opioid Receptor as an Example

The μ-opioid receptor (μOR), a Class A GPCR and main target of analgesic drugs, signals through both the inhibitory G-protein (Gi) and arrestin. Activation of μOR through the first signalling pathway is associated with analgesia and sedation as well as euphoria and addiction, whereas activation through the latter is associated with several adverse effects including constipation and respiratory suppression.[13] Therefore, μOR is a good example to show that investigation of functional selectivity is more than purely basic research and may have implications in drug discovery and therapeutics.[14] In fact, promising results of a phase II clinical trial have been reported for TRV130, a biased ligand of the μOR that activates G protein signalling with little β-arrestin recruitment.[15] These results show that on-target adverse effects can be avoidable if a ligand is found that can drive the receptor conformational ensemble towards a specific signalling pathway. This strategy requires a deep and detailed structural analysis of receptor activation mechanisms to provide a solid basis for accurate structure-based drug design. In this respect, two recent studies on the μOR have provided new insights into GPCR activation through G proteins.[16,17] In the first study,[16] the authors reported a 2.1 Å X-ray crystal structure of the murine μOR bound to the high-affinity agonist BU72 and a G protein mimetic camelid antibody fragment (Figure 5.3A). The study identified a conserved amino acid triad (F6.44, P5.50, and I3.40) which was shown to be conformationally linked to the orthosteric site by molecular dynamics simulations (Figure 5.3B).[16] In addition, an extensive polar network between the orthosteric site and the cytoplasmic domain was found. The conformational changes upon receptor activation at the cytoplasmic

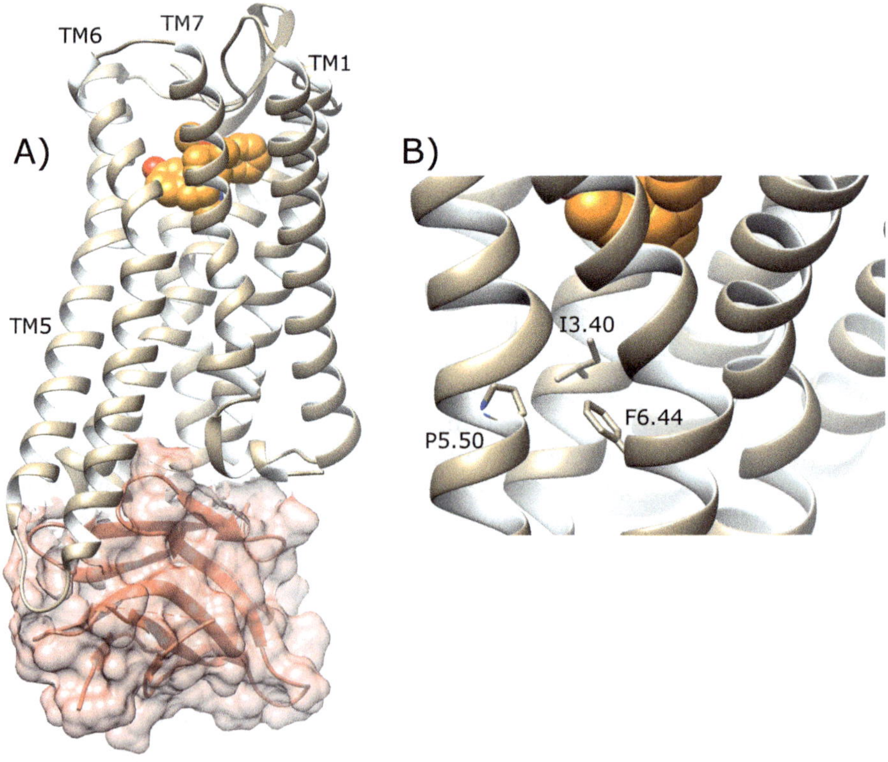

Figure 5.3 (A) Crystal structure of μOR (PDB id: 5C1M, beige ribbon) with bound agonist (orange) and bound nanobody (terracotta). (B) Close-up of an important residue-triad that is conformationally linked to the orthosteric site above.

region consisted of the typical large outward movement of transmembrane helix 6 (TM6) and a smaller inward movement of TM5 and TM7 [16] (Figure 5.4). In the second study,[17] the authors reported solution-state nuclear magnetic resonance (NMR) data on a μOR mouse construct in an amphiphile membrane-like environment. The experiments were made with the free receptor, and in the presence of the agonist BU72 alone, or with BU72 and a G protein mimetic nanobody. The results suggest a mechanism of receptor activation in which agonist binding induces conformational changes in intracellular loop 1 (ICL1) and helix 8 (H8), which facilitate G protein binding and the subsequent movements of TM5, TM6, and TM7 that lead to full receptor activation.[17] As the authors showed, the conformational movements of TM5 and TM6 necessary for G protein binding require the presence of both the agonist and the G protein mimetic nanobody. This result, which is consistent with what was found in β_2-adrenergic receptor NMR[18,19] and computational[20] studies, as well as in the analyses of Class A GPCR crystal structures,[1,21] is an indication of the weak allosterism between the orthosteric site and the G protein binding site.[17]

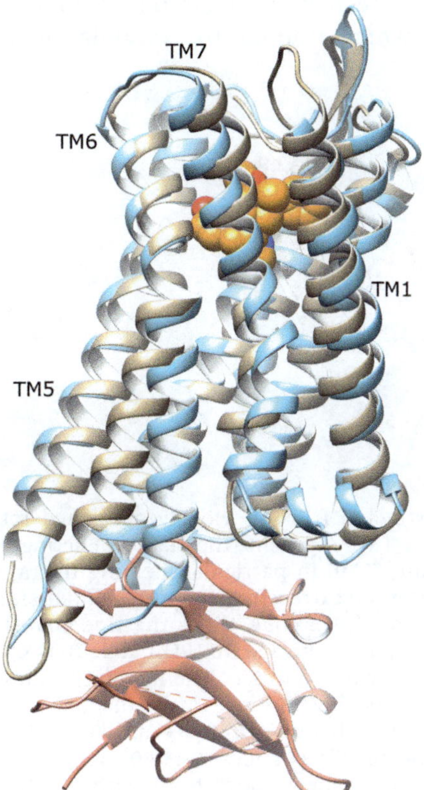

Figure 5.4 Comparison of the crystal structure of μOR in active state (PDB id: 5C1M, beige ribbon) with bound agonist (orange) and bound nanobody (terracotta) with crystal structure of antagonist-bound μOR in inactive state (PDB id: 4DKL, blue ribbon).

Different attempts have been made to develop mechanistic models for GPCR activation through G protein signalling that could be of general application to this receptor superfamily.[22,23] It is of note that in the above μOR studies the allosteric involvement of the F6.44, P5.50 and I3.40 triad[16] or a sequential movement including changes first in ICL1 and H8 and later in TM5 and TM6[17] for G protein binding were proposed. As a complementary approach, a recent statistical analysis of GPCR crystal structures comparing inactive and active states has identified a helix segment of TM3 that includes residues L/I3.40 to L3.43 (the H3 core) as a hinge of movement that could transmit a conformational signal from the extracellular to the intracellular side of the receptor, thus triggering the conformational changes in the cytoplasmic side of the TM domain (TMD) in Class A GPCRs[24] (Figure 5.5).

5.2.2 Dancing with Another Partner: The β-Arrestin Signalling Pathway

The β-arrestin signalling pathway competes with the G-protein signalling pathway in the GPCR mechanism of action. In contrast with the G-protein signalling pathway, which implies the modulation of second-messenger

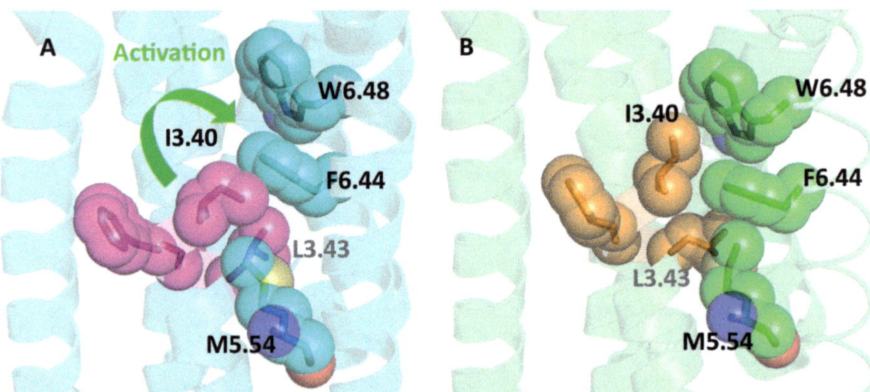

Figure 5.5 Image extracted from previously published ref. 24, with permission, showing a TM3 segment (the H3 core) acting as a hinge with respect to TM5 and TM6 in particular, helping to transmit a conformational signal from the extracellular to intracellular side of Class A GPCRs. (A) Inactive state (PDB code 4EIY); the side chains of H3 core are shown in magenta. (B) Active state (PDB code 3QAK); the side chains of H3 core are shown in orange. The agonist binding changes the interaction profile of the H3 core. In the inactive state (A) L/I3.40 precludes the interaction between I/M5.54 and F6.44. In the active state (B) the interaction profile changes: L/I3.40 moves to the other side of F6.44 where it makes new interactions with W6.48 (green arrow in Panel A), thus allowing the interaction of I/M5.44 with F6.44. Reproduced from Journal of Structural Biology, 192, Isaias Lans, James A. R. Dalton, Jesús Giraldo, Helix 3 acts as a conformational hinge in Class A GPCR activation: An analysis of interhelical interaction energies in crystal structures, Copyright 2015 with permission from Elsevier.[24]

molecules by activated downstream effector enzymes, the β-arrestin signalling pathway includes phosphorylation of the activated receptor by GPCR kinases (GRKs), which facilitates β-arrestin binding, receptor desensitization and internalization, and ultimately β-arrestin signalling.[25]

Two sequential steps have been suggested for the interactions of β-arrestin with activated GPCRs. The first step involves electrostatic interactions between basic residues on the N-terminal domain of β-arrestin and phosphate groups on the phosphorylated C-terminus of the receptor. The second step involves interactions of β-arrestin with the TMD of the receptor.[26] This proposal has been confirmed by recently described structures of complexes of β-arrestin with β_2-adrenergic receptor[27] and rhodopsin[28] (Figure 5.6A). Interestingly, the arrestin–rhodopsin complex has allowed the identification of conformational differences in TM1, TM4, TM5 and TM7 of the arrestin-bound rhodopsin with respect to the G-protein-bound rhodopsin (Figure 5.6B), which may provide new insights into the structural elements necessary for biased signalling.[28]

The following issue then arises: as β-arrestin binding to GPCRs follows receptor activation, if the receptor activated state was just a single state then only one route, which would include one G protein binding state, would be possible. This is consistent with unbiased ligands, but not with biased

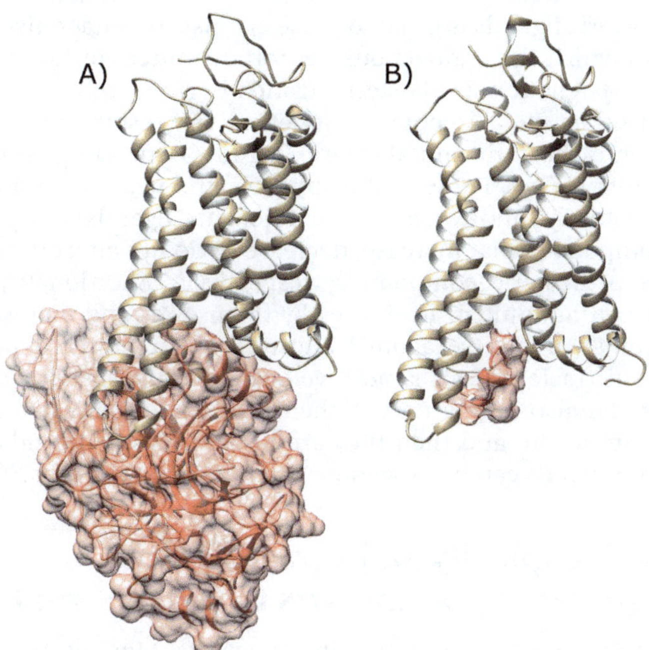

Figure 5.6 (A) Crystal structure (PDB id: 4ZWJ) of rhodopsin (beige ribbon) complexed with arrestin (terracotta). (B) Crystal structure (PDB id: 3PQR) of rhodopsin complexed with a G-protein fragment (terracotta).

ligands. Thus, there must be structural differences between those biased active receptor conformations leading to the G protein signalling pathway and those leading to the β-arrestin signalling pathways, and these biased receptor conformations must also be different from the unbiased active GPCR conformation.[29] The different structural features characterizing each of these active receptor conformations open a large and diverse repertoire of potential ligand–receptor interactions for drug design.[30,31] Also, a question arises: does biased agonism mean that the balance between two conformations is altered or that new active conformations appear? This concern connects with a related conceptual issue appearing in partial agonism. Does a partial agonist induce or select the same conformation as that induced by a full agonist albeit with a lower propensity or alternatively a different conformation to which the G protein or β-arrestin binds with a lower affinity? (discussed elsewhere[32])

5.2.3 This Decision Can be Affected by the Allosteric Interactions Between Orthosteric and Allosteric Ligands

The allosteric interactions between agonists and accessory proteins (G proteins and β-arrestins), which are mediated by the receptor conduit,[11] can be modulated in addition by allosteric ligands.[7,33] Conventionally, the designation of orthosteric ligands is given to those agonists or antagonists that bind to the site to which the endogenous neurotransmitters or hormones bind, whereas the designation as allosteric ligands is given to those compounds that bind at other sites. A cooperativity effect can result from the interactions between orthosteric and allosteric ligands. Depending on the sign of the cooperativity effect, PAMs, NAMs (positive and negative allosteric modulators, respectively) and NALs (neutral allosteric ligands) designate those allosteric compounds that increase, decrease or do not affect the functional response of orthosteric compounds, respectively. Additionally, allosteric compounds can have intrinsic efficacy by themselves and then we say that we have agonist or inverse agonist allosteric compounds, depending on whether the intrinsic efficacy is positive or negative, respectively. Allosteric compounds can also have a part of their total molecular scaffold binding to the orthosteric site and then they are termed bitopic ligands.[34] Finally, allosteric compounds can be exogenous or endogenous ligands.[8,35]

5.3 The Complexity and Versatility that Oligomerisation Imparts to GPCR Signalling

The concept of allosterism was first introduced by Monod *et al.* to explain cooperativity effects in enzymes.[36,37] The model was originally formulated to account for oligomeric proteins, by assuming that the protomers of which they are composed occupy equivalent positions and there is one, and only one, site on each protomer.[38] The term allostery comes from the Greek word "allos" (other), and it is this meaning that has been used in the GPCR context, that is to say, the change in protein reactivity at one site arising from a molecule binding to the protein at another site.[11] Thus, allosterism has been used to describe the change in agonist–receptor affinity after G protein binding to the receptor and, conversely, the change in the receptor–G protein affinity after agonist binding to the receptor. Moreover, the idea of allosterism involving a third component acting at a distance[39] has been used in the GPCR field to denote as allosteric compounds those ligands that modify the affinity or function of ligands binding at the orthosteric site of the receptor by binding to another site: the allosteric site.[40] It is of note that the functional behaviour of GPCRs as oligomeric receptors in many instances recovers the original ideas of Monod's model.

There is controversy in the literature about the need for oligomerisation for GPCR function.[41–46] Against a role for receptor oligomerisation are the results showing that a monomeric receptor is sufficient to activate G proteins.[47–50] In favour of receptor oligomerisation, functional evidence shows that GPCRs can form dimers or higher oligomers[51–57] (see elsewhere[39,58] for review and modelling of cooperativity effects). Some of the processes that have been found to be favoured by receptor oligomerisation are: the interaction with a specific G protein, receptor internalization (reviewed elsewhere[59]), receptor maturation, surface delivery, and function[60] and biosynthesis in the endoplasmic reticulum.[61] Finally, receptor heteromerization provides a new level of complexity to signal transduction by combining signalling pathways of more than one receptor and the interactions between the effects of corresponding ligands.[62–65]

Receptor oligomerisation permits endogenous ligands to have cross-talk effects and tuning of their functional responses by cooperativity effects through receptor protomers where different receptor conformational states depend on the number of protomers that are activated at a time.[66]

To illustrate some of these issues we will comment on the metabotropic glutamate receptors because there are no doubts about their dimeric nature on the basis of both structural and functional data.

5.3.1 Metabotropic Glutamate Receptors: Where Dimerization Meets Allosterism

Glutamate is the major excitatory neurotransmitter in the CNS.[67] Glutamate signalling functions through two distinct kinds of transmembrane receptors: ionotropic cation channels for fast neuronal ionic transmission[68] and metabotropic glutamate (mGlu) receptors for sustained but slower modulation of synaptic activity.[69] mGlu receptors are Class C GPCRs and are classified into eight subtypes, which in turn are clustered into three groups.[70] Group I mGlu receptors (mGlu1 and mGlu5) couple to Gq/G11 whereas groups II (mGlu2 and mGlu3) and III (mGlu4, mGlu6, mGlu7, and mGlu8) couple to Gi or Go. Group I are expressed primarily at postsynaptic sites, group II at both presynaptic and postsynaptic sites and group III at presynaptic sites.[71]

mGlu receptors are suggested to be involved in many neurological disorders and diseases such as schizophrenia, depression, autism spectrum disorders, chronic pain, and Parkinson's disease and because of this they are an important target of current pharmaceutical research.[72–75]

Whereas for some GPCRs there can be doubts about whether the observed receptor oligomerization occurs in native physiological systems or it is a consequence of the experimental design, mGlu receptors are constitutive and strict dimeric receptors.[76,77] Thus, analysis of their functioning can provide confident results when interpreted in a dimeric framework.

In contrast to class A GPCRs, mGlu receptors are characterized by bearing a large extracellular domain (ECD), the bilobate Venus flytrap (VFT), to which glutamate and other orthosteric ligands bind. This leaves the full TMD of mGlu receptors as a region potentially exposed to allosteric binding. The singular arrangement of mGlu receptors as dimeric units with each of them composed of an extracellular VFT domain and a TMD interconnected by a cysteine rich domain (CRD) makes these receptors complex and fine-tuning allosteric machines (Figure 5.7).

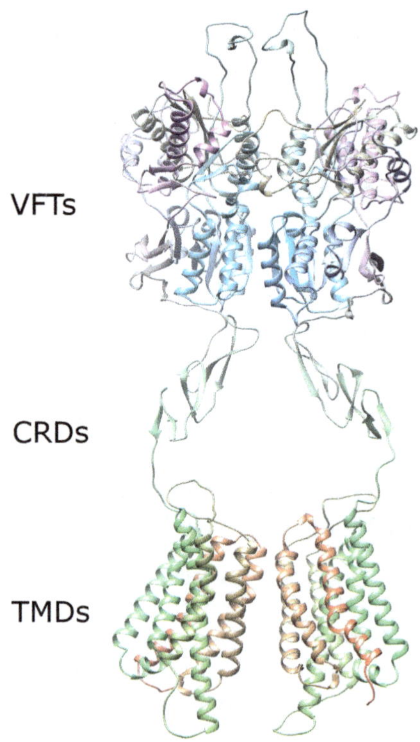

Figure 5.7 A homology model of mGlu5 receptor as a dimeric unit where each protomer is composed of an extracellular Venus Flytrap domain (VFT) and transmembrane domain (TMD), interconnected by a cysteine-rich domain (CRD).

The structural and functional complexity of mGlu receptors has been dissected through the outstanding work of some groups, which have contributed to placing correctly many of the pieces of the puzzle.[69,72,73,78–81] Some of the most relevant structure–function findings are the following:

5.3.1.1 *Structure–Function Insights from X-Ray Crystallographic Structures*

Crystal structures of the ECD of mGlu receptors either free or in the presence of glutamate or antagonists[82,83] have shed light on the relationship between the closure of one or both VFT protomers and the active conformation of the VFT dimer, with the active conformation (A) of the VFT dimer conveying a reorientation of 70° of the two subunits in comparison with the inactive or resting (R) antagonist-bound conformation.[82,83] In the R conformation the lower lobes of the ECDs are far apart whereas in the A conformation they become closer to one another, which is assumed to promote the stabilization of the active state of the TMD (Figure 5.8). It is of note that the free VFT gave rise to open–open structures in a resting conformation (Roo) and closed–open structures in an active conformation (Aco) whereas an Aco conformation was also obtained with VFT in complex with glutamate, indicating that the dimeric ECD is in a dynamic equilibrium between multiple conformations and glutamate stabilizes both protomer closure and ECD active conformation.[82] The Roo and Acc VFT conformations obtained either in the presence of an antagonist or with both glutamate and Gd^{3+} ion, respectively,[83] showed that agonists

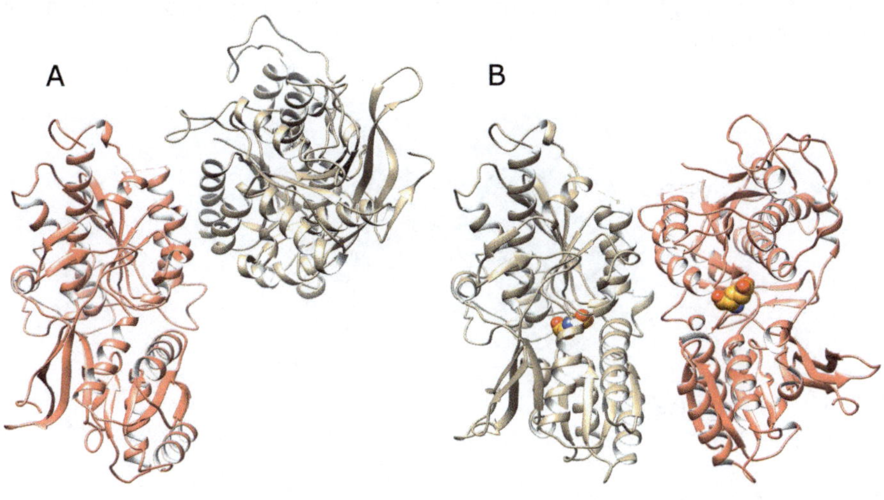

Figure 5.8 (A) The R conformation of the mGlu1 Venus flytrap (VFT) dimer (PDB id: 1ISS, beige and terracotta ribbons). (B) The corresponding A conformation (PDB id: 1EWK, beige and terracotta ribbons) with bound glutamate (in orange).

stabilized both protomer closed states and dimer active conformations and antagonists stabilized both protomer open states and dimer inactive conformations.[82,83] As a result of comparing agonist-bound and antagonist-bound VFT a conformational link between VFT closure, dimer interface, and active conformation was proposed.[83]

Closure of VFT is not a sufficient condition for yielding the active dimer conformation since structures with glutamate bound in both subunits and in a closed state were found in a dimer topology resembling a resting conformation (Rcc).[84]

Inclusion of the CRD in the ECD crystal structures of an mGlu subtype 3 receptor complexed with various agonists suggested a role of the CRD in the transmission of the ligand-induced VFT conformational change to the TMD[84] (Figure 5.9).

Big progress was made towards resolving the structural puzzle when the crystal structures of the TMDs of two mGlu were solved. Crystal structures of the TMDs of mGlu1[85] and mGlu5[86] bound to NAMs show the typical 7TM folding of GPCRs and a number of structural features relevant to mGlu-allosteric-modulator recognition (Figure 5.10).

5.3.1.2 Structure–Function Insights from Dynamic Approaches

The ECD and TMD are connected by the CRD, which is expected to play a role in receptor activation. The specific CRD intersubunit movements responsible for the coupling of the VFT dimer conformational activation with the TMD activation were determined by a combination of approaches including time-resolved Förster resonance energy transfer measurements (trFRET).[87] This study highlights the role of mGlu receptor dimerization in the activation mechanism.[87]

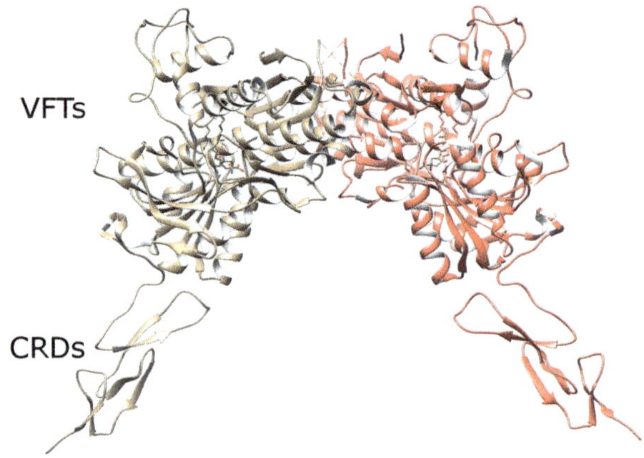

Figure 5.9 Crystal structure (PDB id: 2E4W) of the Venus Flytrap domains (VFTs) of dimeric mGlu3 (terracotta and beige ribbons, respectively) connected to cysteine rich domains (CRDs).

The correlation initially proposed between the open and closed states of the VFT protomers and the inactive and active conformation of the dimeric ECD was challenged as more crystal structures were determined.[88] The paradox generated by static crystal structures was solved when the time parameter was introduced into the experiments through trFRET.[88] The dynamic approach demonstrated that a large movement between the ECDs is associated with receptor and G-protein activation.[88] Importantly, the authors also showed an inherent property of allosterism (mutual interaction): the coupling between the ECD and the TMD domain is mutual.[88] The allosteric interactions between these two domains could explain the "anomalous" agonist-bound Rcc and antagonist-bound Aoo VFT conformations, as a result of the lack of constraints in the ECD movements because of the absence of the TMD.[88] The experiments also shed light on the relationship between ECD dimeric conformation and efficacy and attributed partial agonism to a lower stability of the closed state of the VFTs, with only a fraction of the agonist-bound receptors reaching the A conformation.[88]

Additional kinetic studies provided more information on the relationship between VFT closure and ECD activation. The kinetics of the R↔A equilibrium for the ECD of mGlu receptors was determined by single-molecule FRET.[89] The dynamic transitions between two main populations of resting and active states occur at a sub-millisecond timescale. Agonists, partial agonists, and antagonists shift this equilibrium, and it is the energy difference between the A and R states that determines agonist efficacy. It was shown that partial agonism is associated with a lower ability to shift the equilibrium towards

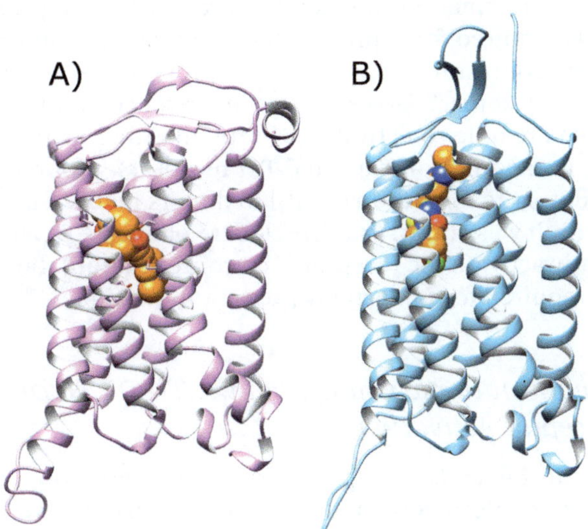

Figure 5.10 (A) Crystal structure (PDB id: 4009) of mGlu5 transmembrane domain (TMD, pink ribbon) with bound negative allosteric modulator (NAM, in orange). (B) Crystal structure of (PDB id: 4OR2) of mGlu1 transmembrane domain (TMD, blue ribbon) with bound NAM (in orange).

the A state in comparison with agonists, and not to stabilization of specific conformations different from A and R.[89] In a new kinetic approach, the activation mechanism of full-length mammalian group II mGlu receptors was analyzed by single-molecule FRET methods.[90] Three states were identified: a resting state and a short-lived intermediate state, representing functionally inactive conformations, and an active state, which were assigned to the Roo, Rco, and Acc conformations, respectively.[90] The authors corroborated the assumed mechanism that activation of mGlu receptors by agonist binding is induced by closure of the ECDs followed by dimer interface reorientation and proposed that efficacy is determined by the fractional occupancy of the Acc conformation.[90]

5.3.1.3 Assessing mGlu Receptor Signalling in a Dimeric Context

The characteristic structural feature of mGlu as dimeric receptors was assessed in some studies. By a combination of trFRET and snap-tag technologies, it was shown that mGlu receptors are strict dimers that do not associate into higher-order oligomers at the cell surface of living cells.[76,77] In this regard, using a combination of biochemical and pharmacological approaches, it was shown that mGlu2 and mGlu4 group III mGlu receptors form heterodimers in native rat and mouse tissue and with a distinct pharmacological profile.[91] Like mGlu homodimers, mGlu heterodimers are also strict dimers.[77]

The issue of monomeric *versus* dimeric mGlu organization for signal transduction is fundamental in mGlu research. A full-length mGlu dimeric arrangement is required for G protein activation by the endogenous agonist glutamate, thus confirming that mGlu receptors signal through a dimeric structure in live systems.[92] However, in the presence of a PAM, monomeric mGlu receptors are able to activate a G protein either as full-length (VFT + CRD + TM) or in the form of truncated TM receptors.[92] Moreover, by using a system that allows the formation of mGlu5 dimers made of a wild-type and a mutant subunit, it was shown that the closure of one ECD per dimer is sufficient to partially activate the receptor (Aco conformation), but the closure of both ECDs is required for full activity (Acc conformation).[93]

5.3.1.4 Conformational Changes at the TMD Determine mGlu Receptor Activation

Investigations at the TMD level provided more information about the link between receptor conformation and activation. By using a system that allows the functional expression of mGlu1 dimeric receptors with specific mutations, it was shown that activation of a G protein by a dimeric mGlu involves the activation of only one TM at a time, thus leading to an asymmetric TMD active state.[94] In addition to this, evidence for mGlu intersubunit rearrangement in the intracellular domains,[95] in the TMDs,[96] or involving a conformational change of TM2 and TM4[97] have been found during receptor activation. These data were complemented by kinetic studies: sequential inter- and intra-subunit TM rearrangements during mGlu1 receptor activation were found, where a relative rearrangement of the two TMs occurs first (time constant of ~35 ms), followed by activation of a single TM within the dimer (time constant of ~50 ms).[98]

As a higher level of both functional and structural complexity we can mention the possibility of multiple signalling and the change in TMD interfaces as a result of mGlu activation. By electrophysiology and immunocytochemistry methods on mutated mGlu1 receptors, it was found that the coupling configuration of mGlu1 for Gq is different from that of mGlu1 for Gi or Go.[99] The authors concluded that multiple G-protein signalling pathways were differentially activated by different intramolecular activation mechanisms of the dimeric mGlu1 receptor.[99] The TM interface of the mGlu2 receptor both in inactive and active states was determined at the cell surface of living cells.[100] The inactive state was characterized by a dimer interface involving TM4 and TM5 whereas the active state was characterized by a dimer interface involving TM6. Thus, agonist-induced activation of dimeric mGlu involves a major rearrangement of the TMD interface[100] (Figure 5.11).

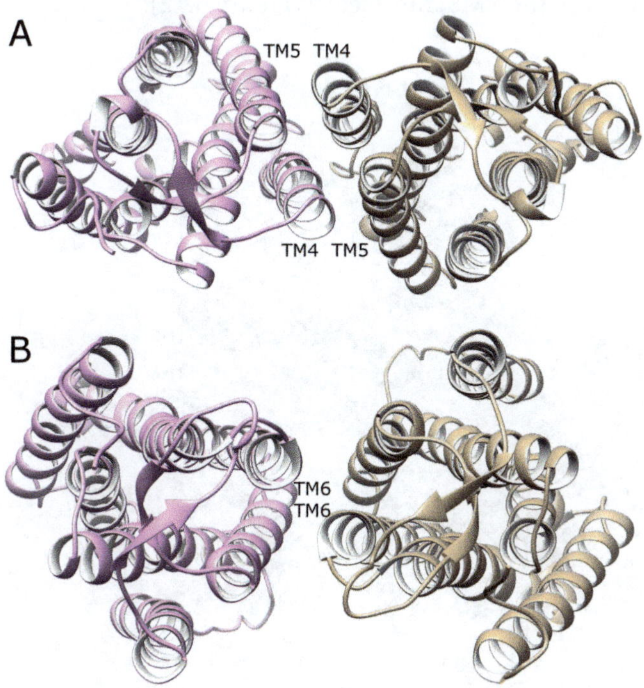

Figure 5.11 The TM interface of the mGlu2 receptor (homology model) in inactive (A) and active (B) states. The inactive state is characterized by a dimer interface involving TM4 and TM5 whereas the active state is characterized by a dimer interface involving TM6.

5.3.1.5 mGlu Receptors Can Form Heteromers with Class A GPCRs

Studying the effects of receptor heteromerization is a big challenge in drug discovery. mGlu receptors can heteromerize with class A GPCRs, thus leading to structurally and functionally distinct pharmacological spaces. Of particular note, a serotonin 2A receptor (5-HT_{2A})–mGlu2 heteromeric complex was identified.[101] The authors proposed that this heteromeric complex integrates serotonin and glutamate signalling to regulate the sensory gating functions of the cortex, a process that is disrupted in psychosis.[101] The Gi- and Gq-dependent signalling through the 5-HT_{2A}–mGlu2 heteromeric complex is affected by ligands of the two partner receptors, thus explaining the psychoactive behavioural effects of a variety of pharmacological compounds[102] and paving the way to a new pharmacology based on the heteromeric concept.[62]

In structural studies and eventually in structure-based drug design with an oligomeric approach, the analysis of the interfaces between the protomers in receptor signalling is fundamental. Within the serotonin 5-HT_{2A}–mGlu2 heteromeric complex, the 5-HT_{2A} receptor interacts with TM4–5 of the mGlu2 receptor.[103] This interaction is consistent with the proposal that the 5-HT_{2A} receptor may stabilize the active state of the mGlu2 receptor which is assumed to involve the TM6 interface[100] (Figure 5.12).

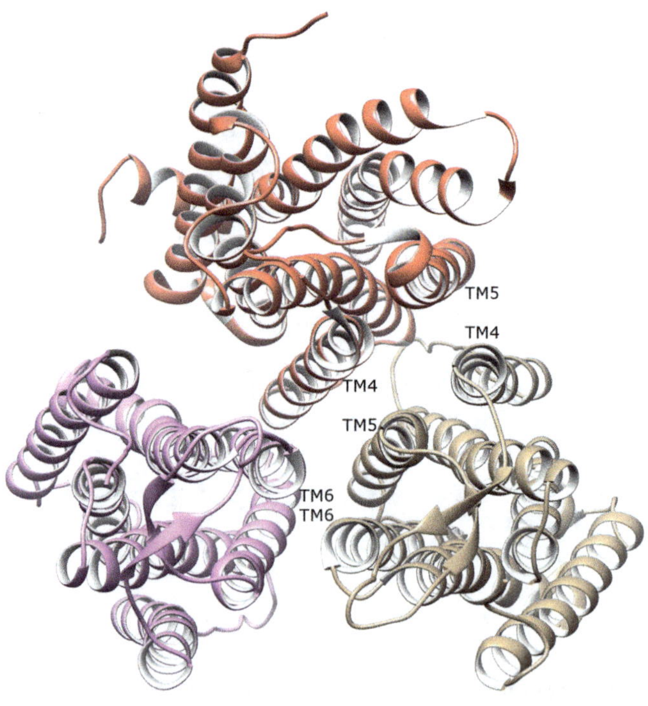

Figure 5.12 Homology model of the serotonin 5-HT_{2A}–mGlu2 heteromeric complex. The 5-HT_{2A} receptor (terracotta) interacts with TM4–5 of mGlu2, potentially stabilizing the active state of mGlu2 (pink and beige ribbons, respectively) *via* a TM6 homodimer interface.

5.4 Getting Help from Mechanistic Mathematical Models: The mGlu Receptor as an Example

Mechanistic mathematical models differ from empirical mathematical models in that the parameters included in mechanistic models have physical meaning.[104,105] Thus, parameter values of mechanistic models may provide help in the understanding of pharmacological receptor signalling. For illustration, mGlu receptors were selected.

5.4.1 Modelling the Transmission of the Signal Through the ECD

A model for the binding and function of mGlu receptors was recently developed.[106] Here, we briefly discuss the functional aspects that this model includes. Figure 5.13 depicts the model, with constants and equations described in Appendix A. The model is shown as an induction approach for the VFT, in which ligands bind to the OO state and induce the closure of protomer subunits. An alternative model can be constructed following a selection approach.[106] These models are equivalent in thermodynamic terms for constitutive active receptors.[107] The model, though designed as simple as possible for simulation and fitting purposes, allows for describing a number of pharmacological features: (1) distinguishing between *cis* and *trans* closure of the VFT subunit: the same or the partner protomer to which the ligand binds, respectively; (2) binding or functional cooperativity depending on whether we consider the binding of the ligand to the OO states or the closure of the VFT subunits; (3) the differential functional capability of the VFT subunits is included by assuming that the equilibrium constants for the activation of the TMD (L constants in Figure 5.13) depend on the VFT state to which the TMD is bound. Thus, the model proposes that any of the OO, CO or CC VFT states can be in either resting or active state and that the propensity of activation is associated with the number of VFT subunits in a closed state and quantified by the relative value of the L constants.

To examine the capability and flexibility of the model, published experimental results[93] were remodelled.[106] Figure 5.14 shows the experimental data and simulated curves under the mechanistic mGlu model (parameter values published previously[106]). Two experimental conditions were contemplated: a wild-type (WT) mGlu5 receptor and a mutated mGlu5 receptor in which a mutation was introduced in the binding pocket of one of the VFT subunits.[93] The consequences of VFT mutation in receptor functionality is reflected in experimental E/[A] curve points through the conversion of the WT sigmoid curve into a biphasic one and a decrease of the maximum response. The estimated parameters of the constants (see Table 5.1) provide a mechanistic interpretation for experimental results. In particular, it is worth mentioning the following changes in the mutated receptor with respect to the WT: (i) the affinity of the agonist for the mutated subunit decreases ($K_2\uparrow$); (ii) the capacity of the mutated subunit for closing decreases ($Y_6\downarrow$); (iii) the capacity of the CC state for activating the TMD decreases ($L_3\downarrow$).

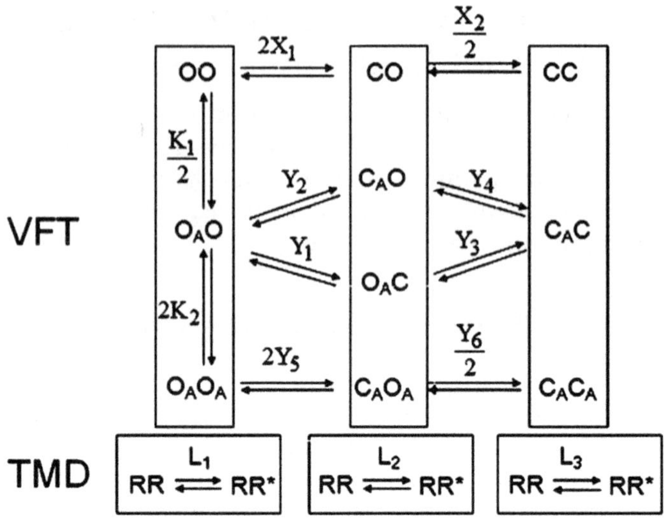

Figure 5.13 The metabotropic glutamate receptor model (adapted from a model published previosuly[106]). The model includes two domains: the extracellular Venus flytrap (VFT) domain to which agonists bind and the transmembrane domain (TMD) to which the signalling G protein binds. The receptor is dimeric. For the VFT domain, open (O) and closed (C) states of the subunits, leading to OO, OC, and CC dimeric states, are considered. The VFT domain can be free, (top row), singly bound (middle row), or doubly bound (bottom row). The model follows a conformational induction approach in which closed states are induced from the ligand binding to OO states. For the dimeric TMD a symmetric disposition of the two protomers (RR) for the inactive state and an asymmetric one (RR*) for the active state is proposed (in agreement with previously published results[94]). The relative populations of active and inactive TMD dimers are determined by an equilibrium constant (L). The binding to the extracellular domain modulates the activity of the transmembrane domain by making the constant L vary according to the VFT state to which the TMD is linked. In the simulations it is assumed that the proportion of active TMD (RR*) increases with the number of closed subunits of the VFT domain (in agreement with previously published work[93]). The definition of the constants and functional response are included in Appendix A.

A useful property of the mathematical mGlu model is in providing the distribution of receptor species over agonist concentration.[106] The analysis is particularly interesting for the intermediate and upper plateaus depicted in the functional curve for the mutated receptor (Figure 5.14). The intermediate plateau corresponds to singly occupied VFT states with one closed receptor at most. The spanning of this plateau over 2 log[A] units reflects the difficulty of the added agonist in both binding the mutated protomer and promoting its closure. On the other hand, the upper plateau corresponds to the doubly occupied VFT states CC, CO, and OO, in decreasing concentration order, whereas in the WT receptor the upper plateau is fully represented by the doubly occupied CC.[106]

The mathematical mGlu model[106] allows a quantitative description of ligand action at the ECD orthosteric sites. Yet what can be said about the

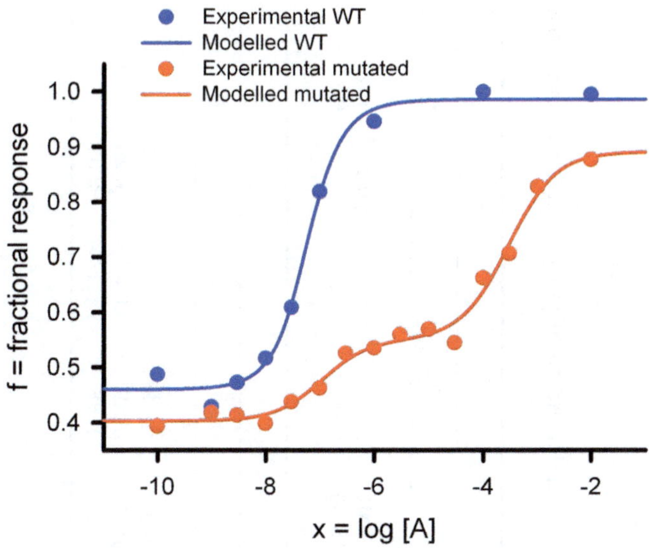

Figure 5.14 Experimental data points and theoretical curves of WT (blue) and single-mutated (red) mGlu5 receptor.[93] Theoretical curves were constructed under fractional response function (Appendix A3) with parameter values listed in Table 1 (best run of 100 using an in-house evolutionary algorithm).[130]

allosteric interactions between orthosteric and allosteric ligands which bind to the ECD and TMD, respectively?

5.4.2 Modelling Cooperativity Effects Between the ECD and the TMD

It is worth recapitulating on the complex interconnection between the ECD and TMD and the dimeric nature of mGlu receptors. Monomeric mGlu2, either as truncated TMDs or as full-length protein, can activate G-proteins upon direct activation by a PAM; however, activation of the receptor by agonists requires a dimeric arrangement of the full receptor.[92] Different speculative hypotheses can be raised by these results: did nature select the ECD for mGlu receptors (and class C GPCRs, in general) as an alternative to the TMD for endogenous binding in order to provide more efficient differentiation between receptors? Since dimerization is required for agonist but not for PAM mGlu activation, does this imply that there are no endogenous mGlu PAMs? If that were true, is dimerization a structural organization devised by nature to substitute the absence of endogenous PAMs?

There is currently no mechanistic model for mGlu receptors including, in a dimeric context, the allosteric interactions between orthosteric ligands in the VFT and allosteric ligands in the TMD. Generally, the operational model of allosterism[108–112] is used, which does not take into account receptor

Table 5.1 Interpretation of the experimental data included in a previous publication[93] under the mGlu model (fractional functional response in Appendix A3).

	Parameter values											
	X_1	X_2	K_1	K_2	Y_1	Y_2	Y_3	Y_5	Y_6	L_1	L_2	L_3
WT												
Mean	−5.07	−4.30	−6.02	−4.07	−4.56	0.84	1.52	−0.01	2.79	−0.37	−0.21	1.81
SD	1.90	1.50	0.08	0.82	1.50	0.76	0.97	0.69	0.86	0.02	0.57	0.35
Best run	−6.23	−3.70	−6.10	−3.46	−5.03	0.63	1.15	0.00	3.45	−0.37	0.02	1.55
	Parameter values											
	X_1	X_2	K_1	K_2	Y_1	Y_2	Y_3	Y_5	Y_6	L_1	L_2	L_3
Single-mutated VFT												
Mean	−6.85*	−5.95*	−5.90	−2.41*	−4.63	0.42*	1.30	0.03	1.28*	−0.47*	−0.17	0.71*
SD	1.24	1.56	0.36	0.54	1.49	0.51	1.16	0.46	0.49	0.00	0.20	0.09
Best run	−6.92	−5.92	−6.16	−2.27	−3.95	0.12	1.60	0.31	0.88	−0.47	−0.09	0.72

dimerization. The operational model of allosterism uses commonly eqn (5.1) for data fitting or simulation.[112]

$$E = \frac{E_m\left(\tau_A[A](K_B + \alpha\beta[B]) + \tau_B[B]K_A\right)^n}{\left([A]K_B + K_AK_B + K_A[B] + \alpha[A][B]\right)^n + \left(\tau_A[A](K_B + \alpha\beta[B]) + \tau_B[B]K_A\right)^n} \tag{5.1}$$

Where E_m is the maximum effect of the system; [A] and [B], the concentrations of the agonist A and the allosteric modulator B, respectively; τ_A and τ_B, the operational efficacies of A and B, respectively; K_A and K_B, the dissociation constants of A and B, respectively; α and β, the cooperativity parameters governing binding and function, respectively; and n, a parameter related with the slope of the curves. It is worth noting that in the case of mGlu receptors τ_A and τ_B, though reflecting both the activation of G-proteins, refer to ligand binding sites located in the ECD and TMD, respectively. Those allosteric modulators which do not have efficacy by themselves and thus are not able to activate the receptor in the absence of the agonist will yield $\tau_B = 0$.

Eqn (5.1) can be simplified to eqn (5.2) under several assumptions (see Appendix B), in which a parameter (Basal) accounting for ligand-independent system responses has been additionally included.[113]

$$E = \text{Basal} + \frac{(E_m - \text{Basal})\left([A](K_B + \alpha\beta[B]) + \tau_B[B][EC_{50}]\right)^n}{[EC_{50}]^n(K_B + [B])^n + \left([A](K_B + \alpha\beta[B]) + \tau_B[B][EC_{50}]\right)^n} \tag{5.2}$$

The lesser number of parameters in eqn (5.2) makes this model more efficient for curve fitting, but some caution is needed because of the assumptions made (discussed elsewhere[114]).

In a recent study, the cooperativity effects between the TMD and ECD of the mGlu4 receptor were examined using a combination of mGlu4 biosensors and second-messenger assays.[115] Functional data was analysed using the operational model of allosterism.[112,113] Interestingly, it was found that the rank order of PAM cooperativities provided by the sensor was only partly in agreement with that produced by the operational model.[115] As the authors noticed[115] and also discussed elsewhere[114] the operational β parameter includes in its definition, in addition to cooperativity, a component reflecting the intrinsic efficacy of the allosteric modulator. Therefore, it may happen (and this may explain the discrepancies between sensor results and mathematical modelling) that those PAMs demonstrating high efficacy (high τ) will also yield overestimated β values.[115]

Eqn (5.3) is a variant of the operational model in which functional cooperativity does not include efficacy effects[116–118] (see Appendix C).

$$E = \frac{E_m\chi^n\left(K_AK_B + \varepsilon_AK_B[A] + \varepsilon_BK_A[B] + \alpha\varepsilon_A\varepsilon_B\delta[A][B]\right)^n}{\left(K_AK_B + K_B[A] + K_A[B] + \alpha[A][B]\right)^n + \chi^n\left(K_AK_B + \varepsilon_AK_B[A] + \varepsilon_BK_A[B] + \alpha\varepsilon_A\varepsilon_B\delta[A][B]\right)^n} \tag{5.3}$$

[A] and [B] are the agonist (A) and the allosteric modulator (B) concentrations, respectively; K_A and K_B, the dissociation constants of A and B for the receptor; ε_A and ε_B, the intrinsic efficacies of A and B; α, the binding cooperativity between A and B; and δ, the activation cooperativity between A and B. As can be seen in Appendix C, δ does not suffer from the mixing of efficacy and cooperativity concepts included in the definition of β. Values of α and δ greater than, equal to, and lower than one indicate positive, absent, and negative binding and activation cooperativities, respectively. In eqn (5.3), constitutive activity of the receptor is included in the model definition.

5.5 Concluding Remarks and Future Work

The mechanism of GPCR activation is progressively being elucidated by knowledge gained from many disciplines. Crystallographic analysis,[21,24,119] NMR experiments,[120,121] fluorescence- and bioluminescence-based approaches,[122,123] mathematical modelling,[105,117,124,125] and molecular dynamics simulations[20,126,127] are fundamental in the understanding of the functioning of these complex allosteric machines. In particular, much work is needed at the computational level to provide insightful mechanistic information. Mechanistic mathematical models for the mGlu receptor incorporating allosteric modulators at the TMD and mathematical models for the analysis of functional responses through heteromeric receptors would be of great help. For the latter aspect, computational molecular dynamics simulations analyzing the most stable interfaces between protomers and how they change during the activation process are fundamental to understanding GPCR allosterism in an oligomeric context.

Appendix

Appendix A. The Metabotropic Glutamate Receptor Model

The metabotropic glutamate receptor model is shown in Figure 5.13.

Appendix A1. The Equilibrium Constants of the Model for the VFT Domain

$$\mathrm{OO} \xleftrightarrow{2X_1} \mathrm{OC};\quad 2X_1 = \frac{[\mathrm{OC}]}{[\mathrm{OO}]}$$

$$\mathrm{OC} \xleftrightarrow{\frac{X_2}{2}} \mathrm{CC};\quad \frac{X_2}{2} = \frac{[\mathrm{CC}]}{[\mathrm{OC}]}$$

$$\mathrm{A} + \mathrm{OO} \xleftrightarrow{\frac{K_1}{2}} \mathrm{O_AO};\quad \frac{K_1}{2} = \frac{[\mathrm{OO}][\mathrm{A}]}{[\mathrm{O_AO}]}$$

$$O_AO \xleftrightarrow{Y_1} O_AC;\ Y_1 = \frac{[O_AC]}{[O_AO]}$$

$$O_AO \xleftrightarrow{Y_2} C_AO;\ Y_2 = \frac{[C_AO]}{[O_AO]}$$

$$O_AC \xleftrightarrow{Y_3} C_AC;\ Y_3 = \frac{[C_AC]}{[O_AC]}$$

$$C_AO \xleftrightarrow{Y_4} C_AC;\ Y_4 = \frac{[C_AC]}{[C_AO]}$$

$$A + O_AO \xleftrightarrow{2K_2} O_AO_A;\ 2K_2 = \frac{[O_AO][A]}{[O_AO_A]}$$

$$O_AO_A \xleftrightarrow{2Y_5} O_AC_A;\ 2Y_5 = \frac{[O_AC_A]}{[O_AO_A]}$$

$$O_AC_A \xleftrightarrow{\frac{Y_6}{2}} C_AC_A;\ \frac{Y_6}{2} = \frac{[C_AC_A]}{[O_AC_A]}$$

Appendix A2. The Equilibrium Constants of the Model for the TMD

The equilibrium between inactive (RR) and active (RR*) TMD

$$RR \xleftrightarrow{2L} RR^*$$

where R and R* stand for the inactive and active TMD within the dimer, respectively, and $2L = \frac{[RR^*]}{[RR]}$ is the macroscopic equilibrium constant for the equilibrium between TM dimer states.

The equilibrium constant α depends on the state of the VFT domain. Thus, three apparent constants for the equilibrium between inactive (RR) and active (RR*) TMs are defined:

$$2L_1 = \frac{[RR^*_{oo_T}]}{[RR_{oo_T}]};\ 2L_2 = \frac{[RR^*_{oc_T}]}{[RR_{oc_T}]};\ 2L_3 = \frac{[RR^*_{cc_T}]}{[RR_{cc_T}]}$$

where $\{[RR^*_{ooT}], [RR^*_{ocT}]$, and $[RR^*_{ccT}]\}$ and $\{[RR_{ooT}], [RR_{ocT}]$, and $[RR_{ccT}]\}$ stand for total active and inactive TMD connected to open–open (OO), open–closed (OC), and closed–closed (CC) VFT, respectively.

Appendix A3. Functional Response: Fraction of Active Receptors

$$f = \frac{[(RR)]_{Active}}{[(RR)]_T} = 2\frac{c_1 + c_2[A] + c_3[A]^2}{c_4 + c_5[A] + c_6[A]^2} = 2\frac{a_1 + a_2[A] + a_3[A]^2}{a_4 + a_5[A] + [A]^2} \quad (5.4)$$

Where

$$[(\mathrm{RR})]_{\mathrm{Active}} = [\mathrm{RR}^{*}_{\mathrm{OO_T}}] + [\mathrm{RR}^{*}_{\mathrm{OC_T}}] + [\mathrm{RR}^{*}_{\mathrm{CC_T}}]$$

$$[(\mathrm{RR})]_{\mathrm{T}} = [\mathrm{RR}^{*}_{\mathrm{OO_T}}] + [\mathrm{RR}_{\mathrm{OO_T}}] + [\mathrm{RR}^{*}_{\mathrm{OC_T}}] + [\mathrm{RR}_{\mathrm{OC_T}}] + [\mathrm{RR}^{*}_{\mathrm{CC_T}}] + [\mathrm{RR}_{\mathrm{CC_T}}]$$

$$c_1 = K_1K_2(L_1 + 2L_2X_1 + L_3X_1X_2)$$

$$c_2 = 2K_2(L_1 + L_2(Y_1 + Y_2) + L_3Y_2Y_4)$$

$$c_3 = L_1 + 2L_2Y_5 + L_3Y_5Y_6$$

$$c_4 = K_1K_2(2L_1 + 1 + 2(2L_2 + 1)X_1 + (2L_3 + 1)X_1X_2)$$

$$c_5 = 2K_2(2L_1 + 1 + (2L_2 + 1)(Y_1 + Y_2) + (2L_3 + 1)Y_2Y_4)$$

$$c_6 = 2L_1 + 1 + 2(2L_2 + 1)Y_5 + (2L_3 + 1)Y_5Y_6;$$

$$\text{and } a_i = \frac{c_i}{c_6} \text{ for } i = 1 \text{ to } 6$$

Appendix A4. The Asymptotes of the Functional Curves: Extracting Pharmacological Information from Geometric Descriptors

- Left asymptote: Basal activity

$$\underset{\text{for [A]=0}}{f} = 2\frac{a_1}{a_4} = \frac{1}{1 + \dfrac{1}{2}\dfrac{1 + 2X_1 + X_1X_2}{L_1 + 2L_2X_1 + L_3X_1X_2}} \tag{5.5}$$

- Right asymptote: efficacy

$$\lim_{[\mathrm{A}]\to\infty} f_{\mathrm{R}^*} = 2a_3 = \frac{1}{1 + \dfrac{1}{2}\dfrac{1 + 2Y_5 + Y_5Y_6}{L_1 + 2L_2Y_5 + L_3Y_5Y_6}} \tag{5.6}$$

Appendix B. The Operational Model of Allosterism

The operational model of agonism and allosterism[112] proposes the following equation:

$$E = \frac{E_{\mathrm{m}}\left(\tau_{\mathrm{A}}[\mathrm{A}](K_{\mathrm{B}} + \alpha\beta[\mathrm{B}]) + \tau_{\mathrm{B}}[\mathrm{B}]K_{\mathrm{A}}\right)^n}{([\mathrm{A}]K_{\mathrm{B}} + K_{\mathrm{A}}K_{\mathrm{B}} + K_{\mathrm{A}}[\mathrm{B}] + \alpha[\mathrm{A}][\mathrm{B}])^n + \left(\tau_{\mathrm{A}}[\mathrm{A}](K_{\mathrm{B}} + \alpha\beta[\mathrm{B}]) + \tau_{\mathrm{B}}[\mathrm{B}]K_{\mathrm{A}}\right)^n} \tag{5.7}$$

Where E_m is the maximum effect of the system; A and B, the agonist and the allosteric modulator, respectively; τ_A and τ_B, the operational efficacies of A and B, respectively; K_A and K_B, the dissociation constants of A and B, respectively; α and β, the cooperativity parameters governing binding and function, respectively; and n, a parameter related to the slope of the curves.

Appendix B1. The Asymptotes of the Functional Response

- Left asymptote of E in eqn (5.7) in the absence of A and B (Basal response or receptor constitutive activity: E for [A] = 0 and [B] = 0)

$$\text{Basal} = 0 \tag{5.8}$$

The basal response is 0. Receptor constitutive activity is not included in the model.

- Left asymptote in the absence of A (Bottom: E for [A] = 0)

$$\text{Bottom} = \frac{E_m}{1 + \left(\dfrac{K_B + [B]}{\tau_B [B]}\right)^n} \tag{5.9}$$

Bottom is higher as τ_B increases (increased efficacy of the allosteric modulator).

- Right asymptote, the asymptotic E value as [A] increases (Top: $\lim_{[A]\to\infty} E$):

$$\text{Top} = \frac{E_m}{1 + \dfrac{1}{\tau_A^n}\left(\dfrac{K_B + \alpha[B]}{K_B + \alpha\beta[B]}\right)^n} \tag{5.10}$$

where if the allosteric modulator is absent ([B] = 0) then

$$\text{Top} = \frac{E_m}{1 + \dfrac{1}{\tau_A^n}} \tag{5.11}$$

Comparison between eqn (5.10) and (5.11) shows that there is a change of the maximum response in the presence of the allosteric modulator if β is different from 1, and the effect is either positive or negative if this parameter is greater or lower than one, respectively. It is worth noting that τ_B is not included in eqn (5.10), when it should logically appear. The reason for this is that the efficacy of the allosteric modulator B is implicitly included in the definition of β (see below and work published previously[114] for further discussion).

Appendix B2. A Simplified Operational Model

Eqn (5.7) can be simplified to eqn (5.12), in which a parameter (Basal) accounting for ligand-independent system responses has been additionally included,[113]

$$E = \text{Basal} + \frac{\left(E_m - \text{Basal}\right)\left([A]\left(K_B + \alpha\beta[B]\right) + \tau_B[B][EC_{50}]\right)^n}{[EC_{50}]^n\left(K_B + [B]\right)^n + \left([A]\left(K_B + \alpha\beta[B]\right) + \tau_B[B][EC_{50}]\right)^n} \qquad (5.12)$$

under the following assumptions

(i) $[A] \ll K_A$, and then $[A] + K_A \approx K_A$
(ii) $\alpha[A] \ll K_A$, and then $\alpha[A] + K_A \approx K_A$
(iii) $\tau_A{}^n \gg 1$, and then $[EC_{50}] = \dfrac{K_A}{\left(2+\tau_A^n\right)^{1/n} - 1}$ in the operational model of agonism[128] simplifies to $[EC_{50}] \approx \dfrac{K_A}{\tau_A}$

The three assumptions relay on A being a full agonist. In this regard, it can bee seen that the asymptotic top value is equal to the maximum response of the system, E_m, in all cases. Thus, this modelling approach cannot be used neither for partial agonists A nor for allosteric modulators B which increase or decrease the maximum response yielded by compounds A.[114] Additionally, assumption (ii) includes the property that the binding cooperativity α cannot be much greater than 1. Eqn (5.12) reduces the number of parameters present in eqn (5.7) by considering jointly K_A and τ_A through $[EC_{50}]$ and α and β through their product $\alpha\beta$. The usefulness of this procedure in mGlu5 was evaluated in previous studies.[129]

Appendix B3. The Relation Between Efficacy and Functional Cooperativity in the Operational Model

The intrinsic efficacy of the receptor complexed with both the orthosteric agonist A and the allosteric compound B was defined previously[112] as $\varepsilon_{AB} = \varepsilon_A\beta$, where ε_A denotes the intrinsic efficacy of A and β is a coupling factor that describes the ability of B to alter the signalling capacity of A when bound to the same receptor in a ternary ARB complex. This definition of ε_{AB} is pertinent in the construction of the model in order not to increase the number of parameters of the system. However, it makes β not a pure cooperativity factor but a hybrid parameter embodying both intrinsic efficacy and functional cooperativity.[114,115]

Appendix C. The Operational Model of Allosterism Including Constitutive Receptor Activity

Eqn (5.13) is a variant of the operational model which includes receptor constitutive activity in its definition.[116–118]

$$E = \frac{E_m \chi^n \left(K_A K_B + \varepsilon_A K_B[A] + \varepsilon_B K_A[B] + \alpha\varepsilon_A\varepsilon_B\delta[A][B]\right)^n}{\left(K_A K_B + K_B[A] + K_A[B] + \alpha[A][B]\right)^n + \chi^n\left(K_A K_B + \varepsilon_A K_B[A] + \varepsilon_B K_A[B] + \alpha\varepsilon_A\varepsilon_B\delta[A][B]\right)^n} \tag{5.13}$$

E_m is the maximum effect of the system; A and B, the agonist and the allosteric modulator, respectively; K_A and K_B, the dissociation constants of A and B for the receptor; ε_A and ε_B, the intrinsic efficacies of A and B; α, the binding cooperativity between A and B; and δ, the activation cooperativity between A and B. Values of α and δ greater than, equal to, and lower than one indicate positive, absent, and negative binding and activation cooperativities, respectively. In this model functional cooperativity (δ) is uncoupled from the intrinsic efficacy of the allosteric modulator because the intrinsic efficacy of the receptor complexed with both the orthosteric agonist A and the allosteric compound B is defined as $\varepsilon_{AB} = \varepsilon_A\varepsilon_B\delta$.[116–118]

Appendix C1. The Asymptotes of the Functional Response

Left asymptote of E in eqn (5.13) in the absence of A and B (Basal response or receptor constitutive activity: E for [A] = 0 and [B] = 0)

$$\text{Basal} = \frac{E_m}{1 + \dfrac{1}{\chi^n}} \tag{5.14}$$

The basal response is greater as E_m and χ increase.

- Left asymptote in the absence of A (Bottom: E for [A] = 0)

$$\text{Bottom} = \frac{E_m}{1 + \dfrac{1}{\chi^n}\left(\dfrac{M + [B]}{M + \varepsilon_B[B]}\right)^n} \tag{5.15}$$

Values of ε_B greater than, equal to, or lower than one make bottom greater than, equal to, or lower than basal response, respectively, making allosteric compound B an agonist, neutral antagonist, or inverse agonist, respectively.

- Right asymptote, the asymptotic E value as [A] increases (Top: $\lim_{[A]\to\infty} E$)

$$\text{Top} = \frac{E_m}{1 + \dfrac{1}{\chi^n\varepsilon_A^n}\left(\dfrac{M + \alpha[B]}{M + \varepsilon_B\alpha\delta[B]}\right)^n} \tag{5.16}$$

where if the allosteric modulator is absent ([B] = 0) then

$$\text{Top} = \frac{E_m}{1 + \dfrac{1}{\chi^n\varepsilon_A^n}} \tag{5.17}$$

Comparison between eqn (5.16) and (5.17) shows that there is a change of the maximum response in the presence of the allosteric modulator if either of the functional parameters ε_B or δ is different from 1, and the effect is either positive or negative if these parameters are greater or less than one, respectively.

Equilibrium constants (see Appendix A for definitions) are expressed in logarithmic form. An in-house evolutionary algorithm was used for parameter optimization.[130] 100 independent runs were performed for both WT and mutated receptors; mean and standard deviation (SD) are shown; the best (closest to the mean parameters) run is also shown.[106] The parameter values for the latter run were used for graphic representation in Figure 14.

Statistical comparison between WT and mutated receptor samples for each of the parameters was done by Student's *t*-test including Šidák correction for multiple testing, $*P < 0.05$.

The Y_4 equilibrium constant (see Figure 13) has not been included explicitly in the table because it is determined by the values of Y_1, Y_2, and Y_3.

Acknowledgements

This study was supported in part by Ministerio de Economía y Competitividad (ERA-NET NEURON PCIN-2013-018-C03-02 and SAF2014-58396 R). B.Z. was the recipient of a CSC fellowship.

References

1. V. Katritch, V. Cherezov and R. C. Stevens, *Annu. Rev. Pharmacol. Toxicol.*, 2013, **53**, 531–556.
2. C. P. Lawler, C. Prioleau, M. M. Lewis, C. Mak, D. Jiang, J. A. Schetz, A. M. Gonzalez, D. R. Sibley and R. B. Mailman, *Neuropsychopharmacology*, 1999, **20**, 612–627.
3. J. D. Kilts, H. S. Connery, E. G. Arrington, M. M. Lewis, C. P. Lawler, G. S. Oxford, K. L. O'Malley, R. D. Todd, B. L. Blake, D. E. Nichols and R. B. Mailman, *J. Pharmacol. Exp. Ther.*, 2002, **301**, 1179–1189.
4. M. B. Jarpe, C. Knall, F. M. Mitchell, A. M. Buhl, E. Duzic and G. L. Johnson, *J. Biol. Chem.*, 1998, **273**, 3097–3104.
5. S. Galandrin, G. Oligny-Longpre and M. Bouvier, *Trends Pharmacol. Sci.*, 2007, **28**, 423–430.
6. T. Kenakin, *Trends Pharmacol. Sci.*, 1995, **16**, 232–238.
7. C. J. Langmead and A. Christopoulos, *Curr. Opin. Cell Biol.*, 2014, **27**, 94–101.
8. E. T. van der Westhuizen, C. Valant, P. M. Sexton and A. Christopoulos, *J. Pharmacol. Exp. Ther.*, 2015, **353**, 246–260.
9. D. L. Hay and A. A. Pioszak, *Annu. Rev. Pharmacol. Toxicol.*, 2016, **56**, 469–487.

10. K. R. Klein, B. C. Matson and K. M. Caron, *Crit. Rev. Biochem. Mol. Biol.*, 2016, 1–7.
11. T. P. Kenakin, *Trends Pharmacol. Sci.*, 2009, **30**, 460–469.
12. R. J. Lefkowitz, *Prog. Mol. Biol. Transl. Sci.*, 2013, **118**, 3–18.
13. K. M. Raehal, J. K. Walker and L. M. Bohn, *J. Pharmacol. Exp. Ther.*, 2005, **314**, 1195–1201.
14. J. D. Violin, A. L. Crombie, D. G. Soergel and M. W. Lark, *Trends Pharmacol. Sci.*, 2014, **35**, 308–316.
15. E. R. Viscusi, L. Webster, M. Kuss, S. Daniels, J. A. Bolognese, S. Zuckerman, D. G. Soergel, R. A. Subach, E. Cook and F. Skobieranda, *Pain*, 2016, **157**, 264–272.
16. W. Huang, A. Manglik, A. J. Venkatakrishnan, T. Laeremans, E. N. Feinberg, A. L. Sanborn, H. E. Kato, K. E. Livingston, T. S. Thorsen, R. C. Kling, S. Granier, P. Gmeiner, S. M. Husbands, J. R. Traynor, W. I. Weis, J. Steyaert, R. O. Dror and B. K. Kobilka, *Nature*, 2015, **524**, 315–321.
17. R. Sounier, C. Mas, J. Steyaert, T. Laeremans, A. Manglik, W. Huang, B. K. Kobilka, H. Demene and S. Granier, *Nature*, 2015, **524**, 375–378.
18. R. Nygaard, Y. Zou, R. O. Dror, T. J. Mildorf, D. H. Arlow, A. Manglik, A. C. Pan, C. W. Liu, J. J. Fung, M. P. Bokoch, F. S. Thian, T. S. Kobilka, D. E. Shaw, L. Mueller, R. S. Prosser and B. K. Kobilka, *Cell*, 2013, **152**, 532–542.
19. A. Manglik, T. H. Kim, M. Masureel, C. Altenbach, Z. Yang, D. Hilger, M. T. Lerch, T. S. Kobilka, F. S. Thian, W. L. Hubbell, R. S. Prosser and B. K. Kobilka, *Cell*, 2015, **161**, 1101–1111.
20. R. O. Dror, D. H. Arlow, P. Maragakis, T. J. Mildorf, A. C. Pan, H. Xu, D. W. Borhani and D. E. Shaw, *Proc. Natl. Acad. Sci. U. S. A.*, 2011, **108**, 18684–18689.
21. J. A. Dalton, I. Lans and J. Giraldo, *BMC Bioinf.*, 2015, **16**, 124, DOI: 10.1186/s12859-015-0567-3.
22. B. Trzaskowski, D. Latek, S. Yuan, U. Ghoshdastider, A. Debinski and S. Filipek, *Curr. Med. Chem.*, 2012, **19**, 1090–1109.
23. B. G. Tehan, A. Bortolato, F. E. Blaney, M. P. Weir and J. S. Mason, *Pharmacol. Ther.*, 2014, **143**, 51–60.
24. I. Lans, J. A. Dalton and J. Giraldo, *J. Struct. Biol.*, 2015, **192**, 545–553.
25. J. L. Benovic, *Nature*, 2015, **523**, 538–539.
26. V. V. Gurevich and E. V. Gurevich, *Trends Pharmacol. Sci.*, 2004, **25**, 105–111.
27. A. K. Shukla, G. H. Westfield, K. Xiao, R. I. Reis, L. Y. Huang, P. Tripathi-Shukla, J. Qian, S. Li, A. Blanc, A. N. Oleskie, A. M. Dosey, M. Su, C. R. Liang, L. L. Gu, J. M. Shan, X. Chen, R. Hanna, M. Choi, X. J. Yao, B. U. Klink, A. W. Kahsai, S. S. Sidhu, S. Koide, P. A. Penczek, A. A. Kossiakoff, V. L. Woods, Jr., B. K. Kobilka, G. Skiniotis and R. J. Lefkowitz, *Nature*, 2014, **512**, 218–222.
28. Y. Kang, X. E. Zhou, X. Gao, Y. He, W. Liu, A. Ishchenko, A. Barty, T. A. White, O. Yefanov, G. W. Han, Q. Xu, P. W. de Waal, J. Ke, M. H. Tan, C. Zhang, A. Moeller, G. M. West, B. D. Pascal, E. N. Van, L. N. Caro, S. A.

Vishnivetskiy, R. J. Lee, K. M. Suino-Powell, X. Gu, K. Pal, J. Ma, X. Zhi, S. Boutet, G. J. Williams, M. Messerschmidt, C. Gati, N. A. Zatsepin, D. Wang, D. James, S. Basu, S. Roy-Chowdhury, C. E. Conrad, J. Coe, H. Liu, S. Lisova, C. Kupitz, I. Grotjohann, R. Fromme, Y. Jiang, M. Tan, H. Yang, J. Li, M. Wang, Z. Zheng, D. Li, N. Howe, Y. Zhao, J. Standfuss, K. Diederichs, Y. Dong, C. S. Potter, B. Carragher, M. Caffrey, H. Jiang, H. N. Chapman, J. C. Spence, P. Fromme, U. Weierstall, O. P. Ernst, V. Katritch, V. V. Gurevich, P. R. Griffin, W. L. Hubbell, R. C. Stevens, V. Cherezov, K. Melcher and H. E. Xu, *Nature*, 2015, **523**, 561–567.
29. A. K. Shukla, G. Singh and E. Ghosh, *Trends Biochem. Sci.*, 2014, **39**, 594–602.
30. K. A. Jacobson, *Biochem. Pharmacol.*, 2015, **98**, 541–555.
31. Z. Rankovic, T. F. Brust and L. M. Bohn, *Bioorg. Med. Chem. Lett.*, 2016, **26**, 241–250.
32. G. Swaminath, X. Deupi, T. W. Lee, W. Zhu, F. S. Thian, T. S. Kobilka and B. Kobilka, *J. Biol. Chem.*, 2005, **280**, 22165–22171.
33. A. Christopoulos and T. Kenakin, *Pharmacol. Rev.*, 2002, **54**, 323–374.
34. J. R. Lane, P. M. Sexton and A. Christopoulos, *Trends Pharmacol. Sci.*, 2013, **34**, 59–66.
35. P. R. Gentry, P. M. Sexton and A. Christopoulos, *J. Biol. Chem.*, 2015, **290**, 19478–19488.
36. J. Monod and F. Jacob, *Cold Spring Harbor Symp. Quant. Biol.*, 1961, **26**, 389–401.
37. J. Monod, J. Changeux and F. Jacob, *J. Mol. Biol.*, 1963, **6**, 306–329.
38. J. Monod, J. Wyman and J. P. Changeux, *J. Mol. Biol.*, 1965, **12**, 88–118.
39. J. Giraldo, *Prog. Mol. Biol. Transl. Sci.*, 2013, **115**, 349–373.
40. A. Christopoulos, J. P. Changeux, W. A. Catterall, D. Fabbro, T. P. Burris, J. A. Cidlowski, R. W. Olsen, J. A. Peters, R. R. Neubig, J. P. Pin, P. M. Sexton, T. P. Kenakin, F. J. Ehlert, M. Spedding and C. J. Langmead, *Pharmacol. Rev.*, 2014, **66**, 918–947.
41. M. Chabre and M. le Maire, *Biochemistry*, 2005, **44**, 9395–9403.
42. M. Bouvier and T. E. Hebert, *J. Physiol.*, 2014, **592**, 2439–2441.
43. N. A. Lambert and J. A. Javitch, *J. Physiol.*, 2014, **592**, 2443–2445.
44. M. Bouvier and T. E. Hebert, *J. Physiol.*, 2014, **592**, 2447.
45. J. R. James, M. I. Oliveira, A. M. Carmo, A. Iaboni and S. J. Davis, *Nat. Methods*, 2006, **3**, 1001–1006.
46. M. Bouvier, N. Heveker, R. Jockers, S. Marullo and G. Milligan, *Nat. Methods*, 2007, **4**, 3–4.
47. B. H. Meyer, J. M. Segura, K. L. Martinez, R. Hovius, N. George, K. Johnsson and H. Vogel, *Proc. Natl. Acad. Sci. U. S. A.*, 2006, **103**, 2138–2143.
48. M. R. Whorton, M. P. Bokoch, S. G. Rasmussen, B. Huang, R. N. Zare, B. Kobilka and R. K. Sunahara, *Proc. Natl. Acad. Sci. U. S. A.*, 2007, **104**, 7682–7687.
49. T. H. Bayburt, A. J. Leitz, G. Xie, D. D. Oprian and S. G. Sligar, *J. Biol. Chem.*, 2007, **282**, 14875–14881.

50. J. F. White, J. Grodnitzky, J. M. Louis, L. B. Trinh, J. Shiloach, J. Gutierrez, J. K. Northup and R. Grisshammer, *Proc. Natl. Acad. Sci. U. S. A.*, 2007, **104**, 12199–12204.
51. C. D. Rios, B. A. Jordan, I. Gomes and L. A. Devi, *Pharmacol. Ther.*, 2001, **92**, 71–87.
52. G. Milligan, *Mol. Pharmacol.*, 2004, **66**, 1–7.
53. S. Terrillon and M. Bouvier, *EMBO J.*, 2004, **5**, 30–34.
54. B. Jastrzebska, D. Fotiadis, G. F. Jang, R. E. Stenkamp, A. Engel and K. Palczewski, *J. Biol. Chem.*, 2006, **281**, 11917–11922.
55. J. J. Carrillo, J. F. Lopez-Gimenez and G. Milligan, *Mol. Pharmacol.*, 2004, **66**, 1123–1137.
56. L. Albizu, M. Cottet, M. Kralikova, S. Stoev, R. Seyer, I. Brabet, T. Roux, H. Bazin, E. Bourrier, L. Lamarque, C. Breton, M. L. Rives, A. Newman, J. Javitch, E. Trinquet, M. Manning, J. P. Pin, B. Mouillac and T. Durroux, *Nat. Chem. Biol.*, 2010, **6**, 587–594.
57. J. R. Lane, P. Donthamsetti, J. Shonberg, C. J. Draper-Joyce, S. Dentry, M. Michino, L. Shi, L. Lopez, P. J. Scammells, B. Capuano, P. M. Sexton, J. A. Javitch and A. Christopoulos, *Nat. Chem. Biol.*, 2014, **10**, 745–752.
58. J. Y. Springael, E. Urizar, S. Costagliola, G. Vassart and M. Parmentier, *Pharmacol. Ther.*, 2007, **115**, 410–418.
59. G. Milligan, *Br. J. Pharmacol.*, 2008, **153**, S216–S229.
60. J. F. López-Giménez, M. Canals, J. D. Pediani and G. Milligan, *Mol. Pharmacol.*, 2007, **71**, 1015–1029.
61. H. Kobayashi, K. Ogawa, R. Yao, O. Lichtarge and M. Bouvier, *Traffic*, 2009, **10**, 1019–1033.
62. S. Ferre, R. Baler, M. Bouvier, M. G. Caron, L. A. Devi, T. Durroux, K. Fuxe, S. R. George, J. A. Javitch, M. J. Lohse, K. Mackie, G. Milligan, K. D. Pfleger, J. P. Pin, N. D. Volkow, M. Waldhoer, A. S. Woods and R. Franco, *Nat. Chem. Biol.*, 2009, **5**, 131–134.
63. H. F. Vischer, M. Castro and J. P. Pin, *Mol. Pharmacol.*, 2015, **88**, 561–571.
64. N. J. Smith and G. Milligan, *Pharmacol. Rev.*, 2010, **62**, 701–725.
65. N. J. Birdsall, *Trends Pharmacol. Sci.*, 2010, **31**, 499–508.
66. X. Rovira, J. P. Pin and J. Giraldo, *Trends Pharmacol. Sci.*, 2010, **31**, 15–21.
67. D. E. Featherstone, *ACS Chem. Neurosci.*, 2010, **1**, 4–12.
68. R. Dingledine, K. Borges, D. Bowie and S. F. Traynelis, *Pharmacol. Rev.*, 1999, **51**, 7–61.
69. P. Rondard and J. P. Pin, *Curr. Opin. Pharmacol.*, 2015, **20**, 95–101.
70. P. J. Conn and J. P. Pin, *Annu. Rev. Pharmacol. Toxicol.*, 1997, **37**, 205–237.
71. D. E. O'Brien and P. J. Conn, *Int. J. Neuropsychopharmacol.*, 2015, pyv133, DOI: 10.1093/ijnp/pyv133.
72. F. Nicoletti, V. Bruno, R. T. Ngomba, R. Gradini and G. Battaglia, *Curr. Opin. Pharmacol.*, 2014, **20C**, 89–94.
73. J. M. Wieronska, S. H. Zorn, D. Doller and A. Pilc, *Pharmacol. Ther.*, 2016, **157**, 10–27.
74. P. J. Conn, C. W. Lindsley, J. Meiler and C. M. Niswender, *Nat. Rev. Drug Discovery*, 2014, **13**, 692–708.

75. F. Acher and C. Goudet, *Curr. Opin. Pharmacol.*, 2015, **20**, 64–72.
76. D. Maurel, L. Comps-Agrar, C. Brock, M. L. Rives, E. Bourrier, M. A. Ayoub, H. Bazin, N. Tinel, T. Durroux, L. Prezeau, E. Trinquet and J. P. Pin, *Nat. Methods*, 2008, **5**, 561–567.
77. E. Doumazane, P. Scholler, J. M. Zwier, E. Trinquet, P. Rondard and J. P. Pin, *FASEB J.*, 2011, **25**, 66–77.
78. H. Jingami, S. Nakanishi and K. Morikawa, *Curr. Opin. Neurobiol.*, 2003, **13**, 271–278.
79. K. A. Bennett, A. S. Dore, J. A. Christopher, D. R. Weiss and F. H. Marshall, *Curr. Opin. Pharmacol.*, 2014, **20C**, 1–7.
80. Y. Kubo and M. Tateyama, *Curr. Opin. Neurobiol.*, 2005, **15**, 289–295.
81. C. W. Lindsley, K. A. Emmitte, C. R. Hopkins, T. M. Bridges, K. J. Gregory, C. M. Niswender and P. J. Conn, *Chem. Rev.*, 2016, **116**, 6707–6741.
82. N. Kunishima, Y. Shimada, Y. Tsuji, T. Sato, M. Yamamoto, T. Kumasaka, S. Nakanishi, H. Jingami and K. Morikawa, *Nature*, 2000, **407**, 971–977.
83. D. Tsuchiya, N. Kunishima, N. Kamiya, H. Jingami and K. Morikawa, *Proc. Natl. Acad. Sci. U. S. A.*, 2002, **99**, 2660–2665.
84. T. Muto, D. Tsuchiya, K. Morikawa and H. Jingami, *Proc. Natl. Acad. Sci. U. S. A.*, 2007, **104**, 3759–3764.
85. H. Wu, C. Wang, K. J. Gregory, G. W. Han, H. P. Cho, Y. Xia, C. M. Niswender, V. Katritch, J. Meiler, V. Cherezov, P. J. Conn and R. C. Stevens, *Science*, 2014, **344**, 58–64.
86. A. S. Dore, K. Okrasa, J. C. Patel, M. Serrano-Vega, K. Bennett, R. M. Cooke, J. C. Errey, A. Jazayeri, S. Khan, B. Tehan, M. Weir, G. R. Wiggin and F. H. Marshall, *Nature*, 2014, **511**, 557–562.
87. S. Huang, J. Cao, M. Jiang, G. Labesse, J. Liu, J. P. Pin and P. Rondard, *Proc. Natl. Acad. Sci. U. S. A.*, 2011, **108**, 15480–15485.
88. E. Doumazane, P. Scholler, L. Fabre, J. M. Zwier, E. Trinquet, J. P. Pin and P. Rondard, *Proc. Natl. Acad. Sci. U. S. A.*, 2013, **110**, E1416–E1425.
89. L. Olofsson, S. Felekyan, E. Doumazane, P. Scholler, L. Fabre, J. M. Zwier, P. Rondard, C. A. Seidel, J. P. Pin and E. Margeat, *Nat. Commun.*, 2014, **5**, 5206.
90. R. Vafabakhsh, J. Levitz and E. Y. Isacoff, *Nature*, 2015, **524**, 497–501.
91. S. Yin, M. J. Noetzel, K. A. Johnson, R. Zamorano, N. Jalan-Sakrikar, K. J. Gregory, P. J. Conn and C. M. Niswender, *J. Neurosci.*, 2014, **34**, 79–94.
92. D. El Moustaine, S. Granier, E. Doumazane, P. Scholler, R. Rahmeh, P. Bron, B. Mouillac, J. L. Baneres, P. Rondard and J. P. Pin, *Proc. Natl. Acad. Sci. U. S. A.*, 2012, **109**, 16342–16347.
93. J. Kniazeff, A. S. Bessis, D. Maurel, H. Ansanay, L. Prezeau and J. P. Pin, *Nat. Struct. Mol. Biol.*, 2004, **11**, 706–713.
94. V. Hlavackova, C. Goudet, J. Kniazeff, A. Zikova, D. Maurel, C. Vol, J. Trojanova, L. Prezeau, J. P. Pin and J. Blahos, *EMBO J.*, 2005, **24**, 499–509.
95. M. Tateyama, H. Abe, H. Nakata, O. Saito and Y. Kubo, *Nat. Struct. Mol. Biol.*, 2004, **11**, 637–642.

96. C. Brock, N. Oueslati, S. Soler, L. Boudier, P. Rondard and J. P. Pin, *J. Biol. Chem.*, 2007, **282**, 33000–33008.
97. T. Yamashita, A. Terakita, T. Kai and Y. Shichida, *J. Neurochem.*, 2008, **106**, 850–859.
98. V. Hlavackova, U. Zabel, D. Frankova, J. Batz, C. Hoffmann, L. Prezeau, J. P. Pin, J. Blahos and M. J. Lohse, *Sci. Signaling*, 2012, **5**, ra59.
99. M. Tateyama and Y. Kubo, *Neuropharmacology*, 2011, **61**, 832–841.
100. L. Xue, X. Rovira, P. Scholler, H. Zhao, J. Liu, J. P. Pin and P. Rondard, *Nat. Chem. Biol.*, 2015, **11**, 134–140.
101. J. González-Maeso, R. L. Ang, T. Yuen, P. Chan, N. V. Weisstaub, J. F. López-Gimenez, M. Zhou, Y. Okawa, L. F. Callado, G. Milligan, J. A. Gingrich, M. Filizola, J. J. Meana and S. C. Sealfon, *Nature*, 2008, **452**, 93–97.
102. M. Fribourg, J. L. Moreno, T. Holloway, D. Provasi, L. Baki, R. Mahajan, G. Park, S. K. Adney, C. Hatcher, J. M. Eltit, J. D. Ruta, L. Albizu, Z. Li, A. Umali, J. Shim, A. Fabiato, A. D. MacKerell, Jr., V. Brezina, S. C. Sealfon, M. Filizola, J. Gonzalez-Maeso and D. E. Logothetis, *Cell*, 2011, **147**, 1011–1023.
103. J. L. Moreno, C. Muguruza, A. Umali, S. Mortillo, T. Holloway, F. Pilar-Cuellar, G. Mocci, J. Seto, L. F. Callado, R. L. Neve, G. Milligan, S. C. Sealfon, J. F. Lopez-Gimenez, J. J. Meana, D. L. Benson and J. Gonzalez-Maeso, *J. Biol. Chem.*, 2012, **287**, 44301–44319.
104. J. Giraldo, N. M. Vivas, E. Vila and A. Badia, *Pharmacol. Ther.*, 2002, **95**, 21–45.
105. D. Roche, D. Gil and J. Giraldo, *Adv. Exp. Med. Biol.*, 2014, **796**, 159–181.
106. X. Rovira, D. Roche, J. Serra, J. Kniazeff, J. P. Pin and J. Giraldo, *J. Pharmacol. Exp. Ther.*, 2008, **325**, 443–456.
107. J. Giraldo, *FEBS Lett.*, 2004, **556**, 13–18.
108. F. J. Ehlert, *J. Pharmacol. Exp. Ther.*, 2005, **315**, 740–754.
109. T. Kenakin, *Nat. Rev. Drug Discovery*, 2005, **4**, 919–927.
110. T. Kenakin, *J. Recept. Signal Transduction Res.*, 2007, **27**, 247–259.
111. M. R. Price, G. L. Baillie, A. Thomas, L. A. Stevenson, M. Easson, R. Goodwin, A. McLean, L. McIntosh, G. Goodwin, G. Walker, P. Westwood, J. Marrs, F. Thomson, P. Cowley, A. Christopoulos, R. G. Pertwee and R. A. Ross, *Mol. Pharmacol.*, 2005, **68**, 1484–1495.
112. K. Leach, P. M. Sexton and A. Christopoulos, *Trends Pharmacol. Sci.*, 2007, **28**, 382–389.
113. K. Leach, R. E. Loiacono, C. C. Felder, D. L. McKinzie, A. Mogg, D. B. Shaw, P. M. Sexton and A. Christopoulos, *Neuropsychopharmacology*, 2010, **35**, 855–869.
114. J. Giraldo, *Trends Pharmacol. Sci.*, 2015, **36**, 1–2.
115. X. Rovira, F. Malhaire, P. Scholler, J. Rodrigo, P. Gonzalez-Bulnes, A. Llebaria, J. P. Pin, J. Giraldo and C. Goudet, *FASEB J.*, 2015, **29**, 116–130.
116. R. Slack and D. Hall, *Br. J. Pharmacol.*, 2012, **166**, 1774–1792.
117. D. A. Hall, *Prog. Mol. Biol. Transl. Sci.*, 2013, **115**, 217–290.
118. D. Roche, D. Gil and J. Giraldo, *Br. J. Pharmacol.*, 2013, **169**, 1189–1202.

119. C. L. Piscitelli, J. Kean, C. de Graff and X. Deupi, *Mol. Pharmacol.*, 2015, **88**, 536–551.
120. S. Isogai, X. Deupi, C. Opitz, F. M. Heydenreich, C. J. Tsai, F. Brueckner, G. F. Schertler, D. B. Veprintsev and S. Grzesiek, *Nature*, 2016, **530**, 237–241.
121. J. Okude, T. Ueda, Y. Kofuku, M. Sato, N. Nobuyama, K. Kondo, Y. Shiraishi, T. Mizumura, K. Onishi, M. Natsume, M. Maeda, H. Tsujishita, T. Kuranaga, M. Inoue and I. Shimada, *Angew. Chem., Int. Ed. Engl.*, 2015, **54**, 15771–15776.
122. P. Scholler, J. M. Zwier, E. Trinquet, P. Rondard, J. P. Pin, L. Prezeau and J. Kniazeff, *Prog. Mol. Biol. Transl. Sci.*, 2013, **113**, 275–312.
123. L. A. Stoddart, C. W. White, K. Nguyen, S. J. Hill and K. D. Pfleger, *Br. J. Pharmacol.*, 2015, , DOI: 10.1111/bph.13316.
124. T. Kenakin, *Chem. Rev.*, 2016, , DOI: 10.1021/acs.chemrev.5b00561.
125. R. S. Stein and F. J. Ehlert, *J. Recept. Signal Transduction Res.*, 2014, 1–15.
126. S. Bhattacharya and N. Vaidehi, *Biophys. J.*, 2014, **107**, 422–434.
127. Y. Shang, H. R. Yeatman, D. Provasi, A. Alt, A. Christopoulos, M. Canals and M. Filizola, *ACS Chem. Biol.*, 2016, **11**, 1220–1229.
128. J. W. Black and P. Leff, *Proc. R. Soc. London, Ser. B*, 1983, **220**, 141–162.
129. K. J. Gregory, M. J. Noetzel, J. M. Rook, P. N. Vinson, S. R. Stauffer, A. L. Rodriguez, K. A. Emmitte, Y. Zhou, A. C. Chun, A. S. Felts, B. A. Chauder, C. W. Lindsley, C. M. Niswender and P. J. Conn, *Mol. Pharmacol.*, 2012, **82**, 860–875.
130. D. Gil, D. Roche, A. Borràs and J. Giraldo, *Comput. Optim. Appl.*, 2015, **61**, 489–515.

CHAPTER 6

A Unifying Approach to the Duality of "Energetic" Versus "Conformational" Formulations of Allosteric Coupling: Mechanistic Implications for GPCR Allostery

H. ONGUN ONARAN*[a] AND TOMMASO COSTA*[b]

[a]Ankara University Faculty of Medicine, Dept. Pharmacology, Sihhiye, 06100, Ankara, Turkey; [b]Dept. of Pharmacology, Istituto Superiore di Sanità, Rome, Italy
*E-mail: Ongun.Onaran@medicine.ankara.edu.tr, tommaso.costa@iss.it

6.1 Introduction

Allostery represents a very general biological phenomenon. Allosteric effects include all the molecular mechanisms that proteins use in adapting activity to changes in the environment, such as association and dissociation of surrounding molecules. Thus, allostery is at the core of fundamental life

RSC Drug Discovery Series No. 56
Allosterism in Drug Discovery
Edited by Dario Doller

Published by the Royal Society of Chemistry, www.rsc.org

processes, including metabolic regulation, control of enzyme catalysis, membrane transport, intracellular trafficking and signal transduction.

A wealth of experiments have expanded the biological range in which allosteric regulation is known to be involved,[1] but two seminal concepts in protein physics and chemistry have dramatically changed our views on the nature of this phenomenon. One is the logic of linked function analysis[2,3] in describing the allosteric behaviour of macromolecules.[4,5] According to this unified theoretical formalism, allosteric effects and their most common manifestation, *i.e.*, cooperativity in ligand binding, depend on how multiple external interactions add up in shifting the global energetic equilibria of the macromolecule. Thus, allostery can be quantified only within the mechanism-independent framework of protein energetics.[6,7] However, with regard to detailed molecular events and mechanistic principles there is nothing general in allostery. A myriad of structurally diverse molecular arrangements or dynamically different processes can result in identical allosteric effects. Despite overenthusiastic claims of many papers in the field, there is no common rule encoded in the structure of a protein that may unlock a "universal" molecular mechanism of allosteric regulation.

The second concept emerged from breakthrough studies: cryospectroscopic experiments on ligand binding kinetics in heme proteins at diminishing levels of thermal energy.[8,9] The interpretation of that initial work now rests on a broader range of experimental evidence,[10–15] including single-molecule experiments.[16–19] This has led to radically new views on protein structure and dynamics and the relationship between protein architecture and biological activity. Protein structure consists of a large number of non-identical conformations, which are organized as a hierarchy of states and substates. The energy of this system is described by a rugged multidimensional landscape, where minima are separated by heterogeneously distributed barriers.[20] Random thermal jumps across substates result in conformational fluctuations of the protein. Such fluctuations, particularly those relevant to functional activity, depend on and cross-interact with those occurring on the water shell of the molecule.[21,22] Thus, protein motion is best described as waves propagating through an inhomogeneous matrix. Literally, as "protein quakes".[23,24]

The far reaching implications of these ideas for protein science and the recognition of biomolecules as a new target for the physics of complexity have been reviewed.[22,25] Here, we wish to emphasize the impact on the understanding of allosteric regulation. The new window opened regarding protein complexity breaks the protein representations that were dominant during a large part of the past century. One is the view of a protein as a crystal-like molecular structure existing in one or few unique conformations, isolated from the environment and essentially motionless. Another is the picture of functional activity as a smooth transition that generates active or inactive states of the macromolecule.

One additional important implication is that functional activity cannot be resolved by identifying active or inactive states of the protein by high-resolution X-ray imaging of the molecule. In fact, a cooperative interplay of external and internal fluctuations is what most probably determines function. Thus, conformational differences alone cannot predict or explain the presence of an allosteric effect. What matters is how the hierarchy of conformational states and substates of the protein reacts to different perturbations. Such motions may or may not result in visible changes of the collective molecular behaviour that show as biological function. Thus, in protein function dynamics is more important than structure and information on the latter, however rich, is of limited help in predicting the former. Indeed, system complexity implies that even small differences in the atomic structure of homologous proteins can produce detectable differences of their average behaviour; conversely, relatively large molecular divergences may have invisible effects.

All of what is said above can be condensed into two simple messages: (a) There is no general molecular mechanism that can explain allostery, except for the common tendency of all proteins to restore their highly degenerate ground state following perturbations. (b) Dynamics, far more than structure, may underlie function and allosteric behaviour.[26] Thus, hunting allosteric conformations in the resolved molecular structures of a protein may lead nowhere.

The above ideas have had a troubled acceptance among biologists, especially in the mechanism-hungry circles of biochemists and pharmacologists. For many years, allostery and allosteric transitions have been considered as distinctive properties of oligomeric proteins, given the pioneering experimental work[27–29] and the elegant models[30,31] that have mostly contributed to the spread of the concept. More recently, however, the so called "new view" of allostery[32,33] and the growth of experiments on protein dynamics are contributing to popularize a more general idea.[34–38] Important notions are gaining increasing acceptance. For example, allostery occurs in both monomeric and multimeric proteins; allosteric effects may be generated without visible conformational changes;[37,39] protein motion has a key role in the understanding of allosteric effects.[26,40,41] However, the ambition that drives the work of many investigators still is identifying a "general structure" or a "general mechanism" that may explain allosteric behaviour in all proteins.

It may be argued that protein complexity only appears when substates of the molecule are trapped into discrete wells of the energy landscape, *e.g.*, by sub-freezing temperatures or ultrafast spectroscopy. Likewise, only single-molecule studies may reveal the aperiodic and disordered nature of protein dynamics. However, in the physiological environment of cells the collective behaviour of the protein is still largely consistent with a much simpler view. Thus, the notion of protein complexity in describing allostery may bring needless complications more than insight. This argument explains why protein activity and allosteric behaviour can effectively be modelled using elementary principles, *e.g.*, mass action law, as discussed in the next section. However, the issue of complexity arises when we try to explain or correlate the results of those models with the available information about structure and dynamics.

6.2 Dualism in the Definitions of Allostery

As mentioned, relatively simple models based on thermodynamic equilibrium and kinetics have been successfully used to represent the allosteric behaviour of proteins. Such models influence design and interpretation of experiments on allosteric regulation, but also the biochemical definition of allostery. As reviewed in the past,[42] there is a dualism in the current definitions of the allosteric effect. Two apparently conflicting views are used.

One is the definition of allostery as a free energy coupling between the bindings of ligands at two non overlapping sites of the protein.[43] The factor that describes this coupling, which we call α here, represents a reciprocal cooperativity between binding events: an identical effect that the two ligands exert on each other when they are bound to their respective sites.[44] If the free energy underlying $\log\alpha$ is ≈ 0, there is no visible allostery in the protein. But in all other cases there will be an allosteric effect manifested as negative ($\log\alpha < 0$) or positive cooperativity ($\log\alpha > 0$), which unfolds as a decrease or increase in the binding constant of each ligand upon addition of the other.[45]

The second definition is based on the idea of protein conformational changes driven by a ligand-binding process. If a protein exists in two or more conformational states at equilibrium, the binding of ligands with different affinities across states may alter that equilibrium. Depending on the number of postulated states and the way in which the pre-existing equilibria are shifted upon ligand binding, the system may or may not show visible allosteric effects. In this view, the terms "allosteric states" and "allosteric transition" are often used to emphasize the role of conformational changes in generating allostery.

It is important to understand that this model leads to exactly the same cooperative behaviour as the energetic coupling just discussed. Consider two ligands that bind to separate protein spots. If in doing so both ligands shift the equilibrium towards the same state or collection of states, a positive cooperativity will be observed between the two binding events: one ligand will enhance the binding of the other. Or, if the ligands induce different protein states, the binding of one ligand will decrease the binding of the second, which will show as negative cooperativity.

The important question is whether these two definitions of allosteric effect reflect different theories and provide alternative concepts for data interpretation. The answer is that they do not. Both views are perfectly equivalent descriptions of allostery, since both are rooted in exactly the same principles of linkage thermodynamics.[4] In fact, they represent two possible different ways in which the general concept of allosteric binding potential[5] may be dressed in the language of mass action law-based biochemical reactions.

However, since models based on such representations are used to benchmark mechanisms and their experimental consistency, the difference can have significant consequences; particularly for our thinking about the structure–activity relationships of allosteric effects.

The first view, *i.e.*, cooperativity between binding sites, is phenomenological and has all the benefits that this attribute implies. It defines allosteric effect as a purely energetic quantity[45] without any assumption about underlying mechanisms or structural organizations. Thus, free energy coupling exactly predicts the functional expression of an allosteric effect and allows its quantification in experiments, regardless of the molecular nature of protein and ligands under investigation. In fact, the concept of free energy coupling was adopted early in receptor research, as the ternary complex model (TCM) of receptor–G protein interaction,[46] and also is widely used in the study of allosteric regulation of enzymes[42,47,48] or in the analysis of the effect of mutagenesis on proteins.[49–52]

The second view, *i.e.*, ligand-induced conformational change, implies a more mechanistic interpretation. The allosteric effect and the resulting cooperativity are attributed to the existence of protein conformations in equilibrium. Since neither the equilibrium constants nor the ligand binding affinities of such allosteric states are accessible to experimental measurement, this approach says literally nothing about the magnitude of allosteric effects that can be observed in experiments. Yet, this definition of allostery is far more popular than the first.

One reason is the commonly adopted assumption that a change of functional activity in a protein implies a change of conformation. Thus, if we attach "functional" and "non functional" attributes to each allosteric conformation, it is evident that the idea of allosteric states can lead to a powerful and easily understandable representation of how ligands can control protein function. A ligand that preferentially binds to the functional conformation can activate the protein; another with more affinity for the non-functional conformation can turn it off. The second reason that makes more popular the idea of allosteric state is its apparently closer connection with structural data. If we know that a ligand can activate the protein by stabilizing the active allosteric conformation, then a comparison of the ligand-bound and ligand-free structures of the protein should in principle yield useful information about the mechanism underlying the allosteric effect exerted by the ligand.

However, it is important to understand that in allosteric theory we are really discussing models of states not the physical states of the protein. Thus, it is neither true that a defined protein function can be identified with a specific allosteric conformation nor can a specific X-ray image of the protein ever correspond exactly to a model conformation. In addition, the comparison of ligand-bound *vs.* unbound protein structures tells virtually nothing about the allosteric effect of the ligand. In fact, allostery has little to do with the structural changes that are evoked by a single ligand binding to the protein.

It has a lot to do, instead, with the mutual interactions taking place when more than one ligand combines with different sites of the protein.

This confusion between model and mechanisms, which is often encouraged by the idea of allosteric states, is responsible for many misconceptions that hamper progress in the field. For example, a very common belief is that under special conditions it may be possible to determine from ligand binding or functional responses the binding constant of a ligand for the active state of a protein. This is hopeless, since allosteric theory shows that any infinitesimal change in ligand binding corresponds to an identical change in distribution of the allosteric states of the protein. Ligand association and change across allosteric states are inextricably linked events, which cannot be separated conceptually or experimentally.

However, the above and additional distortions about the possibility of isolating individual properties of allosteric states are widespread, particularly in the recent pharmacological literature on bias signalling. Collectively this leads to a vision of the protein as an essentially predictable mechanical object: something like a Rubik's cube, which different ligands can manipulate in order to fix stable "conformations" with different functions. Such a view may seem a useful way to simplify difficult concepts. Really it is instead a mystification, as it ignores that disorder and stochastic behaviour are what may drive the molecular mechanisms behind allostery and function in proteins.

It may be argued that the idea of allosteric state transition is necessary if we deal with systems in which a ligand appears to change the intrinsic activity of a protein without any obvious link with any additional binding process. However, as we have pointed out previously,[53] any protein function can be related to existing perturbations occurring between binding events at distant sites of the molecule. For example, the opening of an ion channel requires volume expansion and hydration of the conducting pore (*i.e.*, translocation of water from the bulk solution), which can be seen as water binding in osmotic equilibrium. Likewise, the intramolecular charge transfer or the formation or breaking of hydrogen bonds underlying enzyme catalysis can be regarded as specific binding events between catalytically critical residues and their intramolecular surroundings. An elegant demonstration that allosteric regulation can always be described as energetic coupling between distant binding sites was given for haemoglobin,[54] which represents the archetypal model of allostery described as conformational change.

To resolve the dualism in the definition of allostery we have proposed in the past,[55,56] and further do so here, an integrative approach. It is based on the notion that the most important difference between the two definitions lies in their diverse connection with functional macroscopic observations. Thus, we can use a typical black-box approach to relate the purely phenomenological view of free energy coupling to the more mechanistic view of allosteric state transition. To exemplify this relationship, we use the simplest and best known models that are derived from the two concepts. Both models describe the binding of two ligands (X, Y) to different sites of a receptor protein R (Figure 6.1).

Outside of the "black box" is the ternary complex model (TCM). This is represented by the thermodynamic cycle of the outer reaction schemes (Figure 6.1A–C), which involves two affinity constants (K and M) and the coupling constant α, all of which are experimentally measurable. K and M are directly obtained from the individual binding of each ligand to the receptor; α is acquired as a ratio from the shift in binding constant that each ligand shows when the second ligand is absent or present at saturating concentration. Thus, the parameters of this model are all macroscopically observable quantities, where the free energy coupling factor α quantifies the overall allosteric effect in the system.

Inside the "black box" is the two-state allosteric model, represented by the inner cubic reaction scheme of Figure 6.1A. It describes exactly the same reaction, but with the additional explicit assumption that R can exist in equilibrium between two allosteric conformations (r_0 and r_1). None of the parameters describing this system, *i.e.*, the equilibrium constant of receptor states (j), the ligand binding constants (k's and m's) and the allosteric effects of ligands on those constants (β and γ), are accessible to experimental measurement. Thus, these constants represent macroscopically non-observable quantities. However, they are responsible for the changes in macroscopic equilibrium constants that we measure in the TCM. In fact, the receptor forms that we observe in the external reaction are simply the sums of the conformational states at equilibrium that we assume to be present in the allosteric model of the internal scheme (shaded arrows, Figure 6.1A).

It is evident that this simple conceptual framework resolves the existing duplicity in the definition of allostery, as makes it clear that the two definitions reflect different intents. One is meant to quantify the visible energetic consequence of the allosteric effect, the other to explain how that energy may be generated by an equilibrium representation of the internal energetics of the protein. The first tells what we can see, the second what we shall hardly ever see.

An additional important notion inherent in this approach is that a model based on the concept of free energy coupling, such as the TCM, is always the most general macroscopic representation of a vast complexity of conformational allosteric transitions. In fact, Figure 6.1B shows that the essential correspondence between inner and outer parameters of the models remains true even if we expand the allosteric model inside the black box to an indefinitely large number of possible allosteric states. This graph underscores a conclusion also reachable by intuitive reasoning: implicit in the TCM there is a degenerate allosteric model, in which we imagine the protein R interconverting through a nearly continuous range of potential conformations at any change of ligands occupancy. Under this "analogical" view of conformational changes, keeping track of ligand effects on the individual equilibria between states makes obviously no sense. Thus, the only parameter that can meaningfully describe the allostery of the system is the macroscopic cooperative effect embodied in α.

	A Two State	B Multi State	C Probabilistic
Ternary complex	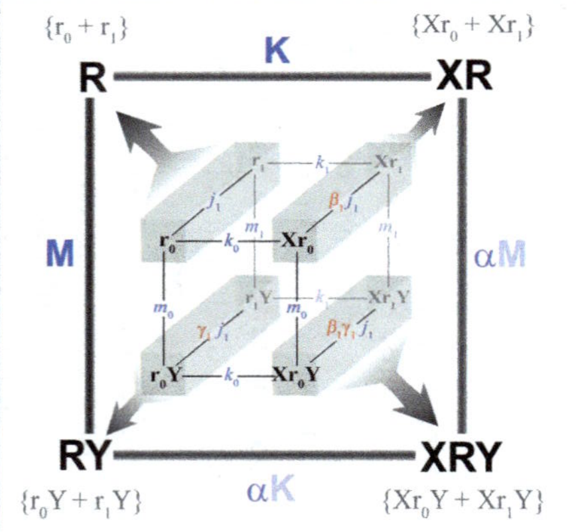	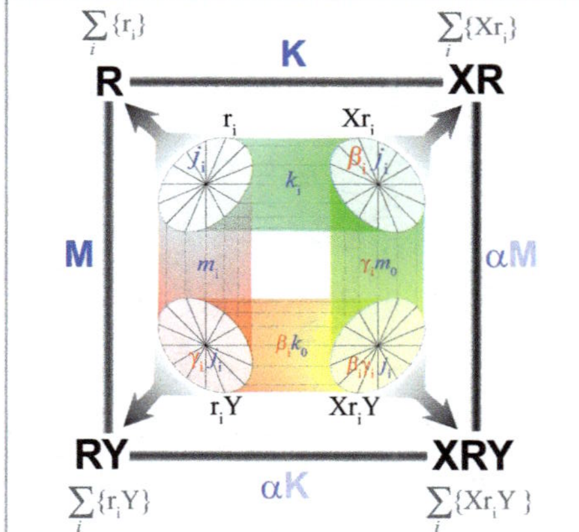	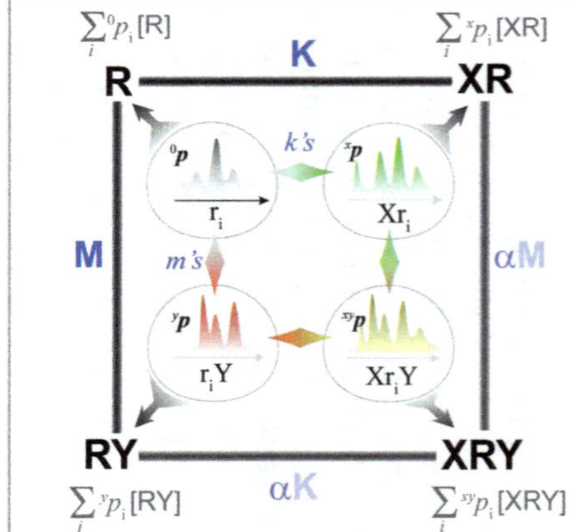
Macroscopic parameters	$\mathbf{K} = k_0 \frac{1+\beta_1 j_1}{1+j_1}$ $\mathbf{M} = m_0 \frac{1+\gamma_1 j_1}{1+j_1}$ $\alpha = \frac{(1+j_1)(1+\beta_1\gamma_1 j_1)}{(1+\beta_1 j_1)(1+\gamma_1 j_1)}$	$\mathbf{K} = k_0 \frac{1+\sum_i \beta_i j_i}{1+\sum_i j_i}$ $\mathbf{M} = m_0 \frac{1+\sum_i \gamma_i j_i}{1+\sum_i j_i}$ $\alpha = \frac{\left(1+\sum_i j_i\right)\left(1+\sum_i \beta_i\gamma_i j_i\right)}{\left(1+\sum_i \beta_i j_i\right)\left(1+\sum_i \gamma_i j_i\right)}$	$\mathbf{K} = \sum {}^0p_i k_i$ $\mathbf{M} = \sum {}^0p_i m_i$ $\alpha = \sum \frac{{}^xp_i\, {}^yp_i}{{}^0p_i}$

Figure 6.1 Allosteric coupling between two ligands X and Y that can bind to distant sites on a receptor protein R. The relationship between the ternary complex reaction cycle (TCM) and possible microscopic models at different level of complexity that explicitly involve receptors states, are shown. In each case the TCM cycle is drawn as an outer reaction box with three parameters K, M and α, which correspond to the macroscopic affinities of X or Y for R and the allosteric coupling factor between the two binding events, respectively. Different microscopic models, as labelled at the top of each case, are shown inside the TCM cycle In both outside and inside schemes, connecting lines indicate reversible reactions in equilibrium. The lower panels show how the three macroscopic parameters of the TCM cycle are uniquely determined by the parameters of the corresponding models shown in the inner reaction boxes. In each case, the relationships between the four macroscopically observable ligation-states of the TCM (*i.e.*, R, XR, RY and XRY) and the model states of the inner reactions are symbolized with gray arrows and explicitly indicated as sums of corresponding species (gray text next to the four corners of the TCM). *Inner-box reaction schemes*: (A) the receptor can be in two allosteric states r_0 and r_1 in equilibrium, with an equilibrium constant j_1. Binding of X or Y modifies this equilibrium with a factor of β_1 or γ_1, respectively. Affinities of X or Y for the allosteric states are given as k_i or m_i, respectively. Note that k_0 and k_1 or m_0 and m_1 are not independent. In fact, $k_1 = \beta k_0$ and $m_1 = \gamma m_0$. (B) the receptor can be in multiple allosteric states $r_0, r_1, \ldots, r_n$ in equilibrium, with equilibrium constants j_i defined with respect to r_0 (*i.e.*, $j_i = [r_i]/[r_0]$). All the other parameters (*i.e.*, β_i, γ_i, k_i and m_i; $i = 1,2,\ldots n$) have exactly the same meaning as in panel A. Internal equilibria between r_0 and r_i species are symbolized with radial lines in ellipses (r_0 being in the centre). Four ellipses correspond to four ligation states, and the internal equilibrium constants at each ellipse are given inside the ellipses. Each state in a given ellipse is in ligand binding equilibrium with the corresponding state in every other ellipse as depicted. Note that k_i and m_i are not independent from k_0 and m_0 and they are uniquely determined as $k_0\beta_i$ and $m_0\gamma_i$, respectively. (C) a probabilistic representation of the case described in panel B. ${}^0p_i$, ${}^xp_i$ and ${}^yp_i$ are the probabilities of encountering the state r_i in equilibrium when the receptor is unligated, X-bound or Y-bound, respectively. In this case, the states can be assumed to be any physically possible equilibrium states of the protein. See text for more details. The mathematical derivation of the probabilistic model was reported previously.[53,56,57]

The seamless transformation of states implied in the TCM led us to propose an even more general model of conformational changes. This view, which we call the probabilistic model of allosteric regulation,[53,56,57] postulates that the energetic coupling α results from probability distributions of the protein conformational states. Both ligands on binding may perturb the distributions, thus giving raise to a visible allosteric effect revealed as $\alpha \neq 1$. As shown in the simplified cartoon of Figure 6.1C, all four macroscopic receptor species of the TCM can be related to distinct probability distributions of all possible receptor states, under the assumption that all the states are at equilibrium in the ensemble. The generality of this model lies in the fact that no mechanistic assumption other than the idea that ligands can alter the distribution of receptor states is needed to account for the macroscopic cooperativity.

The theoretical framework illustrated in Figure 6.1 can predict the macroscopically observable cooperative behaviour of the protein from changes occurring in non-accessible parameters of models that describe intramolecular protein contributions. In fact, both the two-state or multi-state allosteric models and the probabilistic model are connected by simple mathematical rules to the macroscopic parameters of the TCM reaction scheme (see the equations underlying each reaction scheme in Figure 6.1). In the last section of this chapter we show some examples of how this logic can be adopted in understanding recent structural information regarding G protein coupled receptors.

6.3 Structural Changes and Receptor Allostery

To further discuss the relationship between allostery and structure we focus on the family of G protein coupled receptors (GPCRs).[58–60] This abundant group of monomeric transmembrane proteins[61,62] has a biological peculiarity. "Function" – defined as a measurable intrinsic property of the protein– is not in the molecule itself but elsewhere. The biological activity of these receptors, *i.e.*, triggering several signalling pathways, depends on external proteins, starting from the transduction proteins, like G proteins, arrestins and possibly others, to which GPCRs couple and transfer perturbations in response to the binding of extracellular ligands.[63–65] Thus, the only function we may consider resident in this class of molecules is the ability to generate an allosteric effect, when both a ligand and the transduction partner are bound to the receptor.[60,66]

The size of that allosteric effect is very important in physiology and therapeutics. It explains why a variety of hormones, neurotransmitters and similar physiological regulators can control cellular activity,[59,62,63,65] but also determines whether drugs designed to mimic the activity of those regulators may act as agonists or antagonists.[67] In fact, the allosteric effect between ligands and transduction proteins in GPCRs closely matches the idea of "ligand efficacy", which expresses the power of a drug to promote the signalling activity of a receptor. The exact formulation of this concept has historically obsessed, and still does, the mind of many pharmacologists.[67–70]

With regard to models derived from allosteric theory, it is evident that the TCM,[46] which we have discussed in the previous section (Figure 6.1), represents the most frugal and yet effective way in which we can describe the allosteric effect at the heart of GPCR function. In fact, if we assign X to a GPCR ligand and Y to the transduction protein in the closed thermodynamic box of Figure 6.1, then the coupling constant α and the cooperative effect that it quantifies clearly tells whether a ligand can shift the equilibrium toward or away from ternary complex stabilization. Therefore, $\alpha > 1$ (positive cooperativity) indicates agonism and $\alpha \leq 1$ (no cooperativity or negative cooperativity) indicates antagonism, by definition.[67,69]

The frugality of the TCM turns into oversimplification if we think that this model may represent all the molecular steps that initiate signal transduction in the living cell. The G protein itself is a heterotrimeric assembly of proteins, the dissociation of which is regulated by a guanine nucleotide. In addition, at the concentrations of guanine nucleotides existing in the cell, the XRY complex is only transient, which makes the terms equilibrium or steady state meaningless. In fact, from a mechanistic point of view, dissociation of the G protein subunits (which follows XRY formation) is a necessary step for activating effectors, such as ion channels or enzymes, that propagate intracellular signalling.[63] This is the reason why some general reviews on allostery[32] broadcast the rather confusing notion that it is negative cooperativity (*i.e.*, ligand-induced dissociation of G protein subunits) that describes the key allosteric effect of agonists for GPCR activation.

However, the TCM is not meant to simulate the signalling process. It is an equilibrium representation useful for learning and measuring something about the energetics of the driving forces that make a GPCR capable of initiating the chain of signal transduction. Thus, the free energy behind α indicates changes occurring inside the receptor, given a specified ligand and a specified transducer. The structure and dynamic mechanisms underlying this allosteric coupling is the basis of how ligands control intracellular signals. However, as we said previously the definition of α does not imply any mechanism. This argument is true even at the thermodynamic level, because as a thermodynamic quantity α cannot be dissected in a unique way into the enthalpic (H) and entropic (S) contributions (*i.e.*, $\Delta H - T\Delta S$) that underlie free energy. In fact, a specific α value can be in principle obtained by infinitely many different combinations of conformational and entropic changes. This means that many different types of conformational changes with appropriate entropic compensations may result in the same level of allosteric coupling in each specific case.

Nonetheless, it is clear that visualizing and understanding the structural mechanisms behind the allosteric effect in GPCRs is a challenging problem in protein chemistry, biochemistry and pharmacology, and the solution is expected to have significant impact on drug design.[71,72] One main structural question concerning allosteric coupling is what may determine the necessary symmetry of the energetic effect. As schematized in the TCM reaction box of Figure 6.1, the effect of X binding on the binding affinity of R $\leftrightarrows$ Y (*i.e.*, α) is exactly the same as the effect imparted by Y binding on the affinity of

R $\leftrightarrows$ X (*i.e.*, α, again). Such a symmetry exists, although the chemical nature of X and Y, and thus the configurations of the binding sites that are connected through the reciprocal effect, are very different.

Therefore, any structural mechanism useful to explain allosteric coupling should be sought in the process that determines the *interplay* between the individual effects exerted by the two ligands from non overlapping sites of the protein, not in any individual ligand effect by itself. In other words, we must understand both: (1) the structural–dynamical effect that each ligand separately exerts on the protein and (2) the structural–dynamical basis of the "interaction" between those effects when the two ligands are simultaneously bound. In summary, the structural comparisons of interest in relating the underlying thermodynamics of each side of the TCM box to the opposite side, involve at least three bound receptor forms, *i.e.*, XR, RY and XRY (Figure 6.1).

During the past 15 years a wealth of ligand-bound GPCR crystal structures have become available,[73–75] Technical hurdles have so far prevented the obtaining of a GPCR structure free of ligands (except for opsin[76]). However, several of the resolved X-ray images allow comparison of the agonist-bound and antagonist-bound forms of the same receptor,[76–90] from which we can learn something about the structural differences caused by agonist-induced activation. Most importantly, the breakthrough resolutions of a receptor in complex with the trimeric G protein[83,84] and, more recently, of a GPCR associated with visual arrestin,[90] have shed light on the conformational changes of the receptor in ternary complex form.

Thus, while data on RY are still missing, information about XR, XRY and the extent to which they might differ from R, are currently available. The emerging picture is still fuzzy, but clearly indicates that bound agonists are not generally able to stabilize the receptor in the "active" or "target" configuration. This configuration can only be shown in a receptor that is concurrently bound to agonist and G protein (or a mimicking antibody). The photoreceptor rhodopsin represents one notable exception to this trend.

Such results came as a disappointment. Particularly to those who may have expected, according to the view of allostery as protein conversion to a functional conformation, that the dissimilarities between agonist- and antagonist-bound receptors would have easily revealed the secret of agonist efficacy. The data, however, provide an interesting glimpse into the potential variety of structure–function relationships that may underlie allosteric coupling in GPCRs.

To illustrate concisely the results mentioned above, we drew 3D ribbon diagrams from publically available atomic coordinates corresponding to several structures of four different GPCRs (Figure 6.2 upper panel).[76–87] The "inactive" (top) and "active" (bottom) forms are compared using false colours to indicate the largest structural differences. In the lower part of Figure 6.2 we show the same differences in a more quantitative way, by plotting the absolute distances of the α-carbon atom of each residue in "active" and "inactive" forms of the receptor as a function of residue number. The

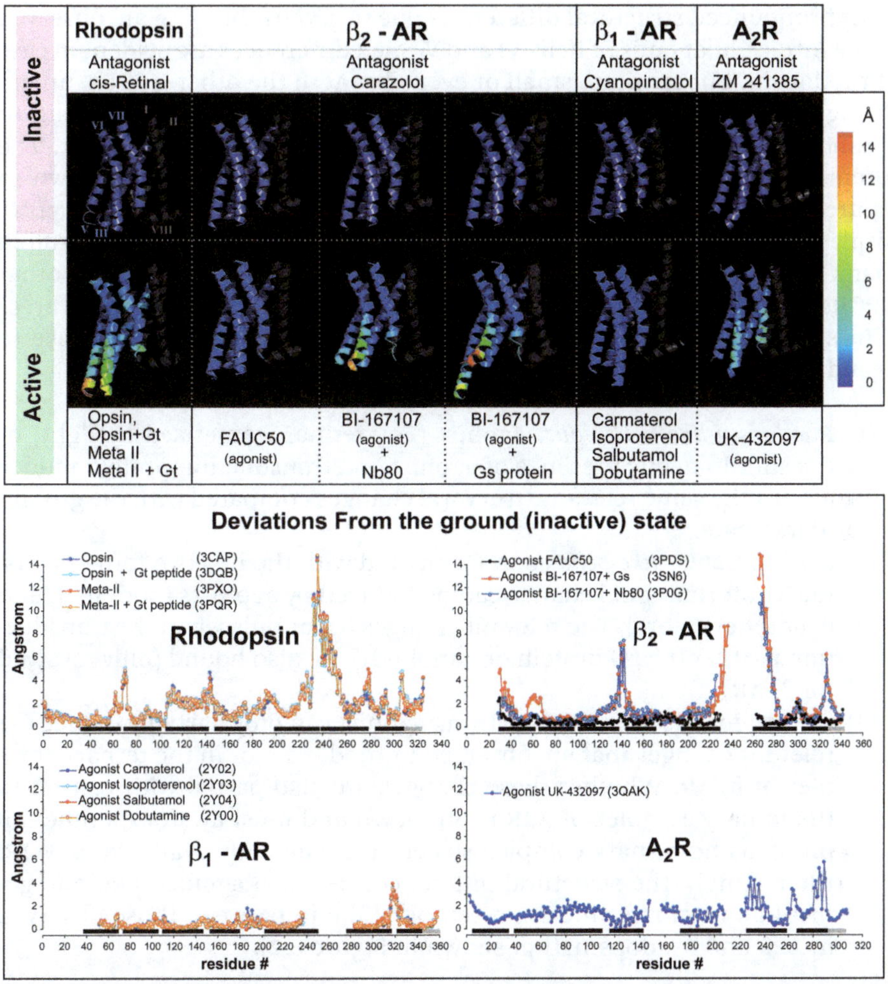

Figure 6.2 Structures of different GPCRs in their different ligation states. Upper panel: X-ray crystallographic structures of indicated GPCRs in their indicated ligation states are shown as ribbon representations. Antagonist-bound "inactive" states and the corresponding agonist-bound "active" states are shown in the first and second rows, respectively, as indicated in the figure. Transmembrane helices (except TM4, which is not visible in this orientation) are labelled in the first picture (upper left corner). Helices 3, 5, 6 and 7 in the "active" structures are coloured with a rainbow scale according to the deviations of the α-carbons from the corresponding inactive structures (given at the right as Ångstroms). The pictures were rendered from the atomic coordinates deposited in the Research Collaboratory for Structural Bioinformatic (RCSB) protein data bank by using PyMol (DeLano Scientific LLC, South San Francisco, California, USA). See the lower panel for the PDB ID codes of the structural data. Lower panel: Deviations from the corresponding inactive structures are shown as a function of residue numbers for the indicated cases (as in the upper panel). Transmembrane helices are indicated by

most pronounced structural differences due to agonist binding are observed in the intracellular ends of helix VI and V. Such differences are evident in case of rhodopsin, but are quite small or even absent in the other GPCRs, unless the agonist-bound receptor (*e.g.*, β_2AR) is associated with the G protein or a nanobody (Nb80) that occupies the same G protein binding pocket. The recently published structure[90] of arrestin-bound rhodopsin (not shown in Figure 6.2) reveals minimal deviations from the averaged "active" structure displayed in Figure 6.2. Although arrestin is a different transduction protein than G protein, this is a further clue that rhodopsin deviates from the other GPCRs with regard to the lack of difference between XR and XRY structures.

In summary, three different types of structural changes in response to ligand binding can be distinguished from the data listed in Figure 6.2:

(1) *Rhodopsin-like behaviour*: Agonist (*i.e.*, retinal isomerised by light) or the simultaneous presence of agonist and transduction protein induce almost the same relevant structural changes compared with the ground (dark) state.
(2) *β-adrenoceptor-like behaviour*: Compared with the inactive form, no visible conformational change can be induced by agonists (both in β_1 and β_2 adrenoceptors). The relevant changes occur only when the transduction partner (*i.e.*, G protein or nanobody) are also bound (only resolved for β_2-AR).
(3) *A2 receptor-like behaviour*: A strong agonist can only partially induce the relevant changes that are observed in rhodopsin or in the ternary complex of β_2-AR. Whether these changes are also partial with respect to the ternary complex of A2R is unknown and must await experimental proof, as no ternary complex structure is currently available for A2R. Interestingly, the structural differences between agonist- and antagonist-bound muscarinic M2 receptor[88,89] lie in between those observed in A2R and rhodopsin, (not shown in Figure 6.2).

The above results are certainly not compatible with the simple idea of a two-state allosteric model (*e.g.*, the inner cubic reaction scheme in Figure 6.1A), whereby one "active" conformation is stabilized by either agonist or transducer binding, and the inactive conformation may be stabilized by inverse agonists. Such a conclusion is further supported by new information

---

black bars next to the residue number axes (1–7 from left to right). Helix 8 is indicated by a gray bar. The PDB ID codes for the active forms of the receptors (*i.e.*, agonist-bound or agonist + transducer-bound forms of the receptors) are given in parenthesis. The used PDB ID codes for the inactive sates of rhodopsin, β2-AR, β1-AR and A2R are: 1U19, 2RH1, 2VT4 and 3EML, respectively. Note that agonist binding to β1 or β2 adrenoceptor causes practically no structural changes in the receptor, whereas the active-like signature arises only when the transducer G protein (or its nanobody mimic Nb80) is also bound in β2-AR. For more details, see the references that are associated with the structures shown.[76–81,83–87] about dynamics. Particularly interesting are results from recent studies based on ^{19}F-nuclear magnetic resonance (NMR) shifts analysis and double electron–electron resonance (DEER) experiments in β_2AR.[91] This work has elucidated the local dynamics of the cytoplasmic end of helix VI and its relative distances from helix IV in the absence and presence of ligands and G protein mimicking nanobody. The results not only confirm that the agonist alone cannot induce the target state that predominates in the concomitant presence of nanobody. They also demonstrate that at least four states, two active and two inactive configurations, can be resolved from the analysis of the dynamics of the receptor.[91]

6.4 Allosteric Coupling as the Result of Probability Distributions of Receptor States

In short, the current structural data for GPCRs describe two opposite patterns. One in which R and XR have very similar structures but XRY sharply differs from XR (*i.e.*, β_2AR), another in which R and XR show clearly dissimilar structures but XRY and XR have almost identical ones (*i.e.*, rhodopsin). Within such bounds, there may be a full range of intermediate patterns to which other GPCRs belong. In addition, the dynamics of the cytosolic helical edges causing the largest structural differences unveil far more conformational states than X-ray structure by itself would suggest. Here we analyze how this disconnection between physical, structural and functional states may be deciphered using the unified framework that we sketched in previous sections. The question is what model of conformational changes can best rationalize the variegated display of structural differences that are observed in the macroscopic receptor species summarized above. As already mentioned, the simple two-state allosteric model (*e.g.*, Figure 6.1A) is out of the question, except possibly in the case of rhodopsin. Yet even so, it would not make any sense at all to conclude that rhodopsin activation can be described with only two conformations, whereas the other GPCRs require a larger number of conformational changes. Especially if we consider that all this family of receptor proteins share an incredible similarity in the structural properties of the transmembrane bundle.[74,75,92]

However, even the multi-state allosteric model (*e.g.*, Figure 6.1B) implies conceptual problems that are difficult to overcome. As long as the number of states is two, allosteric conformations and the postulated active and inactive conformations perfectly match. But, as the number of conformations exceeds two, the allosteric transition model reveals all the indeterminacy that is inherent in the idea of distributing function as a fractional quantity across different states. For example, let us apply a multi-state allosteric model to the four receptor conformations recently identified in the β_2AR. NMR analysis indicates that the active state S3, which is only partially induced upon agonist binding, still differs from the active state S4 that is predominant when both the agonist and the G-protein-mimicking nanobody are bound.[91] Does this mean that we should distinguish active and sub-active states among the functional conformations? Likewise, the existence of two different inactive conformations implies that the stability of non-functional states may either be reduced or left unchanged upon binding of different agonists. How to distinguish then non-responsive and agonist-quenchable states among the inactive conformations that populate the unbound receptor? Evidently the root of the problem is that physical and allosteric states are different things. This is so even if both are defined through similar ensemble averaging principles of statistical thermodynamics. In other words, although we may use the same postulates to derive allosteric conformations from function (*e.g.*, ligand binding) or to derive real conformational states from physical signals (*e.g.*, NMR shifts), still the two representations cannot coincide. The reason is that the physics of complexity implies that in a protein, structure, dynamics and function cannot be related according to simple deterministic relationships.

To overcome these limitations, we propose to abolish the partitions among discrete allosteric conformations and to consider all physical conformations of the receptor as the whole distribution of possible states. As we discussed in the introduction, the naturally folded form of a receptor in thermal equilibrium can assume a large number of conformational substates.[20] Hence, the internal equilibria of the receptor can be reasonably described as a probability distribution over its conformational space.[53,56,57] A perturbation such as ligand binding is expected to disturb this probability distribution. As exemplified in Figure 6.1C, different perturbations of probability distribution may result from the binding of ligands and from the binding of transduction proteins. Their joint effect explains the observed allosteric couplings between binding events. In fact, the macroscopic allosteric coupling (*i.e.*, α) can be defined in terms of the relevant probability distributions as:

$$\alpha = \sum_i \frac{{}^{y}p_i \, {}^{x}p_i}{{}^{0}p_i} \tag{6.1}$$

where ${}^{0}p$, ${}^{x}p$ and ${}^{y}p$ respectively represent the probability distributions of receptor states in unbound, ligand-bound or transducer-bound forms[56,57] and the summation runs over the entire conformational space. Note that the probability distribution for the dual bound receptor (${}^{xy}p$) is not independent and is uniquely determined by the other three distributions as:

$${}^{xy}p_i = \frac{1}{\alpha} \frac{{}^{x}p_i \, {}^{y}p_i}{{}^{0}p_i} \tag{6.2}$$

Therefore, three independent probability distributions representing the internal equilibrium of the unbound receptor and bound to each individual ligand determine unambiguously the value of α. Eqn (6.1) has interesting implications:

1. In order for the ligand and the transducer to generate a visible allosteric effect (*i.e.*, $\alpha \neq 1$), they both must disturb internal states of the receptor on binding (*i.e.*, ${}^{0}p \neq {}^{x}p$ AND ${}^{0}p \neq {}^{y}p$).

2. Many different ligand-induced changes can lead to the same observable α value. For example, given a specified transduction protein, lack of allosteric coupling (*i.e.*, $\alpha = 1$) can result from changing receptor state distributions in many different ways. Ligands can be "neutral" (*i.e.*, producing no cooperative effects on transducer binding) even if they modify receptor states to a varying and divergent extent (*e.g.*, see here ref. 93). This follows from eqn (6.1), which shows that for any fixed pair of ${}^0p$ and ${}^yp$ (*i.e.*, unbound and G-protein-bound receptor distributions), an infinite number of different ${}^xp$ distributions can yield the same solution, $\alpha = 1$. Similarly, equal levels of agonism ($\alpha > 1$) and inverse agonism ($\alpha < 1$) can be achieved through a vast array of diverse perturbations. This essentially means that ligand efficacy is a predictable quantity only as a macroscopically observable parameter (*i.e.*, α). It may not be predicted from specific microscopic parameters derived from the physics of ligand-induced receptor perturbations (*e.g.*, state lifetimes, height of energy barriers, *etc.*)
3. The magnitude of the cooperative effect that ultimately determines formation of the functional ternary complex XRY may be to a great extent dependent on conformational substates that are "unlikely" (that is, rarely populated) in the unbound (R) or ligand-bound (XR) receptor. Conversely substates that are most abundant in either of these receptor forms most probably determine the detectable structural differences between R and XR, even if their contribution to the size of α might be minimal. Consequently, the structural difference between agonist-bound receptor and antagonist-bound (or empty) receptor does not necessarily allow prediction of the extent of agonism and the allosteric coupling that the ligand may generate with a given transducer. This conclusion also derives from eqn (6.1). For example, consider a single very unlikely substate (say ${}^0p_i = 10^{-9}$) present in R. Suppose that this substate becomes more probable although still scarcely represented in both XR and RY (*i.e.*, ${}^xp_i = {}^yp_i = 10^{-3}$). Such a change will determine 1000 units of increase in α value, despite a probably negligible contribution to the difference in average structure between R and XR.

The above predictions are exemplified in Figure 6.3. We use arbitrary probability distributions on a conformational space consisting of 200 substates. The panels show alternative ways in which changes of receptor state distribution may occur in a receptor (R) bound to a ligand H (HR), the G protein G (RG) or both (HRG). The resulting effect on α is shown in each panel.

Figure 6.3A and B show the case in which the same α value (~9000), indicating strong agonism, is achieved by very different perturbations of the probability distributions existing in empty receptor. Note that in panel A, either H or G increase the probabilities of rare substates that still remain underpopulated in the receptor form to which they are bound (see the bumps around position 170 of the states coordinates); yet those rare substates become the predominant ones in HRG. In panel B, both H and G increase the probability of the same collection of unlikely states, but the substates that are most likely

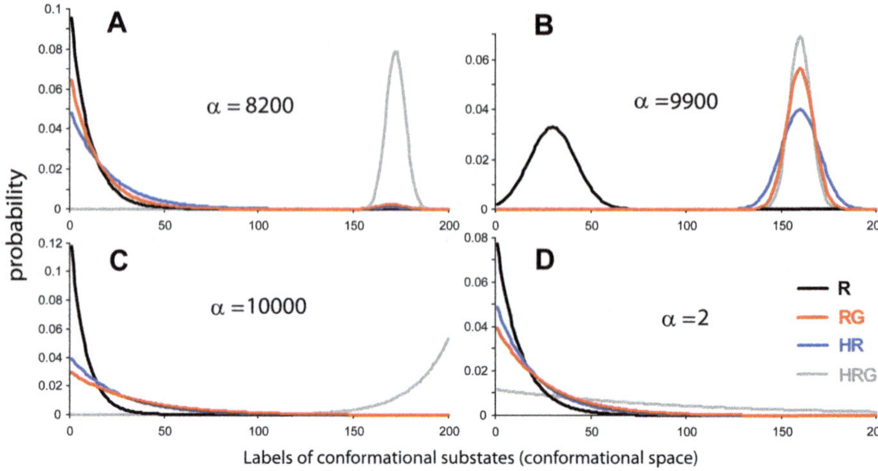

Figure 6.3 A schematic example of the probabilistic view of the allosteric coupling between ligand (H) and G protein (G) binding to the receptor protein (R). In thermal equilibrium, the receptor is assumed to possess a conformational space of size 200. In each panel, arbitrary, but different, probability distributions (represented as coloured solid lines) are attributed to the empty (R), H-bound (HR) or G-protein-bound (RG) forms of the receptor (panels A–D) as indicated in the picture. Distributions for the double-bound receptor species (HRG) are calculated according to eqn (6.2). The allosteric coupling factors (α's) generated by the probability distributions are calculated according to eqn (6.1) and indicated in the picture for each case. Note that the contributions of the states to the final value of α are proportional to their probabilities in the double-bound form of the receptor, as implied by eqn (6.2). See text for more details.

in R virtually disappear in RG or HR. Yet the effect on the distribution in HRG is similar to that seen in panel A and so is the effect on the macroscopic cooperativity value. If we suppose that crystals obtained from the various receptor forms might reflect an average structure of the most abundant substate conformations that populate the receptor in each form, then panel A and B delineate two radically different patterns. In fact, in A, HR would show little structural differences from R but significant differences from HRG, whereas in B, HR and HRG would be similar and both quite different from R.

In contrast, Figure 6.3C and D illustrate an opposite case. Relatively small differences in the ways ligands perturb the probability distribution of the empty receptor can lead to a major difference in the distribution of substates existing in the HRG form. Consequently, the macroscopic coupling α can either indicate a strong allosteric effect (C) or a nearly undetectable effect (D).

To illustrate in a plain way how the same level of macroscopic cooperativity may emerge from diverse substate distributions and average structures, we introduce an oversimplified paradigm. Imagine a hinged rod device as drawn in the top panel of Figure 6.4. Assume that this setup, although obviously

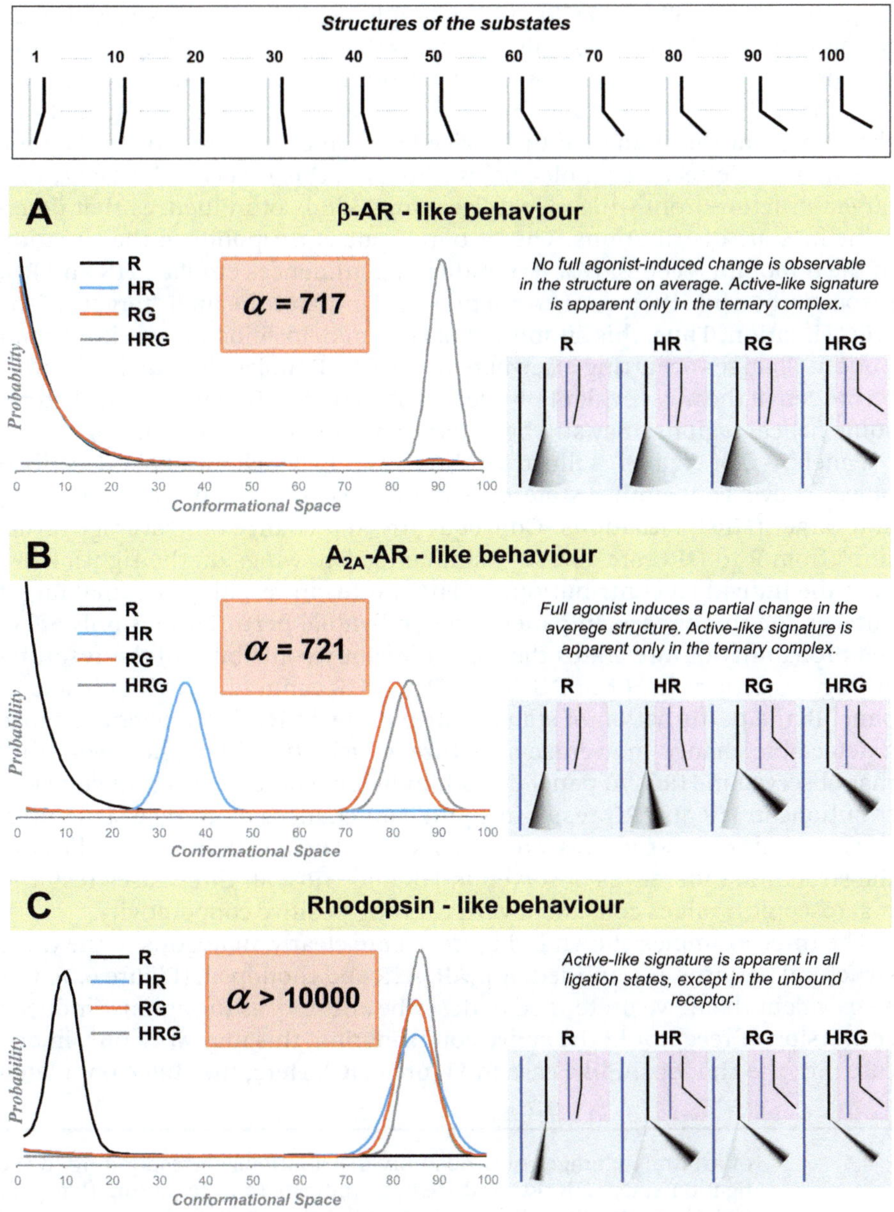

Figure 6.4 *Examples of agonism with different microscopic structural patterns.* A conformational space for the receptor (R) consisting of 100 substates is considered. To each substate in that space is also assigned a specific configuration of a hinge-like structure, representing a simplified analogy of the transmembrane helix 6 of GPCRs. The hinge angle (*i.e.*, the configurational coordinate) can change from state number 1 to 100 in equal steps as shown in the top panel (only one in every ten structures is shown for simplicity). This schematic model is then subjected to sets

inspired by helix VI of GPCRs, has a simple conformational space limited to the 2D plane. It includes only the states that are generated by changes of the hinge angle at regular steps labelled from 1 to 100 (schematically shown in Figure 6.4, top). Such states can be described by probability distributions that represent the chance of the device being in each state. To emulate the role of a whole receptor molecule, we imagine this object to be a part of a larger structure from which it receives three kinds of influences that determine its state distributions. The "ground" one corresponds to the situation of the unbound receptor (R). Two individual influences emulate HR and RG, respectively, while their concurrent presence imitates the final "target" of the HRG situation. Thus, this simplified paradigm helps illustrate an important notion. Changes occurring elsewhere in the GPCR molecule, which may thus be hidden in the most evident physical or structural differences identified for some helical regions, may still be determinant for allosteric coupling.

Panels A–C of Figure 6.4 illustrate three cases in which the above described object shows very similar state probability distributions under ground (R) and target (HRG) situations. Consequently, the changes in "average structure" from R to HRG are also very similar (see drawings on the right). However, the individual contributions of HR and RG in reaching the final target differ in the three cases. In panel A, the individual perturbations only affect the probability of rare states that insignificantly contribute to the resulting average structures in HR or RG. Thus, R and HR cannot be distinguished. In panel B, the perturbation of states distributions in HR is sufficient to cause a detectable change in average structure, which still differs, however, from that observed in HRG. In panel C, both individual perturbations of state distributions in HR and RG result in significant changes of average structures; thus very similar structures are observed in HR and HRG. Regardless of the structural differences observed in HR and HRG, all three cases result in macroscopic α values consistent with a strong positive cooperativity.

The three examples shown in Figure 6.4 are clearly analogous to the three structural patterns exemplified by β_2AR, A2R and rhodopsin (Figure 6.2). It is also evident that if we pretended to describe allostery as the agonist-induced conversion of receptor in the active conformation, this may work only in one case (*i.e.*, the rhodopsin-like case in Figure 6.4C). Here, the states populated

of arbitrary changes of probability distributions corresponding to the ligation states HR, RG and HGR, as indicated in the picture (left panels). Corresponding distributions of the structures at different ligation states are shown in the right panels along with the average structures (pink boxed insets). The distributions of structures on the right are drawn with a gray scale, where lowest and highest probabilities correspond to white and black coloured hinge lines (the individual lines are not visible but produce a smooth gray scale). Three types of behaviour are imitated as indicated and commented on in the picture (from A–C). For each case, allosteric coupling factor α between H and G implied by the corresponding probability distributions is calculated using eqn (6.1) and is shown in the picture.

on agonist or G protein binding result in average structures that are almost identical to that observed in HRG. Altogether these structures (*i.e.*, HR, RG and HRG) identify a unique "active" allosteric conformation, such that the simple two-state model becomes virtually equivalent to the probabilistic model in describing the system. In all other cases, however, the idea of unique agonist-induced active conformation leads nowhere and cannot explain how the agonist promotes receptor–G-protein coupling.

6.5 Conclusions

The touchstone concepts that currently dominate in modelling allosteric regulation, *i.e.*, "allosteric coupling between binding processes" and "interconversion between allosteric conformations", are two sides of the same medal. Although the latter concept may seem better suited for correlating protein function with structural information, this is simply an illusion. In fact, allosteric conformations are not physical states but idealized representations of an equilibrium blueprint of the protein which does not exist in real life. We have shown that the two concepts may be merged into a unified framework, if we recognize that the observable energetic coupling is related by a set of specific rules to microscopic changes postulated to occur inside the protein. Accordingly, function is the macroscopic outcome of a reorganization involving the whole repertoire of protein states, not the specific property of any individual state by itself. We also suggest that the classical view of allosteric regulation as transitions between protein conformations each carrying a specified quota of functional activity should be replaced by a vision of the protein as probability distributions of all its possible physical states. We believe that this probabilistic view is more useful in shaping our imagination regarding the molecular mechanisms underlying the mutual dependence between the separate binding events that give raise to allosteric coupling.

Acknowledgements

H. O. Onaran's research is supported by Turkish Scientific and Technical Research Council (TUBITAK) grant number 113S909. T. Costa acknowledges support from the Italian Ministry of Health funding program "*Ricerca Finalizzata*" grant no.: RF-2011-02351158.

References

1. Q. Shen, G. Wang, S. Li, X. Liu, S. Lu, Z. Chen, K. Song, J. Yan, L. Geng, Z. Huang, W. Huang, G. Chen and J. Zhang, *Nucleic Acids Res.*, 2016, **44**, D527–D535.
2. J. Wyman Jr, *Adv. Protein Chem.*, 1948, **4**, 407–531.
3. J. Wyman Jr, *Adv. Protein Chem.*, 1964, **19**, 223–286.
4. J. Wyman, *J. Mol. Biol.*, 1965, **11**, 631–644.

5. J. Wyman, *J. Am. Chem. Soc.*, 1967, **89**, 2202–2218.
6. G. Weber, *Protein interactions*, Chapman and Hall, 1992.
7. J. Wyman and S. J. Gill, *Binding and linkage: Functional chemistry of biological macromolecules*, University Science Books, U.S., United States, 1990.
8. R. H. Austin, K. W. Beeson, L. Eisenstein, H. Frauenfelder and I. C. Gunsalus, *Biochemistry*, 1975, **14**, 5355–5373.
9. D. Beece, L. Eisenstein, H. Frauenfelder, D. Good, M. C. Marden, L. Reinisch, A. H. Reynolds, L. B. Sorensen and K. T. Yue, *Biochemistry*, 1980, **19**, 5147–5157.
10. H. Frauenfelder, G. A. Petsko and D. Tsernoglou, *Nature*, 1979, **280**, 558–563.
11. H. Hartmann, F. Parak, W. Steigemann, G. A. Petsko, D. R. Ponzi and H. Frauenfelder, *Proc. Natl. Acad. Sci. U. S. A.*, 1982, **79**, 4967–4971.
12. I. E. Iben, D. Braunstein, W. Doster, H. Frauenfelder, M. K. Hong, J. B. Johnson, S. Luck, P. Ormos, A. Schulte, P. J. Steinbach, A. H. Xie and R. D. Young, *Phys. Rev. Lett.*, 1989, **62**, 1916–1919.
13. H. Keller and P. G. Debrunner, *Phys. Rev. Lett.*, 1980, **45**, 68–71.
14. J. R. Lewandowski, M. E. Halse, M. Blackledge and L. Emsley, *Science*, 2015, **348**, 578–581.
15. T. R. M. Barends, L. Foucar, A. Ardevol, K. Nass, A. Aquila, S. Botha, R. B. Doak, K. Falahati, E. Hartmann, M. Hilpert, M. Heinz, M. C. Hoffmann, J. Kofinger, J. E. Koglin, G. Kovacsova, M. Liang, D. Milathianaki, H. T. Lemke, J. Reinstein, C. M. Roome, R. L. Shoeman, G. J. Williams, I. Burghardt, G. Hummer, S. Boutet and I. Schlichting, *Science*, 2015, **350**, 445–450.
16. H. S. Chung and W. A. Eaton, *Nature*, 2013, **502**, 685–688.
17. A. N. Gupta, A. Vincent, K. Neupane, H. Yu, F. Wang and M. T. Woodside, *Nat. Phys.*, 2011, **7**, 631–634.
18. M. Tsytlonok, S. M. Ibrahim, P. J. E. Rowling, W. Xu, M. J. Ruedas-Rama, A. Orte, D. Klenerman and L. S. Itzhaki, *Structure*, 2015, **23**, 190–198.
19. H. Neuweiler, C. M. Johnson and A. R. Fersht, *Proc. Natl. Acad. Sci. U. S. A.*, 2009, **106**, 18569–18574.
20. H. Frauenfelder, S. G. Sligar and P. G. Wolynes, *Science (New York, NY)*, 1991, **254**, 1598–1603.
21. P. W. Fenimore, H. Frauenfelder, B. H. McMahon and R. D. Young, *Proc. Natl. Acad. Sci. U. S. A.*, 2004, **101**, 14408–14413.
22. H. Frauenfelder, G. Chen, J. Berendzen, P. W. Fenimore, H. Jansson, B. H. McMahon, I. R. Stroe, J. Swenson and R. D. Young, *Proc. Natl. Acad. Sci. U. S. A.*, 2009, **106**, 5129–5134.
23. A. Ansari, J. Berendzen, S. F. Bowne, H. Frauenfelder, I. E. Iben, T. B. Sauke, E. Shyamsunder and R. D. Young, *Proc. Natl. Acad. Sci. U. S. A.*, 1985, **82**, 5000–5004.
24. O. Miyashita, J. N. Onuchic and P. G. Wolynes, *Proc. Natl. Acad. Sci. U. S. A.*, 2003, **100**, 12570–12575.
25. H. Frauenfelder, *Nat. Struct. Biol.*, 1995, **2**, 821–823.
26. D. Kern and E. R. Zuiderweg, *Curr. Opin. Struct. Biol.*, 2003, **13**, 748–757.
27. J. P. Changeux, *J. Mol. Biol.*, 1962, **4**, 220–225.

28. J. P. Changeux, *Nat. Rev. Mol. Cell Biol.*, 2013, **14**, 819–829.
29. J. Monod, J. P. Changeux and F. Jacob, *J. Mol. Biol.*, 1963, **6**, 306–329.
30. D. E. Koshland Jr, G. Nemethy and D. Filmer, *Biochemistry*, 1966, **5**, 365–385.
31. J. Monod, J. Wyman and J. P. Changeux, *J. Mol. Biol.*, 1965, **12**, 88–118.
32. C. J. Tsai, A. Del Sol and R. Nussinov, *Mol. BioSyst.*, 2009, **5**, 207–216.
33. M. S. Formaneck, L. Ma and Q. Cui, *Proteins*, 2006, **63**, 846–867.
34. Q. Cui and M. Karplus, *Protein Sci.*, 2008, **17**, 1295–1307.
35. V. J. Hilser, E. B. Garcia-Moreno, T. G. Oas, G. Kapp and S. T. Whitten, *Chem. Rev.*, 2006, **106**, 1545–1558.
36. H. N. Motlagh, J. O. Wrabl, J. Li and V. J. Hilser, *Nature*, 2014, **508**, 331–339.
37. C. J. Tsai, A. del Sol and R. Nussinov, *J. Mol. Biol.*, 2008, **378**, 1–11.
38. C. J. Tsai and R. Nussinov, *PLoS Comput. Biol.*, 2014, **10**, e1003394.
39. A. Cooper and D. T. Dryden, *Eur. Biophys. J.*, 1984, **11**, 103–109.
40. K. Gunasekaran, B. Ma and R. Nussinov, *Proteins*, 2004, **57**, 433–443.
41. R. Nussinov and C. J. Tsai, *Curr. Opin. Struct. Biol.*, 2015, **30**, 17–24.
42. A. W. Fenton, *Trends Biochem. Sci.*, 2008, **33**, 420–425.
43. G. Weber, *Biochemistry*, 1972, **11**, 864–878.
44. D. A. Kolb and G. Weber, *Biochemistry*, 1975, **14**, 4471–4476.
45. G. Weber, *Adv. Protein Chem.*, 1975, **29**, 1–83.
46. A. De Lean, J. M. Stadel and R. J. Lefkowitz, *J. Biol. Chem.*, 1980, **255**, 7108–7117.
47. J. L. Johnson and G. D. Reinhart, *Biochemistry*, 1997, **36**, 12814–12822.
48. G. D. Reinhart, S. B. Hartleip and M. M. Symcox, *Proc. Natl. Acad. Sci. U. S. A.*, 1989, **86**, 4032–4036.
49. E. di Cera, *Adv. Protein Chem.*, 1998, **51**, 59–119.
50. A. Horovitz, *J. Mol. Biol.*, 1987, **196**, 733–735.
51. G. Schreiber and A. R. Fersht, *J. Mol. Biol.*, 1995, **248**, 478–486.
52. C. Ambrosio, P. Molinari, S. Cotecchia and T. Costa, *Mol. Pharmacol.*, 2000, **57**, 198–210.
53. H. O. Onaran and T. Costa, *Curr. Protein Pept. Sci.*, 2009, **10**, 110–115.
54. M. F. Colombo, D. C. Rau and V. A. Parsegian, *Science*, 1992, **256**, 655–659.
55. H. O. Onaran and T. Costa, *Ann. N. Y. Acad. Sci.*, 1997, **812**, 98–115.
56. H. O. Onaran, A. Scheer, S. Cotecchia and T. Costa, in *Handbook of Experimental Pharmacology*, ed. T. P. Kenakin, Springer, United States, 2000, vol. 148, pp. 217–259.
57. H. O. Onaran and T. Costa, in *Biomedical Applications of Computer Modelling*, ed. A. Christopoulos, CRC Press, Washngton D.C., U.S.A., 2000, ch. 109–134.
58. J. Bockaert and J. P. Pin, *EMBO J.*, 1999, **18**, 1723–1729.
59. K. L. Pierce, R. T. Premont and R. J. Lefkowitz, *Nat. Rev. Mol. Cell Biol.*, 2002, **3**, 639–650.
60. D. M. Rosenbaum, S. G. Rasmussen and B. K. Kobilka, *Nature*, 2009, **459**, 356–363.
61. T. K. Bjarnadottir, D. E. Gloriam, S. H. Hellstrand, H. Kristiansson, R. Fredriksson and H. B. Schioth, *Genomics*, 2006, **88**, 263–273.
62. K. Kristiansen, *Pharmacol. Ther.*, 2004, **103**, 21–80.

63. W. M. Oldham and H. E. Hamm, *Q. Rev. Biophys.*, 2006, **39**, 117–166.
64. A. K. Shukla, K. Xiao and R. J. Lefkowitz, *Trends Biochem. Sci.*, 2011, **36**, 457–469.
65. N. Wettschureck and S. Offermanns, *Physiol. Rev.*, 2005, **85**, 1159–1204.
66. B. K. Kobilka and X. Deupi, *Trends Pharmacol. Sci.*, 2007, **28**, 397–406.
67. T. Costa, Y. Ogino, P. J. Munson, H. O. Onaran and D. Rodbard, *Mol. Pharmacol.*, 1992, **41**, 549–560.
68. T. Kenakin, *A Pharmacology Primer: Techniques for More Effective and Strategic Drug Discovery*, Elsevier Science, 2014.
69. H. O. Onaran and T. Costa, *Nat. Chem. Biol.*, 2012, **8**, 674–677.
70. H. O. Onaran, S. Rajagopal and T. Costa, *Trends Pharmacol. Sci.*, 2014, **35**, 639–647.
71. B. K. Kobilka, *Trends Pharmacol. Sci.*, 2011, **32**, 213–218.
72. B. K. Shoichet and B. K. Kobilka, *Trends Pharmacol. Sci.*, 2012, **33**, 268–272.
73. V. Katritch, V. Cherezov and R. C. Stevens, *Annu. Rev. Pharmacol. Toxicol.*, 2013, **53**, 531–556.
74. B. Kobilka and G. F. Schertler, *Trends Pharmacol. Sci.*, 2008, **29**, 79–83.
75. D. Zhang, Q. Zhao and B. Wu, *Mol. Cells*, 2015, **38**, 836–842.
76. J. H. Park, P. Scheerer, K. P. Hofmann, H.-W. Choe and O. P. Ernst, *Nature*, 2008, **454**, 183–187.
77. T. Okada, M. Sugihara, A.-N. Bondar, M. Elstner, P. Entel and V. Buss, *J. Mol. Biol.*, 2004, **342**, 571–583.
78. V. Cherezov, D. M. Rosenbaum, M. A. Hanson, S. G. F. Rasmussen, F. S. Thian, T. S. Kobilka, H. J. Choi, P. Kuhn, W. I. Weis, B. K. Kobilka and R. C. Stevens, *Science*, 2007, **318**, 1258–1265.
79. V. P. Jaakola, M. T. Griffith, M. A. Hanson, V. Cherezov, E. Y. T. Chien, J. R. Lane, A. P. Ijzerman and R. C. Stevens, *Science*, 2008, **322**, 1211–1217.
80. P. Scheerer, J. H. Park, P. W. Hildebrand, Y. J. Kim, N. Krauß, H.-W. Choe, K. P. Hofmann and O. P. Ernst, *Nature*, 2008, **455**, 497–502.
81. T. Warne, M. J. Serrano-Vega, J. G. Baker, R. Moukhametzianov, P. C. Edwards, R. Henderson, A. G. W. Leslie, C. G. Tate and G. F. X. Schertler, *Nature*, 2008, **454**, 486–491.
82. H.-W. Choe, Y. J. Kim, J. H. Park, T. Morizumi, E. F. Pai, N. Krauß, K. P. Hofmann, P. Scheerer and O. P. Ernst, *Nature*, 2011, **471**, 651–655.
83. S. G. F. Rasmussen, H.-J. Choi, J. J. Fung, E. Pardon, P. Casarosa, P. S. Chae, B. T. DeVree, D. M. Rosenbaum, F. S. Thian, T. S. Kobilka, A. Schnapp, I. Konetzki, R. K. Sunahara, S. H. Gellman, A. Pautsch, J. Steyaert, W. I. Weis and B. K. Kobilka, *Nature*, 2011, **469**, 175–180.
84. S. G. F. Rasmussen, B. T. DeVree, Y. Zou, A. C. Kruse, K. Y. Chung, T. S. Kobilka, F. S. Thian, P. S. Chae, E. Pardon, D. Calinski, J. M. Mathiesen, S. T. A. Shah, J. A. Lyons, M. Caffrey, S. H. Gellman, J. Steyaert, G. Skiniotis, W. I. Weis, R. K. Sunahara and B. K. Kobilka, *Nature*, 2011, **477**, 549–555.
85. D. M. Rosenbaum, C. Zhang, J. A. Lyons, R. Holl, D. Aragao, D. H. Arlow, S. G. F. Rasmussen, H.-J. Choi, B. T. DeVree, R. K. Sunahara, P. S. Chae, S. H. Gellman, R. O. Dror, D. E. Shaw, W. I. Weis, M. Caffrey, P. Gmeiner and B. K. Kobilka, *Nature*, 2011, **469**, 236–240.

86. T. Warne, R. Moukhametzianov, J. G. Baker, R. Nehmé, P. C. Edwards, A. G. W. Leslie, G. F. X. Schertler and C. G. Tate, *Nature*, 2011, **469**, 241–244.
87. F. Xu, H. Wu, V. Katritch, G. W. Han, K. A. Jacobson, Z. G. Gao, V. Cherezov and R. C. Stevens, *Science*, 2011, **332**, 322–327.
88. K. Haga, A. C. Kruse, H. Asada, T. Yurugi-Kobayashi, M. Shiroishi, C. Zhang, W. I. Weis, T. Okada, B. K. Kobilka, T. Haga and T. Kobayashi, *Nature*, 2012, **482**, 547–551.
89. A. C. Kruse, A. M. Ring, A. Manglik, J. Hu, K. Hu, K. Eitel, H. Hübner, E. Pardon, C. Valant, P. M. Sexton, A. Christopoulos, C. C. Felder, P. Gmeiner, J. Steyaert, W. I. Weis, K. C. Garcia, J. Wess and B. K. Kobilka, *Nature*, 2013, **504**, 101–106.
90. Y. Kang, X. E. Zhou, X. Gao, Y. He, W. Liu, A. Ishchenko, A. Barty, T. A. White, O. Yefanov, G. W. Han, Q. Xu, P. W. de Waal, J. Ke, M. H. E. Tan, C. Zhang, A. Moeller, G. M. West, B. D. Pascal, N. Van Eps, L. N. Caro, S. A. Vishnivetskiy, R. J. Lee, K. M. Suino-Powell, X. Gu, K. Pal, J. Ma, X. Zhi, S. Boutet, G. J. Williams, M. Messerschmidt, C. Gati, N. A. Zatsepin, D. Wang, D. James, S. Basu, S. Roy-Chowdhury, C. E. Conrad, J. Coe, H. Liu, S. Lisova, C. Kupitz, I. Grotjohann, R. Fromme, Y. Jiang, H. Yang, J. Li, M. Wang, Z. Zheng, D. Li, N. Howe, Y. Zhao, J. Standfuss, K. Diederichs, Y. Dong, C. S. Potter, B. Carragher, M. Caffrey, H. Jiang, H. N. Chapman, J. C. H. Spence, P. Fromme, U. Weierstall, O. P. Ernst, V. Katritch, V. V. Gurevich, P. R. Griffin, W. L. Hubbell, R. C. Stevens, V. Cherezov, K. Melcher and H. E. Xu, *Nature*, 2015, **523**, 561–567.
91. A. Manglik, T. H. Kim, M. Masureel, C. Altenbach, Z. Yang, D. Hilger, M. T. Lerch, T. S. Kobilka, F. S. Thian, W. L. Hubbell, R. S. Prosser and B. K. Kobilka, *Cell*, 2015, **161**, 1101–1111.
92. V. Isberg, C. de Graaf, A. Bortolato, V. Cherezov, V. Katritch, F. H. Marshall, S. Mordalski, J. P. Pin, R. C. Stevens, G. Vriend and D. E. Gloriam, *Trends Pharmacol. Sci.*, 2015, **36**, 22–31.
93. M. Zocher, J. J. Fung, B. K. Kobilka and D. J. Muller, *Structure*, 2012, **20**, 1391–1402.

CHAPTER 7

mGlu2 Receptor Positive Allosteric Modulators

A. A. TRABANCO*[a], J. M. CID[a] AND G. TRESADERN[b]

[a]Neuroscience Medicinal Chemistry, Janssen Research & Development, Janssen–Cilag SA, C/Jarama 75A, Toledo, 45007, Spain; [b]Discovery Sciences, Janssen Research & Development, Janssen–Cilag SA, C/Jarama 75A, Toledo, 45007, Spain
*E-mail: atrabanc@its.jnj.com

7.1 Introduction

Glutamate is the major excitatory neurotransmitter in the central nervous system of vertebrates. It elicits and modulates synaptic responses through the activation of ionotropic and metabotropic glutamate receptors.[1,2] The metabotropic glutamate receptor (mGlu) family encompasses eight members, which have been classified into three different subgroups based on sequence similarity, pharmacology and preferential signal transduction pathway.[3] The mGlu2 receptor belongs to the Group II subclass along with the mGlu3 receptor and has been extensively studied in neuropharmacology. mGlu2 is preferentially located on presynaptic nerve terminals where it negatively regulates glutamate release as an autoreceptor.[4] mGlu2 receptors are widely expressed in the brain, and in particular in regions known to be implicated in neuropsychiatric disorders such as prefrontal cortex, hippocampus, striatum, thalamus and amygdala.[5–7] It is therefore proposed that activation of mGlu2 receptors constitutes an attractive approach for the treatment of

RSC Drug Discovery Series No. 56
Allosterism in Drug Discovery
Edited by Dario Doller

Published by the Royal Society of Chemistry, www.rsc.org

disorders associated with an excessive glutamatergic tone, such as anxiety and schizophrenia.[8–11]

Pharmaceutical companies and academic groups have primarily focused on the identification and pharmacological evaluation of mixed mGlu2/3 receptor agonists. Mixed orthosteric agonists such as **1** (LY354740), **2** (LY544344) and **3** (LY2140023; pomaglumetad methionil), the methionine amide prodrug of **4** (LY404039) have shown activity in a variety of preclinical animal models of anxiety and schizophrenia (Figure 7.1).[12,13]

Several of these compounds have advanced into clinical trials to date. Thus, the anxiolytic potential of **1** was confirmed in healthy volunteers, showing activity in fear-potentiated startle and panic induction after CO_2 inhalation challenge.[14] Compound **3** demonstrated improvements to positive and negative symptoms compared with placebo in schizophrenic patients.[15] However pomaglumetad methionil (**3**) failed to meet the primary endpoint based on the positive and negative syndrome scale in phase III clinical trials and the originating company announced the decision to stop further development of **3** for the treatment of patients suffering from schizophrenia.[16–18]

Despite the interesting clinical findings with mixed mGlu2–3 orthosteric agonists, and the recent examples of subtype-selective agonists,[19] there remain hurdles in developing clinically useful compounds.[20]

7.2 mGlu2 Receptor Positive Allosteric Modulators

Given the promise offered by mGlu2 receptor activation, there is increased interest in identifying small molecules able to activate the receptor. One alternative approach to mixed orthosteric agonists like **1–4** is the use of sub-type selective positive allosteric modulators (PAMs). Molecules that act *via* an allosteric site may offer several advantages.[21] mGlu2 receptor PAMs are unlikely to be amino-acid-like-molecules and therefore less polar, with improved central nervous system (CNS) penetration. Furthermore, the lower evolutionary pressure on allosteric sites may allow for improved mGlu receptor subtype selectivity. Finally, mGlu2 PAMs will typically not activate the receptor alone, but only enhance its function in the presence of the endogenous agonist glutamate. Thus, mGlu2 PAMs may be less prone to causing receptor desensitization and would respond to physiological

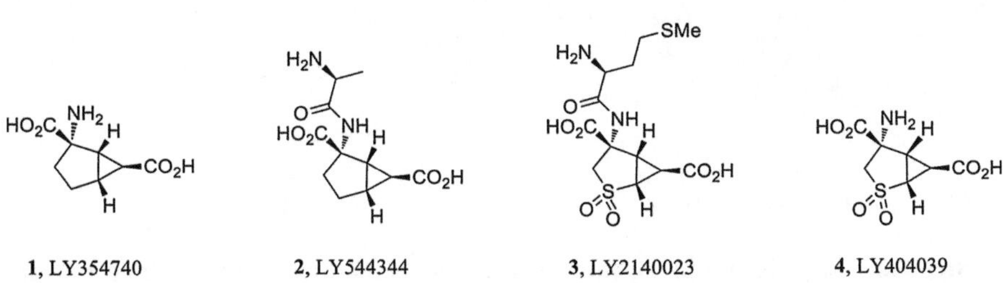

1, LY354740 **2**, LY544344 **3**, LY2140023 **4**, LY404039

Figure 7.1 Representative mixed mGlu2–3 receptor agonists.

glutamate fluctuations.[22] To exploit all these potential advantages, the number of reported selective mGlu2 receptor PAMs has increased in recent years.[23–25]

7.2.1 Medicinal Chemistry of mGlu2 Receptor PAMs

Researchers at Eli Lilly and Merck & Co., Inc. (Merck) disclosed the first examples mGlu2 receptor PAMs in the early 2000's. Medicinal chemistry optimization efforts at both companies led to the identification of potent and subtype-selective mGlu2 receptor PAMs **5** (LY487379)[26] and **6** (biphenylindanone A (BINA)), shown in Figure 7.2. Compound **5** was found to be active in animal models predictive of antipsychotic and anxiolytic activity, providing the first preclinical proof-of-concept that mGlu2 receptor PAMs can mimic the effects of mixed mGlu2/3 receptor orthosteric agonists.[27,28] Short duration of action and minimal efficacy at low doses limited further development of this sulfonamide series.

Scientists at Merck extensively characterized **6** as a selective and potent mGlu2 modulator with long-lasting effects in a number of animal models used to predict potential antipsychotic- and anxiolytic-like properties in mice. Furthermore, the behavioural effects of **6** were blocked by the mixed mGlu2–3 receptor antagonist (2*S*)-2-amino-2-[(1*S*,2*S*)-2-carboxycycloprop-1-yl]-3-(xanth-9-yl) propanoic acid (LY341495) supporting the hypothesis that the *in vivo* effects observed were mediated by increased activation of the mGlu2 receptor.[29] These seminal reports spurred on leading academic groups and pharmaceutical companies in the search for more potent and selective mGlu2 PAMs, resulting in the identification of multiple novel mGlu2 receptor PAM chemotypes and two compounds that advanced into clinical trials.

Herewith we provide an overview of the state of the art of medicinal chemistry for mGlu2 receptor PAMs. The available data have been classified on the basis of reported structure chemotypes of described mGlu2 receptor PAMs. Examples have been chosen from each report to highlight the chemical matter within, or simply because the corresponding authors have emphasised them and provided additional data.

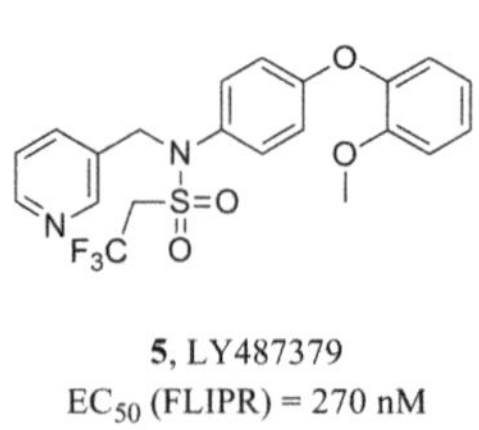

5, LY487379
EC_{50} (FLIPR) = 270 nM

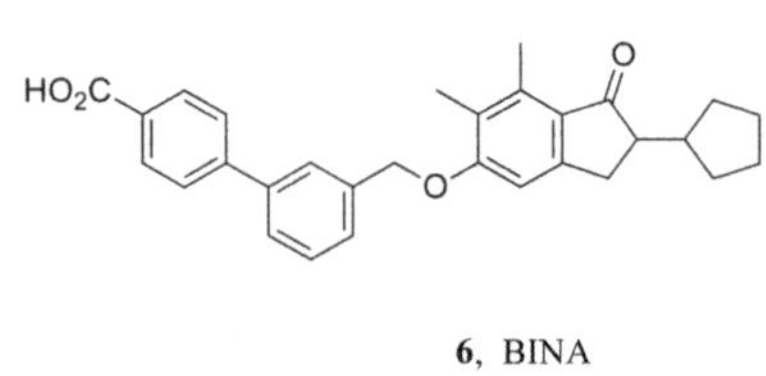

6, BINA
EC_{50} (GTPγS) = 111 nM

Figure 7.2 Early examples of mGlu2 receptor PAMs. EC_{50} (half maximal effective concentration) FLIPR (fluorescent imaging plate reader), GTPγS (guanosine 5′-*O*-[gamma-thio]triphosphate).

7.2.1.1 *Acetophenones and Indoles*

A series of acetophenones with mGlu2 PAM activity were identified by scientists at Merck in the early 2000s. Initial prototype **7**[30] was further optimized to increase poor central penetration leading to thiopyridine **8**,[31] having much higher brain penetration (Figure 7.3).

Scientists at Eli Lilly also described a series of similar acetophenones,[32] from which compound **9** [*N*-(4-((2-(trifluoromethyl)-3-hydroxy-4-(isobutyryl) phenoxy)methyl)benzyl)-1-methyl-1*H*-imidazole-4-carboxamide (THIIC)] was identified as a potent and selective mGlu2 PAM, active *in vivo* in models predictive for antidepressant and anxiolytic activity.[33,34] More recently, Sanford-Burnham Medical Research Institute described analogues of **7** having alternative replacements of the tetrazole ring, with the carboxylic acid derivative **10** being active in a model of cocaine dependence.[35,36]

A series of indole derivatives was developed by Merck in an attempt to improve the suboptimal pharmacokinetic (PK) properties of mGlu2 PAM acetophenones masking the hydroxyketone motif.[37,38] From this exploration compound **11** was shown to reduce ketamine-induced hyperactivity in rats, suggesting potential efficacy of mGlu2 receptor PAMs in schizophrenia.

7.2.1.2 *Isoindolones*

This series was initially disclosed by AstraZeneca/NPS in 2006.[39] Since then, several groups have worked on this chemotype with over 1200 examples disclosed.[40–43] AstraZeneca identified the isoindolone derivative **12** (AZD8529), the first mGlu2 receptor PAM that advanced into clinical trials (Figure 7.4). However, in early 2011, AstraZeneca announced the discontinuation of

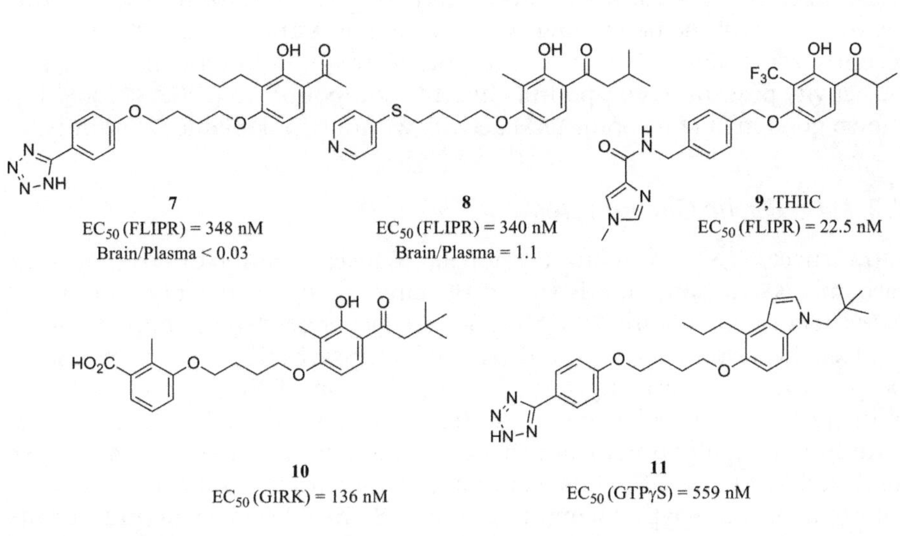

Figure 7.3 Representative mGlu2 receptor PAMs **10–11** from the acetophenone class.

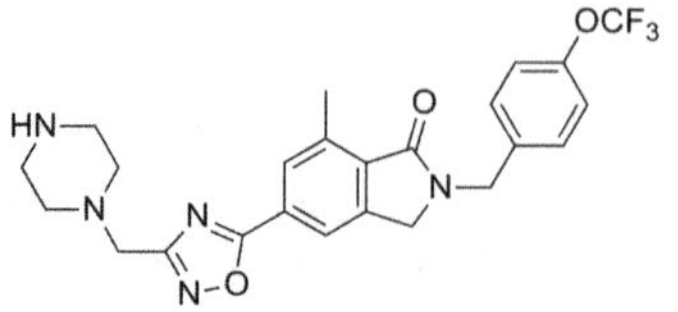

12, AZD8529
EC_{50} (GTPγS) = 231 nM

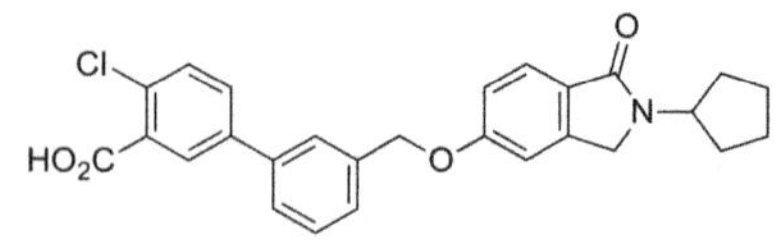

13
EC_{50} (thallium flux) = 50 nM

Figure 7.4 Representative mGlu2 receptor PAMs **12–13** from the isoindolone class.

AZD8529 in a phase IIa proof of concept (POC) study in schizophrenia, as it showed no effect at a dose of 40 mg every other day, while active control (Risperidone) showed good activity ($P < 0.001$) on the primary outcome—change in Positive and Negative Syndrome Scale (PANSS) total score from baseline to week 4.[44]

The Sanford-Burnham Medical Research Institute followed up on previous work on indanones, in particular BINA (**6**). Their exploration resulted in the potent mGlu2 PAM molecule **13**, which was shown to have favourable PK properties and activity in a rat model of nicotine self-administration, therefore indicating potential utility for the treatment of nicotine dependence in humans.[45]

7.2.1.3 Benzimidazoles

Benzimidazoles with mGlu2 receptor PAM activity were identified by Pfizer[46–51] and GSK[52] *via* high-throughput screens (HTS), with compounds **14** and **15** being among the preferred ones (Figure 7.5). A subseries of tricyclic benzimidazoles **16**, where a cycloalkyl ring is fused from the imidazole nitrogen back to the benzo ring, was reported by AstraZeneca/NPS.[53,54] More recently, scientists at GSK reported on their progress in reducing off-target D2 activity present in compounds like **14**. Compound **17** (GSK1331268) has shown good mGlu2 receptor PAM activity without appreciable D2 affinity.

7.2.1.4 Cyclic Carbamates

AstraZeneca/NPS[55,56] and Pfizer[57] pioneered the exploration of series of cyclic carbamates with compounds **18** and **19** being among the most potent mGlu2 PAMs identified (Figure 7.6). Compound **19** reduced methamphetamine-induced hyperlocomotion (predictive of antipsychotic activity) in mice.[58] Despite these encouraging results, the reduction of the high overall lipophilicity of the molecules remains a challenge for their further development.

Additional compounds from this cyclic carbamate series have also been disclosed by Merck.[59–61] Thus, compound **20** was found to be active in a rat ketamine-induced hyperlocomotion model. Some relevant structure–activity relationships (SAR) showed that the (*R*)-configuration of position C5 is crucial for activity, which was also the case for previously described oxazolidinone

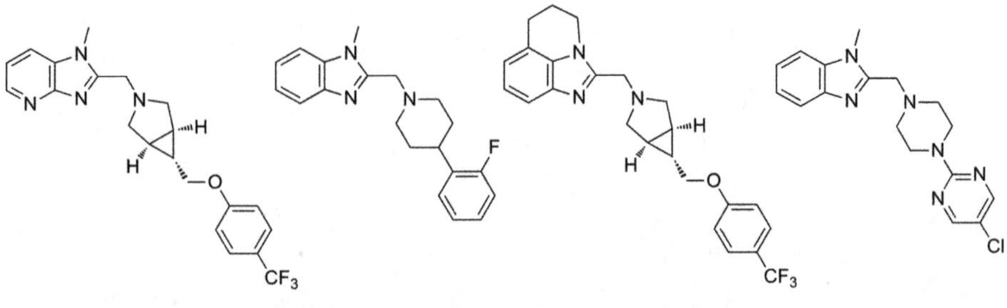

Figure 7.5 Representative mGlu2 receptor PAMs **14–17** from the benzimidazole series.

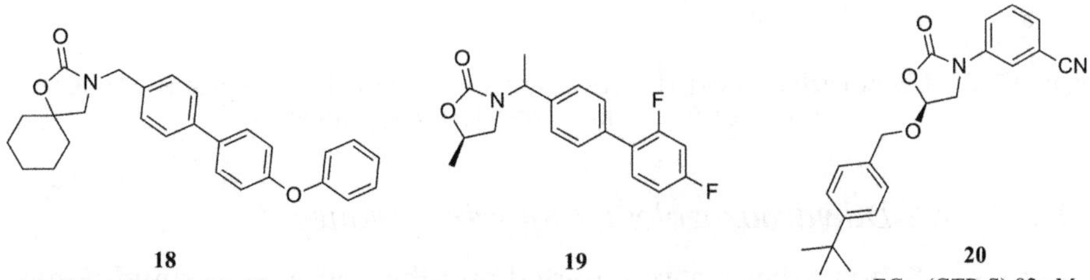

Figure 7.6 Representative cyclic carbamates **18–20** as mGlu2 receptor PAMs.

series. Interestingly, substitution at the *ortho* position of the phenyl ring is not permitted for activity, which suggests that coplanarity between the aryl ring and the oxazolidinone core is beneficial for achieving good activity. This hypothesis was later validated by the preparation of conformationally restricted compounds such as the novel oxazolobenzimidazole chemotype (**B**, Figure 7.7) that, according to Merck scientists, successfully addressed the inherent limitations of the oxazolidinone core.

7.2.1.5 1,2-Dihydrooxazolo[2,3-a]benzimidazoles

A series of dihydrooxazolobenzimidazoles was disclosed by Merck researchers in 2009.[62,63] As previously described, compounds in this series (**B**) can be seen as a conformationally restricted analogues of *N*-phenyloxazolidinones (**A**) due to cyclization of the phenyl ring onto the C2 position of the oxazolidinone core. This cyclization locks a coplanar conformation which is suggested to be preferred for mGlu2 receptor PAM activity. The most interesting compound, 2(*S*)-(6-*tert*-butylpyridin-3-yloxymethyl)-2,3-dihydrooxazolo[3,2-a]benzimidazole-7-carbonitrile (TBPCOB) (**21**), showed robust efficacy in inhibiting phencyclidine (PCP)-induced locomotor activity, a model for schizophrenia, in rats.[64]

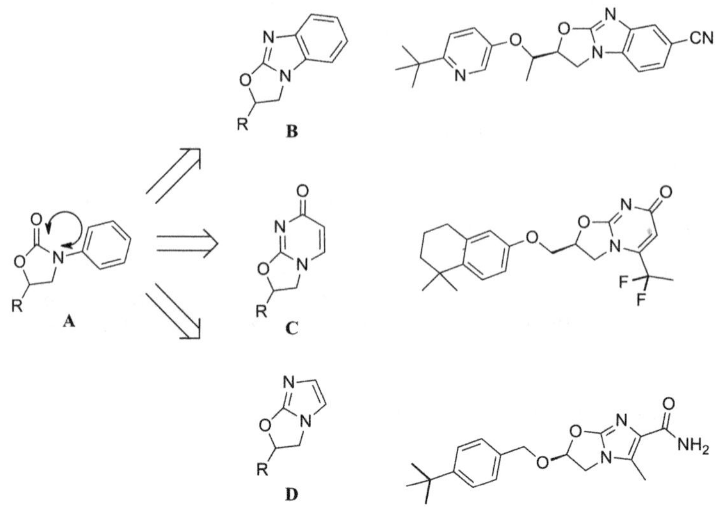

21, TBPCOB
EC_{50} (FLIPR) 28 nM

22
EC_{50} (GTPγS) 15 nM

23; TASP0433864
EC_{50} (GTPγS) 180 nM

Figure 7.7 Cyclic carbamates (**A**) as the origin of novel mGlu2 receptor PAM chemotypes **B–D** *via* conformational restriction strategies.

*7.2.1.6 2,3-Dihydrooxazolo[3,2-*a*]pyrimidinones*

Scientists at Sanofi-Aventis have reported on dihydrooxazolopyrimidinones as a novel mGlu2 receptor PAM chemotype. This novel scaffold can be seen as resulting from another conformational restriction around the *N*-phenyloxazolidinone series (**A**), where the acceptor properties of the oxazolidinone carbonyl group are now embedded into a pyrimidone amide-like function (**C**).[65–69] A representative compound from this chemical class (**22**) is shown in Figure 7.7.

*7.2.1.7 2,3-Dihydroimidazo[2,1-*b*]oxazoles*

Dihydroimidazooxazoles **D** have been described by Taisho.[70–72] In a recent publication, compound **23** (TASP0433864)[73] was identified as a selective mGlu2 PAM with antipsychotic activity, and its ability to attenuate the excess of glutamatergic neurotransmission in several mechanistic *in vitro* and *in vivo* models has also been proven (Figure 7.7).

7.2.1.8 Imidazoimidazolones

The most recently reported scaffold with mGlu2 receptor PAM activity has been disclosed by Taisho in a patent application describing compounds with overall potencies between 3.4 micromolar and nanomolar ranges.[74] The most potent compound exemplified is **24** with a half maximal effective concentration (EC_{50}) of 11 nM (Figure 7.8). Most of the examples maintain the *N*-alkyl substitution and the terminal 2-pyridyl substituent as illustrated with compound **24**.

7.2.1.9 Pyridones

Series of pyridones[75] and benzopyridones[76,77] were identified by Janssen/Addex after an HTS. Optimization from HTS hits led to the identification JNJ40068782, **25**[78] as an initial lead (Figure 7.9). Extensive *in vitro* and *in vivo* characterization has been recently reported for **25**.[79] For instance, **25** influenced rat sleep–wake organization, decreasing rapid eye movement (REM) sleep at an oral dose of 3 mg kg^{-1}. In addition, **25** reversed PCP-induced hyperlocomotion in mice with a half maximal effective dose (ED_{50}) of 5.7 mg kg^{-1} administered subcutaneously (s.c.).

Further optimization carried out on **25** culminated with the identification of JNJ40411813/ADX71149 (**26**),[80] to date, the most advanced mGlu2 receptor PAM in clinical studies. It has been disclosed that, in an exploratory phase IIa study in schizophrenia, **26** proved to meet the primary objectives of safety and tolerability. Moreover, patients treated with antipsychotics who experience residual negative symptoms were identified as the subgroup of patients who may potentially benefit from add-on treatment with **26**, although this is yet to be established in a formal POC study.[81,82] Additionally, a phase II POC study with **26** as adjunctive therapy in patients with major depressive disorder (MDD) with significant anxiety symptoms has been completed.[83] Overall, **26** was well-tolerated, and treatment emergent adverse events reported were similar to those seen in previous clinical studies. On the basis of a preliminary analysis of the primary efficacy end point, the Six-Item Hamilton Anxiety Subscale (HAM-A6), **26** did not meet the criterion for efficacy signal detection *versus* placebo. Despite a lack of efficacy in the primary outcome

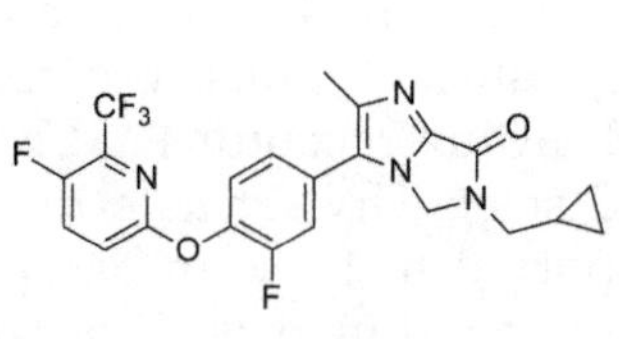

24
EC_{50} (GTPγS) 11 nM

Figure 7.8 Representative imidazoimidazolone **24** as mGlu2 receptor PAM.

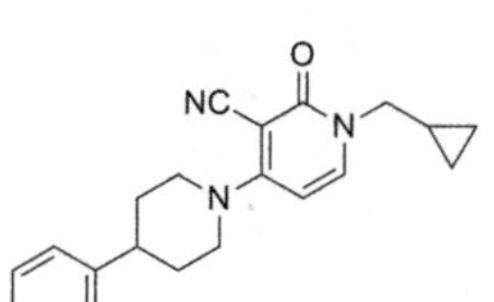

25, JNJ40068782
EC_{50} (GTPγS) 143 nM

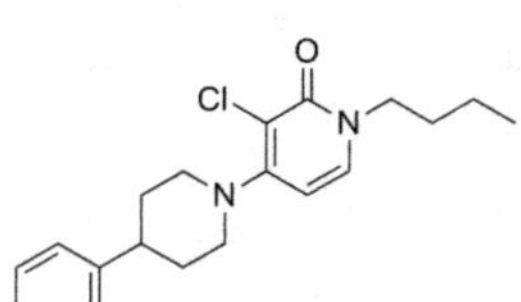

26, JNJ40411813/ADX71149
EC_{50} (GTPγS) 147 nM

Figure 7.9 Representative mGlu2 receptor PAMs **25–26** from the pyridone class.

measure, treatment with **26** showed efficacy signals on several anxiety measures (Hamilton Depression Rating Scale 17 (HDRS17) anxiety somatization factor, Inventory of Depressive Symptomatology for Clinicians 30 (IDS-C30) anxiety subscale) and on all depression measures (HDRS17, HAM-D6 and IDS-C30).[84] Although efficacy signals were evident, the data overall did not support the further development of **26** for treatment of anxious depression. Further exploration of **26** for other therapeutic indications remains of potential interest.

7.2.1.10 Imidazopyridines

Janssen/Addex have identified a class of imidazo[1,2-*a*]pyridine derivatives as PAMs of the mGlu2 receptor.[85] This novel imidazopyridine scaffold (**F**) was designed *via* medicinal chemistry-driven SAR and computational techniques. The imidazopyridine scaffold provided a similar spatial distribution of the pendent substituents as well as similar electrostatic surface potential around the central core compared with the corresponding C4-substituted pyridone (**E**).[86] Compound **27** has been reported to be one of the most interesting representatives within the series combining good mGlu2 PAM activity with selectivity and modulation of REM sleep variables in a rat sleep model, a mechanism of action consistent with mGlu2 receptor activation (Figure 7.10).[87]

7.2.1.11 Triazolopyridines

1,2,4-Triazolopyridine scaffolds (**G**) were identified by Janssen/Addex as suitable replacements for the imidazopyridine core (**F**, Figure 7.11). The additional nitrogen atom in the bicyclic core provides improved metabolic stability to the compounds while retaining or slightly improving the primary activity. Very potent compounds were disclosed in a patent-application-claiming series of triazolopyridines.[88] Thus, the recently described JNJ42153605 (**28**)[89] displays good mGlu2 receptor PAM activity (EC_{50} = 17 nM) and remarkable over 50-fold selectivity *versus* other mGlu receptors (Figure 7.11). Likewise, **28** is reported to have an acceptable PK profile in rodent and non-rodent species, combined with *in vivo* central activity in sleep-wake electroencephalograms (sw-EEG), showing suppressed

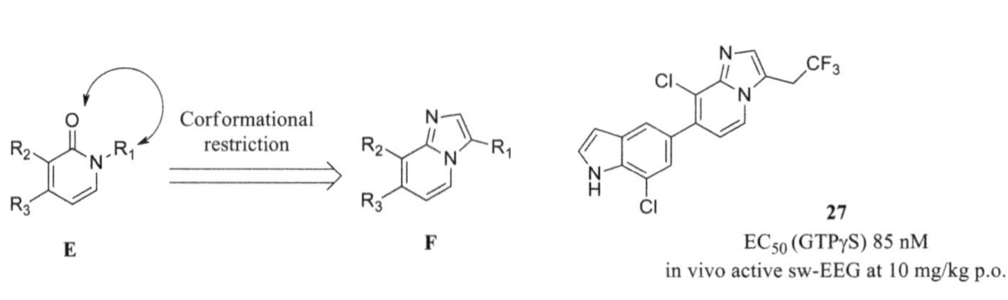

Figure 7.10 Representative mGlu2 receptor PAM imidazopyridine **27**.

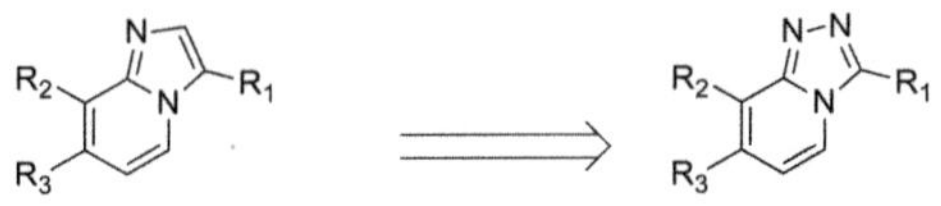

F G

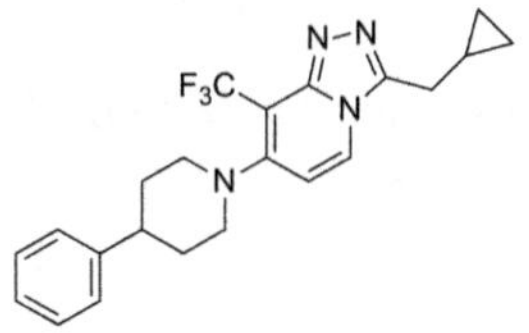

28, JNJ42153605
EC_{50} (GTPγS) 17 nM
ED_{50} (PCP-LMA, sc) = 5.4 mg/kg
ED_{50} (CAR, sc) = 2.35 mg/kg

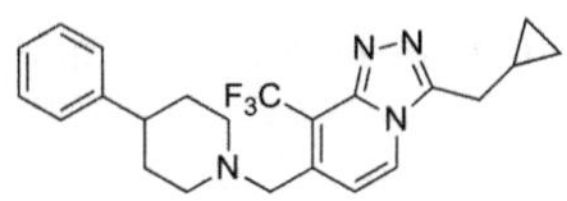

29, JNJ46281222
EC_{50} (GTPγS) 17 nM
ED_{50} (CAR, sc) = 5 mg/kg

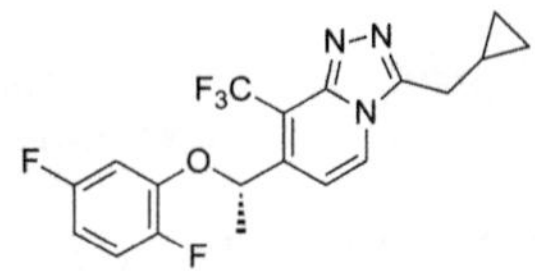

30
EC_{50} (GTPγS) 5.6 nM

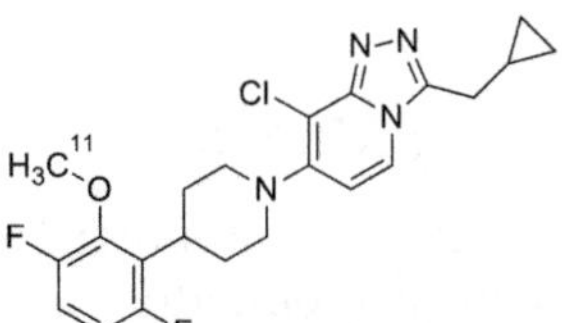

31, [^{11}C]-JNJ42491293
EC_{50} (GTPγS) 9.2 nM

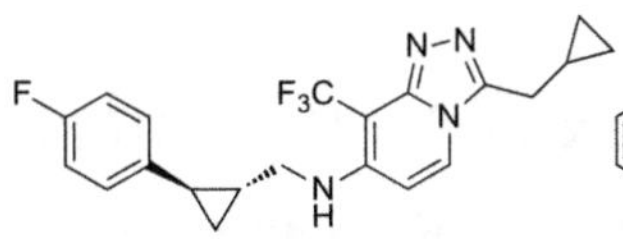

32
EC_{50} (cAMP) 1 nM

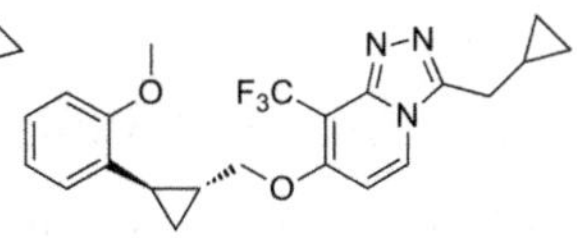

33
EC_{50} (cAMP) 0.2 nM

Figure 7.11 Representative triazolopyridines **28–33** as mGlu2 receptor PAMs.

REM sleep duration. Elongation of the phenylpiperidine moiety as in **29** (JNJ46281222) is well tolerated for potency. A more recent patent application discloses series of analogues having the piperidine ring replaced by a branched ethoxy linker (**30**).[90] Janssen scientists were also able to develop radiolabeled mGlu2 receptor PAM positron-emission tomography (PET) ligands from this chemical class.[91] Thus, [^{11}C]-JNJ42491293 (**31**) has been reported as a PET tracer for measuring mGlu2 receptor availability in humans that may be suitable for assessing mGlu2 receptor occupancy of mGlu2 PAM drug candidates.[92]

Finally, Bristol-Myers Squibb has disclosed similar 1,2,4-triazolopyridines having either secondary amines (**32**)[93] or ethers (**33**)[94] at the C7 position, leading to molecules with nanomolar activity in a cAMP biological assay.

7.2.1.12 Benzotriazoles

A family of 1,2,3-benzotriazoles was disclosed as potent mGlu2 receptor PAMs in a series of eight patent applications from Merck published between 2010 and 2012.[95–102] This benzotriazole series is closely related to a series of benzazole mGlu2 receptor potentiators also disclosed by Merck in 2006. Representative examples **34–35** within this class are shown in Figure 7.12.

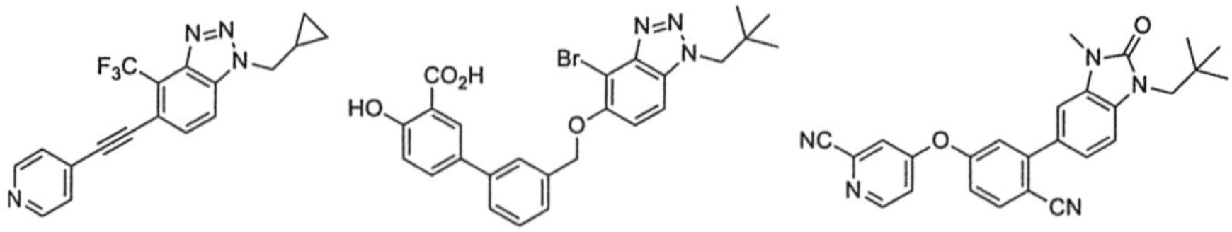

34
EC_{50} (GTPγS) 12 nM

35
EC_{50} (GTPγS) 3.5 nM

36
EC_{50} (FLIPR) = 8 nM

Figure 7.12 Representative benzotriazole **34–35** and benzoimidazolone **36** mGlu2 receptor PAMs.

7.2.1.13 (Aza)Benzimidazolones

In 2011 scientists at Merck published a series of patent applications detailing series of benzimidazolone and aza-benzimidazolone derivatives as mGlu2 receptor PAMs. A representative example is compound **36** in Figure 7.12.[103–105]

7.2.2 mGlu Structure and mGlu2 PAM Binding Modes

The crystal structures of the mGlu1 receptor and mGlu5 receptor 7-transmembrane (7-TM) domains bound with negative allosteric modulators were published in mid-2014 (Figure 7.13).[106,107] More recently additional structures of mGlu5 have been published.[108] These structures confirmed the binding site of allosteric modulators in the 7-TM, in the region of evolutionary conservation seen for binding of orthosteric ligands in class A GPCRs. The ligands form few polar interactions, and instead make multiple hydrophobic and lipophilic contacts, and provide steric complementarity with the receptor. Mavoglurant binds much deeper in mGlu5 than fluoro-*N*-[4-[6-(isopropylamino)pyrimidin-4-yl]-1,3-thiazol-2-yl]-*N*-methylbenzamide (FITM) at mGlu1. This is due to an unusual amino acid, P655 in mGlu5 (S688 in mGlu1), which opens a narrow channel suitable for the aryl acetylene motif seen in mGlu5 allosteric modulators.

The significant sequence identity of mGlu1 and mGlu5 to mGlu2, approximately 50% in each case, permits the relatively accurate building of homology models and subsequent further simulation. Such modelling has been used in conjunction with mutagenesis, radiolabelled ligand competition assay data and understanding of SAR to derive hypotheses for the binding mode of a number of mGlu2 receptor PAMs. Firstly, an extensive mutagenesis and modelling study led to the proposed binding mode of examples from multiple series including acetophenone **9** (THIIC) and pyridone **25** (JNJ40068782).[109] Homology modelling and docking approaches provided a consistent binding mode for **28** (JNJ42153605), an example from the triazolopyridine series.[110] Around the same time a more sophisticated modelling approach was used to study **29** (JNJ46281222), another mGlu2 PAM from the triazolopyridine class.[111] A homology model of an mGlu2 active state monomer, with G-protein bound, was studied using molecular dynamics simulations in the presence of ligand, solvent and lipid membrane. The modelled binding mode was

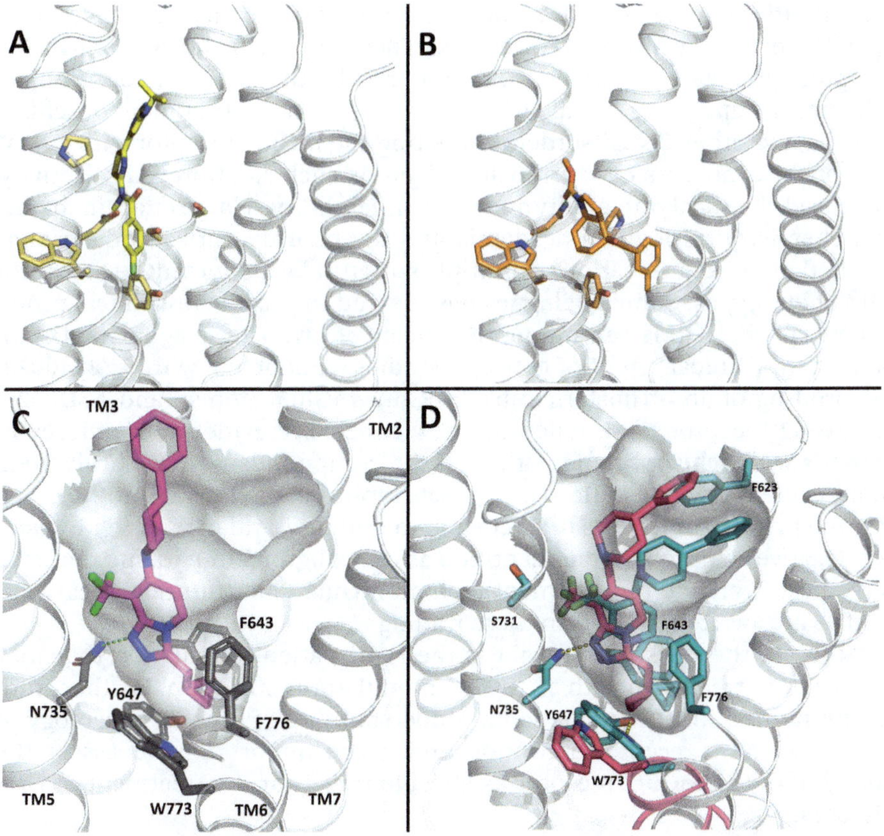

Figure 7.13 Binding modes for mGlu allosteric modulators. (A) Negative allosteric modulator FITM bound at mGlu1, PDB 4OR2. (B) NAM Mavoglurant bound at mGlu5, PDB 4OO9. (C) Modelled binding mode of **28** (JNJ42153605). (D) Modelled binding mode of **29** (JNJ46281222) before (magenta) and after MD simulation (blue).

consistent with SAR and mutagenesis data. The simulations revealed a stable binding mode for the ligand throughout the 500 ns of simulation and the movement of amino acids around the ligand, some of which may play a role in receptor activation.

In summary, the availability of close structural templates, mutagenesis data and extensive knowledge of SAR, now permits reliable modelling hypotheses for the binding mode for mGlu2 PAMs. Such approaches can be incorporated into molecular design for future PAM series.

7.3 Conclusions

The first decade of the twenty-first century has seen a significant discovery and early development effort around modulation of the mGlu2 receptor. The first generation of mGlu2 receptor activators were non-selective conformationally restricted analogues of glutamate acting as agonists at both mGlu2 and mGlu3 receptors. These mixed orthosteric mGlu2–3 receptor agonists generated encouraging results that served to accumulate evidence to support potential utility of the target for the treatment of schizophrenia and other CNS disorders. Targeting the mGlu2 receptor with a PAM may offer advantages over orthosteric ligands such as improved selectivity, increased chemical tractability and better tolerability. A large effort in medicinal chemistry throughout academia and the pharmaceutical industry has resulted in the identification of multiple structurally distinct mGlu2 receptor PAM chemotypes. Some analogues have served to provide preclinical proof-of-concept that PAMs can mimic the effects of orthosteric agonists. Disappointingly, clinical proof-of-concept studies conducted with LY2140023, the prodrug of an orthosteric mixed mGlu2–mGlu3 agonist, and AZD8529, an mGlu2 receptor PAM, failed to show conclusive evidence of efficacy in patients with schizophrenia, although the compounds were well tolerated. In an exploratory phase IIa study in schizophrenia, JNJ40411813, however, met the primary objectives of safety and tolerability, and patients with residual negative symptoms were identified as the subgroup of patients that may serve best from add-on treatment with an mGlu2 PAM; future clinical POC studies are awaited to confirm these findings.

All in all, the question of what is the right patient population that may benefit from treatment with an mGlu2 modulator is still open. Further studies are needed for the fundamental understanding of the pharmacology of mGlu2 selective receptor activation and, more importantly, to assess the potential of the mGlu2 receptor as a viable target for the treatment of CNS diseases.

References

1. A. Nakanishi, *Science*, 1992, **258**, 597.
2. J. N. Kew and J. A. Kemp, *Psychopharmacology*, 2005, **179**, 4.
3. C. M. Niswender and P. J. Conn, *Annu. Rev. Pharmacol. Toxicol.*, 2010, **50**, 295.
4. J. Cartmel and D. D. Schoepp, *J. Neurochem.*, 2000, **75**, 889.
5. H. Ohishi, R. Shigemoto, S. Nakanishi and N. Mizuno, *Neuroscience*, 1993, **53**, 1009.
6. G. Richards, J. Messer, P. Malherbe, R. Pink, M. Brockhaus, H. Stadler, J. Wichmann, H. Schaffhauser and V. Mutel, *J. Comp. Neurol.*, 2005, **487**, 15.
7. R. A. Wright, B. G. Johnson, C. Zhang, C. Salhoff, A. E. Kingston, D. O. Calligaro, J. A. Monn, D. D. Schoepp and G. J. Marek, *Neuropharmacology*, 2013, **66**, 89.
8. C. J. Swanson, M. Bures, M. P. Johnson, A. M. Linden, J. A. Monn and D. D. Schoepp, *Nat. Rev. Drug Discovery*, 2005, **4**, 131.
9. P. J. Conn, C. W. Lindsley and C. K. Jones, *Trends Pharmacol. Sci.*, 2009, **30**, 25.

10. F. Nicoletti, J. Bockaert, G. L. Collingridge, P. J. Conn, F. Ferraguti, D. D. Schoepp, J. T. Wroblewski and J. P. Pin, *Neuropharmacology*, 2011, **60**, 1017.
11. C. Shigeyuki and H. Hiroiko, *Curr. Pharm. Des.*, 2011, **17**, 94.
12. M. J. Fell, D. L. MacKinzie, J. A. Monn and K. A. Svensson, *Neuropharmacology*, 2012, **62**, 1473–1483.
13. D. D. Schoepp and G. J. Marek, *Curr. Drug Targets: CNS Neurol. Disord.*, 2002, **1**, 215.
14. D. D. Schoepp, R. A. Wright, L. R. Levine, B. Gaydos and W. Z. Potter, *Stress*, 2003, **6**, 189.
15. S. T. Patil, L. Zhang, F. Martenyi, S. L. Lowe, K. A. Jackson, B. V. Andreev, A. S. Avedisova, L. M. Bardenstein, I. Y. Gurovich, M. A. Morozova, M. A. Mosolov, N. G. Neznanov, A. M. Reznik, A. B. Smulevich, V. A. Tochilov, B. G. Johnson, J. A. Monn and D. D. Schoepp, *Nat. Med.*, 2007, **13**, 1102.
16. B. J. Kinon, L. Zhang, B. A. Millen, O. O. Osuntokun, J. E. Williams, S. Kollac-Walker, K. Jackson, L. Kryzhanovskaya and N. Jankova, *J. Clin. Psychopharmacol.*, 2011, **31**, 349.
17. V. L. Stauffer, B. A. Millen, S. Andersen, B. J. Kinon, L. Lagrandeur, J. P. Lindemayer and J. C. Gomez, *Schizophr. Res.*, 2013, **150**, 434.
18. D. H. Adams, L. Zhang, B. A. Millen, B. J. Kinon and J.-C. Gomez, *Schizophr. Res. Treat.*, 2014, **2014**, 758212.
19. J. A. Monn, L. Prieto, L. Taboada, J. Hao, M. R. Reinhard, S. S. Henry, C. D. Beadle, L. Walton, T. Man, H. Rudyk, B. Clark, D. Tupper, S. R. Baker, C. Lamas, C. Montero, A. Marcos, J. Blanco, M. Bures, D. K. Clawson, S. Atwell, F. Lu, J. Wang, M. Russell, B. A. Heinz, X. Wang, J. H. Carter, B. G. Getman, J. T. Catlow, S. Swanson, B. G. Johnson, D. B. Shaw and D. L. McKinzie, *J. Med. Chem.*, 2015, **58**, 7526.
20. M.-L. Li, X.-Q. Hu and W.-J. Gao, *Prog. Neuro-Psychopharmacol. Biol. Psychiatry*, 2015, **60**, 66.
21. J. P. Conn, A. Christopoulos and C. W. Lindsley, *Nat. Rev. Drug Discovery*, 2009, **8**, 41.
22. T. Gjoni and S. Urwyler, *Neuropharmacology*, 2008, **55**, 1293.
23. A. A. Trabanco, J. M. Cid, H. Lavreysen, G. J. Macdonald and G. Tresadern, *Curr. Med. Chem.*, 2011, **18**, 47.
24. A. A. Trabanco and J. M. Cid, *Expert Opin. Ther. Pat.*, 2013, **23**, 629.
25. G. Szabó and G. M. Keserű, *Curr. Top. Med. Chem.*, 2014, **14**, 1771.
26. M. P. Johnson, M. Baez, G. E. Jagdmann, T. C. Britton, T. H. Large, D. O. Callagaro, J. P. Tizzano, J. A. Monn and D. D. Schoepp, *J. Med. Chem.*, 2003, **46**, 3189.
27. M. P. Johnson, D. Barda, T. C. Britton, R. Emkey, W. J. Hornback, G. E. Jadmann, D. L. Mckinzie, E. S. Nisenbaum, J. P. Tizzano and D. D. Schoepp, *Psychopharmacology*, 2005, **179**, 271.
28. R. Galici, N. G. Echemendia, A. L. Rodriguez and P. J. Conn, *J. Pharmacol. Exp. Ther.*, 2005, **315**, 1181.
29. R. Galici, C. K. Jones, K. Hemstapat, Y. Nong, N. G. Echemendia, L. C. Williams, T. de Paulis and P. J. Conn, *J. Pharmacol. Exp. Ther.*, 2006, **318**, 173.

30. A. B. Pinkerton, J. M. Vernier, H. Schaffhauser, B. A. Rowe, U. C. Campbell, D. E. Rodriguez, D. S. Lorrain, C. S. Baccei, L. P. Daggett and L. J. Bristow, *J Med Chem.*, 2004, **47**, 4595.
31. R. V. Cube, J. M. Vernier, J. H. Hutchinson, M. F. Gardner, J. K. Joyce, B. A. Rowe, H. Schaffhauser, L. Dagget and A. B. P. Pinkerton, *Bioorg. Med. Chem. Lett.*, 2005, **15**, 2389.
32. A. Khilevich, B. Liu, D. R. Mayhugh, J. M. Schkeryantz and Z. Deyi, WO 2010009062 A1, 2010.
33. M. J. Fell, J. M. Witkin, J. F. Falcone, J. S. Katner, K. W. Perry, J. Hart, L. Rorick-Kehn, C. D. Overshiner, K. Rasmussen, S. F. Chaney, M. J. Benvenga, X. Li, D. L. Marlow, L. K. Thompson, S. K. Luecke, K. A. Wafford, W. F. Seidel, D. M. Edgar, A. T. Quets, C. C. Felder, X. Wang, B. A. Heinz, A. Nikolayev, M.-S. Kuo, D. Mayhugh, A. Khilevich, D. Zhang, P. J. Ebert, J. A. Eckstein, B. L. Ackermann, S. P. Swanson, J. T. Catlow, R. A. Dean, K. Jackson, S. Tauscher-Wisniewski, G. J. Marek, J. M. Schkeryantz and K. A. Svensson, *J. Pharmacol. Exp. Ther.*, 2012, **336**, 165.
34. P. L. Johnson, S. D. Fitz, E. A. Engleman, K. A. Svensson, J. M. Schkeryantz and A. Shekhar, *J. Psychopharmacol.*, 2013, **27**, 152.
35. R.-P. Dhanya, D. J. Sheffler and N. D. P. Cosford, WO 2015157187 A1, 2015.
36. R.-P. Dhanya, D. J. Sheffler, R. Dahl, M. Davis, P. S. Lee, L. Yang, H. H. Nickols, H. P. Cho, L. H. Smith, M. S. D′Souza, J. P. Conn, A. Der-Avakian, A. Markou and N. D. P. Cosford, *J. Med. Chem.*, 2014, **57**, 4154.
37. S. P. Govek, C. Bonnefous, J. H. Hutchinson, T. Kamenecka, J. McQuiston, R. Pracitto, L. X. Zhao, M. F. Gardner, J. K. Joyce, L. P. Daggett, B. A. Rowe, H. Schaffhauser, L. J. Bristow, U. C. Campbell, D. E. Rodriguez and J.-M. Vernier, *Bioorg. Med. Chem. Lett.*, 2005, **15**, 4068.
38. S. P. Govek, J. M. Vernier, T. Kamenecka, J. H. Hutchinson and R. Pracitto, WO 2006091496 A2, 2006.
39. J. Clayton, F. Ma, B. Van Wagenen, R. Ukkiramapandian, I. Egle, J. Empfield, I. Methvin, A. Slassi, G. Steelman, R. Urbanek and S. Walsh, WO 2006020879 A1, 2006.
40. H. Geneste, D. Sauer, W. Braje, W. Amberg, M. Mezler and M. H. M. Bakker, WO 2008145616 A1, 2008.
41. J. Adam, J. Gillespie, S. Laats, J. K. F. Maclean and R. D. McArthur, WO 2011051490 A2, 2011.
42. N. D. P. Cosford, D. R. Panickar and S. Sidique, WO 2011116356 A2, 2011.
43. R.-P. Dhanya, S. Sidique, D. L. Sheffler, H. H. Nickols, A. Herath, Y. Li, D. Russell, R. Ardecky, S. Semenova, A. Markou, J. P. Conn and N. D. P. Cosford, *J. Med. Chem.*, 2011, **54**, 342.
44. R. E. Litman, *AZD8529, A positive allosteric modulator at the mGluR2 receptor, does not improve symptoms in schizophrenia: a proof of principle study; NCDEU An Annual Meeting of the aSCP Westin Diplomat*, Hollywood Florida, US, May 28–31, 2013.

45. S. Sidique, R.-P. Dhanya, D. L. Sheffler, H. H. Nickols, L. Yang, R. Dahl, A. Mangravita-Novo, L. H. Smith, M. S. D´Souza, S. Semenova, J. P. Conn, A. Markou and N. D. P. Cosford, *J. Med. Chem.*, 2012, **55**, 9434.
46. L. Zhang, B. N. Rogers, A. J. Duplantier, S. F. McHardy, I. Efremov, H. Berke, W. Qian, A. Q. Zhang, N. Maklad, J. Candler, A. C. Doran, J. T. Lazzaro and A. H. Ganong, *Bioorg. Med. Chem. Lett.*, 2009, **18**, 5493.
47. L. Zhang, M. A. Brodney, J. Candler, A. C. Doran, A. J. Duplantier, I. V. Efremov, E. Evrad, K. Kraus, A. H. Ganong, J. A. Haas, A. N. Hanks, K. Jenza, J. T. Lazzaro, N. Makland, S. A. McCarthy, W. Qian, B. N. Rogers, M. D. Rottas, C. J. Schmidt, J. A. Siuciak, F. D. Tingley and A. Q. Zhang, *J. Med. Chem.*, 2011, **54**, 1724.
48. H. Berke, A. J. Duplantier, I. Efremov, S. F. Mchardy, B. N. Rogers, W. Qian, L. Zhang and Q. Zhang, WO 2007135527 A2, 2007.
49. I. V. Efremov, B. N. Rogers, A. J. Duplantier., L. Zhang, Q. Zhang and N. S. Maklad, WO 2008012622 A2, 2008.
50. I. V. Efremov, B. N. Rogers, A. J. Duplantier., L. Zhang, Q. Zhang, N. S. Maklad, E. V. Evrard and M. A. Brodney, WO 2008012623 A1, 2008.
51. H. Berke, A. J. Duplantier, I. V. Efremov, S. F. Mchardy, W. Qian, B. N. Rogers, L. Zhan and Q. Zhang, WO 2007135529 A2, 2007.
52. P. L. D'Alessandro, C. Corti, A. Roth, A. Ugolini, A. Sava, D. Montanari, F. Bianchi, S. L. Garland, B. Powney, E. L. Koppe, M. Rocheville, G. Osborne, P. Perez, J. de la Fuente, M. De Los Frailes, P. W. Smith, C. Branch, D. Nash and S. P. Watson, *Bioorg. Med. Chem. Lett.*, 2009, **20**, 759.
53. M. Isaac, A. Slasi, I. Egle and F. Ma, US20070032469 A1, 2007.
54. I. Egle, M. Isaac, R. Urbanek, F. M. Mclaren, S. B. Walsh, G. B. Steelman, D. G. Brown, D. Nugiel, D. W. Chen, A. Slassi and F. Ma, WO 2007115077 A2, 2007.
55. A. Slassi, B. Joseph, F. Ma, I. Egle, J. Clayton, M. Isaac and K. Swierczek, WO 2007078523 A2, 2007.
56. M. Isaac, A. Slassi, I. Egle, J. Clayton, B. Joseph and F. Ma, WO 2008032191 A2, 2008.
57. A. J. Duplantier, I. V. Efremov, L. Zhang, N. S. Maklad and T. J. ÓSullivan, WO 2009004430 A1, 2009.
58. A. J. Duplantier, I. Efremov, J. Candler, A. C. Doran, A. H. Ganong, J. A. Haas, A. N. Hanks, K. G. Kraus, J. T. Lazzaro, J. Lu, N. Maklad, S. A. McCarthy, T. O´Sullivan, B. N. Rogrs, J. A. Siuciak, D. K. Spracklin and L. Zhang, *Bioorg. Med. Chem. Lett.*, 2009, **19**, 2524.
59. E. J. Brnardic, M. Fraley, R. Garbaccio and M. Layton, WO 2009094265 A1, 2009.
60. E. J. Brnardic, M. Fraley, M. Layton and R. Garbaccio, US 20100292241 A1, 2010.
61. P. L. D'Alessandro, C. Corti, A. Roth, A. Ugolini, A. Sava, D. Montanari, F. Bianchi, S. L. Garland, B. Powney, E. L. Koppe, M. Rocheville, G. Osborne, P. Perez, J. de la Fuente, M. De Los Frailes, P. W. Smith, C. Branch, D. Nash and S. P. Watson, *Bioorg. Med. Chem. Lett.*, 2010, **20**, 759.

62. E. J. Brnardic, A. Converso, M. Fraley, R. M. Garbaccio and S. Y. Shaei, WO 2009140163 A1, 2009.
63. E. J. Brnardic, M. E. Fraley and R. M. Garbaccio, WO 2009140166 A2, 2009.
64. R. M. Garbaccio, E. J. Brnardic, M. E. Fraley, G. D. Hartman, P. H. Hutson, J. A. ÓBrien, B. C. Magliaro, J. M. Uslander, S. L. Huszar, K. L. Fillgrove, J. H. Samll, C. Tang, Y. Kuo and M. A. Jacobson, *ACS Med. Chem. Lett.*, 2010, **1**, 406.
65. B. Cao, V. Gurunian, S. Kongsamut, R. W. Kosley Jr., R. Sher and R. E. Hartung, WO 2008112483 A2, 2008.
66. R. Sher, WO 2009110901 A1, 2009.
67. P. George, D. Hall, R. Hartung, R. W. Kosley Jr. and A. C. Scotese, WO 2011034828 A1, 2011.
68. R. W. Kosley Jr. and R. Sher, WO 2011034830 A1, 2011.
69. R. W. Kosley Jr., R. Sher, K. W. Neuenschwander and V. Gurunian, WO 2011034832 A1, 2011.
70. H. Urabe, R. Shimono, T. Tamita, H. Nobukata, K. Sakagami, Y. Matsuda, A. Yasuhara, S. Matsuda, K. Konishi, T. Nakamura and S. Yamamoto, WO 2013062079 A1, 2013.
71. H. Urabe, R. Shimono, T. Minda, N. Hattori, K. Sakagami, Y. Matsuda, A. Yasuhara, S. Masuda, I. Konishi, T. Nakamura and S. Yamamoto, JP 2014214130 A, 2014.
72. H. Urabe, R. Nishikawa, T. Minda, N. Hattori, K. Sakagami, Y. Matsuda and A. Yasuhara, WO 2013062079 A1, 2013.
73. T. Hiyoshi, T. Marumo, H. Hikichi, Y. Tomishima, H. Urabe, T. Tamita, I. Iida, A. Yasuhara, J. Karasawa and S. Chaki, *J. Pharmacol. Exp. Ther.*, 2014, **351**, 642.
74. R. Shimono, K. Sakagami, T. Tamita, S. Yamamoto, S. Masuda, Y. Matsuda, N. Hattori, M. Yoshinaga and K. Otake, JP 2014214124 A, 2014.
75. J. M. Cid, G. Duvey, P. Cluzeau, V. Nhem, K. Macary, A. Raux, N. Poirier, J. Muller, C. Bolea, T. Finn, I. Urios, M. Epping-Jordan, E. Chamelot, F. Derouet, F. Girad, G. J. Macdonald, J. A. Vega, A. I. de Lucas, E. Matesanz, H. Lavreysen, M. L. Linares, D. Oehlrich, J. Oyarzabal, G. Tresadern, A. A. Trabanco, J. I. Andrés, E. Le Poul, H. Imogai, R. Lutjens and J.-P. Rocher, *ACS Med. Chem. Lett.*, 2010, **1**, 788.
76. A. A. Trabanco, G. Duvey, J. M. Cid, G. J. Macdonald, P. Cluzeau, V. Nhem, R. Furnari, N. Behaj, G. Poulain, T. Finn, S. Poli, H. Lavreysen, A. Raux, Y. Thollon, N. Poirier, D. D´Adonna, J. I. Andrés, R. Lutjens, E. Le Poul, H. Imogai and J.-P. Rocher, *MedChemComm*, 2011, **2**, 132.
77. A. A. Trabanco, G. Duvey, J. M. Cid, G. J. Macdonald, P. Cluzeau, V. Nhem, R. Furnari, N. Behaj, G. Poulain, T. Finn, H. Lavreysen, S. Poli, A. Raux, Y. Thollon, N. Poirier, D. D´Adonna, J. I. Andrés, R. Lutjens, E. Le Poul, H. Imogai and J.-P. Rocher, *Bioorg. Med. Chem. Lett.*, 2011, **21**, 971.
78. J. M. Cid-Núñez, A. A. Trabanco-Suárez, G. J. Macdonald, G. A. J. Duvey and R. J. Luetjens, WO 2008107479 A1, 2008.

79. H. Lavreysen, X. Langlois, A. Ahnaou, W. Drinkenburg, P. te Riele, I. Biesmans, I. Van der Linden, L. Peters, A. Megens, C. Wintmolders, J. M. Cid, A. A. Trabanco, J. I. Andrés, F. M. Dautzenberg, R. Lutjens, G. Macdonald and J. R. Atack, *J. Pharmacol. Exp. Ther.*, 2013, **346**, 514.
80. J. M. Cid, G. Tresadern, G. Duvey, R. Lutjens, T. Finn, J.-P. Rocher, S. Poli, J. A. Vega, A. I. de Lucas, E. Matesanz, M. L. Linares, J. I. Andrés, J. Alcazar, J. M. Alonso, G. J. Macdonald, D. Oehlrich, H. Lavreysen, A. Ahnaou, W. Drinkenburg, C. Mackie, S. Pype, D. Gallacher and A. A. Trabanco, *J. Med. Chem.*, 2014, **57**, 6495.
81. *Addex reports top-line data from a successful phase 2a clinical study with ADX71149 in schizophrenia patients*, http://www.addextherapeutics.com/investors/press-releases/news-details.
82. H. Lavreysen, *Discovery and early clinical development of novel mGlu2 receptor PAMs, American College of Neuropsychopharmacology 52nd Annual Meeting*, Hollywood, Florida, Dec 8–12, 2013.
83. *A study of JNJ-40411813 as supplementary treatment to an antidepressant in adults with depression and anxiety symptoms*, http://clinicaltrials.gov/show/NCT01582815.
84. *Addex Reports top-line data from ADX71149 phase 2a study in patients with major depressive disorder (MDD) with significant anxiety symptoms*, http://www.addextherapeutics.com/investors/pressreleases/news-details/article/addex-reports-top-line-data-fromadx71149-phase-2a-study-in-patients-with-major-depressive-disorder/.
85. A. A. Trabanco-Suárez, G. J. Tresadern, J. A. Vega-Ramiro and J. M. Cid-Núñez, WO 2009062676 A2, 2009.
86. G. Tresadern, J. M. Cid, G. J. Macdonald, J. A. Vega, A. I. de Lucas, A. García, E. Matesanz, M. L. Linares, D. Oehlrich, H. Lavreysen and I. Biesmans, *Bioorg. Med. Chem. Lett.*, 2010, **20**, 175.
87. A. A. Trabanco, G. Tresadern, G. J. Macdonald, J. A. Vega, A. I. de Lucas, E. Matesanz, A. García, M. L. Linares, S. A. Alonso de Diego, J. M. Alonso, D. Oehlrich, A. Ahnaou, W. Drinkenburg, C. Mackie, J. I. Andrés, H. Lavreysen and J. M. Cid, *J. Med. Chem.*, 2012, **55**, 2688.
88. J. M. Cid-Núñez, D. Oehlrich, A. A. Trabanco-Suárez, G. J. Tresadern, J. A. Vega-Ramiro and G. J. Macdonald, WO2010130424 A1, 2010.
89. J. M. Cid, G. Tresadern, J. A. Vega, A. I. de Lucas, E. Matesanz, L. Iturrino, M. L. Linares, A. García, J. I. Andrés, G. J. Macdonald, D. Oehlrich, H. Lavreysen, A. Megens, A. Ahnaou, W. Drinkenburg, C. Mackie, S. Pype, D. Gallacher and A. A. Trabanco, *J. Med. Chem.*, 2012, **55**, 8770.
90. J. M. Cid-Núñez, A. A. Trabanco-Suárez, H. Lavreysen and M. Ceusters, WO 2015032790 A1, 2015.
91. J. I. Andrés, J. Alcazar, J. M. Cid, M. De Angelis, L. Iturrino, X. Langlois, H. Lavreysen, A. A. Trabanco, S. Celen and G. Bormans, *J. Med. Chem.*, 2012, **55**, 8685.
92. M. Schmidt, *J. Cereb. Blood Flow Metab.*, 2012, **32**, S13–S40, DOI: 10.1038/jcbfm.2012.77.

93. R. J. Mattson and Z. Meng, WO 2013138687 A1, 2013.
94. L. R. Marcin, M. A. Higgins, J. J. Bronson, F. C. Zusi, J. E. Macor and M. Ding, WO2015042243 A1, 2015.
95. V. Dudkin, M. E. Fraley, C. Wang and J. Steen, WO 2010114726 A1, 2010.
96. D. C. Beshore, V. Dudkin, S. D. Kuduk, J. W. Skudlarek and C. Wang, WO 2010141360 A1, 2010.
97. V. Y. Dudkin, M. E. Fraley, C. Wang, R. M. Garbaccio, D. C. Beshore, S. D. Kuduk and J. W. Skudlarek, WO 2011022312 A1, 2011.
98. D. C. Beshore, V. Dudkin, R. M. Garbaccio, A. W. Johnson, S. D. Kuduk, J. W. Skudlarek, C. Wang and M. E. Fraley, US 20120135977 A1, 2012.
99. D. C. Beshore, R. M. Garbaccio, S. D. Kuduk, A. W. Johnson and J. W. Skudlarek, WO 2012151136 A1, 2012.
100. D. C. Beshore and S. D. Kuduk, WO 2012151140 A1, 2012.
101. D. C. Beshore and S. D. Kuduk, WO 2012151138 A1, 2012.
102. S. D. Kuduk and J. W. Skudlarek, 2012151139 A1, 2012.
103. K. L. Arrington, V. Dudkin, M. E. Layto, J. E. Pero and A. J. Reif, WO 2011156245 A1, 2011.
104. V. Dudkin, M. E. Fraley, K. L. Arrington, M. E. Layton, K. J. Rodzinak, J. E. Pero and A. J. Alexander, WO2012021382 A1, 2012.
105. M. E. Layton and M. J. Kelly, WO2011137046 A1, 2011.
106. A. S. Doré, K. Okrasa, J. C. Patel, M. Serrano-Vega, K. Bennett, R. M. Cooke, J. C. Errey, A. Jazayeri, S. Khan, B. Tehan, M. Weir, G. R. Wiggin and F. H. Marshall, *Nature*, 2014, **511**, 557.
107. H. Wu, C. Wang, K. J. Gregory, G. W. Han, H. P. Cho, Y. Xia, C. M. Niswender, V. Katrich, J. Meiler, V. Cherezov, P. J. Conn and R. C. Stevens, *Science*, 2014, **344**, 58.
108. J. A. Christopher, S. J. Aves, K. A. Bennett, A. S. Doré, J. C. Errey, A. Jazayeri, F. H. Marshall, K. Okrasa, M. J. Serrano-Vega, B. G. Tehan, G. R. Wiggin and M. Congreve, *J. Med. Chem.*, 2015, **58**, 6653.
109. A. Farinha, H. Lavreysen, L. Peeters, B. Russo, S. Masure, A. A. Trabanco, J. M. Cid and G. Tresadern, *Br. J. Pharmacol.*, 2015, **172**, 2383.
110. A. Ahnaou, H. Lavreysen, G. Tresadern, J. M. Cid and W. H. I. M. Drinkenburg, *PLoS One*, 2015, **10**, e0144017.
111. M. L. J. Doornbos, L. Pérez-Benito, G. Tresadern, T. Mulder Krieger, I. Biesmans, A. A. Trabanco, J. M. Cid, H. Lavreysen, A. P. IJzerman and L. H. Heitman, *Br. J. Pharmacol.*, 2016, **173**, 588–600.

CHAPTER 8

Muscarinic Receptors Allosteric Modulation

BRUCE J. MELANCON[a] AND COREY R. HOPKINS*[b]

[a]Department of Chemistry and Biochemistry, Warren Family Research Center for Drug Discovery and Development, University of Notre Dame, Notre Dame, IN 46556, USA; [b]Department of Pharmaceutical Sciences, College of Pharmacy, University of Nebraska Medical Center, Omaha, NE, 68198 USA

*E-mail: corey.hopkins@unmc.edu

8.1 Introduction

In molecular pharmacology of receptors, the binding site for the endogenous ligand is termed the orthosteric site. An allosteric site is one that is topographically distinct from the orthosteric site, usually potentiating or inhibiting the binding and/or signaling of the endogenous ligand.[1–6] Allosteric ligands can possess a number of modes of pharmacology including potentiation (positive allosteric modulators (PAMs)) of the receptor signals, ablation (negative allosteric modulators (NAMs)) of the receptor signals, and having no effect on the receptor signals (neutral allosteric ligands (NALs), formerly silent allosteric modulators (SAMs)) but block the binding of other PAMs or NAMs.[2] The use of allosteric ligands imparts an exquisite level of selectivity, as many allosteric compounds bind on sites of a receptor that differ from

RSC Drug Discovery Series No. 56
Allosterism in Drug Discovery
Edited by Dario Doller

Published by the Royal Society of Chemistry, www.rsc.org

those of the closest family members in the same class. Allosteric modulators often operate under a dependence on the native ligand, exploiting the spatial and temporal signaling that occurs during binding. This results in less desensitization of the receptor to the ligand, less down-regulation of the receptor (decreased signaling), and often limits the off-target activity as compared with an orthosteric compound. Conversely, many orthosteric compounds suffer from a lack of selectivity due to homology of the binding site within a receptor family, and promiscuous binding at off-target or "out-of-family" receptor classes. It is no surprise that many small-molecule discovery programs that target receptors have endeavored to identify allosteric compounds to address selectivity and desensitization issues associated with their orthosteric counterparts. With recent advances in screening technologies and pharmacology, allosteric compounds have been developed for a variety of receptor types including kinases, ion channels, phospholipases, and G-protein coupled receptors (GPCRs).[7]

G-Protein coupled receptors (GPCRs) are also known as seven-transmembrane-spanning receptors (7TMRs) and represent the largest class of cell surface receptors, comprising the targets of 40% of all marketed drugs.[2] Upon binding of the signaling molecule, the receptor conformation is altered, activating signaling cascades by coupling to G-proteins and other secondary proteins. GPCRs are commonly divided into three families or classes (families A, B, and C), based on their amino acid sequences; family B differs from family A by the presence of a larger extracellular loop. Family C possesses a bi-lobed N-terminal Venus Fly Trap (VFT) domain.[1–6] The orthosteric bind sites also differ in their locations as well. The orthosteric binding site for many family A GPCRs are located in the transmembrane domains while family B binding sites are in the extracellular loop. Family C binding sites are in the VFT domain. The endogenous ligand also differs between families; family A and C ligands are neurotransmitters acetylcholine (ACh) and glutamate, respectively, while family B ligands are peptides of over 30 amino acids.

Family A GPCRs are comprised of many types of receptors including those for adenosine and cannabinoids, dopaminergic, muscarinic, and serotonin.[8] Of these, muscarinic acetylcholine receptors (mAChRs) are stimulated by the neurotransmitter acetylcholine. There are five subtypes, defined as M_1–M_5, which are further delineated by their coupling to signal transduction pathways. When stimulated by ACh, M_1, M_3, and M_5 induce release of intracellular calcium stores through the activation of the phospholipase C *via* $G\alpha_q$. M_2 and M_4 couple to $G\alpha_{i/o}$ to regulate adenylyl cyclase and many ion channels.[9–13] These receptors are localized in various regions of the body, where M_1 and M_4 are predominately expressed in the central nervous system (CNS), in addition to lower levels of M_5. M_2 is localized in the cardiac tissues and M_3 is highly expressed in the gut.[9,10] Historically, it is the localization of M_1 and M_4, which attracted much attention due to the promise of therapeutic potential for cognitive disorders. Many preclinical studies with nonselective muscarinic agonists showed improvements in cognitive function in rat and mouse, and clinical studies also showed improvement in humans. Acetylcholinesterase (AChE) inhibitors displayed enhancements in cholinergic transmission in patients suffering from Alzheimer's disease (AD). While muscarinic receptors held much promise as druggable targets, these programs were plagued with compounds reported to be selective, efficacious ligands but which were in fact orthosteric compounds with off-target activity. This was especially problematic for muscarinic receptors due to the high conservation of sequence homology between receptor subtypes. In both preclinical and clinical studies, nonselective muscarinic agonists displayed adverse effects mediated through M_2 and M_3 receptors (bradycardia, gastrointestinal (GI) distress, excessive salivation and sweating, and emesis). The development of more selective orthosteric compounds to avoid these adverse effects is a sound principle, but this often proved a difficult objective to achieve. Allosteric compounds often provide higher selectivity, but until recently were not widely identified for muscarinic receptors. Renewed interest of academic researchers in muscarinic receptors has flourished due to many investigational efforts into new, selective orthosteric and allosteric compounds. This research will be aided by the recent disclosure of the first crystal structures of the M_1 and M_4 subtypes.[14] These structures show similarities between the orthosteric binding sites, as predicted, and Christopoulos and coworkers also indicate potential allosteric binding sites. Targeting these sites with new high-throughput and virtual screens promises to provide new undisclosed scaffolds with improvements and challenges compared with their disclosed counterparts.

This chapter will describe both allosteric and bitopic compounds reported in the primary literature identified as selective for either M_1 or M_4, regardless of pharmacology. As with any preclinical program, *in vivo* efficacy in established models will also be an established benchmark for consideration. There have been a number of excellent reviews over the past five years, and as such, this chapter will outline only recent advances over the past few years.[8,15–19]

8.2 Recent Advances with M_1 and M_4 Bitopic Ligands

Compounds with dual M_1 and M_4 activity, while not selective for a single receptor, have some therapeutic value. In fact, a central question in the area of schizophrenia therapies is whether M_1 or M_4 mAChR is responsible for the beneficial outcomes seen in clinical trials using muscarinic pan-agonists. It has also been suggested that compounds which may provide the best therapeutic value those that possess dual M_{1-4} agonist activity. A series of partial agonists, disclosed by Dainippon Sumitomo Pharma in 2013, displayed good selectivity for M_1 and M_4 over the other muscarinic subtypes (Figure 8.1).[20] Compound **1** was a 29 nM agonist and shown to be centrally penetrant and reversed methamphetamine- and apomorphine-induced psychotic-like behaviors in mice with minimal side effects. This compound was also evaluated against xanomeline in mice and while the treated animals still displayed GI distress when treated at dose levels that reversed psychotic-like behaviors, these effects were lower than those with xanomeline at the same dose.[20] Similarly, 2, also from Sumiyoshi and coworkers, showed good activation for M_1 and M_4 (hM_1 = 93% activation, hM_4 = 134% activation) with a better selectivity profile against M_2, M_3 and M_5. It was also classified as a weak human ether-a-go-go related gene (hERG) inhibitor (concentration giving 50% of maximum inhibition (IC_{50}) = 2.8 μM)). Compound 2 was orally bioavailable and reversed methamphetamine-induced psychotic-like behaviors in rats.[21] Further optimization led to 3 (concentration giving 50% of maximum effectiveness (EC_{50}) for hM_1 = 170 nM, EC_{50} for hM_4 = 47 nM), whose origin from 1 and 2 is evident and addresses the hERG liabilities of the first- and second-generation compounds. Incorporation of a pyridine in the indoline ring system and subsequent acylation to give tertiary amides (*e.g.*, 3) ablated any intramolecular hydrogen bonding, and lowered hERG inhibition while also lowering cLogP.[22]

Bitopic ligands can often be confounding in muscarinic research and can display unusual pharmacology for the various subtypes. Dependence on receptor reserve and the loss of subtype selectivity are common reasons why bitopic agonists languish in the preclinical development stages. However, there remains inherent value in the study of their *in vitro* pharmacology and electrophysiological effects, since muscarinic receptor density is not uniform throughout the CNS. An example of a bitopic ligand that highlights these challenges is compound 4, from Piergentili and coworkers (Figure 8.2).[23] Compound 4 is an M_1–M_4 hybrid bitopic ligand with structural

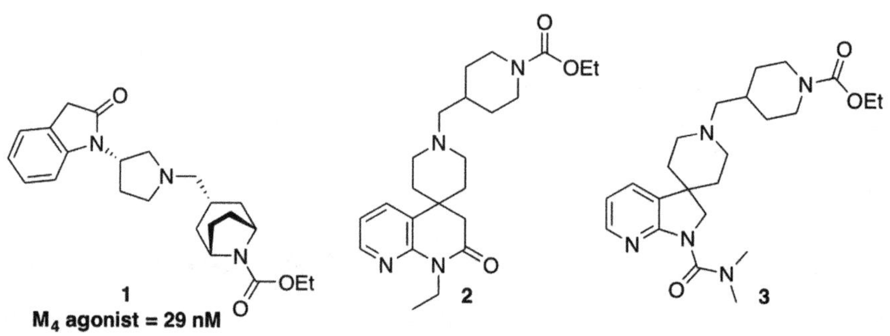

Figure 8.1 Bitopic M_1–M_4 mAChR agonists.

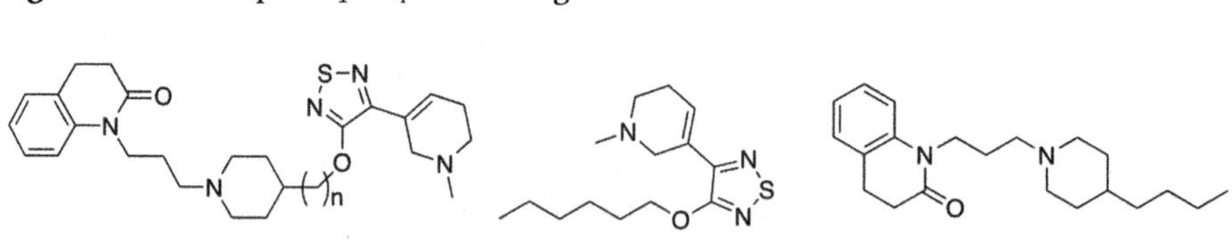

Figure 8.2 An M_1–M_4 hybrid agonist, 4, with Xanomeline, 5, an orthosteric agonist and 77-LH-28-1, 6, an allosteric agonist.

components from the orthosteric pan-agonist, xanomeline, **5**, and the M_1 allosteric agonist 77-LH-28-1, **6**. It was postulated that this type of hybrid may possess better affinity for the M_1 and M_4 receptors, and perhaps a synergistic effect would be observed. However, **4** and other close structural analogs showed pan-antagonist activity at M_1–M_5. Piergentili and coworkers were able to report on interesting results related to signal bias for these compounds; some structural analogs of **4** preferentially activated G-protein signaling while others recruited *β*-arrestin in a robust manner.[23] The details of these studies indicated the depth of the challenges related to bitopic ligands as development candidates, which often prove too difficult to overcome.

8.3 Recent Advances with M_4 Allosteric Ligands

The role of muscarinic receptors as novel targets for the treatment of symptoms of schizophrenia is a highly active area of research in recent years.[7,24–26] By potentiating the activity of the M_4 receptor, many investigators demonstrated the ability to reverse psychotic-like effects in preclinical models in rats and mice. Additionally, M_4 has been implicated in improving cognitive deficits,[27] modulating circadian activity rhythms,[28] impacting sleep–wake architecture,[29] and improving motor and synaptic deficits in a mutant mouse model of Huntington's disease.[30,31]

8.3.1 Current Efforts Using LY2033298

In the context of positive allosteric modulators, LY2033298 (**7**) represents the inspiration for discovery programs to investigate small-molecule potentiators of M_4 (Figure 8.3).[24] Disclosed by Eli Lilly in the late 2000's, LY2033298 as a potent M_4 PAM (hM_4 EC_{50} = 65 nM), it also showed some potentiation of M_2 (hM_2 EC_{50} = 1.5 μM). In combination with its high clearance and off-target activity, this resulted in it being discontinued as a development candidate. However, Christopoulos and coworkers have employed it as a probe to analyze allosteric modulation of muscarinic signaling. Through several studies of the M_2 mAChR and the cooperativity of LY2033298 and orthosteric compounds such as xanomeline they determined that LY2033298 adopts a different orientation of binding in the allosteric site of M_2 *versus* M_4, and

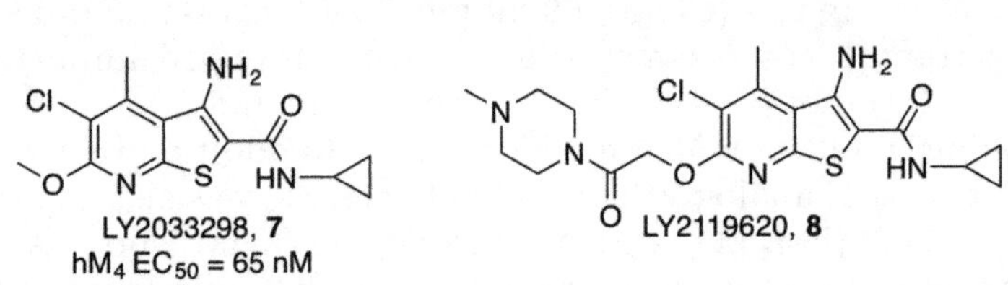

Figure 8.3 M_4 PAMs from Eli Lilly and Co.

that the selectivity of LY2033298 for M_4 arises from a difference in cooperativity with ACh across all mAChRs. This type of probe dependence has relevance to many programs in this area. Allosteric effects may be missed if neutral cooperativity is occurring and results can be confounding.[32] Felder and coworkers have also developed radiolabeled analogs of LY2033298, specifically $[^3H]$LY2119620 (**8**).[33] This compound was shown to bind allosterically to both M_2 and M_4 through $[^3H]$*N*-methylscopolamine assays and was displaced by compounds of similar chemical structure, notably VU0152100 (**9**) and LY2033298.[6] Autoradiography in cynomolgus monkey brain sections showed high concentrations of radiolabeled compound present where M_4 is expressed. These results related to this scaffold give important context under which the compounds from the Vanderbilt Center for Neuroscience Drug Discovery (VCNDD) were developed.

8.3.2 M_4 Positive Allosteric Modulators Developed at Vanderbilt University

The VCNDD has worked in the area of M_4 allosteric modulation since 2011. Through the Molecular Libraries Probe Production Centers Network, a part of the National Institutes of Health (NIH) Molecular Libraries Roadmap and through sponsored research with external pharmaceutical companies, the VCNDD developed allosteric modulators for M_1, M_4, and M_5 muscarinic receptors. This section will cover a series of M_4 PAMs developed by the VCNDD for the treatment of schizophrenia (Figure 8.4).

In the development of novel M_4 PAMs with efficacy in preclinical models of schizophrenia, several compounds from the VCNDD have served as excellent tool compounds for validating the hypothesis that an M_4 PAM could reverse these effects given the correct pharmacokinetic profile. Encouraged by the disclosure of LY2033298, **7**, a structure–activity relationship (SAR) campaign resulted in the first selective rat M_4 probe VU0152100 (ML108, **9**).[34] Compound **9** was a potent M_4 PAM (rM_4 EC_{50} = 150 nM); however, this compound displayed a species preference for the rat receptor (hM_4 EC_{50} = 560 nM), a finding common in M_4 PAM work, and was over 100-fold selective against the other muscarinic subtypes. Compound **9** was rapidly cleared both in microsome experiments (rCL_{HEP} = 65 ml min^{-1} kg^{-1} and hCL_{HEP} = 18 ml min^{-1} kg^{-1}, where CL_{HEP} is the predicted hepatic clearance) and observed clearance *in vivo* in rats (CL_{obs} = 58 ml min^{-1} kg^{-1}).[35] Compound **9** was centrally penetrant (brain : plasma ratio = 0.66) and reversed amphetamine-induced hyperlocomotion as well as amphetamine-induced contextual fear conditioning in rats at a 56.6 mg kg^{-1} dose.[34] In addition to compound **9**, its close structural analog VU0448088 (ML253, **10**) was also reported to be a potent M_4 PAM (hM_4 EC_{50} = 56 nM and rM_4 = 176 nM) and was > 100-fold selective against the other muscarinic subtypes.[36] Similarly to **9**, **10** suffered from high observed clearance *in vivo* in rats (CL_{obs} = 103 ml min^{-1} kg^{-1})

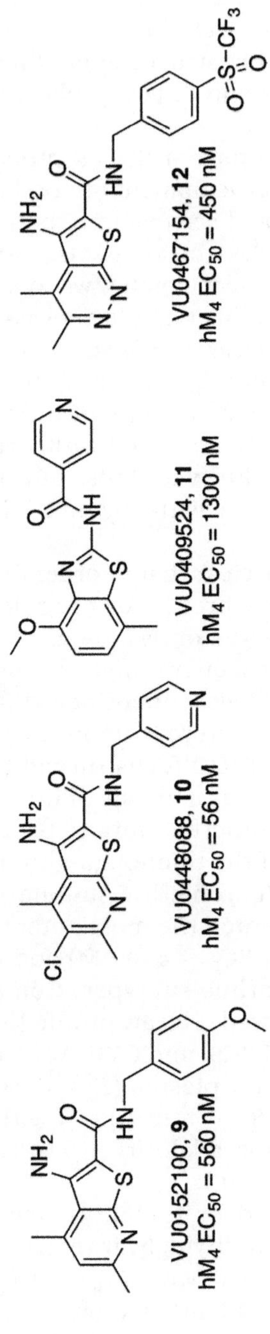

Figure 8.4 M_4 PAMs from the Vanderbilt Center for Neuroscience Drug Discovery based on the LY2033298 core scaffold.

and a very short half-life ($t_{1/2}$ = 17 min). Despite the high clearance, compound **10** was centrally penetrant (brain : plasma = 0.88) and reversed amphetamine-induced hyperlocomotion in rats (53% reversal at 56.6 mg kg^{-1} dose).[36] Neither of these scaffolds could overcome the pharmacokinetic (PK) limitations with other structural modifications. While they serve as excellent tool compounds and *in vivo* probes, they were abandoned as drug development candidates.

Notwithstanding the limitations of the PK properties of **9** and **10**, there was still great interest in the development of novel M_4 tool compounds with improved properties. Additional screens were performed at Vanderbilt and revealed a novel structural class of M_4 PAMs, exemplified by VU0409524 (ML293, **11**).[37] A modest decrease in potency was observed for this compound (hM_4 EC_{50} = 1.3 μM); however, this scaffold showed an improved PK profile. Compound **11** displayed low clearance *in vivo* in rats (CL_{obs} = 11.5 ml min^{-1} kg^{-1}) and an improved half-life ($t_{1/2}$ = 57 min) and showed excellent brain exposure (brain : plasma = 0.85).[37] In spite of the robust PK improvements for **11**, its potency could not be improved with structural modifications in order to bring its activity into line with those of compounds **9** and **10**. This resulted in its exclusion from use with any preclinical models of psychotic-like behavior.

At this point in time, it was clear that in order to develop a potent, selective M_4 PAM with a robust PK profile for drug development, a new scaffold would need to be discovered or designed. After mining the screening results from additional subsequent high-throughput screening (HTS) campaigns, no new scaffolds were identified. Efforts then turned to *de novo* synthesis of scaffolds to improve upon the properties of known M_4 PAMs (*i.e.*, **9** and **10**). The VCNDD focus turned toward the thieno[2,3-*b*] pyridine ring of VU0152100, **9**, as a means of improving both potency and PK properties. A limited synthetic effort returned a novel thieno[2,3-*c*] pyridazine moiety in place of the thieno[2,3-*b*]pyridine scaffold. Coupling to an interesting benzyltrifluoromethyl sulfone resulted in VU0467154, **12**.[38] This compound represented the most potent rat selective M_4 PAM to date (rM_4 EC_{50} = 18 nM; hM_4 EC_{50} = 450 nM) and was over 100-fold selective against the other muscarinic subtypes. Compound **12** was analyzed for P450 inhibition and showed a clean profile (>30 μM against four isoforms; CYP1A2, CYP2D6, CYP2C9, and CYP3A4). VU0467154, **12**, had good unbound concentrations in rat plasma (F_u = 0.031) and was a low-clearance compound (rCL_{obs} = 7.8 ml min^{-1} kg^{-1}) with a good half-life ($t_{1/2}$ = 342 min). It had good oral bioavailability (F = 61%) and had modest brain penetrance (brain : plasma = 0.46).

VU0467154, **12**, was evaluated for its ability to reverse psychotic-like symptoms in preclinical models. Gratifyingly, it reversed amphetamine-induced hyperlocomotion in rats with an *in vivo* EC_{50} of 710 nM (42% reversal at 56.6 mg kg^{-1}).[38] In the presence of MK-801, a known *N*-methyl-D-aspartate receptor (NMDAR) antagonist, **12**, reversed the MK-801-induced deficits in mice.

Compound **12** was also evaluated for its ability to improve cognitive impairments. Using a touchscreen-based cognitive task, mice were trained and then impaired with MK-801. Compound **12** reversed these impairments as well as MK-801-induced impairments for a context-conditioned fear model.[9,38] When these data are taken in context with other compounds reported from the muscarinic community **12** represents the best-in-class rodent tool compound for the study of M_4 mAChR activation as a therapy for psychotic symptoms of schizophrenia.

8.4 Recent Advances in M_1 Positive Allosteric Modulators

8.4.1 Quinolinone and Quinolizidinone Scaffolds

With the pioneering clinical studies with xanomeline, **5**, an M_1–M_4-preferring agonist, researchers have been trying to dissect the contributions of the M_1 and M_4 receptors in multiple disease states.[39–42] BQCA (1-(4-methoxybenzyl)-4-oxo-1,4-dihydroquinoline-3-carboxylic acid; benzyl quinolone carboxylic acid), **13**, was identified *via* a high-throughput screen of the Merck Research Laboratories (MRL) sample collection and has served as a starting point for multiple efforts and iterations over the past several years in order to develop selective M_1 positive allosteric modulators (Figure 8.5).[43] BQCA has served as a valuable research tool, however, it has several limitations which hinder future development (low potency and CNS penetration). The group at MRL has made several modifications to the scaffold over the past 7 years—identifying the carboxylic acid as a key residue for potency at the M_1 receptor.[44–46] More recently, the group has identified a quinolinone replacement as the core scaffold—quinolizidinone (**14–16**). An original scaffold of the quinolizidinone was reported in 2010 (**14**)[47] and further elaborated to the cyano piperidine compound, **15**.[48] Compound **15** (also known as 1-((4-*cyano*-4-(pyridine-2-yl)piperidin-1-yl) methyl-4-oxo-4*H*-quinolizine-3-carboxylic acid (PQCA)) was more potent (M_1 potentiator inflection point (Pot IP) = 135 nM) and displayed significantly better free fraction and *in vivo* PK properties, including brain penetration, although the brain : plasma ratio remained low (<0.3).[48] PQCA, **15**, has been extensively studied *in vivo* and has been shown to have efficacy in a number of pre-clinical animal models, including, improved performance of memory and attention in rodents and non-human primates[49,50] as well as improvements in learning and memory in a Tg2576 Alzheimer's disease mouse model.[51] Next, the group looked for replacements for the carboxylic acid group that was previously shown to be required for activity. They were able to identify a series of carboxamides, namely the *trans*-1S,2S-2-aminocyclohexanol, with similar potency to the carboxylic acid (*e.g.*, **16**), however, these cyclohexanol amides showed varying degrees of agonist activity.[52]

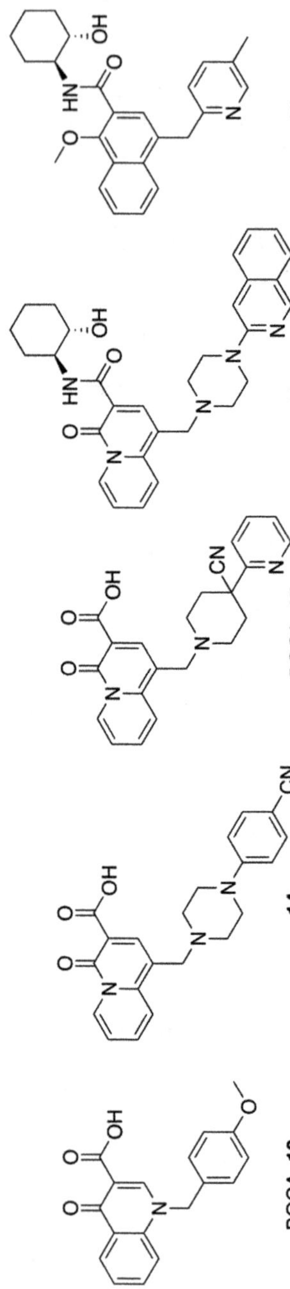

BQCA, **13**
M_1 Pot IP = 660 nM

14
M_1 Pot IP = 520 nM

PQCA, **15**
M_1 Pot IP = 135 nM

16
M_1 Pot IP = 110 nM

17
M_1 Pot IP = 121 nM

Figure 8.5 Structures of Merck's BQCA and quinolizidinone analogs.

Compound **16** was shown to be a selective M_1 positive allosteric modulator, and unlike the other 2-aminocyclohexanol analogs, was not shown to be a P-glycoprotein (P-gp) substrate. Compound **16** provided efficacy in a mouse contextual fear model of episodic memory.[52] The Merck team next looked for alternative scaffolds to the quinolinone or quinolizidinone moiety. The key component to a replacement would be to keep the internal hydrogen bond between the upper carbonyl and the amide NH present. They were able to successfully navigate the scaffold change by incorporating a methoxy naphthalene moiety, **17**.[53] In addition to maintaining the potency (M_1 Pot IP = 121 nM), they were able to address the P-gp liabilities associated with the quinolizidinone scaffold.

8.4.2 Tricyclic Positive Allosteric Modulators

Based on the knowledge and desire to maintain the internal hydrogen bond, the Merck group further elaborated this concept by cyclizing the moiety into a tricyclic core scaffold (*e.g.*, **18**) (Figure 8.6).[54] Cyclization of the upper moiety was a significant advancement because it showed that the quinolone or quinolizidinone carbonyl was not required for activity, nor was the methoxy of the naphthalene compound, highlighting a key structural feature that was assumed to be required for some time. Compound **18** was also shown to demonstrate high cerebrospinal fluid levels and limited P-gp liabilities, and was active in the mouse contextual fear model.[54] The compound did show agonist activity at higher concentrations. In addition to the Merck group, the group at the VCNDD also reported on a tricyclic scaffold as a selective M_1 PAM. Their efforts initiated with a screen of the NIH Molecular Libraries Small Molecule Repository (MLSMR) compound collection (~300 000 compounds). From this, dibenzyl-2*H*-pyrazolo[4,3-*c*]quinolin-3(5*H*)-one (DBPQ) (**19**), was identified as a novel structure for the M_1 receptor.[55] This scaffold incorporates a similar architecture to the BQCA scaffold—keeping the para-relationship between the quinolone nitrogen and the pyrazolo nitrogen. The VCNDD

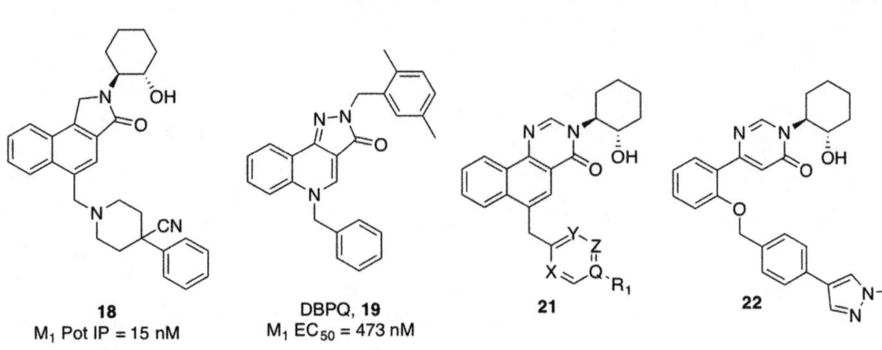

Figure 8.6 Tricyclic scaffold analogs as M_1 PAMs.

has reported several key SAR points about this molecule and although it has modest potency, this represents a key novel scaffold within the literature. A more recent report is from the group at Monash University of a key scaffold that is taken from the Merck patent literature.[56] The Monash group reports on a pyrimidin-4(3*H*)one, **21**,[57] scaffold which is found in a number of Merck's patents (*e.g.*, **20**)—albeit in the patent the molecule is tricyclic in nature. The Monash group breaks a bond in the naphthalene core and replaces this with a benzyl ether moiety.

8.4.3 Indole–Oxindole Scaffolds

Previously, the VCNDD group reported the lead optimization of an isatin-containing M_1 selective PAM.[58] This same group has recently developed a series of compounds that has removed the isatin moiety of the molecule, revealing a series of substituted or spirocyclic oxindole compounds to be selective M_1 PAMs.[59,60] The highlighted compound, **22**, showed modest potency as an M_1 PAM, however, this represented a novel chemical scaffold for further optimization (Figure 8.7). From a screen of the MLSMR library, the VCNDD group identified a series of 3-sulfonyl indole compounds, **23**.[61,62] The initial HTS structure was a 3-sulfonyl indole compound,[61] and further optimization led to the incorporation of the 4,6-difluoro atoms which, when coupled with heteroaryl moieties (pyridyl) in the southern portion, imparted significant improvements in potency.[62] Evaluation of **23** showed that this compound was able to shift the amyloid protein precursor (APP) processing toward a nonamyloidogenic pathway. This process is measured by an increase in soluble APP (APPsα) release in TREx293-hM_1 cells,[62] suggesting that M_1 PAMs may be disease-modifying in Alzheimer's disease. Lastly, a group from Pfizer recently disclosed their efforts towards the discovery and development of a novel azaindole amide M_1 PAM.[63] This group used design elements from the Merck aminocyclohexanol amides

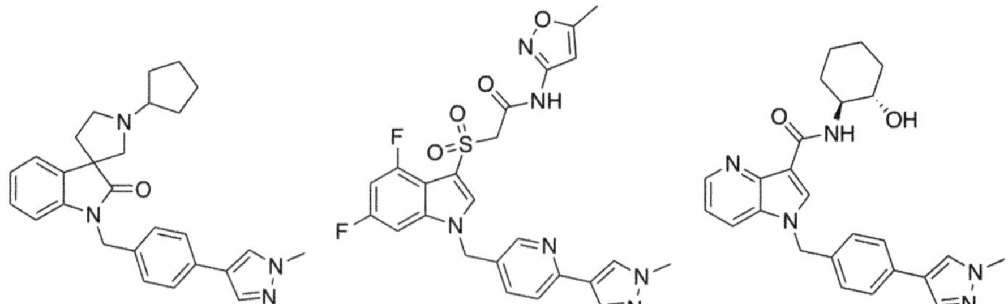

22
M_1 EC_{50} = 1.8 μM

23
M_1 EC_{50} = 310 nM

24
M_1 EC_{50} = 55 nM

Figure 8.7 Oxindole and indole M_1 PAMs.

as well as the indole core scaffolds first disclosed by the VCNDD. Merging these two scaffolds yielded the advanced lead compound, **24**, which is a potent and selective M_1 PAM. Compound **24** had modest brain penetration (unbound brain concentration (Cb_u): unbound plasma concentration (Cp_u): 0.2) and was active in an amphetamine-stimulated locomotor activity study in mice.[63]

8.5 Recent Advances in M_5 Negative and Positive Allosteric Modulation

Unlike M_1 and M_4, the muscarinic acetylcholine receptor subtype 5 (M_5) has received less attention due to the lack of understanding of localization and expression, as well as of subtype-selective ligands. However, recently, M_5 has started to become more prominent due to the results of genetic studies in M_5-knockout (KO) mice linking the M_5 receptor and addiction.[64–66] Thus, to better understand the role of M_5 in a variety of disease areas, there was a need to develop selective tool compounds. To this end, the VCNDD group performed a high-throughput screen of the MLSMR compound library and identified a novel tetrahydro-5*H*-imidazo[2,1-*a*]isoindol-5-one scaffold as a weak M_5 NAM (Figure 8.8). The group performed a lead optimization campaign and identified compound **25** as the first, selective NAM of the M_5 receptor with high CNS penetration.[67] Further optimization of **25**, led to the discovery of a close analog, **26**.[68]

In addition to the dearth of selective M_5 NAMs, there were, until recently, no reports of selective M_5 PAMs. This in spite of KO mice data showing the animals display a decrease in prepulse inhibition (a recognized model of psychosis) as well as CNS neuronal and cerebrovascular irregularities, leading to cognitive shortfalls which could have implications for conditions such as Alzheimer's disease, schizophrenia, and ischemic stroke.[69,70] In addition to the discovery of the first selective M_5 NAMs, the VCNDD group has been the

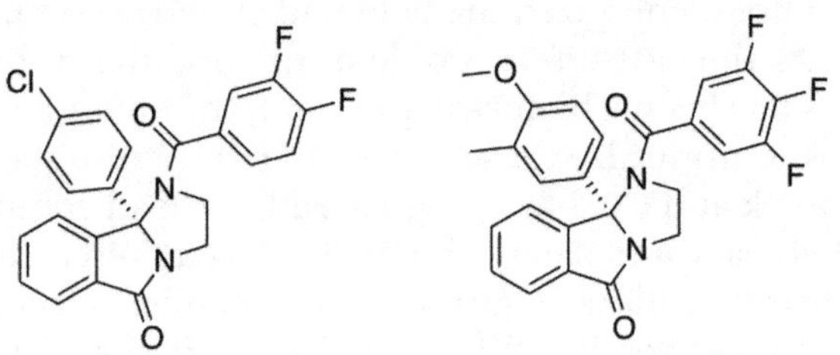

25
hM_5 IC_{50} = 300 nM

26
hM_5 IC_{50} = 900 nM

Figure 8.8 Recently reported M_5 NAMs.

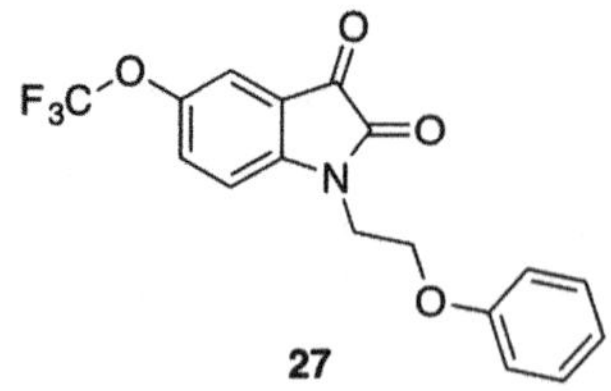

27
hM_5 EC_{50} = 409 nM

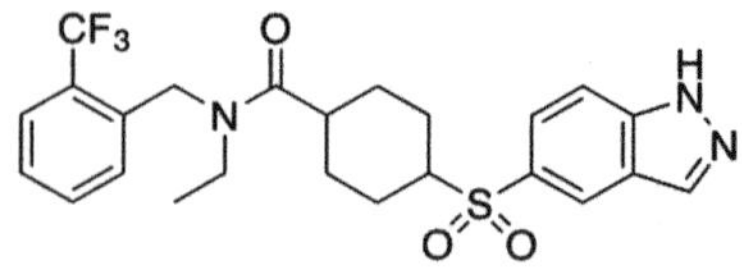

28
hM_5 EC_{50} = 190 nM

Figure 8.9 Selective M_5 PAMs.

leader in the discovery and development of selective M_5 PAMs (Figure 8.9). The first M_5 PAM from the Vanderbilt group is from an isatin scaffold that originated from a high-throughput screen and was originally reported as a non-selective muscarinic modulator. The group changed the *N*-linked benzyl group to *N*-ethoxyphenyl and was able to identify a potent and selective M_5 PAM, **27**. Lastly, from the same HTS of the MLSMR compound library that led to the identification of the selective M_5 NAMs (*vide supra*), the Vanderbilt group identified a novel piperidine sulfonamide scaffold as a weak M_5 PAM. Further optimization of this scaffold lead to the identification of **28**, a potent, selective and CNS-penetrant M_5 PAM.

8.6 Conclusions

There has been significant activity in the development of novel, selective and brain-penetrant M_1, M_4, and M_5 positive and negative allosteric modulators over the past several years. The M_1–M_4 field has progressed significantly since the disclosure of xanomeline as an M_1–M_4-preferring agonist with advancements in the basic biological understanding of the receptor class which has been enabled by the discovery of multiple classes of selective tool compounds. The M_5 field has lagged behind the other receptor subtypes due to a lack of selective tool compounds; however, recent disclosures obtained using KO mice have reenergized the field and selective compounds are starting to emerge. As the side-effect profile of the original orthosteric agonists and antagonists has been a historical problem, the emerging allosteric modulators will look to minimize these issues. In addition to these selective tool compounds, Merck and Co. have progressed their lead compound, MK-7622 into a Phase 2 efficacy and safety clinical trial as adjunct therapy in participants with Alzheimer's disease. Although the chemical structure of MK-7622 has not been disclosed yet, it is thought to belong to the tricyclic quinazoline series, **20**. The future of drug discovery and development of allosteric modulators of the acetylcholine muscarinic receptors looks bright and the next several years will be pivotal as these compounds move through clinical trials towards the market.

References

1. T. Kenakin and L. J. Miller, *Pharmacol. Rev.*, 2010, **62**, 265.
2. P. J. Conn, A. Christopoulos and C. W. Lindsley, *Nat. Rev. Drug Discovery*, 2009, **8**, 41.
3. T. M. Bridges and C. W. Lindsley, *ACS Chem. Biol.*, 2008, **3**, 530.
4. A. Christopoulos and T. Kenakin, *Pharmacol. Rev.*, 2002, **54**, 323.
5. A. Christopoulos, *Nat. Rev. Drug Discovery*, 2002, **1**, 198.
6. T. Kenakin, *Trends Pharmacol. Sci.*, 2009, **30**, 460.
7. B. J. Melancon, C. R. Hopkins, M. R. Wood, K. A. Emmitte, C. M. Niswender, A. Christopoulos, P. J. Conn and C. W. Lindsley, *J. Med. Chem.*, 2012, **55**, 1445.
8. B. J. Melancon, J. C. Tarr, J. D. Panarese, M. R. Wood and C. W. Lindsley, *Drug Discovery Today*, 2013, **18**, 1185.
9. A. Levey, *Life Sci.*, 1993, **52**, 441.
10. P. Abrams, K.-E. Andersson, J. J. Buccafusco, C. Chapple, W. C. de Groat, A. D. Fryer, G. Kay, A. Laties, N. M. Nathanson, P. J. Pasricha and A. J. Wein, *Br. J. Pharmacol.*, 2006, **148**, 565.
11. L. A. Volpicelli and A. I. Levey, *Prog. Brain Res.*, 2004, **145**, 59.
12. C. J. Langmead, J. Watson and C. Reavill, *Pharmacol. Ther.*, 2008, **117**, 232.
13. J. M. Gold, *Schizophr. Res.*, 2004, **72**, 21.
14. D. M. Thal, B. Sun, D. Feng, V. Nawaratne, K. Leach, C. C. Felder, M. G. Bures, D. A. Evans, W. I. Weis, P. Bachhawat, T. S. Kobilka, P. M. Sexton, B. K. Kobilka and A. Christopoulos, *Nature*, 2016, **531**, 335.
15. T. M. Bridges, E. P. LeBois, C. R. Hopkins, M. R. Wood, C. K. Jones, P. J. Conn and C. W. Lindsley, *Drug News Perspect.*, 2010, **23**, 229–240.
16. D. J. Foster, D. L. Choi, P. J. Conn and J. M. Rook, *Neuropsychiatr. Dis. Treat.*, 2014, **10**, 183.
17. S. D. Kuduk and D. C. Beshore, *Expert Opin. Ther. Pat.*, 2012, **22**, 1385.
18. S. D. Kuduk and D. C. Beshore, *Curr. Top. Med. Chem.*, 2014, **14**, 1738.
19. C. W. Lindsley, K. A. Emmitte, C. R. Hopkins, T. M. Bridges, D. J. Gregory, C. M. Niswender and P. J. Conn, *Chem. Rev.*, 2016, **116**, 6707–6741.
20. T. Sumiyoshi, T. Enomoto, K. Takai, Y. Takahashi, Y. Konishi, Y. Uruno, K. Tojo, A. Suwa, H. Matsuda, T. Nakako, M. Sakai, A. Kitamura, Y. Uematsu and A. Kiyoshi, *ACS Med. Chem. Lett.*, 2013, **4**, 244.
21. K. Takai, Y. Inoue, Y. Konishi, A. Suwa, Y. Uruno, H. Matsuda, T. Nakako, M. Sakai, H. Nishikawa, G. Hashimoto, T. Enomoto, A. Kitamura, Y. Uematsu, A. Kiyoshi and T. Sumiyoshi, *Bioorg. Med. Chem. Lett.*, 2013, **23**, 4644.
22. K. Takai, Y. Inoue, Y. Konishi, A. Suwa, Y. Uruno, H. Matsuda, T. Nakako, M. Sakai, H. Nishikawa, G. Hashimoto, T. Enomoto, A. Kitamura, Y. Uematsu, A. Kiyoshi and T. Sumiyoshi, *Bioorg. Med. Chem. Lett.*, 2014, **24**, 3189.
23. A. Bonifazi, H. Yano, F. Del Bello, A. Farande, W. Quaglia, R. Petrelli, R. Matucci, M. Nesi, G. Vistoli, S. Ferré and A. Piergentilli, *J. Med. Chem.*, 2014, **57**, 9065.

24. W. Y. Chan, D. L. McKinzie, S. Bose, S. N. Mitchell, J. M. Witkin, R. C. Thompson, A. Christopoulos, S. Lazareno, N. J. M. Birdsall, F. P. Bymaster and C. C. Felder, *Proc. Natl. Acad. Sci. U. S. A.*, 2008, **105**, 10978.
25. A. C. Kruse, B. K. Kobilka, D. Gautam, P. M. Sexton, A. Christopoulos and J. Wess, *Nat. Rev. Drug Discovery*, 2014, **13**, 549.
26. C. K. Jones, N. Byun and M. Bubser, *Neuropsychopharm. Rev.*, 2012, **37**, 16.
27. E. T. Tzavara, F. P. Bymaster, C. C. Felder, M. Wade, J. Gomeza, J. Wess, D. L. McKinzie and G. G. Nomikos, *Mol. Psychiatry*, 2003, **8**, 673.
28. R. A. Hut and E. A. Van der Zee, *Behav. Brain Res.*, 2011, **221**, 466.
29. R. E. Brown, R. Basheer, J. T. McKenna, R. E. Strecker and R. W. McCarley, *Physiol. Rev.*, 2012, **92**, 1087.
30. K. W. Lange, F. Javoy-Agid, Y. Agid, P. Jenner and C. D. Marsden, *J. Neurol.*, 1992, **239**, 103.
31. T. Pancani, D. J. Foster, M. S. Moehle, T. J. Bichell, E. Bradley, T. M. Bridges, R. Klar, M. Poslusney, J. M. Rook, J. S. Daniels, C. M. Niswender, C. K. Jones, M. R. Wood, A. B. Bowman, C. W. Lindsley, Z. Xiang and P. J. Conn, *Proc. Natl. Acad. Sci. U. S. A.*, 2015, **112**, 14078.
32. C. Valant, C. C. Felder, P. M. Sexton and A. Christopoulos, *Mol. Pharmacol.*, 2012, **81**, 41.
33. D. A. Schober, C. H. Croy, H. Xiao, A. Christopoulos and C. C. Felder, *Mol. Pharmacol.*, 2014, **86**, 116.
34. N. E. Byun, M. Grannan, M. Bubser, R. L. Barry, A. Thompson, J. Rosanelli, R. Gowrishankar, N. D. Kelm, S. Damon, T. M. Bridges, B. J. Melancon, J. C. Tarr, J. T. Brogan, M. J. Avison, A. Y. Deutch, J. Wess, M. R. Wood, C. W. Lindsley, J. C. Gore, P. J. Conn and C. K. Jones, *Neuropsychopharmacology*, 2014, **39**, 1578.
35. M. L. Lewis, T. M. Bridges, C. M. Niswender, C. D. Weaver and C. W. Lindsley, Discovery of a Highly Selective in vitro and in vivo M4 Positive Allosteric Modulator (PAM), 2010, Probe Reports from the NIH Molecule Libraries Program, ID: NBK50687.
36. U. Le, B. J. Melancon, T. M. Bridges, P. N. Vinson, T. J. Utley, A. Lamsal, A. L. Rodriguez, D. Venable, D. J. Sheffler, C. K. Jones, A. L. Blobaum, M. R. Wood, J. S. Daniels, P. J. Conn, C. M. Niswender, C. W. Lindsley and C. R. Hopkins, *Bioorg. Med. Chem. Lett.*, 2013, **23**, 346.
37. J. M. Salovich, P. N. Vinson, D. J. Sheffler, A. Lamsal, T. J. Utley, A. L. Blobaum, T. M. Bridges, U. Le, C. K. Jones, M. R. Wood, J. S. Daniels, P. J. Conn, C. M. Niswender, C. W. Lindsley and C. R. Hopkins, *Bioorg. Med. Chem. Lett.*, 2012, **22**, 5084.
38. M. Bubser, T. M. Bridges, D. Dencker, R. W. Gould, M. Grannan, M. J. Noetzel, A. Lamsal, C. M. Niswender, J. S. Daniels, M. S. Poslusney, B. J. Melancon, J. C. Tarr, F. W. Byers, J. Wess, M. E. Duggan, J. Dunlop, M. W. Wood, N. J. Brandon, M. R. Wood, C. W. Lindsley, P. J. Conn and C. K. Jones, *ACS Chem. Neurosci.*, 2014, **5**, 920.

39. N. C. Bodick, W. W. Offen, A. I. Levey, N. R. Cutler, S. G. Gauthier, A. Satlin, H. E. Shannon, G. D. Tollefson, K. Rasmussen, F. P. Bymaster, D. J. Hurley, W. Z. Potter and S. M. Paul, *Arch. Neurol.*, 1997, **54**, 465.
40. N. R. Mirza, D. Peters and R. G. Sparks, *CNS Drug Rev.*, 2003, **9**, 159.
41. F. P. Bymaster, C. A. Whitesitt, H. E. Shannon, N. DeLapp, J. S. Ward, D. O. Calligaro, L. A. Shipley, J. L. Buelke-Sam, N. C. Bodick, L. Farde, M. J. Sheardown, P. H. Olesen, K. T. Hansen, P. D. Suzdak, M. D. B. Swedberg, P. Sauerberg and C. H. Mitch, *Drug Dev. Res.*, 1997, **40**, 158.
42. A. Shekhar, W. Z. Potter, J. Lienemann, K. Sundblad, J. Lightfoot, J. Herrara, F. Unverzagt, F. P. Bymaster and C. Felder, *40th Annual Meeting of American College of Neuropsychopharmacology*, Hawaii, Hawaii, 2001.
43. L. Ma, M. A. Seager, M. Wittmann, M. Jacobson, D. Bickel, M. Burno, K. Jones, V. K. Graufelds, G. Xu, M. Pearson, A. McCampbell, R. Gaspar, P. Shughrue, A. Danziger, C. P. Regan, R. Flick, D. Pascarella, S. Garson, S. Doran, C. Kreatsoulas, L. Veng, C. W. Lindsley, W. Shipe, S. Kuduk, C. Sur, G. Kinney, G. R. Seabrook and W. J. Ray, *Proc. Natl. Acad. Sci. U. S. A.*, 2009, **106**, 15950.
44. F. V. Yang, W. D. Shipe, J. L. Bunda, M. B. Nolt, D. D. Wisnoski, Z. Zhao, J. C. Barrow, W. J. Ray, L. Ma, M. Wittman, M. A. Seager, K. A. Koeplinger, G. D. Hartman and C. W. Lindsley, *Bioorg. Med. Chem. Lett.*, 2010, **20**, 531.
45. S. D. Kuduk, C. N. Di Marco, V. Cofre, D. R. Pitts, W. J. Ray, L. Ma, M. Wittman, M. Seager, K. Koeplinger, C. D. Thompson, G. D. Hartman and M. T. Bilodeau, *Bioorg. Med. Chem. Lett.*, 2010, **20**, 657.
46. S. D. Kuduk, C. N. Di Marco, V. Cofre, D. R. Pitts, W. J. Ray, L. Ma, M. Wittmann, L. Veng, M. A. Seager, K. Koeplinger, C. D. Thompson, G. D. Hartman and M. T. Bilodeau, *Bioorg. Med. Chem. Lett.*, 2010, **20**, 1334.
47. S. D. Kuduk, R. K. Chang, C. N. Di Marco, W. J. Ray, L. Ma, M. Wittman, M. A. Seager, K. A. Koeplinger, C. D. Thompson, G. D. Hartman and M. T. Bilodeau, *ACS Med. Chem. Lett.*, 2010, **1**, 263.
48. S. D. Kuduk, R. K. Chang, C. N. Di Marco, D. R. Pitts, T. J. Greshock, L. Ma, M. Wittmann, M. A. Seager, K. A. Koeplinger, C. D. Thompson, G. D. Hartman, M. T. Bilodeau and W. J. Ray, *J. Med. Chem.*, 2011, **54**, 4773.
49. J. M. Uslaner, D. Eddins, V. Puri, C. E. Cannon, J. Sutcliffe, C. S. Chew, M. Pearson, J. A. Vivian, R. K. Chang, W. J. Ray, S. D. Kuduk and M. Wittmann, *Psychopharmacology*, 2013, **225**, 21.
50. H. S. Lange, C. E. Cannon, J. T. Drott, S. D. Kuduk and J. M. Uslaner, *J. Pharmacol. Exp. Ther.*, 2015, **355**, 442.
51. V. Puri, X. Wang, J. D. Vardigan, S. D. Kuduk and J. M. Uslaner, *Behav. Brain Res.*, 2015, **287**, 96–99.
52. S. D. Kuduk, R. K. Chang, T. J. Greshock, W. J. Ray, L. Ma, M. Wittmann, M. A. Seager, K. A. Koeplinger, C. D. Thompson, G. D. Hartman and M. T. Bilodeau, *ACS Med. Chem. Lett.*, 2012, **3**, 1070.
53. S. D. Kuduk, C. N. Di Marco, J. R. Saffold, W. J. Ray, L. Ma, M. Wittmann, K. A. Koeplinger, C. D. Thompson, G. D. Hartman, M. T. Bilodeau and D. C. Beshore, *Bioorg. Med. Chem. Lett.*, 2014, **24**, 1417.

54. Z.-Q. Yang, Y. Shu, L. Ma, M. Wittmann, W. J. Ray, M. A. Seager, K. A. Koeplinger, C. D. Thompson, G. D. Hartman, M. T. Bilodeau and S. D. Kuduk, *ACS Med. Chem. Lett.*, 2014, **5**, 604.
55. C. Han, A. Chatterjee, M. J. Noetzel, J. Panarese, E. Smith, P. Chase, P. Hodder, C. M. Niswender, P. J. Conn, C. W. Lindsley and S. R. Stauffer, *Bioorg. Med. Chem. Lett.*, 2015, **25**, 384.
56. S. D. Kuduk, D. C. Beshore, C. Di Marco and T. J. Greshock, Aryl methyl benzoquinazolinone M1 receptor positive allosteric modulators, WO2010059773, 2010.
57. S. N. Mistry, M. Jörg, H. Lim, N. B. Vinh, P. M. Sexton, B. Capuano, A. Christopoulos, J. R. Lane and P. J. Scammells, *J. Med. Chem.*, 2016, **59**, 388.
58. T. M. Bridges, J. P. Kennedy, M. J. Noetzel, M. L. Breininger, P. R. Gentry, P. J. Conn and C. W. Lindsley, *Bioorg. Med. Chem. Lett.*, 2010, **20**, 1972.
59. B. J. Melancon, M. S. Poslusney, P. R. Gentry, J. C. Tarr, D. J. Sheffler, M. E. Mattmann, T. M. Bridges, T. J. Utley, J. S. Daniels, C. M. Niswender, P. J. Conn, C. W. Lindsley and M. R. Wood, *Bioorg. Med. Chem. Lett.*, 2013, **23**, 412.
60. M. S. Poslusney, B. J. Melancon, P. R. Gentry, D. J. Sheffler, T. M. Bridges, T. J. Utley, J. S. Daniels, C. M. Niswender, P. J. Conn, C. W. Lindsley and M. R. Wood, *Bioorg. Med. Chem. Lett.*, 2013, **23**, 1860.
61. P. R. Reid, T. M. Bridges, D. J. Sheffler, H. P. Cho, L. M. Lewis, E. Days, J. S. Daniels, C. K. Jones, C. M. Niswender, C. D. Weaver, P. J. Conn, C. W. Lindsley and M. R. Wood, *Bioorg. Med. Chem. Lett.*, 2011, **21**, 2697.
62. J. C. Tarr, M. L. Turlington, P. R. Reid, T. J. Utley, D. J. Sheffler, H. P. Cho, R. Klar, T. Pancani, M. T. Klein, T. M. Bridges, R. D. Morrison, A. L. Blobaum, Z. Xiang, J. S. Daniels, C. M. Niswender, P. J. Conn, M. R. Wood and C. W. Lindsley, *ACS Chem. Neurosci.*, 2012, **3**, 884.
63. J. E. Davoren, S. V. O'Neil, D. P. Anderson, M. A. Brodney, L. Chenard, K. Dlugolenski, J. R. Edgerton, M. Green, M. Garnsey, S. Grimwood, A. R. Harris, G. W. Kauffman, E. LaChapelle, J. T. Lazzaro, C.-W. Lee, S. M. Lotarski, D. M. Nason, R. S. Obach, V. Reinhart, R. Salomon-Ferrer, S. J. Steyn, D. Webb, J. Yan and L. Zhang, *Bioorg. Med. Chem. Lett.*, 2016, **26**, 650.
64. A. Fink-Jensen, I. Fedorova, G. Wörtwein, D. P. D. Woldbye, T. Rasmussen, M. Thomsen, T. G. Bolwig, K. M. Knitowski, D. L. McKinzie, M. Yamada, J. Wess and A. Basile, *J. Neurosci. Res.*, 2003, **74**, 91.
65. M. Thomsen, D. P. D. Woldbye, G. Wörtwein, A. Fink-Jensen, J. Wess and S. B. Caine, *J. Neurosci.*, 2005, **25**, 8141.
66. A. S. Basile, I. Fedorova, A. Zapata, X. Liu, T. Shippenberg, A. Duttaroy, M. Yamada and J. Wess, *Proc. Natl. Acad. Sci. U. S. A.*, 2002, **99**, 11452.
67. P. R. Gentry, M. Kokubo, T. M. Bridges, N. R. Kett, J. M. Harp, H. P. Cho, E. Smith, P. Chase, P. S. Hodder, C. M. Niswender, J. S. Daniels, P. J. Conn, M. R. Wood and C. W. Lindsley, *J. Med. Chem.*, 2013, **56**, 9351.

68. H. Kurata, P. R. Gentry, M. Kokubo, H. P. Cho, T. M. Bridges, C. M. Niswender, F. W. Byers, M. R. Wood, J. S. Daniels, P. J. Conn and C. W. Lindsley, *Bioorg. Med. Chem. Lett.*, 2015, **25**, 690.
69. M. Thomsen, G. Wörtwein, A. Fink-Jensen, D. P. D. Woldbye, J. Wess and S. B. Caine, *Psychopharmacology*, 2007, **192**, 97.
70. R. Araya, T. Noguchi, M. Yuhki, N. Kitamura, M. Higuchi, T. C. Saido, K. Seki, S. Itohara, M. Kawano, K. Tanemura, A. Takashima, K. Yamada, Y. Kondoh, I. Kanno, J. Wess and M. Yamada, *Neurobiol. Dis.*, 2006, **24**, 334.

CHAPTER 9

Positive Allosteric Modulators of Opioid Receptors

KRISTIN L. ROCKWELL† AND ANDREW ALT*[a]

[a]Leads Discovery and Optimization, Bristol–Myers Squibb, Wallingford, Connecticut, USA
*E-mail: andrew.alt@gmail.com

9.1 Opioid Receptors and Pain

Pain is the most common ailment for which people seek medical attention. In the United States, the cost of pain-related healthcare and lost work time is estimated at $100 billion per year.[1] Morphine and its derivatives remain the gold standard for treatment of acute and chronic pain. These drugs produce their therapeutic effect by acting as agonists at opioid receptors.[2] Opioid receptors belong to the class A family of heterotrimeric guanine nucleotide binding protein (G protein)-coupled receptors (GPCRs). Four opioid receptor types exist: μ, δ, κ, and opioid receptor-like 1 (ORL1). These four receptors share *ca.* 60% amino acid homology (primarily in the transmembrane domains). While all four opioid receptor types are known to produce some antinociceptive effects,[3–5] activation of the μ-opioid receptor produces the most powerful analgesia, and the analgesic effects of opioid drugs are primarily mediated through the μ receptor. Due to their clinical importance,

† Current address: Pfizer Global Research and Development, Pfizer Inc., Groton, Connecticut, USA.

RSC Drug Discovery Series No. 56
Allosterism in Drug Discovery
Edited by Dario Doller

Published by the Royal Society of Chemistry, www.rsc.org

μ-opioid receptors are among the most studied and well characterized GPCRs.

While highly efficacious for the treatment of pain, morphine and other μ-opioid receptor agonists are also associated with a number of untoward side effects. Respiratory depression and constipation are among the most serious side effects of μ-opioid receptor agonists. Additionally, μ-opioid agonists have considerable abuse potential and addiction liability, and this dramatically exacerbates the safety issues associated with μ-opioid agonists, particularly respiratory depression. The current epidemic of prescription opioid drug abuse has led to a more than fourfold increase in the number of accidental overdose deaths since 1999.[6,7] Finally, long-term use of μ-opioid agonists also leads to significant tolerance to their analgesic effects, resulting in escalation of the doses required for analgesic efficacy over time, again exacerbating the addiction and safety liabilities associated with their use.

Clearly, safer pain medications are urgently needed. To this end, a wide variety of structurally diverse opioid agonists have been developed which have defined selectivity between the various opioid receptor types,[8] and which exhibit varying degrees of intrinsic efficacy for producing receptor activation.[9] Combination therapy with two agonists has also been explored,[10] as well as alternative formulations designed to control the systemic release of opioid agonists and/or make their abuse more difficult.[11] To date, however, the success of these approaches has been limited. These diverse approaches share a single commonality in that they all target the orthosteric agonist binding site (the site where endogenous opioid peptides and all current opioid medications bind to the receptor). An alternative strategy is the discovery and development of allosteric opioid receptor modulators. Allosteric modulators bind and exert their effects at a site on the receptor that is distinct from the orthosteric ligand binding site.

9.2 Allosteric Modulation

Allosteric ligands can produce a variety of effects on receptor activity.[12] Allosteric agonists directly activate the receptor, and are in many ways similar to orthosteric agonists except that they bind to a distinct site on the protein. Positive allosteric modulators (PAMs) do not activate the receptor directly, but increase the ability of an orthosteric agonist to activate the receptor, usually by increasing the orthosteric agonist's affinity for the receptor and/or increasing the functional efficacy of the orthosteric agonist. It should be noted that many allosteric ligands that function as PAMs at lower concentrations can also have direct allosteric agonist activity at higher concentrations.[13–15] Such compounds are frequently referred to as "PAM–agonists" or "ago-PAMs."[16] Negative allosteric modulators (NAMs) inhibit activation of the receptor by reducing the affinity and/or the functional efficacy of the orthosteric ligand. Finally, silent allosteric modulators (SAMs), also referred to as neutral allosteric ligands (NALs), bind to allosteric sites but exert no functional effect on receptor activation.

Allosteric modulators have a number of advantages over their orthosteric counterparts.[12] One advantage of allosteric ligands is the potential for increased selectivity. Orthosteric binding sites are highly evolutionarily conserved, making it difficult to design orthosteric drugs that confer selectivity between various receptor subtypes that recognize the same endogenous agonist. Free from such evolutionary constrains, allosteric sites often have greater variance in their amino acid sequence compared with the orthosteric binding pocket, giving allosteric ligands greater potential to achieve selectivity at different subtypes of GPCRs. In the case of opioid receptors, agonists that are highly selective between the various opioid receptor types already exist.[17,18] Furthermore, the side effects associated with μ-opioid receptor agonists do not represent "off-target" effects (*i.e.*, effects mediated by other receptor types), and instead result from activation of the μ receptor in tissues or brain regions where receptor activation is undesirable.[19] Therefore, the potential for increased selectivity may be less important in the case of opioid receptors, as side effects are unlikely to be avoided by the development of more selective agonists.

Allosteric modulators have other potentially important advantages over orthosteric agonists and antagonists. PAMs, in particular, are very different from direct-acting agonists in that PAMs lack intrinsic agonist efficacy and only exert an effect when bound to the receptor in the presence of an orthosteric agonist. Therefore, in the absence of an exogenously administered agonist, PAMs are expected to selectively increase receptor activity only in tissues or regions where endogenous agonists are present, allowing the spatial and temporal characteristics of native signaling to remain intact.[12,20] This functional selectivity of PAM effects may be expected to produce fewer "on-target" (*i.e.*, μ-opioid receptor mediated) side effects than direct-acting agonists, which indiscriminately activate all of their receptors throughout the body, irrespective of the patterns of native signaling (Figure 9.1).

Another problem with traditional direct-acting agonists is the sustained receptor activation that they produce, which can trigger receptor down-regulation and desensitization, as well as other compensatory mechanisms leading to tolerance and dependence. This is another feature which may be improved by the use of PAMs instead of direct agonists of the receptor. Because PAMs exert an effect only when the endogenous agonist is present, the temporal fidelity and pulsatile nature of native signaling patterns is preserved.[12,20] This is in sharp contrast to surrogate agonists such as morphine or oxycodone, which cause continuous activation of receptors until the drug is cleared (Figure 9.2). This difference between agonists and PAMs may be especially important in the case of opioid receptors, where the development of tolerance and dependence significantly exacerbates safety and addiction liabilities and undermines the therapeutic utility of direct-acting agonists.[21]

A final advantage of PAMs over direct agonists comes from the inherent limitation of PAM effects – PAMs have finite cooperativity ($\alpha\beta$)[22] values, and once the allosteric site is saturated, increasing the concentration of the PAM produces no further effect.[12,20] The cooperativity value can be modified

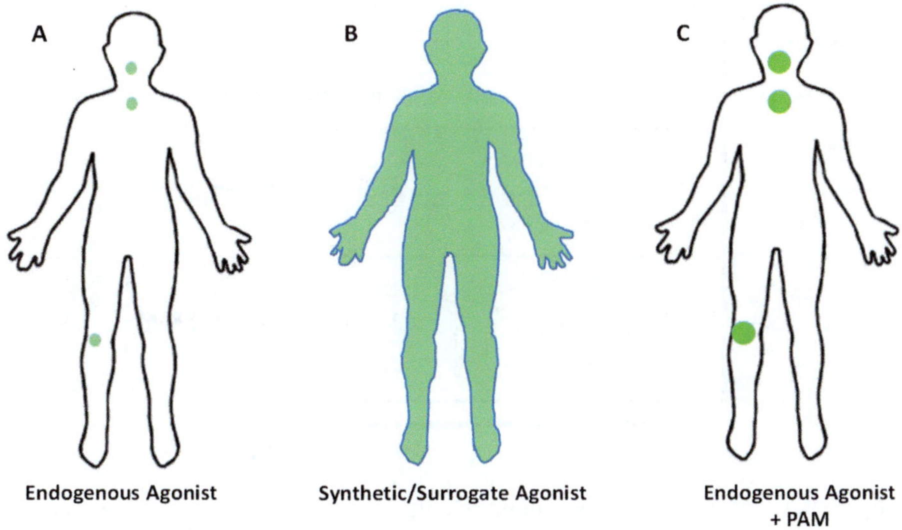

Figure 9.1 PAMs maintain the spatial fidelity of native signaling. The individual depicted in the cartoon may have injured his knee, resulting in pain. Endogenous opioid agonists are released at the area of the injury (where they will act at peripheral opioid receptors), and at specific locations within the brain and spinal cord involved in pain modulation (A). Exogenous opioid agonists such as morphine or oxycodone are distributed systemically and produce indiscriminate activation of opioid receptors throughout the body (B). PAMs do not activate the receptor by themselves, but amplify endogenous opioid signaling in areas where it occurs naturally (C). Therefore, activation of opioid receptors in other areas, such as the enteric nervous system (leading to constipation) and in brain regions mediating reward and addiction pathways or respiratory depression, may be avoided by the use of a PAM, resulting in improvements in abuse and addiction liability and side effect profile. (Adapted with permission from ref. 77, © 2014 The British Pharmacological Society).

through structure–activity relationship (SAR) studies, and the ideal level of cooperativity for a PAM of a given receptor target can be determined empirically *in vivo*, allowing researchers to find the level of cooperativity that yields the optimal balance of safety and efficacy.[23,24] In the case of opioid receptors, this advantage of PAMs may also be quite important in improving therapeutic windows as well as improving safety in the event of overdose.

9.3 Potential Utility of Opioid Receptor PAMs for Pain Management

There are two fundamental ways in which opioid receptor PAMs might be used therapeutically for pain management: (1) an opioid PAM used in conjunction with a direct opioid agonist, such as morphine, could reduce the

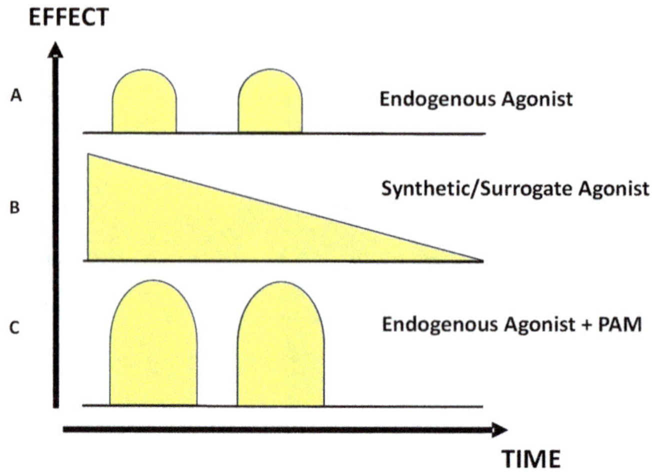

Figure 9.2 PAMs maintain temporal fidelity of native signaling. Endogenous agonists are released in a pulsatile fashion, and can be cleared quickly (A). Exogenous agonists such as morphine or oxycodone produce prolonged receptor activation, leading to receptor down-regulation and other intracellular compensatory mechanisms such as the up-regulation of adenylyl cyclase in nociceptive neurons. These effects ultimately lead to opioid tolerance and dependence (B). PAMs can enhance the effects of endogenous opioid agonists while still maintaining the temporal fidelity of native signaling (C); therefore, the hypothesis has been proposed that PAMs may produce significantly less tolerance and dependence than direct agonists.

dose of opioid agonist required for analgesic efficacy, and therefore may result in an improved side effect profile, or (2) an opioid PAM administered alone could provide analgesia by enhancing endogenous opioid signaling. As discussed above, the untoward side effects of μ-opioid receptor agonists represent "on-target" effects (*i.e.*, these effects are also mediated by μ receptor activation). Therefore, although using an opioid PAM may decrease the dose of morphine (or other opioid agonist) needed for pain relief, there is no *a priori* reason to assume that the unwanted side effects would not also occur similarly at lower doses. The major potential advantages of a PAM approach, maintenance of the spatial and temporal fidelity of native signaling (described above), would apply only if the PAM were administered in the absence of a synthetic agonist. Under these conditions, the PAM would be expected to augment the functional effects of endogenously released opioid peptides such as β-endorphin, methionine-enkephalin (met-enkephalin) and leucine-enkephalin (leu-enkephalin). The argument that such a PAM approach to drugging opioid receptors would provide functional selectivity to the drug's effects, and therefore result in a safer analgesic medicine, is predicated upon two major assumptions: first, that endogenous opioids are released *in vivo* at levels (tone) sufficient for opioid receptor PAMs to have an effect (*i.e.*, that there is enough native signaling for the PAM to amplify); and second, that under pain conditions this native opioid signaling selectively occurs in tissues that are involved in the modulation of pain signaling (as opposed to other areas that modulate gastrointestinal motility, respiratory rate, or reward and addiction pathways). The scientific evidence to support these assumptions is described below.

9.4 Endogenous Opioid Signaling

A number of endogenous peptides act as opioid receptor agonists.[4] Among these, dynorphins, endorphins (primarily β-endorphin), endomorphins, and the enkephalins (met-enkephalin and leu-enkephalin) may be of primary importance. Dynorphins are selective for κ-opioid receptors, and probably do not play a major role in native μ-opioid receptor signaling. While the enkephalins are slightly selective for binding to δ-opioid receptors, they also act as full agonists at μ-opioid receptors[25] and probably act through both μ- and δ-opioid receptors physiologically. Endorphin peptides such as β-endorphin are somewhat selective for binding to μ-opioid receptors, and were initially suggested to be endogenous morphine-like molecules (endorphin is a contraction for endogenous morphine). Another class of opioid peptides termed endomorphins, which are much more selective for μ-receptors, was discovered more recently.[26]

It is well documented that the expression of μ-opioid receptors and the release of endogenous opioid peptides are increased in circumstances of stress or injury.[27,28] Many studies have reported an increase in endogenous opioid synthesis and signaling in the spinal cord following induction of peripheral inflammation with the painful inflammatory agent complete Freund's adjuvant (CFA).[29–32] For example, Mousa *et al.*[33] used intraplantar injection of CFA to elicit unilateral hindpaw inflammation in rats, which was characterized by swelling and hyperalgesia in the injected paw only. These authors observed an increase in endomorphin levels in tissue from the inflamed paw but not the contralateral paw. An increase in μ-opioid receptor expression was also observed within laminae I and II of the dorsal horn of the lumbar spinal cord in areas receiving innervation from the inflamed paw, but not in other regions. Other animal studies using capsaicin and other noxious stimuli support the idea that endogenous opioid signaling is activated in specific areas of the brain (such as the periaqueductal gray) and spinal cord in response to pain.[27,34,35] In a rodent model of rheumatoid polyarthritis, up-regulation of both opioid receptors and endogenous opioid peptides was observed in the dorsal horn of the spinal cord, an effect which lasted for over two months.[36] A human study using positron emission tomography reported that sustained experimentally-induced masseter (jaw) muscle pain caused the release of endogenous μ-opioid agonists in the brain.[37] This endogenous opioid release was highly regionally selective, being observed in the dorsal anterior cingulate cortex, lateral prefrontal cortex, insular cortex, thalamus, hypothalamus, and amygdala, and not in other brain regions.

Combined, the results of these studies suggest that up-regulation of μ-opioid receptors and increased endogenous opioid signaling serve as part of the body's natural response to acute, chronic, and inflammatory pain.[28,35] If this hypothesis is correct, it is reasonable to assume that blocking endogenous opioid signaling through the use of an opioid antagonist would cause a reversal of this effect and result in hyperalgesia and/or allodynia. Supporting this prediction are the results of animal studies, which demonstrate that administration of competitive opioid receptor antagonists (naloxone or naltrexone) increases pain behaviors in animals with experimentally-induced pain conditions, but not in control animals.[36] Similarly, results from numerous human studies have shown that administration of naloxone enhances pain in post-surgical patients.[38–42] Importantly, these studies utilized subjects who were not receiving any opioid medications, and therefore the increase in pain scores produced by naloxone can be inferred to be the result of blocking endogenous opioid signaling. When naloxone was given to control subjects (who hadn't experienced surgery), pain scores remained unchanged.

9.5 Enkephalinase Inhibitors

As described above, there is considerable evidence to suggest that significant endogenous opioid signaling occurs in brain and spinal centers involved in pain modulation when humans and animals are in pain. Therefore, amplification of this endogenous signaling by an opioid receptor PAM is expected to produce antinociceptive effects. A conceptually similar approach is to enhance endogenous opioid activity by preventing the degradation of endogenous opioids, effectively increasing their concentration. This approach has been explored through the use of enkaphalinase inhibitors. Enkephalins are degraded by the concomitant action of two metallopeptidases, neutral endopeptidase (NEP) and aminopeptidase N (APN). Synthetic peptides which inhibit the activity of these enzymes have been described. These dual NEP–APN inhibitors have been shown to produce antinociceptive effects in animal pain models.[43] Unfortunately, the development of enkephalinase inhibitors has been limited by a lack of drug-like pharmacokinetic and pharmacodynamic properties as well as a lack of selectivity in their effects, as NEP and APN inhibition affects the endogenous regulation of many other native peptides in addition to the enkephalins. However, the results of exploratory studies using enkephalinase inhibitors provide further evidence supporting the hypothesis that potentiation of endogenous opioid signaling is sufficient to produce analgesic effects.

9.6 Discovery and Characterization of μ-Opioid Receptor PAMs

In 2006, cannabidiol (a CB_1 cannabinoid receptor agonist) was reported to negatively modulate agonist binding to δ- and μ-opioid receptors through an allosteric interaction.[44] Salvinorin A has also been reported to function as a NAM at μ-opioid receptors.[45] These reports suggested the existence of one or more allosteric binding sites on these receptors. It wasn't until 2013, however, that the first opioid receptor PAMs were reported.[46] These compounds were discovered at Bristol–Myers Squibb from a high-throughput screen which measured β-arrestin recruitment in human osteosarcoma cells (U2OS) using an enzyme fragment complementation approach (PathHunter technology, DiscoveRx Corp., Fremont, CA, USA).[47] The cells used recombinantly express both μ- and δ-opioid receptors; and test compounds were screened in the presence of both a concentration producing a 10% maximal effect (EC_{10}) of the highly μ-selective orthosteric agonist endomorphin-1 and the EC_{10} of the (relatively) δ-selective orthosteric agonist leu-enkephalin, allowing the identification of both μ- and δ-opioid receptor PAMs and agonists from a single high-throughput screen.[48]

In order to differentiate PAMs from direct-acting agonists, hits from this high-throughput screen were tested both alone (agonist-detection mode) and in the presence of an EC_{10} concentration of the μ-selective orthosteric agonist, endomorphin-1 (PAM-detection mode). Hits were then evaluated for receptor selectivity by testing concentration–response curves (both in PAM and agonist mode) using U20S cells recombinantly expressing the μ-opioid receptor alone (U20S-OPRM1 cells), and in U20S cells expressing the δ-opioid receptor alone (U2OS-OPRD1 cells) in the β-arrestin assay. Two μ-selective PAMs were identified: (*E*)-*N*-(2,6-dichlorophenyl)-4-(4-nitrophenyl)-2,3-dihydro-1,3-thiazol-2-imine (BMS-986121, **1**), and 2-(3-bromo-4-methoxyphenyl)-3-(4-chlorobenzenesulfonyl)-1,3-thiazolidine (BMS-986122, **2**) (Figure 9.3), which exhibited "pure PAM" behavior (meaning no significant β-arrestin recruitment occurred when compounds were run in "agonist detection mode;" see Sections 9.2 and 9.11) in this assay (Figure 9.4). Both compounds were selective for the μ-opioid receptor in the β-arrestin assay with no appreciable agonist or PAM activity detected in U2OS-OPRD1 cells. Both compounds exhibited similar cooperativity, producing a sevenfold leftward shift in the potency of endomorphin-1 ($\alpha\beta = 7$) in the β-arrestin recruitment assay (Figure 9.5).

Opioid receptor activation leads to the activation of G proteins (primarily the G inhibitory (G_i) and G other (G_o) G protein subtypes), as well as the recruitment of β-arrestin. The analgesic effects of opioid receptor activation are believed to be mediated primarily by activation of G proteins rather than β-arrestin.[49] Therefore, it was also critical to investigate the effects of the PAMs on G-protein-mediated signaling pathways. To this end, the effects of BMS-986121 and BMS-986122 on G_i- or G_o-mediated inhibition of adenylyl cyclase activity were tested in CHO cells recombinantly expressing μ-opioid receptors.[46] As shown in Figure 9.6, both of these μ-PAMs were found to potentiate the G-protein-mediated decrease in cAMP accumulation produced by an EC_{10} of endomorphin-1. Interestingly, both BMS-986121 and BMS-986122 behaved as mixed "PAM–agonists" (see Sections 9.2 and 9.11) in this assay, exhibiting some direct agonist activity at concentrations approximately 4–5 times higher than those required for PAM effects, although the agonist activity of BMS-986121 was low-efficacy (maximal

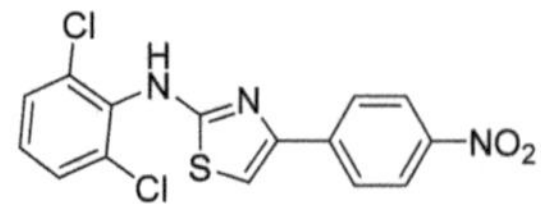

1, BMS-986121
μ-PAM

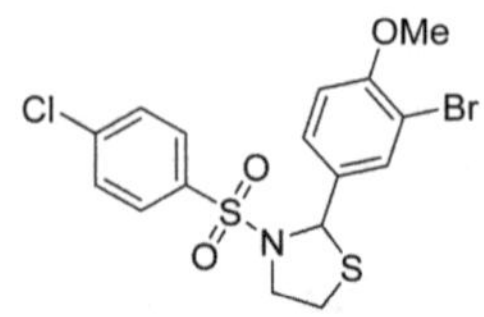

2, BMS-986122
μ-PAM

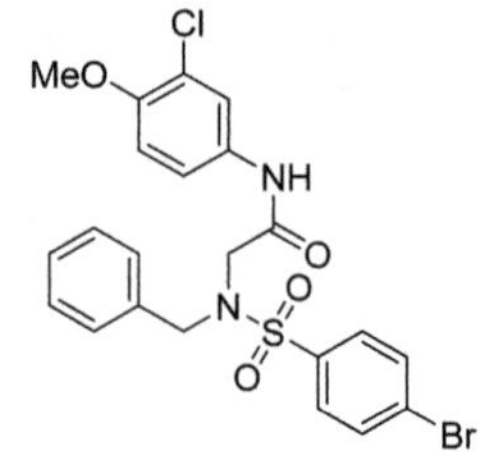

3, MS1
μ-PAM

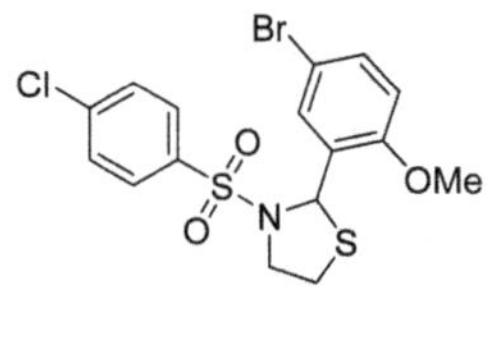

4, BMS-986124
μ-SAM

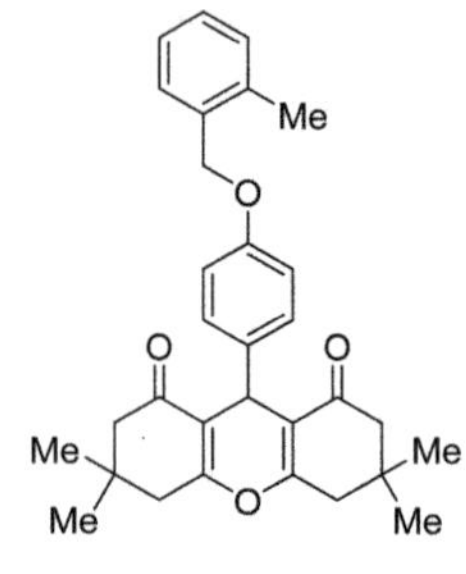

5, BMS-986187
δ-selective PAM

Figure 9.3 Representative structures from currently known opioid receptor PAM chemotypes.

response (E_{max}) ~ 36%) and not consistently reproducible between experiments. While these results confirm that these PAMs potentiate the ability of the μ-opioid receptor to activate G-protein-mediated signaling pathways as well as β-arrestin, they are discrepant with results from the β-arrestin assay in that no direct agonist activity was seen for β-arrestin recruitment. As discussed in Section 9.11, the physiological relevance of the direct agonist activity observed for many PAMs is questionable, and may in many cases result from recombinant receptor expression or highly amplified assay systems.[13,14,16,50,51] Indeed, the recombinant receptor system used for the cAMP experiments is highly amplified relative to the β-arrestin assay system, as endomorphin-1 exhibited more than 1000-fold greater potency in the cAMP assay (half maximal effective concentration (EC_{50}) = 76 pM) than the β-arrestin assay (EC_{50} = 86 nM), which has been reported to represent a relatively unamplified assay system with little if any receptor reserve.[52]

The effects of BMS-986122 were further characterized in an assay measuring the ability of compounds to stimulate binding of the poorly hydrolyzed guanosine-5′-triphosphate analog guanosine-5′-*O*-(3-[^{35}S]thio)triphosphate ([^{35}S]GTPγS) in membranes from C6 glioma cells recombinantly expressing rat μ-opioid receptors (C6μ cells). This assay represents a very direct

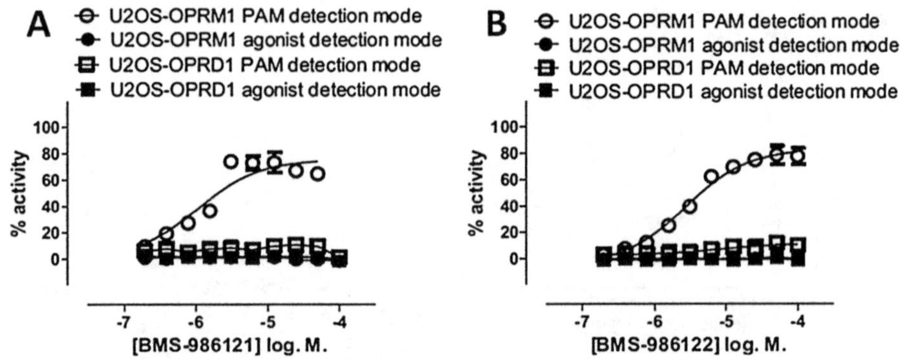

Figure 9.4 Effect of μ-PAMs on β-arrestin recruitment. Two μ-opioid receptor selective PAMs, BMS-986121 (A) and BMS-986122 (B), were assayed at varying concentrations in agonist-detection mode (with compound alone) and in PAM-detection mode (compound in the presence of a low concentration (~EC_{10}) of orthosteric agonist, either endomorphin-1 (20 nM) for μ-expressing cells (U2OS-OPRM1) or leu-enkephalin (0.4 nM) for δ-expressing cells (U2OS-OPRD1)). Data are represented as mean ± standard error of the mean (SEM) of three experiments. In agonist detection mode, 0 and 100% activity represent basal activity and a maximal response (E_{max}) of endomorphin-1 (in U2OS-OPRM1, a sixfold signal) or leu-enkephalin (in U2OS-OPRD1, a fourfold signal), respectively. In PAM detection mode, 0% activity is normalized to the low concentration (~EC_{10}) of orthosteric agonist; 100% activity represents the response to an E_{max} concentration of orthosteric agonist. (Reproduced with permission from ref. 46).

measurement of G protein activation. The potency of the synthetic peptide μ-receptor agonist [D-Ala[2], *N*-MePhe[4], Gly-ol]-enkephalin (DAMGO) for stimulating [^{35}S]GTPγS binding was increased sevenfold in the presence of BMS-986122 (10 μM), again confirming that the μ-PAM potentiates G protein activation as well as β-arrestin recruitment (Figure 9.7A). A very similar effect was seen in a [^{35}S]GTPγS binding assay utilizing membranes from mouse brain homogenates (Figure 9.7B) rather than recombinant cells, with BMS-986122 producing a fivefold increase in DAMGO's potency in this assay system.[46] This finding is important, as it confirms that the μ-PAM is active in natively expressed receptor systems as well as recombinant systems. In contrast to its effects in the cAMP accumulation assay, BMS-986122 showed no direct agonist activity in the [^{35}S]GTPγS binding assays in membranes from either recombinant cells or mouse brain homogenates. This again is consistent with the hypothesis that the direct agonist activity observed in the cAMP assay is a result of testing using highly amplified assay systems. Like the β-arrestin assay system,[52] the [^{35}S]GTPγS binding assay has been reported to exhibit no significant amplification (*i.e.*, has little if any receptor reserve) when performed using short [^{35}S]GTPγS incubation times (1 min),[53] as was the case for these experiments (5 min).[46]

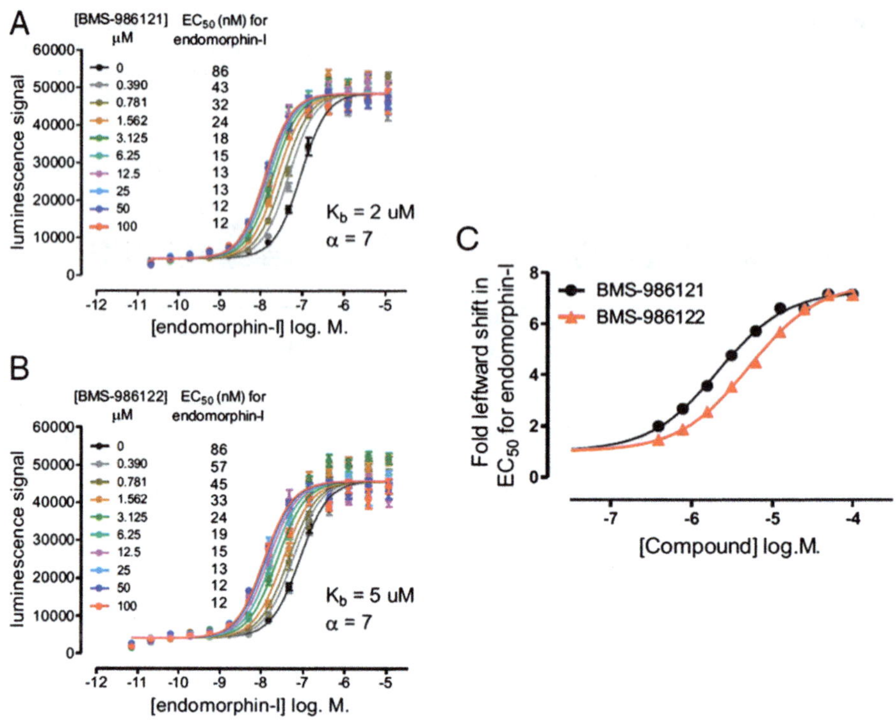

Figure 9.5 Effect of μ-PAMs BMS-986121 and BMS-986122 on endomorphin-1 stimulated β-arrestin recruitment in U2OS-OPRM1 cells. Both BMS-986121 (A) and BMS-986122 (B) produced concentration-dependent leftward shifts in the β-arrestin recruitment response to the agonist endomorphin-1. The data were analyzed simultaneously using an allosteric ternary complex model to provide binding constant (K_b) and cooperativity factor (α; also commonly referred to as $\alpha\beta$; Kenakin, 2009[22]) values for each compound. Calculated EC_{50} values (nM) for endomorphin-1 at each concentration of the PAM are shown in the legend for each graph. The fold leftward shift in EC_{50} values for endomorphin-1 in the presence of increasing concentrations of PAM compound is presented (C). Data are represented as mean ± SEM of four experiments. (Reproduced with permission from ref. 46).

In order to further dissect the exact molecular mechanism of potentiation produced by BMS-986122, its effect on orthosteric ligand affinity was examined. In saturation binding experiments, BMS-986122 was shown to have no significant effect on the affinity of the radiolabeled opioid receptor antagonist [^{3}H]diprenorphine (Figure 9.8A). However, BMS-986122 increased the affinity of the μ-agonist DAMGO as measured by its potency for displacing [^{3}H]diprenorphine (Figure 9.8B) in membrane homogenates from C6μ cells.[46] The magnitude of this increase in DAMGO's affinity (sixfold) very closely matched the sevenfold increase in DAMGO's potency for stimulating [^{35}S]GTPγS binding produced by BMS-986122 in membranes from the

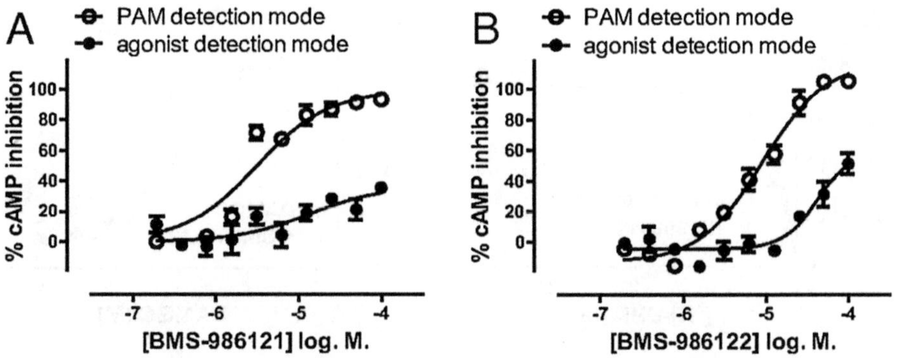

Figure 9.6 Effect of μ-PAMs on inhibition of forskolin-stimulated cAMP accumulation in CHO-μ cells. Both BMS-986121 (A) and BMS-986122 (B) increased the effect of a low (30 pM; ~EC_{10}) concentration of endomorphin-1 (PAM-detection mode) in a concentration-dependent manner. The compounds also showed some agonist activity when added alone (agonist-detection mode). For agonist-detection mode, 0% activity represents vehicle (basal) activity. For PAM-detection mode, 0% is normalized to the response to a 30 pM (~EC_{10}) concentration of endomorphin-1. The 100% response represents the response to an E_{max} concentration of endomorphin-1 (10 nM) in both agonist and PAM detection modes. Data are represented as mean ± SEM of three experiments. (Reproduced with permission from ref. 46).

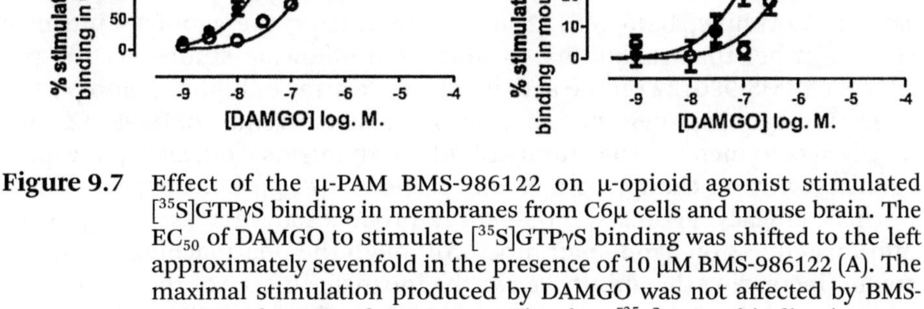

Figure 9.7 Effect of the μ-PAM BMS-986122 on μ-opioid agonist stimulated [^{35}S]GTPγS binding in membranes from C6μ cells and mouse brain. The EC_{50} of DAMGO to stimulate [^{35}S]GTPγS binding was shifted to the left approximately sevenfold in the presence of 10 μM BMS-986122 (A). The maximal stimulation produced by DAMGO was not affected by BMS-986122. The EC_{50} of DAMGO to stimulate [^{35}S]GTPγS binding in membranes from mouse brain was shifted to the left approximately fivefold in the presence of 10 μM BMS-986122 (B). Basal [^{35}S]GTPγS binding (fmol mg^{-1} of protein: 3.2 ± 0.2 in C6μ cells and 4.8 ± 0.4 in mouse brain) was not affected by 10 μM BMS-986122. Shown are the combined mean ± SEM data from three to seven separate assays, each performed in duplicate. (Adapted with permission from ref. 46).

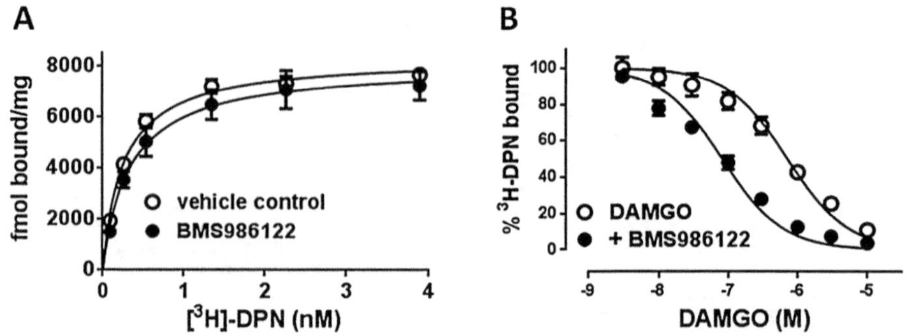

Figure 9.8 Effect of the μ-PAM BMS-986122 on the affinity of the orthosteric antagonist diprenorphine and the high-efficacy orthosteric agonist DAMGO in membranes from C6μ cells. In saturation binding experiments, BMS-986122 (10 μM) had no effect on [^{3}H]diprenorphine ([^{3}H]-DPN) binding affinity (A). BMS-986122 produced a sixfold increase in the affinity of DAMGO as measured by its potency for displacing [^{3}H]-DPN (B). (Adapted with permission from ref. 46).

same cells, implying that most or all of the increase in DAMGO's functional potency produced by the PAM resulted from an increase in DAMGO's affinity for the receptor in the presence of the PAM rather than from changes the intrinsic efficacy of DAMGO for producing receptor activation. A different story emerged, however, when the partial agonist morphine (rather than the full agonist DAMGO) was used as the orthosteric agonist probe. BMS-986122 produced no change in morphine's affinity as measured by its potency for displacing [^{3}H]diprenorphine (Figure 9.9A), but did produce a significant increase in the maximal G protein activation elicited by morphine, as well as a smaller (threefold) increase in morphine's potency for activating G proteins (Figure 9.9B), as measured by stimulation of [^{35}S]GTPγS binding in C6μ cell membrane homogenates.[46,54] This indicates that for the partial agonist morphine, BMS-986122 acts by increasing the agonist's intrinsic efficacy rather than its affinity. Results from follow-up studies testing the effects of BMS-986122 on the activity of a wide array of both full and partial μ receptor agonists suggested that this is a general trend – BMS-986122 primarily acts to increase the affinity of full (high intrinsic efficacy) μ receptor agonists with no change in maximal response (E_{max}); but for partial (low intrinsic efficacy) agonists, BMS-986122 produces an increase in E_{max} with minimal effects on agonist affinity.[54] Although the pharmacological mechanism (increasing the affinity *versus* the intrinsic efficacy of the orthosteric agonist) and the magnitude of BMS-986122's effects vary depending upon the orthosteric agonist used, BMS-986122 has been found to potentiate responses to all orthosteric μ-agonists tested, including the endogenous peptide agonists endomorphin-1, met- and leu-enkephalin, and β-endorphin, as well as synthetic agonists such as morphine, methadone, fentanyl, and several others.[46,54]

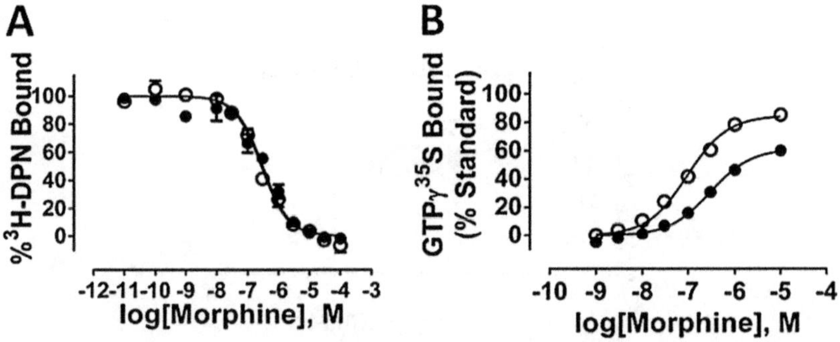

Figure 9.9 Effect of BMS-986122 on binding affinity and G protein activation exhibited by the partial agonist morphine. [^{3}H]-DPN binding to the μ-opioid receptor was determined in membranes from C6μ cells in the presence of increasing concentrations of morphine in the presence of vehicle (filled symbols) or BMS-986122 (open symbols) (A). The ability of increasing concentrations of morphine to elicit [^{35}S]GTPγS binding in C6μ cell membranes was measured in the absence (filled symbols) or presence of 10 μM BMS-986122 (open symbols). Data are presented as % stimulation relative to a maximal concentration (10 μM) of the full agonist standard DAMGO (B). All plotted points are means ± SEM of three to five independent experiments, each in duplicate. (Adapted with permission from ref. 54).

The molecular mechanism of BMS-986122's action may relate to the effect that Na^+ ions have on opioid receptor activity. A specific Na^+ ion binding site is known to exist for opioid receptors and other class A GPCRs. Na^+ binding to this site stabilizes an inactive state of the receptor, leading to a decrease in agonist affinity.[55–61] The magnitude of the affinity shift seen in the presence of Na^+ generally correlates with the intrinsic efficacy of the ligand, with the affinity of high-efficacy agonists being most sensitive to Na^+, and antagonist affinity being unaffected by the presence of Na^+; although this correlation is imperfect. BMS-986122 has been shown to reduce the potency of Na^+ for decreasing the affinity of the high-efficacy agonist DAMGO[54] (Figure 9.10). This effect of BMS-986122 is saturable (*i.e.*, concentrations of BMS-986122 above 10 μM do not produce further inhibition of the Na^+ effect), implying that BMS-986122 inhibits Na^+ binding through an allosteric interaction with Na^+ rather than direct competition for the Na^+ binding site. The proposed relationship between BMS-986122's activity and Na^+ binding is supported by computational chemistry studies using molecular dynamics simulations of the binding of (R)-methadone, Na^+, and BMS-986122 to the μ-receptor.[62] In these simulations, the inclusion of Na^+ (in the absence of BMS-986122) consistently disrupted the interaction between (R)-methadone and a key residue in the orthosteric binding pocket (Asp 3.32), leading the dissociation of (R)-methadone from its binding site. However, the addition of BMS-986122 into the simulation disturbed the position of the Na^+ ion through an allosteric interaction with the Na^+ binding site and completely abolished the effect of Na^+ on (R)-methadone binding.

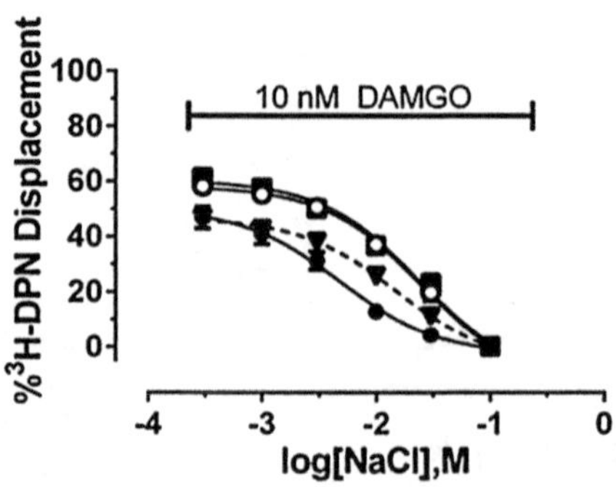

Figure 9.10 BMS-986122 inhibits the effect of Na^+ ions on the binding affinity of the high-efficacy μ-agonist DAMGO. Using C6μ membranes, DAMGO (10 nM) was incubated with 0.2 nM [^{3}H]-DPN and increasing concentrations of NaCl with 10 μM GTPγS in the presence of vehicle (filled circles), 3 μM BMS-986122 (triangles with dashed line), 10 μM BMS-986122 (open circles), or 30 μM BMS-986122 (filled squares). Data shown are means ± SEM of three independent experiments each in duplicate. (Reproduced with permission from ref. 54).

9.7 Structure–Activity Relationship Studies and the Identification of μ-Opioid Receptor SAMs

Several analogs of BMS-986122 were tested in the β-arrestin recruitment assay in order to explore the SAR of this series.[46] As shown in Figure 9.11, all chemical changes to the original compound that were tested led to some loss of activity. This was generally observed as a decrease in the E_{max}, which reflects a decrease in the cooperativity $(\alpha\beta)^{22}$ of the compound.[51] Interestingly, although BMS-986122 itself shows no significant activity at δ-opioid receptors, several of the analogs tested exhibited some very weak (low E_{max}) δ-PAM activity, suggesting that elements of the allosteric site may be conserved between μ- and δ-receptors, and supporting the idea that δ-PAM activity can be selectively engineered in or out through SAR.

Also of interest were two analogs of BMS-986122 (BMS-986123 and BMS-986124, **4**; Figure 9.3) which were identified as SAMs of this allosteric site. These compounds are "silent" binders to the allosteric site, producing no measurable PAM or NAM activity on their own; but can competitively antagonize the effects of BMS-986122 (calculated binding constant (K_b) = 1 and 2 μM, respectively).[46] While SAMs of opioid receptors would have no obvious clinical utility (as no endogenous ligands are known to exist for this allosteric site), SAMs can be very useful as tool molecules.[24]

9.8 Mount Sinai Chemotype

Researchers from the Icahn School of Medicine at Mount Sinai used BMS-986121 and BMS-986122 as starting compounds for structure-guided computational chemistry efforts which resulted in the identification of a novel

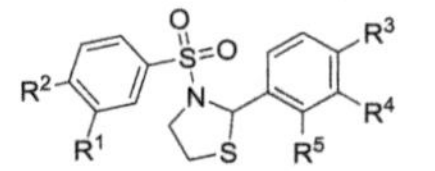

Substance	R^1	R^2	R^3	R^4	R^5	β-arrestin PAM Detection Mode (% E_{max})	β-arrestin PAM Detection Mode (EC_{50}, μM)	Description	β-arrestin SAM Detection Mode (K_b, μM)
BMS-986122	H	Cl	OMe	Br	H	79	3	Full PAM (ref)	
Analog 1	H	NO_2	OMe	Br	H	53	2	Moderate PAM	
Analog 2	H	Me	OMe	Br	H	18	4	Weak PAM	
Analog 3	H	OMe	OMe	Br	H	16	6	Weak PAM	
Analog 4	Me	H	OMe	Br	H	20	9	Weak PAM	
Analog 5	H	Cl	OMe	NO_2	H	45	4	Moderate PAM	
Analog 6	H	Br	OMe	NO_2	H	45	2	Moderate PAM	
Analog 7 (BMS-986123)	H	Me	OMe	NO_2	H	–	–	SAM	1
Analog 8	H	Br	OMe	OMe	H	18	14	Weak PAM	
Analog 9	H	Br	OMe	H	H	44	7	Moderate PAM	
Analog 10	H	H	OMe	H	H	17	23	Weak PAM	
Analog 11	H	Cl	H	Cl	H	65	5	Strong PAM	
Analog 12	H	Cl	Br	H	H	26	6	Weak PAM	
Analog 13 (BMS-986124)	H	Cl	Br	H	OMe	–	–	SAM	2
Analog 14	H	OMe	Br	H	H	3	7	Weak PAM	
Analog 15	H	OMe	OMe	H	H	9	37	Weak PAM	

Figure 9.11 Structure–activity relationship of BMS-986122 and analogs tested in a β-arrestin recruitment assay in U2OS-OPRM1 cells. Compounds exhibiting PAM activity are described based on their cooperativity as full, strong, moderate, or weak PAMs. 0% activity represents the response to a low (~EC_{10}) concentration of endomorphin-1 alone, and 100% activity represents the response to an E_{max} concentration of endomorphin-1. The two compounds that showed no PAM activity were additionally tested in SAM detection mode [inhibition of BMS-986122 (~EC_{80}) response in the presence of a low concentration (~EC_{20}) of endomorphin-1], where the compounds were shown to inhibit the BMS-986122 response. The K_b values provided are calculated from the concentration giving half maximal inhibition (IC_{50}) values for inhibition of the BMS-986122 response. Data represent mean values from three experiments. NA, not active. (Reproduced with permission from ref. 46).

μ-PAM chemotype exemplified by 2-((*N*-benzyl-4-bromophenyl)sulfonamido)-*N*-(3-chloro-4-methoxyphenyl)acetamide (MS1, **3**; Figure 9.3).[63] Exploration of the SAR of this chemotype was performed using endomorphin-1 as the orthosteric agonist and β-arrestin recruitment as the activity readout (Figure 9.12). The most active compounds identified from this series exhibited potency in the 4–9 μM range, with cooperativity ($\alpha\beta$) values between 4 and 8 (*i.e.*, produced 4–8 fold increases in the potency of endomorphin-1 for eliciting β-arrestin recruitment). MS1 produced less pronounced effects on agonist affinity and ability to activate G proteins than were observed for BMS-986122,[63] although MS1 effects were qualitatively similar to those of BMS-986122 in that the largest effects were observed when (R)-methadone was used as the orthosteric agonist.[54]

Interestingly, the MS1 chemical series contained a large number of SAMs.[63] Nearly half of the analogs tested exhibited no measurable activity on their own, but were able to block the PAM activity of BMS-986121 in the β-arrestin-recruitment assay (Figure 9.12). In addition to the obvious value of the

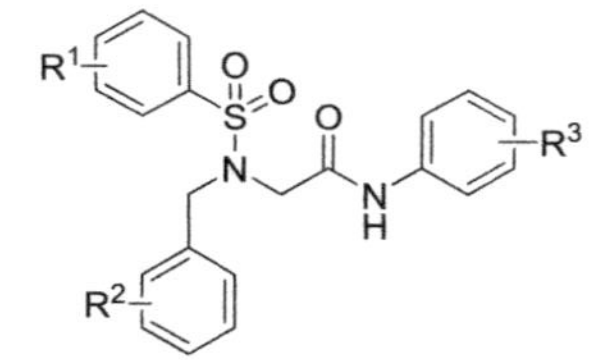

Sample ID	R^1	R^2	R^3	EC_{50}, µM (PAM mode/SAM mode)	PAM mode E_{max} %	Activity
MS1	4-Br	H	3-Cl,4-OMe	6.5	85.1	PAM
MS2	4-Br	H	4-OPh	6.2	109.2	PAM
MS3	4-Cl	3-Me	3-Cl,4-OMe	5.7	44.6	PAM
MS4	4-Br	H	3-Cl	3.9	32.4	PAM
MS5	3-Me,4-OMe	H	3-Cl,4-OMe	4.7	21	PAM
MS6	4-Me	H	3-Cl,4-OMe	5.3	21.8	PAM
MS7	4-Br	H	4-OMe	5.4	22.1	PAM
MS8	4-Cl	H	4-OEt	6.4	31.8	PAM
MS9	4-F	H	3-Cl,4-OMe	7.0	24	PAM
MS10	4-Cl	H	4-OMe	8.9	26.4	PAM
MS11	4-OMe	H	3-Br	14.5	33	PAM
MS12	4-Me	4-Me	3-Cl,4-OMe	21.9	10.3	PAM
MS13	H	H	3-Br	>30	20	PAM
MS14	H	H	3-Cl,4-OMe	76.1	28.4	PAM
MS15	H	2,4-di-Cl	3-Cl,4-OMe	6*	NA	SAM
MS16	4-Cl	H	3-OMe	6.5*	NA	SAM
MS17	H	4-F	3-Cl,4-OMe	6.9*	NA	SAM
MS18	4-Me	H	4-OMe	14.4*	NA	SAM
MS19	H	4-Me	3-Cl	16.2*	NA	SAM
MS20	H	4-Cl	4-OMe	16.6*	NA	SAM
MS21	H	4-Cl	3-Me	21.2*	NA	SAM
MS22	H	4-Br	4-OMe	22.7*	NA	SAM
MS23	4-Cl	H	3-Cl,6-OMe	23.5*	NA	SAM
MS24	H	4-Cl	3-Cl,4-OMe	24.2*	NA	SAM
MS25	H	4-Cl	2-OMe,5-Me	27.7*	NA	SAM
MS26	4-Cl	H	2-OMe,5-Me	>30*	NA	SAM
MS27	H	4-Me	3-Cl,4-OMe		NA	inactive
MS28	4-OMe	H	2-CF_3		NA	inactive

Figure 9.12 Structure–activity relationship of the MS1 μ-PAM chemotype in a β-arrestin recruitment assay. SAM compounds were detected by incubating serial dilution of the compound with cells in the presence of an EC_{20} of endomorphin-1 plus an EC_{80} of the μ-PAM BMS-986121. Under these conditions, the SAM compounds act as antagonists of PAM binding, reducing PAM activity. Control PAMs: BMS-986121 (EC_{50} 2.2 μM, E_{max} 86%) and BMS-986122 (EC_{50} 16.2 μM, E_{max} 108%). NA = not active in PAM mode. (Reprinted with permission from ref. 63. Copyright 2015 American Chemical Society).

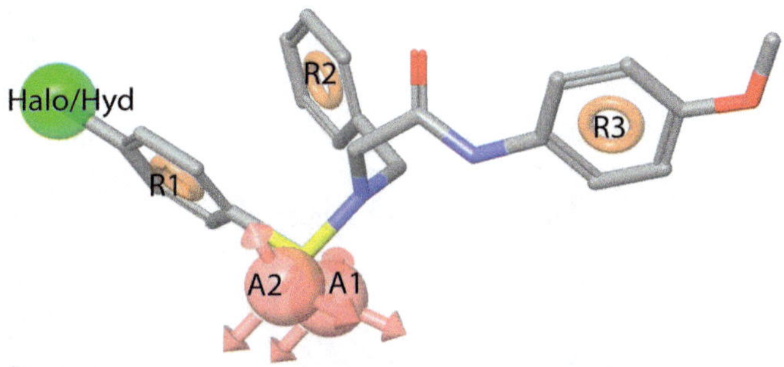

Figure 9.13 Ligand-based 3D pharmacophore model built from μ-PAMs shown in Figure 9.12. The best 3D pharmacophore model that separates μ-PAMs and μ-SAMs includes two H-bond acceptors (*i.e.*, the two oxygen atoms of the sulfur dioxide group (A1 and A2)), one halogen substituent or hydrophobic group ((Halo/Hyd) *i.e.*, R1 = Br, Cl, Me; see Figure 9.12), and three aromatic rings (R1–R3) related by specific distances and angles. (Reprinted with permission from ref. 63. Copyright 2015 American Chemical Society).

identification of a novel μ-PAM series, the results of this work are intriguing in that they suggest that computational approaches may be informative with respect to the chemical determinants of cooperativity for allosteric modulators at this site. The authors identified several physicochemical properties of the MS1 analogs that were predictive of PAM *versus* SAM activity for this series (Figure 9.13). It will be of great interest to see whether these findings will be confirmed upon examination of larger compound sets, and if they are translatable to other μ-PAM scaffolds.

9.9 δ-Opioid Receptor Selective PAMs

A series of δ-opioid receptor selective PAMs has also been reported.[64] This series was discovered from the same high-throughput screening campaign that identified the μ-selective PAMs BMS-986121 and BMS-986122.[48] 3,3,6,6-tetramethyl-9-(4-((2-methylbenzyl)oxy)phenyl)-3,4,5,6,7,9-hexahydro-1*H*-xanthene-1,8(2*H*)-dione (BMS-986187, **5**; Figure 9.3) was identified from initial SAR studies as one of the most active compounds from this series (EC_{50} = 0.03 μM; Figure 9.14). BMS-986187 was shown to potentiate G protein activation, cAMP inhibition, β-arrestin recruitment, and extracellular signal-related kinase (ERK) phosphorylation evoked by the endogenous δ-opioid agonist leu-enkephalin, and to increase the binding affinity of leu-enkephalin as well as other (synthetic) δ-opioid receptor agonists.[64] BMS-986187 exhibits greater cooperativity ($\alpha\beta$ = 15) than the μ-selective PAMs, and also exhibits some direct agonist activity, even in the [^{35}S]GTPγS binding assay (although it exhibits greater agonist activity in the highly amplified cAMP inhibition assay).

PAMs of the δ-opioid receptor have less obvious therapeutic potential than μ-PAMs. Although activation of δ-opioid receptors does produce analgesic effects,[4] δ-receptor mediated analgesia is much weaker than the analgesia

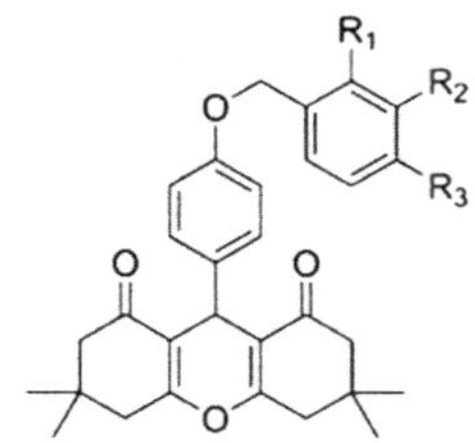

compd	R_1	R_2	R_3	EC_{50} (μM) (% Y_{max})		selectivity μ/δ
				δ	μ	
1	H	H	H	0.2 (126)	6 (130)	30
2 (BMS-986187)	CH_3	H	H	0.03 (124)	3 (121)	100
3	H	CH_3	H	0.2 (136)	3 (120)	15
4	H	H	CH_3	0.3 (92)	4 (65)	13
5	F	H	H	0.1 (120)	5 (107)	50
6	H	F	H	1 (136)	7 (182)	7
7	H	H	F	0.1 (95)	2 (72)	20
8	Cl	H	H	0.3 (68)	>10 (>100)	>33
9	Cl	Cl	H	0.1 (116)	>10 (>87)	>100
10 (BMS-986188)	Br	H	H	0.05 (58)	>10 (>20)	>200
11	$OCHF_2$	H	H	0.2 (109)	3 (80)	15
12	OCF_3	H	H	0.3 (114)	2 (111)	7
13	SO_2CH_3	H	H	0.9 (102)	5 (42)	6
14	CH_2OH	H	H	1 (92)	10 (112)	10
15	CF_3	H	H	>10 (>40)	>10 (>20)	

Figure 9.14 Structure–activity relationship of the δ-PAM chemotype in PathHunter CHO-OPRD1 and CHO-OPRM1 cells in a β-arrestin recruitment assay. No activity was observed in agonist mode (in the absence of orthosteric agonist). In PAM mode (in the presence of an EC_{20} of leu-enkephalin for OPRD1 cells or an EC_{20} of endomorphin-1 for OPRM1 cells), robust responses were observed. The mean EC_{50} values, Y_{max} (maximum response relative to a maximally effective concentration of leu-enkephalin) values, and potency ratio of δ-receptor activity : μ-receptor activity (μ EC_{50} : δ EC_{50}) in PAM mode are reported ($n = 3$). (Reprinted with permission from ref. 64. Copyright 2015 American Chemical Society).

elicited by μ-receptor activation. There is, however, evidence that δ-opioid receptor agonists are particularly effective in animal models of chronic pain, and rodent models using a genetic deletion of the δ receptor support the hypothesis that native opioid signaling at δ receptor mediates an increase in pain threshold.[65] Additionally, δ-opioid agonists have been shown to produce antidepressant-like effects in animal models.[66] Therefore, chronic pain and depression would both represent potential therapeutic indications for δ-agonists or PAMs.

Unfortunately, results from studies in animals, including non-human primates, indicate that δ-opioid receptor agonists have a proconvulsant effect. Results from several studies support the hypothesis that it may be possible to separate the seizurogenic properties from antidepressant-like effects associated with δ-receptor activation. One such study slowed the rate of administration of δ-receptor agonist SNC80 and noted a reduction in convulsions with no effect on the anti-depressant effects.[67] It is possible that a PAM approach may provide greater safety and improved therapeutic windows between efficacy and seizure liabilities than can be attained using direct-acting δ-receptor agonists. Precedence for this idea comes from the metabotropic glutamate receptor subtype 5 (mGluR5), activation of which has procognitive efficacy in animal models but also produces seizures and neurotoxicity.[15,23,68] In recent studies it has been concluded that direct activation of the receptor, or potentiation of receptor activation by PAMs with high cooperativity ($\alpha\beta$) values leads to seizurigenic effects in animals, whereas low-cooperativity PAMs ($\alpha\beta$ <3) do not exhibit procovulsant activity but retain procognitive efficacy in animals.[24]

It is important to point out that BMS-986187 does exhibit significant μ-PAM activity in addition to its δ-PAM activity. BMS-986187 is relatively potent as a δ-PAM (EC_{50} = 0.03 μM); and therefore, even though its activity is ~100-fold selective for δ- *versus* μ-receptors, its PAM potency at μ-receptors (EC_{50} = 3 μM) is actually similar to the μ-PAM potency of the prototype μ-selective PAMs BMS-986121 and BMS-986122. The results of the limited study of SAR that has been performed around this chemotype indicate that potency at μ- and δ-receptors can be independently manipulated through chemical modifications, and the 14 active analogs of this series that have been reported exhibit δ- *versus* μ-receptor selectivity ranging from just sixfold to over 200-fold (Figure 9.14).

9.10 Proposed Binding Site for Opioid Receptor PAMs

Reports from two separate groups were published in early 2016 using *in silico* molecular simulations to predict the binding site for opioid receptor PAMs. The Mount Sinai group and their collaborators at Monash University simulated the binding of the δ-PAM BMS-986187 to the δ-opioid receptor in an explicit lipid–water environment and with the δ-receptor selective agonist SNC80 bound to the orthosteric site.[69] Representative lowest-energy structures from the reconstructed free-energy landscape predicted a binding site for BMS-986187 that primarily involves interactions with transmembrane (TM) helices TM1, TM2, and TM7. To test these *in silico* predictions experimentally, BMS-986187 was studied in functional (ERK phosphorylation) and radioligand binding assays using cells expressing the δ-receptor with various point mutations introduced. The results of these mutagenesis studies supported the proposed binding site, as point mutations in several residues predicted to interact with BMS-986187, including Lys 2.63, Tyr 2.64, Leu 7.35, and His 7.36 (Ballesteros–Weinstein notation[70]) were found to affect the binding affinity and/or cooperativity of BMS-986187 with the orthosteric agonist SNC80.

At about the same time, another group of computational chemists from the Medical University of Lublin, Poland and the University of Eastern Finland published the results of molecular dynamics simulations of the μ-PAM BMS-986122 binding to the μ-opioid receptor in the presence of the μ-agonist (R)-methadone.[62] Very interestingly, based upon the results of these simulations the authors predicted a binding site for the μ-PAM at the μ-receptor that is essentially identical to the binding site for the δ-PAM at the δ-receptor proposed by the Mount Sinai group, with Asn 2.63, Tyr 2.64, Trp 7.35, His 7.36, and Ile 7.39 residues mediating the most important interactions between BMS-986122 and the μ-receptor.

The proposed binding site for both μ- and δ-PAMs is "above" (more extracellular to) and very close to the orthosteric binding pocket, which is similar to allosteric sites that have been identified for other class A GPCRs.[71,72] The idea that both μ-and δ-PAMs bind to a conserved allosteric site on μ- and δ-receptors is consistent with observations that various analogs of the μ-selective thiazolidine PAM chemotype exemplified by BMS-986122 also show some δ-PAM activity,[46] and that many analogs of the δ-selective xanthene-dione PAM chemotype exemplified by BMS-986187 also exhibit μ-PAM activity[64] (Figure 9.14). The fact that this site was identified independently by two different groups, in simulations using different opioid receptors and structurally dissimilar PAMs would seem to strongly suggest that this is in fact the binding pocket targeted by the currently known opioid receptor PAMs, and it may be expected that our emerging understanding of this binding site will greatly facilitate future chemistry efforts to improve the activity of PAMs from the existing opioid PAM scaffolds, as well as aiding in the identification of novel opioid PAM chemotypes.

9.11 Discussion and Future Directions

A strong scientific rationale exists to support the hypothesis that opioid receptor PAMs, especially μ-PAMs, may exhibit analgesic efficacy while providing significant improvement in the abuse liability and side effect profile that limit the therapeutic utility of opioid agonists. Two major questions must be answered experimentally in order to assess the therapeutic potential of opioid PAMs: (1) Will opioid PAMs exert analgesic activity when administered alone *in vivo* (in other words, does sufficient endogenous opioid signaling occur *in vivo* for opioid PAMs to produce significant analgesic effects)?, and (2) Will opioid PAMs administered at analgesic doses exhibit a significant improvement in side effect profile (tolerance, respiratory depression, and constipation) and abuse liability compared with direct-acting opioid agonists? The animal models that will be required to assess analgesic activity, tolerance, and effects on respiration and gastrointestinal motility are straightforward;[73–75] and animal self-administration assays that are predictive of abuse liability in humans (especially for opioid-like mechanisms), though not simple to perform, are very well established.[76] It is likely, however, that better opioid PAM tool molecules may be required.

The small handful of lead molecules that have been reported (especially the μ-selective PAMs BMS-986121 and BMS-986122) suffer from poor potency and metabolic stability, making *in vivo* studies difficult. Efforts to use these molecules for *in vivo* testing in pain models are ongoing. However, the discovery of tool molecules with properties more suitable for *in vivo* testing would be very useful for validating this therapeutic approach. In this regard, the recent discovery of the receptor binding site for μ- and δ-PAMs[62,69] may be expected to facilitate future chemistry efforts to generate more drug-like tool molecules.

The theoretical rationale for expecting opioid PAMs to be safer than direct-acting opioid agonists is predicated upon the idea that opioid PAMs would act only to amplify endogenous opioid signaling rather than activating the receptor directly in the absence of endogenous opioids. This would only be expected to be achieved through the use of "pure PAMs" which do not produce any direct activation of the receptor, as opposed to mixed PAM–agonists (or "ago-PAMs") which can directly activate receptors at concentrations higher than those required for PAM activity.[13–16] Therefore, it will be important to monitor direct agonist activity of new opioid PAMs as they are developed. Mixed PAM and agonist activity is very commonly seen when testing PAMs using *in vitro* assay systems. Unfortunately, this phenomenon is not well understood in the sense that for many PAMs it remains unclear whether direct agonist activity observed *in vitro* is "real" (reflects actual receptor agonism *in vivo*), or an artifact of the *in vitro* assay systems employed, which frequently use overexpression of recombinant receptors and often represent highly amplified and artificial systems. Indeed, for the best-characterized μ-PAM BMS-986122, significant direct agonist activity was observed in a highly amplified cAMP assay using recombinantly expressed receptors. However, no direct agonist activity was observed for this PAM in a β-arrestin recruitment assay also using recombinantly expressed receptors, or in [^{35}S]GTPγS binding assays using membranes either from mouse brain homogenates or from cells recombinantly expressing the μ-receptor. Therefore, there currently is no unequivocal answer to the question "Does BMS-986122 directly activate the receptor *in vivo* or not?" In many cases, direct agonist activity is more likely to be observed with PAMs that also exhibit very high cooperativity with orthosteric agonists. However, this is certainly not a universal correlation, as exemplified by the fact that BMS-986121 exhibits significantly less direct agonist activity than BMS-986122 in the cAMP accumulation assay, despite the fact that both compounds exhibit equal cooperativity ($\alpha\beta$ = 7 for both in the β-arrestin recruitment assay[46]). Gaining a better understanding of the direct agonist activity of opioid receptor PAMs will be important, as a lack of *in vivo* agonism will most probably be desirable for opioid PAM leads that could progress to eventual drug candidates.

The desired level of cooperativity for opioid receptor PAMs also remains unknown. Higher-cooperativity PAMs are not necessarily better. For example, in the case of mGluR5 PAMs, it has been found that very low-cooperativity PAMs ($\alpha\beta$ < 3) may provide the ideal window between efficacy and side effects.[24] A similar case may exist for opioid receptor PAMs, and it is likely that the ideal level of cooperativity for producing analgesia while avoiding unwanted side effects will have to be determined empirically by testing PAMs with a range of cooperativity values *in vivo*.

Acknowledgements

The authors thank Dr Arthur Christopoulos and Dr P. Jeffrey Conn for providing an initial version of Figure 9.2, Dr Dario Doller for critical review of the manuscript, Dr Jonathan O'Connell and Dr Martyn Banks for supporting this research at Bristol-Myers Squibb, and Dr Neil T. Burford for his seminal contributions to the research described within this chapter.

References

1. I. Melnikova, *Nat. Rev. Drug Discovery*, 2010, **9**, 589–590.
2. S. P. Alexander, H. E. Benson, E. Faccenda, A. J. Pawson, J. L. Sharman, M. Spedding, J. A. Peters, A. J. Harmar and C. G. T. P. Collaborators, *Br. J. Pharmacol.*, 2013, **170**, 1797–1867.
3. J. C. Meunier, *Eur. J. Pharmacol.*, 1997, **340**, 1–15.
4. R. Przewlocki and B. Przewlocka, *Eur. J. Pharmacol.*, 2001, **429**, 79–91.
5. L. Toll, M. R. Bruchas, G. Calo, B. M. Cox and N. T. Zaveri, *Pharmacol. Rev.*, 2016, **68**, 419–457.
6. http://www.drugabuse.gov/about-nida/legislative-activities/testimony-to-congress/2016/americas-addiction-to-opioids-heroin-prescription-drug-abuse, (last accessed April 2016).
7. M. Heron, *Natl. Vital. Stat. Rep.*, 2016, **65**, 1–95.
8. N. Dietis, R. Guerrini, G. Calo, S. Salvadori, D. J. Rowbotham and D. G. Lambert, *Br. J. Anaesth.*, 2009, **103**, 38–49.
9. I. K. Khanna and S. Pillarisetti, *J. Pain Res.*, 2015, **8**, 859–870.
10. M. P. Davis, *Expert Opin. Drug Discovery*, 2012, **7**, 165–178.
11. T. J. Cicero and M. S. Ellis, *JAMA Psychiatry*, 2015, **72**, 850–851.
12. A. Christopoulos and T. Kenakin, *Pharmacol. Rev.*, 2002, **54**, 323–374.
13. T. W. Schwartz and B. Holst, *J. Recept. Signal Transduction Res.*, 2006, **26**, 107–128.
14. T. W. Schwartz and B. Holst, *Trends Pharmacol. Sci.*, 2007, **28**, 366–373.
15. J. M. Rook, M. J. Noetzel, W. A. Pouliot, T. M. Bridges, P. N. Vinson, H. P. Cho, Y. Zhou, R. D. Gogliotti, J. T. Manka, K. J. Gregory, S. R. Stauffer,

F. E. Dudek, Z. Xiang, C. M. Niswender, J. S. Daniels, C. K. Jones, C. W. Lindsley and P. J. Conn, *Biol. Psychiatry*, 2013, **73**, 501–509.
16. M. J. Noetzel, J. M. Rook, P. N. Vinson, H. P. Cho, E. Days, Y. Zhou, A. L. Rodriguez, H. Lavreysen, S. R. Stauffer, C. M. Niswender, Z. Xiang, J. S. Daniels, C. K. Jones, C. W. Lindsley, C. D. Weaver and P. J. Conn, *Mol. Pharmacol.*, 2012, **81**, 120–133.
17. S. H. Snyder and G. W. Pasternak, *Trends Pharmacol. Sci.*, 2003, **24**, 198–205.
18. A. D. Corbett, G. Henderson, A. T. McKnight and S. J. Paterson, *Br. J. Pharmacol.*, 2006, **147**(suppl. 1), S153–S162.
19. H. W. Matthes, R. Maldonado, F. Simonin, O. Valverde, S. Slowe, I. Kitchen, K. Befort, A. Dierich, M. Le Meur, P. Dolle, E. Tzavara, J. Hanoune, B. P. Roques and B. L. Kieffer, *Nature*, 1996, **383**, 819–823.
20. A. Christopoulos, *Mol. Pharmacol.*, 2014, **86**, 463–478.
21. E. L. Way, H. H. Loh and F. H. Shen, *J. Pharmacol. Exp. Ther.*, 1969, **167**, 1–8.
22. T. P. Kenakin, *Trends Pharmacol. Sci.*, 2009, **30**, 460–469.
23. S. Parmentier-Batteur, P. H. Hutson, K. Menzel, J. M. Uslaner, B. A. Mattson, J. A. O'Brien, B. C. Magliaro, T. Forest, C. A. Stump, R. M. Tynebor, N. J. Anthony, T. J. Tucker, X. F. Zhang, R. Gomez, S. L. Huszar, N. Lambeng, H. Faure, E. Le Poul, S. Poli, T. W. Rosahl, J. P. Rocher, R. Hargreaves and T. M. Williams, *Neuropharmacology*, 2014, **82**, 161–173.
24. F. Yang, L. B. Snyder, A. Balakrishnan, J. M. Brown, D. V. Sivarao, A. Easton, A. Fernandes, M. Gulianello, U. M. Hanumegowda, H. Huang, Y. Huang, K. M. Jones, Y. W. Li, M. Matchett, G. Mattson, R. Miller, K. S. Santone, A. Senapati, E. E. Shields, F. J. Simutis, R. Westphal, V. J. Whiterock, J. J. Bronson, J. E. Macor and A. P. Degnan, *ACS Med. Chem. Lett.*, 2016, **7**, 289–293.
25. A. Alt, A. Mansour, H. Akil, F. Medzihradsky, J. R. Traynor and J. H. Woods, *J. Pharmacol. Exp. Ther.*, 1998, **286**, 282–288.
26. J. E. Zadina, L. Hackler, L. J. Ge and A. J. Kastin, *Nature*, 1997, **386**, 499–502.
27. M. H. Ossipov, G. O. Dussor and F. Porreca, *J. Clin. Invest.*, 2010, **120**, 3779–3787.
28. B. K. Taylor and G. Corder, *Curr. Top. Behav. Neurosci.*, 2014, **20**, 283–325.
29. M. J. Millan, A. Czlonkowski, B. Morris, C. Stein, R. Arendt, A. Huber, V. Hollt and A. Herz, *Pain*, 1988, **35**, 299–312.
30. R. Dubner and M. A. Ruda, *Trends Neurosci.*, 1992, **15**, 96–103.
31. B. Przewlocka, W. Lason and R. Przewlocki, *Neuroscience*, 1992, **46**, 209–216.
32. M. Pohl, S. Ballet, E. Collin, A. Mauborgne, S. Bourgoin, J. J. Benoliel, M. Hamon and F. Cesselin, *Brain Res.*, 1997, **749**, 18–28.
33. S. A. Mousa, H. Machelska, M. Schafer and C. Stein, *J. Neuroimmunol.*, 2002, **126**, 5–15.
34. O. Nakagawasai, T. Tadano, K. Tan No, F. Niijima, S. Sakurada, Y. Endo and K. Kisara, *Methods Find. Exp. Clin. Pharmacol.*, 1999, **21**, 471–476.
35. C. H. Tambeli, J. D. Levine and R. W. Gear, *Pain*, 2009, **143**, 228–232.

36. L. Calza, M. Pozza, R. Arletti, E. Manzini and T. Hokfelt, *Exp. Neurol.*, 2000, **164**, 333–343.
37. J. K. Zubieta, Y. R. Smith, J. A. Bueller, Y. Xu, M. R. Kilbourn, D. M. Jewett, C. R. Meyer, R. A. Koeppe and C. S. Stohler, *Science*, 2001, **293**, 311–315.
38. M. S. Buchsbaum, G. C. Davis and W. E. Bunney Jr, *Nature*, 1977, **270**, 620–622.
39. J. D. Levine, N. C. Gordon and H. L. Fields, *Nature*, 1979, **278**, 740–741.
40. M. Frid, G. Singer, T. Oei and C. Rana, *Psychopharmacology (Berlin, Ger.)*, 1981, **73**, 116–119.
41. G. Jungkunz, R. R. Engel, U. G. King and H. J. Kuss, *Psychiatry Res.*, 1983, **8**, 13–18.
42. W. S. Anderson, R. N. Sheth, B. Bencherif, J. J. Frost and J. N. Campbell, *Pain*, 2002, **99**, 207–216.
43. F. Noble and B. P. Roques, *Expert Opin. Ther. Targets*, 2007, **11**, 145–159.
44. M. Kathmann, K. Flau, A. Redmer, C. Trankle and E. Schlicker, *Naunyn-Schmiedeberg's Arch. Pharmacol.*, 2006, **372**, 354–361.
45. R. B. Rothman, D. L. Murphy, H. Xu, J. A. Godin, C. M. Dersch, J. S. Partilla, K. Tidgewell, M. Schmidt and T. E. Prisinzano, *J. Pharmacol. Exp. Ther.*, 2007, **320**, 801–810.
46. N. T. Burford, M. J. Clark, T. S. Wehrman, S. W. Gerritz, M. Banks, J. O'Connell, J. R. Traynor and A. Alt, *Proc. Natl. Acad. Sci. U. S. A.*, 2013, **110**, 10830–10835.
47. D. L. Bassoni, W. J. Raab, P. L. Achacoso, C. Y. Loh and T. S. Wehrman, *Methods Mol. Biol.*, 2012, **897**, 181–203.
48. N. T. Burford, T. Wehrman, D. Bassoni, J. O'Connell, M. Banks, L. Zhang and A. Alt, *J. Biomol. Screening*, 2014, **19**, 1255–1265.
49. J. D. Violin, A. L. Crombie, D. G. Soergel and M. W. Lark, *Trends Pharmacol. Sci.*, 2014, **35**, 308–316.
50. T. M. Bridges and C. W. Lindsley, *ACS Chem. Biol.*, 2008, **3**, 530–541.
51. N. T. Burford, J. Watson, R. Bertekap and A. Alt, *Biochem. Pharmacol.*, 2011, **81**, 691–702.
52. S. A. Nickolls, A. Waterfield, R. E. Williams and R. A. Kinloch, *J. Biomol. Screening*, 2011, **16**, 706–716.
53. A. Alt, I. J. McFadyen, C. D. Fan, J. H. Woods and J. R. Traynor, *J. Pharmacol. Exp. Ther.*, 2001, **298**, 116–121.
54. K. E. Livingston and J. R. Traynor, *Proc. Natl. Acad. Sci. U. S. A.*, 2014, **111**, 18369–18374.
55. C. B. Pert, G. Pasternak and S. H. Snyder, *Science*, 1973, **182**, 1359–1361.
56. E. J. Simon and J. Groth, *Proc. Natl. Acad. Sci. U. S. A.*, 1975, **72**, 2404–2407.
57. D. E. Selley, C. C. Cao, Q. Liu and S. R. Childers, *Br. J. Pharmacol.*, 2000, **130**, 987–996.
58. H. Gutierrez-de-Teran, A. Massink, D. Rodriguez, W. Liu, G. W. Han, J. S. Joseph, I. Katritch, L. H. Heitman, L. Xia, A. P. Ijzerman, V. Cherezov, V. Katritch and R. C. Stevens, *Structure*, 2013, **21**, 2175–2185.
59. S. Yuan, H. Vogel and S. Filipek, *Angew. Chem., Int. Ed. Engl.*, 2013, **52**, 10112–10115.

60. G. Fenalti, P. M. Giguere, V. Katritch, X. P. Huang, A. A. Thompson, V. Cherezov, B. L. Roth and R. C. Stevens, *Nature*, 2014, **506**, 191–196.
61. V. Katritch, G. Fenalti, E. E. Abola, B. L. Roth, V. Cherezov and R. C. Stevens, *Trends Biochem. Sci.*, 2014, **39**, 233–244.
62. D. Bartuzi, A. A. Kaczor and D. Matosiuk, *J. Chem. Inf. Model.*, 2016, **56**, 563–570.
63. P. Bisignano, N. T. Burford, Y. Shang, B. Marlow, K. E. Livingston, A. M. Fenton, K. Rockwell, L. Budenholzer, J. R. Traynor, S. W. Gerritz, A. Alt and M. Filizola, *J. Chem. Inf. Model.*, 2015, **55**, 1836–1843.
64. N. T. Burford, K. E. Livingston, M. Canals, M. R. Ryan, L. M. Budenholzer, Y. Han, Y. Shang, J. J. Herbst, J. O'Connell, M. Banks, L. Zhang, M. Filizola, D. L. Bassoni, T. S. Wehrman, A. Christopoulos, J. R. Traynor, S. W. Gerritz and A. Alt, *J. Med. Chem.*, 2015, **58**, 4220–4229.
65. C. Gaveriaux-Ruff and B. L. Kieffer, *Behav. Pharmacol.*, 2011, **22**, 405–414.
66. P. E. Lutz and B. L. Kieffer, *Trends Neurosci.*, 2013, **36**, 195–206.
67. E. M. Jutkiewicz, K. C. Rice, J. R. Traynor and J. H. Woods, *Psychopharmacology (Berlin)*, 2005, **182**, 588–596.
68. S. Conde-Ceide, C. M. Martinez-Viturro, J. Alcazar, P. M. Garcia-Barrantes, H. Lavreysen, C. Mackie, P. N. Vinson, J. M. Rook, T. M. Bridges, J. S. Daniels, A. Megens, X. Langlois, W. H. Drinkenburg, A. Ahnaou, C. M. Niswender, C. K. Jones, G. J. Macdonald, T. Steckler, P. J. Conn, S. R. Stauffer, J. M. Bartolome-Nebreda and C. W. Lindsley, *ACS Med. Chem. Lett.*, 2015, **6**, 716–720.
69. Y. Shang, H. R. Yeatman, D. Provasi, A. Alt, A. Christopoulos, M. Canals and M. Filizola, *ACS Chem. Biol.*, 2016, **11**, 1220–1229.
70. J. A. Ballesteros and H. Weinstein, in *Methods in Neurosciences*, ed. S. C. Sealfon and P. M. Conn, Academic Press, San Diego, CA, 1995, vol. 25, pp. 366–428.
71. A. C. Kruse, A. M. Ring, A. Manglik, J. Hu, K. Hu, K. Eitel, H. Hubner, E. Pardon, C. Valant, P. M. Sexton, A. Christopoulos, C. C. Felder, P. Gmeiner, J. Steyaert, W. I. Weis, K. C. Garcia, J. Wess and B. K. Kobilka, *Nature*, 2013, **504**, 101–106.
72. Q. Tan, Y. Zhu, J. Li, Z. Chen, G. W. Han, I. Kufareva, T. Li, L. Ma, G. Fenalti, J. Li, W. Zhang, X. Xie, H. Yang, H. Jiang, V. Cherezov, H. Liu, R. C. Stevens, Q. Zhao and B. Wu, *Science*, 2013, **341**, 1387–1390.
73. L. Manara, G. Bianchi, P. Ferretti and A. Tavani, *J. Pharmacol. Exp. Ther.*, 1986, **237**, 945–949.
74. D. Le Bars, M. Gozariu and S. W. Cadden, *Pharmacol. Rev.*, 2001, **53**, 597–652.
75. K. T. Pattinson, *Br. J. Anaesth.*, 2008, **100**, 747–758.
76. L. V. Panlilio and S. R. Goldberg, *Addict.*, 2007, **102**, 1863–1870.
77. N. T. Burford, J. R. Traynor and A. Alt, *Br. J. Pharmacol.*, 2015, **172**, 277–286.

CHAPTER 10

mGlu$_4$ PET Ligands as Enablers of Target Biology Understanding

ZHAODA ZHANG[a], PEKKA POUTIAINEN[a] AND ANNA-LIISA BROWNELL*[a]

[a]Athinoula A. Martinos Center for Biomedical Imaging, Department of Radiology, Massachusetts General Hospital, 149, 13th Street, Suite 2301, Charlestown, MA 02129, USA
*E-mail: abrownell@mgh.harvard.edu

10.1 Introduction

L-Glutamate is the most abundant excitatory neurotransmitter in the central nervous system (CNS) of vertebrates and probably mediates more than 50% of all synapses.[1,2] Two major classes of receptors, ionotropic glutamate receptors (iGlus) and metabotropic glutamate receptors (mGlus), as well as transporters are involved in glutamate signaling. iGlus, including the *N*-methyl-D-aspartate (NMDA), α-amino-3-hydroxy-5-methyl-4-isoxazole-propionate (AMPA), and kainite receptors are ligand-gated ion channels that mediate fast synaptic transmission. The mGlus (metabotropic glutamate receptors) modulate the presynaptic glutamate release and/or postsynaptic effects of glutamate. mGlus belong to class C of the G-protein-coupled receptor (GPCR) super family, which can be further divided into three subgroups including eight known receptor

RSC Drug Discovery Series No. 56
Allosterism in Drug Discovery
Edited by Dario Doller

Published by the Royal Society of Chemistry, www.rsc.org

sub-types (group I: $mGlu_1$ and $mGlu_5$, group II: $mGlu_2$ and $mGlu_3$, and group III: $mGlu_4$, $mGlu_6$, $mGlu_7$, and $mGlu_8$) based on their structural similarity, ligand specificity, and preferred coupling mechanism.[3] mGlus are involved in glutamate signaling in almost every excitatory synapse in CNS, and they have distinctive biodistribution in CNS depending on subtypes and subgroups.[4]

10.1.1 Metabotropic Glutamate Receptor 4 ($mGlu_4$)

$mGlu_4$ is expressed at multiple synapses throughout the basal ganglia, mainly localized presynaptically and expressed in the striatum, hippocampus, thalamus, and cerebellum.[3,5,6] $mGlu_4$ participates in presynaptic neurotransmission of excitatory glutamate signal and has received much attention lately due to its potential role in various neuronal diseases such as Parkinson's disease (PD) and other disorders.[7–9] As a group III mGlu, $mGlu_4$ interacts with the $G_{\alpha i/o}$ subunit (an isoform of G_α subunit) of G-protein that negatively couples with adenylate cyclase to inhibit cAMP dependent signal pathways.[10,11] G-proteins have three types of subunits: α, β and γ. Most GPCRs couple to heterotrimeric G-proteins, which comprise a $G_{\beta/\gamma}$ complex and a G_α subunit. The G_α subunits can be further subdivided into $G_{\alpha s}$ (G_s), $G_{\alpha i/o}$ (G_i), $G_{\alpha q/11}$ (G_q) and $G_{\alpha 12/13}$ ($G_{12/13}$) isoforms. Activation of $mGlu_4$ may reduce neurotransmitter release, a mechanism implicated in the pathophysiology of PD.

10.1.2 $mGlu_4$ and Parkinson's Disease

Parkinson's disease is a debilitating disorder with a lifetime risk of 2%, making PD the second most common neurodegenerative disorder after Alzheimer's disease.[12] PD was first described about 200 years ago[13] and there is not yet treatment for it, only palliative approaches for the symptoms. It is believed that the primary pathophysiological change that gives rise to the motor symptoms of PD is the loss of dopaminergic neurons of the substantia nigra pars compacta that is involved in modulating function in the striatum and other basal ganglia structures. This ultimately increases activity of the indirect pathway relative to the direct pathway (Figure 10.1) and leads to the symptoms of PD. The current pharmacotherapies aim to replace the missing dopamine, while the surgical therapies aim to reduce transmission in the indirect pathway. However, the surgical therapies are invasive and expensive, while prolonged dopamine replacement therapy with L-DOPA (L-3,4-dihydroxyphenylalanine) has severe side effects, known as L-DOPA-induced dyskinesia, in more than 40% of the patients. mGlus might provide new approaches for therapies by reducing transmission in the indirect pathways (Figure 10.1).

Part of the reason why PD is so difficult to treat is the late onset of the symptoms. It is postulated that the striatal dopamine content (or dopamine transporter binding) will decrease up to 60–70% before parkinsonian symptoms appear. To save the remaining neurons and abolish progression of degeneration, it is important to find protective mechanisms. It is known that glutamate is involved in basal ganglia function and dysfunction through

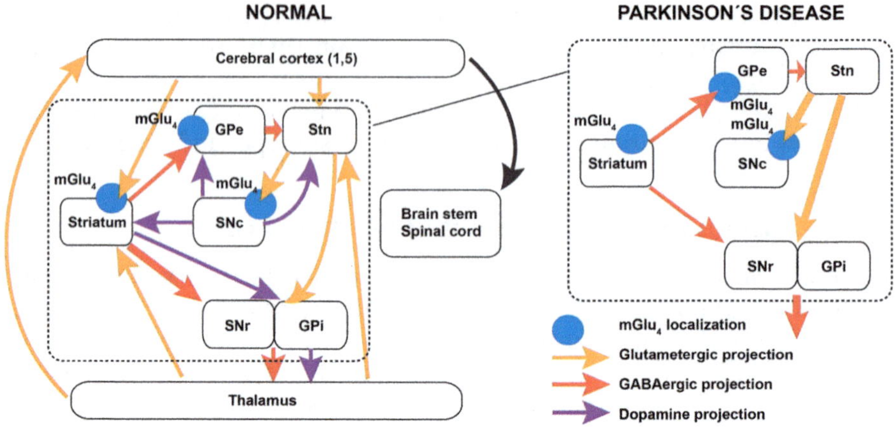

Figure 10.1 Schematic diagram of the proposed dopamine and glutamate signaling pathways. The direct pathway includes striatum, globus pallidus interna (GPi), thalamus, and cortex and the indirect pathway includes striatum, globus pallidus externa (GPe), subthalamic nucleus (Stn), substantia nigra reticulata (SNr), and substantia nigra pars compacta (SNc).

the corticofugal projections to the basal ganglia nuclei and the subthalamic neurons. mGlus affect several intracellular signal transduction mechanisms through G-proteins. The basal ganglia are thought to control movements by a balance of transmission through the two basal ganglia circuits that connect the direct and indirect pathways (Figure 10.1).

Presynaptic location of $mGlu_4$ makes an important contribution to glutamate neurotransmission, since glutamate as well as dopamine is released from the presynaptic site of the neuron and activation of $mGlu_4$ can inhibit the release of neurotransmitters such as glutamate and gamma-aminobutyric acid (GABA) and thus balance neurotransmission through the direct and indirect pathways. Several recent publications propose especially $mGlu_4$ as therapeutic targets for different neurological diseases including Parkinson's disease.[7–9]

10.1.3 Positron Emission Tomography (PET) Imaging

The first clinical studies using positron emission techniques were done in 1952 by Gordon Brownell in Massachusetts General Hospital to localize brain tumors.[14] The success of these studies led to wide use of positron emission imaging as a diagnostic tool for tumors and abscesses in different parts of the body in the following 20 years when film-based X-ray imaging was the only other option. Brownell's team conducted more than 4000 clinical patient studies before computer-tomographic (CT)-imaging came into clinical use. Independently of the development of CT Brownell's team developed software for three-dimensional image reconstruction and turned positron emission imaging into positron emission tomography (PET) by 1971.[15–17] After that came a long silent period for PET imaging. The roadblock for clinical use of PET has been high cost. PET is an expensive technique compared with CT or magnetic resonance imaging (MRI) since in addition to the imaging procedure it needs a radiopharmaceutical agent, which is often prepared right before the specific study. After [^{18}F]fluorodeoxyglucose ([^{18}F]FDG) and its distribution were developed, PET imaging gained a significant clinical role in oncological diagnosis and evaluation of therapeutic response, leading to the concept of FDG–PET.

PET has become an important clinical diagnostic and research modality, and also a valuable technology in drug discovery and development. PET offers picomolar sensitivity and is a fully translational technique that requires specific probes radiolabeled with a usually short-lived positron-emitting radionuclide. Carbon-11 (radioactive half life ($t_{1/2}$) = 20.4 min) and fluorine-18 ($t_{1/2}$ = 109.7 min) are the most commonly used radionuclides in PET imaging. PET has provided the capability of measuring biological processes at the molecular and metabolic levels *in vivo* by the detection of the photons formed as a result of the annihilation of the emitted positrons (Figure 10.2).

As a noninvasive medical and molecular imaging technique and a powerful tool in neurological research, PET offers the possibility of visualizing and analyzing the target receptor expression under physiological and pathophysiological conditions. PET has often been used to detect disease-related biochemical changes before the disease-associated anatomical changes can be found using standard medical imaging modalities.

Moreover, PET tracers serve as invaluable biomarkers during the clinical development of potential neurotherapeutics, in which the receptor

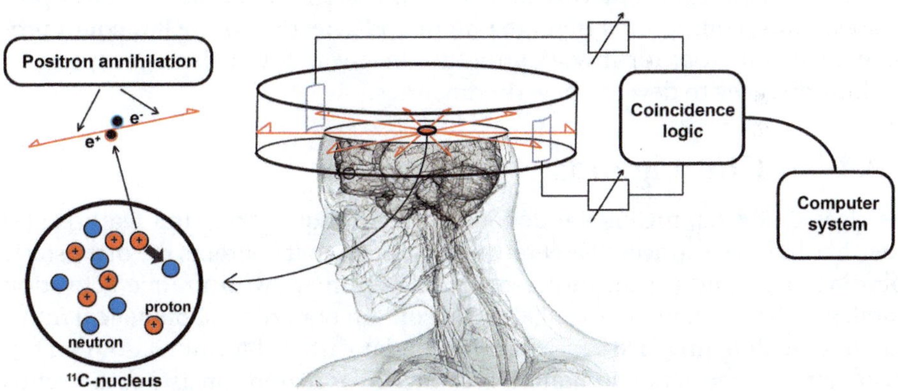

Figure 10.2 Schematic diagram of the processes that lead to PET images starting from a radiolabeled agent. The used isotope, *e.g.* carbon-11, has an extra proton in the nucleus so it is in an excitation state. When the proton in the nucleus turns to neutron it releases a positron what spins off from the nucleus. When the energy of this positron matches its counter particle, an electron, they will join and these particles are converted to energy (annihilation) releasing two photons in almost opposite directions. These photons (511 keV) are recorded by specific detectors of the PET system and converted to images to show the location and intensity of the isotope and simultaneously provide information on the radiolabeled imaging ligand.

occupancy of potential drug candidates in the brain is measured.[18,19] *In vivo* receptor occupancy can help to answer many vital questions in the drug discovery and development process, such as whether potential drugs reach their molecular targets, the relationship between therapeutic dose and receptor occupancy, the correlation between receptor occupancy and plasma drug levels, and the duration of time the drug remains at its target.[18]

Despite the great wealth of information that such probes can provide, the potential of PET strongly depends on the availability of suitable PET radiotracers. Thus, extensive research effort has been directed towards the development of PET radioligands suitable for probing $mGlu_1$ and $mGlu_5$, in which many PET radiotracers have been reported and several of which have been advanced for human clinical trials, such as 4-[^{18}F]fluoro-*N*-methyl-*N*-(4-(6-(methylamino)pyrimidin-4-yl)thiazol-2-yl)benzamide ([^{18}F]FIMX)[20] for $mGlu_1$, and 3-[^{18}F]fluoro-5-(2-pyridinylethynyl)benzonitrile ([^{18}F]FPEB),[21,22] 3-fluoro-5-(2-(2-[^{18}F](fluoromethyl)-thiazol-4-yl)ethynyl)benzonitrile ([^{18}F]SP203)[23] and 3-(6-methyl-pyridin-2-ylethynyl)-cyclohex-2-enone-O-[^{11}C]methyl-oxime ([^{11}C]ABP688)[24] for $mGlu_5$.

To better understand the role of $mGlu_4$ in health and disease conditions, we were interested in developing an $mGlu_4$-selective radiotracer for *in vivo* studies. Imaging of presynaptic $mGlu_4$ function might provide valuable insight into the synaptic transmission of glutamate. Interestingly, combining studies of presynaptic $mGlu_4$ and postsynaptic $mGlu_5$ could potentially provide in-depth information on synaptic transmission of glutamate, which might be useful for evaluating neurotransmission-related pathological processes. Thus, compounds that are potent and selective for $mGlu_4$ could provide a valuable tool for investigating the involvement of these receptors in various diseases to develop new therapeutics.

10.2 $mGlu_4$ Ligands

As class C GPCRs, mGlus are defined by two unique structural features: (1) they have a large extracellular N-terminal domain that contains the orthosteric binding sites; and (2) they form constitutive dimers with unique activation modes.[25] As shown in Figure 10.3, mGlu contain not only a heptahelical transmembrane domain (7TM) and an intracellular carboxyl-terminal domain but also a large extracellular domain. The extracellular region consists of the Venus flytrap domain (VFD), the orthosteric binding site of glutamate, and the cysteine-rich domain (CRD), which relays the conformational change induced by glutamate binding.[26,27] 7TM delivers the conformational change in the extracellular region to the intracellular region of mGlu. The binding locations at the 7TM are the non-competitive binding sites for allosteric modulators of mGlu.[27] The intracellular region comprises the C-terminus where signal transfers occur by G-protein coupling, kinase pathways, and alternative splicing.

Candidate radiotracers are often developed from derivatives of known ligands of the target of interest or even an existing therapeutic agent. Given the structural characteristics of $mGlu_4$, two types of ligands have been identified based on where they bind to the receptor: orthosteric agonists and

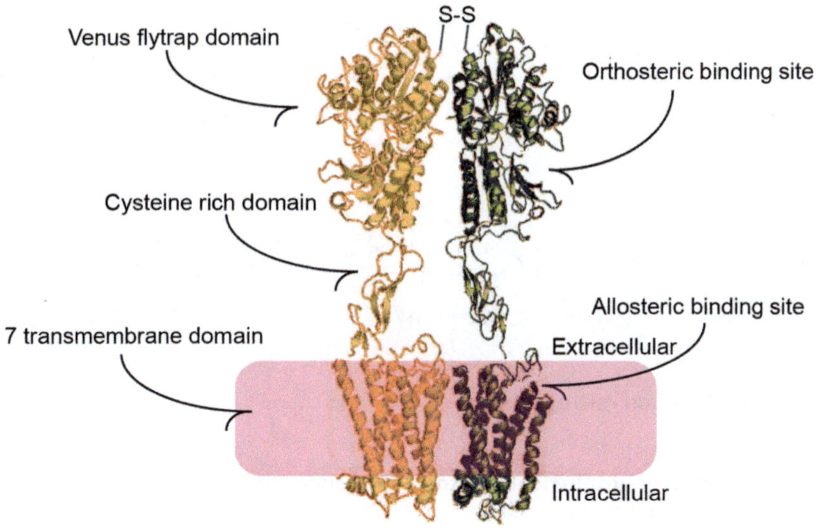

Figure 10.3 General structure of a dimeric mGlu receptor.

antagonists bind competitively at the extracellular glutamate binding site, and positive and negative allosteric modulators (PAMs and NAMs) bind non-competitively at the transmembrane region of the receptor.

PET radiotracers are used in picomolar concentrations to measure the biological process without perturbing the biological system. Therefore, a PET tracer can be developed from a potent ligand of any type without concerns for its biological effect.

10.2.1 Orthosteric Agonists and Antagonists

Glutamate is the endogenous agonist for all mGlu subtypes, and as such, many orthosteric agonists developed earlier for mGlu$_4$ are polar amino acid derivatives bearing additional acidic groups to mimic L-glutamate. However, it has been difficult to develop highly selective ligands for specific mGlu subtypes because of the highly conserved glutamate-binding site across members of this receptor family and the restricted structural requirements for pharmacophores that bind at the orthosteric site. Furthermore, these compounds based on glutamate have difficulty crossing the blood–brain barrier (BBB) because of their general hydrophilicity due to their amino acid origin. Some selected mGlu$_4$ agonists are shown in Figure 10.4.

As Table 10.1 shows, most of the previously developed orthosteric agonists such as compounds **2–4** were capable of discriminating against group I and II of mGlus, but lacked clear sub-type selectivity within group III and BBB penetration.[9] Recently notable examples of selective and brain-penetrant orthosteric agonists, such as LSP4-2022 (**8**) have been reported.[28,29] On the other hand, there is no mGlu$_4$-selective orthosteric antagonist reported so far.[30]

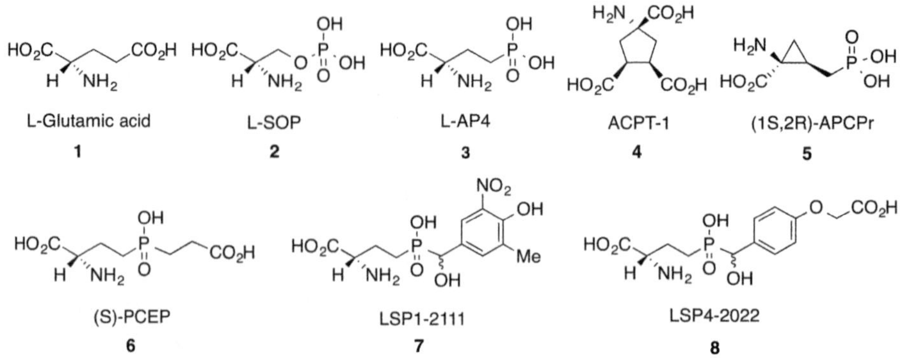

Figure 10.4 Selected mGlu$_4$ agonists.

Table 10.1 Binding affinity (in μM) and BBB penetration for a selection of orthosteric agonists.

Receptor	Group I		Group II		Group III				BBB[a]
Agonist	mGlu$_1$	mGlu$_5$	mGlu$_2$	mGlu$_3$	mGlu$_4$	mGlu$_6$	mGlu$_7$	mGlu$_8$	
1	1–13	3–11	0.3–12	2–9	3–17	5–8	2300	8–10	No
2	n/a	n/a	Antag[b]	Antag[b]	300	>100	>100	>100	No
3	>1000	>1000	>1000	>1000	0.2–1.2	0.6–0.9	160–1200	2	No
4	Antag[b]	>1000	>1000	n/a	7	18.4	n/a	10	No
7	>100	>100	>100	>100	2.2	1.7	53	66	Yes
8	>100	>100	>100	>100	0.11	4.2	12	29	Yes

[a]BBB penetration.
[b]Antagonist.

10.2.2 Allosteric Modulators

Recent efforts have focused on the development of allosteric modulators, which target the seven-transmembrane-spanning domain. In particular, the allosteric modulation of mGlu$_4$ has spurred intense interest after (−)-*N*-Phenyl-7-(hydroxyimino)cyclopropa[*b*]chromen-1a-carboxamide ((−)-PHCCC) (**9**), a partially selective mGlu$_4$ PAM, was discovered and demonstrated neuroprotection in models of PD.[31,32] Since then, there has been substantial progress in identifying PAMs for mGlu$_4$.[7,33,34] Figure 10.5 shows some representative mGlu$_4$ PAMs.[7,33,35–39]

(−)-PHCCC (**9**) and other early disclosed mGlu$_4$ PAMs such as **10–13** are deficient in their BBB penetration.[35] The potencies of these compounds are also relatively low (half maximum effective concentration (EC$_{50}$): 0.65–5.0 μM) and structure–activity relationship (SAR) studies around these structures have given 'flat' results. Significant progress in developing mGlu$_4$ PAMs has been made since 2009. Addex Therapeutics[40] and Vanderbilt University[35] have independently disclosed a series of small arylamide compounds as a new class of mGlu$_4$ PAMs. Engers *et al.* reported the SAR

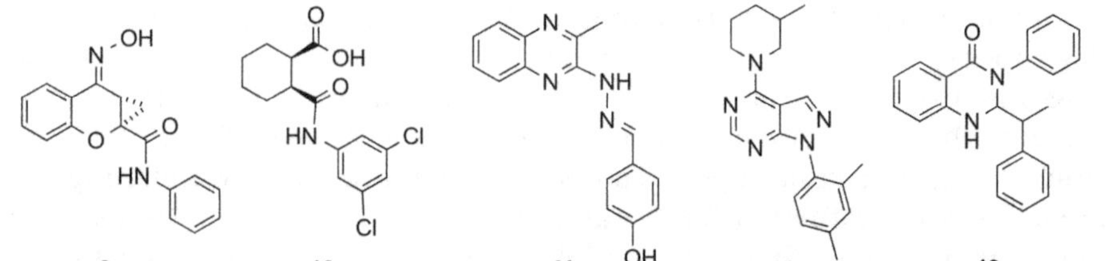

9 (-)-PHCCC EC_{50} = 1.4 μM

10 VU0155041 hEC_{50} = 750 nM

11 VU0001171 EC_{50} = 650 nM

12 VU0080241 EC_{50} = 5.0 μM

13 VU0092145 EC50: 3.0 μM

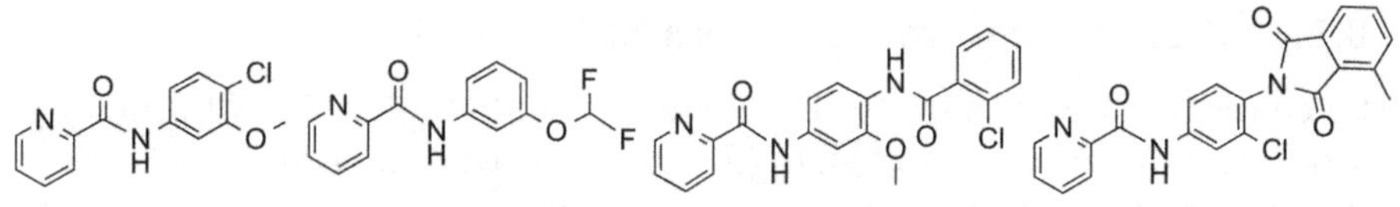

14 ML128 hEC_{50} = 240 nM rEC_{50} = 110 nM

15 DFMPP hEC_{50} = 340 nM rEC_{50} = 80 nM

16 VU0415374 EC_{50} = 100 nM

17 PAM-1 hEC_{50} = 43.6 nM

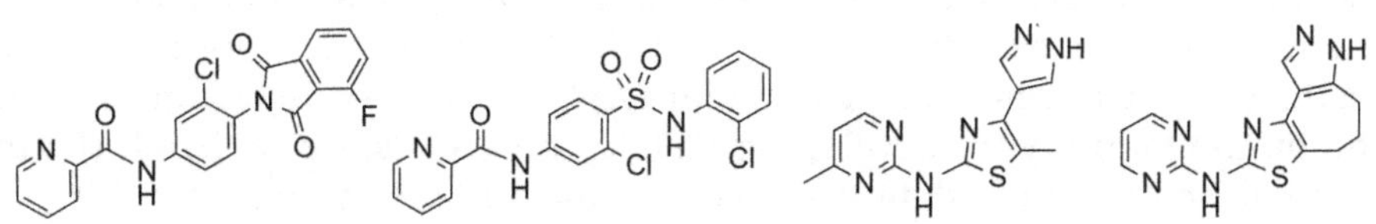

18 hEC_{50} = 50.4 nM

19 VU0364439 hEC_{50} = 20 nM

20 ADX88178 hEC_{50} = 4 nM rEC_{50} = 9 nM

21 hEC_{50} = 9 nM rEC_{50} = nM

Figure 10.5 Some representative $mGlu_4$ PAMs.

study on this series of heterobiaryl amides, in which the most potent compounds were **14** and **15**. The same group also reported a related diamide series as represented by compound **16**, which is brain-penetrant but highly protein-bound.[41] Scientist at Merck described a series of *N*-(phthalimide) phenylpicolinamides, such as **17** and **18**, with improved activity.[42] Engers *et al.* further studied SAR of 4-(phenylsulfamoyl)phenylacetamide derivatives and found that **19** was one of the most potent (20 nM) $mGlu_4$ PAM.[43] However, this class of compounds exhibited high plasma protein binding and poor microsomal stability, thus limiting their utility. Addex has disclosed a series of heteroaromatic compounds as positive allosteric modulators for $mGlu_4$, in which ADX88178 (**20**) is one of most potent and selective compound to date.[37,44,45] Doller and co-workers (Lundbeck Research, USA) have reported a tricyclic thiazolopyrazole compound **21**, which was identified as a very potent and orally available compound with excellent brain penetration and good physicochemical properties.[38] On the other hand, only one $mGlu_4$ NAM has been reported with a half maximum inhibitory concentration (IC_{50}) value of 8.2 μM.[46]

Compared with orthosteric compounds, many allosteric modulators display superior activity, better selectivity among mGlu subtypes and adequate drug-like properties. Subsequent results with PAMs of $mGlu_4$ have further validated the antiparkinsonian activity in animal models of PD,[6,36,45,47–49] in which this approach has opened a new potential avenue for developing non-dopaminergic treatments for PD and for identifying a novel disease-modifying therapeutics. Therefore, we have focused on developing $mGlu_4$ PET tracers from $mGlu_4$ PAMs.

10.2.3 Selection of $mGlu_4$ Ligands

The characteristics of a useful PET ligand include high affinity to the target receptor's concentration (B_{max}), high selectivity over other mGlu subtypes as well as other receptors, suitable pharmacological properties, and chemical structures of the precursor that allow fast labeling at a late stage.

The most of $mGlu_4$ PAMs were identified by using functional assays (calcium mobilization assays for human $mGlu_4$ and thallium flux assays for rat $mGlu_4$) and characterized with respect to EC_{50}, the maximum response, and the fold-shift values.[35] It is known that the EC_{50} value may not always correlate closely with the affinity value for PAMs.[50] It is very important to study the binding affinity for developing PET ligands. Thus, we prepared the tritium-labeled compound **14** ([^{3}H]**14**) (Scheme 10.1) for competitive binding assay.[51]

The selected lead PAM compounds were characterized with competitive binding studies using $mGlu_4$ transfected Chinese hamster ovary (CHO) cells by increasing the concentration of test materials from 0.01 nM to 10 μM in presence of 2 nM of [^{3}H]**14**, in which the binding affinities to $mGlu_4$ were described as IC_{50} values.[52] In Table 10.2, the affinity, activity, and some key structural parameters are given.

10.3 Co-Operative Binding Assay

In vitro assays are important tools for selection of potential PET imaging candidates. It is still common that functional assays are thought to correspond with the binding affinity when new ligands are selected for synthesis as potential PET tracers.[34] To further clarify this issue we developed the first fully characterized $mGlu_4$ binding assay (Figure 10.6).[52]

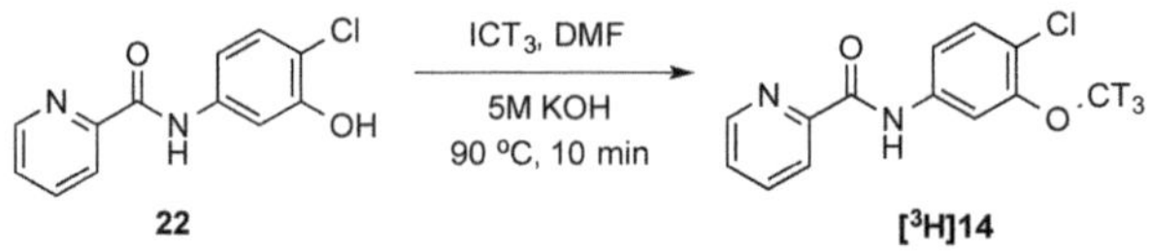

Scheme 10.1 Radiosynthesis of [^{3}H]**14**.

Table 10.2 The binding affinity and structural parameters of lead $mGlu_4$ PAM compounds.

	Activity EC_{50}[a] (nM)	Affinity IC_{50} (nM)	MW[b]	CLogP[c]	tPSA[d]	HBD[e]
14	240 (h); 110 (r)	5.1	262.7	2.8	50.7	1
15	340 (h); 80 (r)	4.6	264.2	2.7	50.7	1
17	43.6 (h)	2.3	391.8	3.6	78.9	1
18	50.4 (h)	2.4	395.8	3.3	78.8	1
20	4 (h); 9 (r)	3.2	272.3	2.0	73.5	2

[a]h = human, r=rat.
[b]Molecular weight.
[c]Calculated log of partition coefficient.
[d]Total polar surface area.
[e]Hydrogen bond donors.

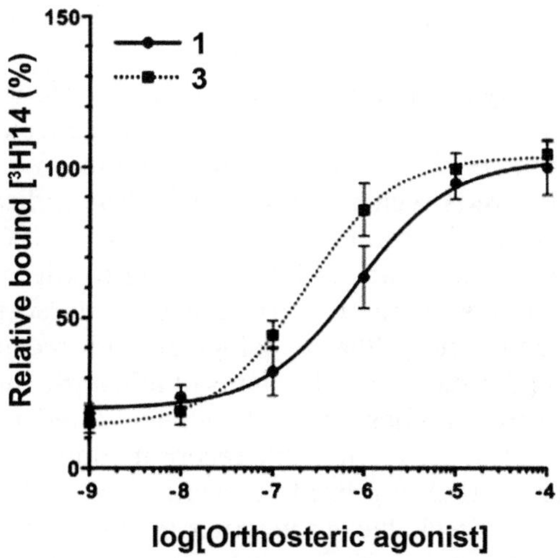

Figure 10.6 The equilibrium analysis of [^{3}H]**14** binding to $mGlu_4$ in CHO cells revealed co-operative binding with orthosteric agonist.

Cell-based assays, either functional or binding, are simplified models to mimic the environment in order to evaluate, screen and estimate drug-like properties of drug candidates in safer, more economical and less time-consuming manner. The presented binding results have been done using a competitive radioligand binding immunoassay (RIA), which is a classical approach, where the drug-target complex is formed and the unbound radiolabeled reference drug is removed from the sample.[52] The method revealed the complex pharmacology of $mGlu_4$ allosteric modulators (Figure 10.6). The results indicate that the characteristics of $mGlu_4$ allosteric modulators should be interpreted with the co-operative factor depending on the concentrations of the drugs and the glutamate in the specific microenvironment.

Typically NAMs are assessed for inhibition of a submaximal concentration of orthosteric agonist, whereas PAMs are assayed for potentiation of a low agonist concentration in functional assays.[53] The allosteric modulators potencies represent the combined contributions of modulator affinity and co-operativity with agonist and are dependent on the surrounding microenvironment.[54] Moreover, published $mGlu_4$ PAMs might actually display both agonist activity and positive co-operativity with orthosteric glutamate.[55] The term agonist–PAM (ago–PAM) refers to compounds that have relatively low co-operativity with glutamate and compounds like **9** and **16** which have more classical PAM characteristics with a strong ability to modulate the binding of glutamate.

The $mGlu_4$ PAMs have at least two binding cavities, although they might be partly overlapping.[56] Compounds **9** and **16** bind to differed cavities of the $mGlu_4$ 7TM domain.[7] Compound **9** has shown strong co-operativity and a larger fold shift in functional assays.[57] It is obvious that the agonist function and co-operativity mechanism have their own unique demands for the ligand–receptor interactions and formation of the complex protein conformations. The binding of class C GPCRs shares common features, where the co-operative $mGlu_4$ PAMs reside in pocket which is indicative of a homologous sodium binding site found from many class A GPCRs.[56,58] Despite a lack of knowledge of the actual sodium binding to the $mGlu_4$ in the 7TM region, this region makes a major contribution to the receptor function and co-operative characteristics. Pharmacological and receptor mutagenesis studies have revealed possible mechanisms of allosteric modulators interactions to the target protein. These studies have suggested a possible common allosteric binding site(s) across multiple receptor subtypes and the presence of multiple and/or overlapping binding cavities in one receptor subtype.[59] There is still a lack of knowledge of the structure-based mechanism of action of $mGlu_4$-targeting drugs.

The co-operativity has an evolutionary background which actually might help to explain the complex network of $mGlu_4$ signal transduction. The dynamic interplay combined with small timescales forms the basis for the reactive processes with changed probability of releasing neurotransmitters in the synaptic cleft.[60] Monitoring, balancing, and fine-tuning of the synaptic signals demand reliable detection of weak signals from random background signals and on the other hand enhancement of the decreased signaling cascade, for example as treatment for neurological dysfunctions.[56] Pure PAMs lack intrinsic agonist activity but rather have the ability to enhance the activity of sub-threshold levels of glutamate.

Overall, the binding profile is an essential feature for optimizing $mGlu_4$ drugs and imaging ligands.[61] The bindings of PAMs cannot be reliably estimated from functional assays. The developed binding assay could be used to explain the behavior of developed $mGlu_4$ PET agents.

10.4 Development of mGlu$_4$ PET Ligands

Although many PET probes have been developed for mGlu$_1$ and mGlu$_5$, mGlu$_4$ still lacks useful PET radioligands for clinical studies. A specific mGlu$_4$ PET radioligand could be an important tool for understanding the role of mGlu$_4$ in healthy and disease conditions, and also for the development of new drugs targeting this receptor.

To develop an mGlu$_4$ PET tracer, it is important to have suitable affinity to mGlu$_4$ as well as sufficient selectivity against other targets. Since poor BBB penetration and high nonspecific binding (NSB) are among the most frequent causes for failure in CNS PET ligand development, it is necessary to consider some important physicochemical parameters at the design stage. It has been proposed that preferred ranges for CNS drugs are molecular weight (MW) <450 Da, calculated log of partition coefficient (cLogP) in the range 2–5, polar surface area (tPSA) <90 Å^2 and less than three hydrogen bond donors (HBD).[62]

10.4.1 *N*-(4-Chloro-3-[^{11}C]methoxyphenyl)picolinamide ([^{11}C]14)

The Vanderbilt University group developed a selective mGlu$_4$ PAM, **14** (ML128, VU0361737), which is potent (rat EC_{50} (rEC_{50}) = 110 nM and human (hEC_{50}) = 240 nM) and centrally penetrating.[35] It was the first mGlu$_4$ PAM to demonstrate efficacy in a preclinical rodent model of motor impairments associated with PD.[7,35] ML128 showed some favorable properties for development as a potential PET radiotracer, which include rapid penetration into rat brain following intraperitoneal injection (time the tracer is present at maximum concentration in serum (T_{max}) for brain: 0.5 h), high brain : plasma partition coefficient (B : P = 4.1), good selectivity over other mGlu subtypes and the appropriate chemical structural attributes of the precursor to allow straightforward and fast radiolabeling. Our competitive binding study showed that **14** had good affinity (IC_{50} = 5.1 nM, Table 10.2). On the other hand, **14** also displayed low microsomal stability, fast clearance and short half-life.

We synthesized carbon-11 labeled [^{11}C]**14** as a PET radiotracer for mGlu$_4$. [^{11}C]**14** was synthesized from **22** using [^{11}C]CH_3I (Scheme 10.2). Total synthesis time was 38 ± 2.2 min (n = 7) from the end of bombardment to the formulation. The radioligand [^{11}C]**14** was obtained in 27.7 ± 5.3% (n = 5)

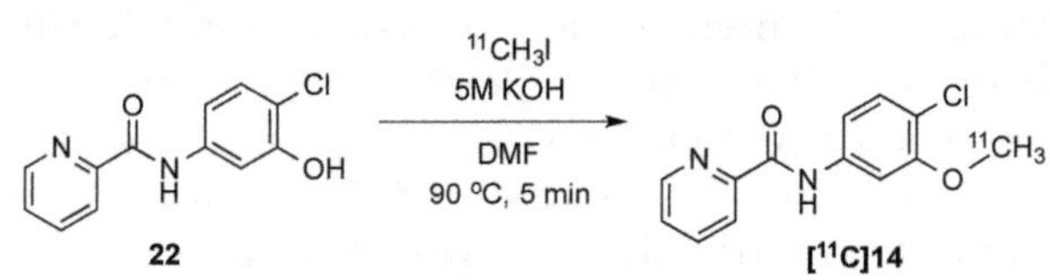

Scheme 10.2 Radiosynthesis of [^{11}C]**14**.

decay-corrected radiochemical yield based on the radioactivity of [^{11}C]CO_2. The radiochemical purity of [^{11}C]**14** was >99%. Specific activity was 188.7 ± 88.8 GBq mol^{-1} (n = 4) at the end of synthesis (EOS).

PET imaging was conducted on normal male Sprague Dawley rats, which showed fast accumulation of [^{11}C]**14** (peak activity between 1 and 3 min) in several brain areas, including the striatum, thalamus, hippocampus, cerebellum, and olfactory bulb, followed by fast washout. Blocking studies with the $mGlu_4$ modulator **15** showed a 22–28% decrease of [^{11}C]**14** accumulation, while studies of selectivity showed only minor decreases, supporting good selectivity over $mGlu_5$. Results from biodistribution studies and blood analyses support fast metabolism.

As the first PET imaging ligand for $mGlu_4$, [^{11}C]**14** was used for imaging its *in vivo* distribution and pharmacokinetics in brain and demonstrated some promising features as a PET radioligand such as fast uptake into brain and specific accumulation in $mGlu_4$-rich regions of the brain. However, in comparison with one of the best $mGlu_5$ PET tracer [^{18}F]FPEB,[21,22] [^{11}C]**14** showed a decreased retention time in the brain, which may affect the quality of the imaging. The results indicate that the affinity and metabolic stability of this class of tracers need further optimization.

10.4.2 *N*-(3-Chloro-4-(4-[^{18}F]fluoro-1,3-dioxoisoindolin-2-yl)phenyl)-2-picolinamide ([^{18}F]18)

Among $mGlu_4$ PAMs, a series of phthalimide derivatives of *N*-phenyl-2-picolinamides have been developed by Merck Research Laboratory and Vanderbilt University.[36,42] Two representative compounds (**17** and **18**) from this series are shown in Figure 10.5. It was also shown that compounds **17** (hEC_{50} = 43.6 nM; IC_{50} = 11 nM) and **18** (hEC_{50} = 50.4 nM; IC_{50} = 36 nM) had enhanced activity compared with **14** (hEC_{50} = 240 nM) and gave superb binding affinity (measured as the IC_{50} value) to $mGlu_4$. Because of their promising affinity to $mGlu_4$, we synthesized the phthalimide derivatives **17** and **18** for further studies. Our results from the *in vitro* competitive binding assay (Table 10.3) confirmed that the phthalimide derivatives, **17** (2.3 nM) and **18** (2.4 nM), gave somewhat improved affinity compared with **14** (5.1 nM).[51] This result shows that **17** and **18** have similar affinity while Merck's patent[42] presents that **17** (11 nM) has better affinity than **18** (36 nM), which may be caused by the use of cells expressing different $mGlu_4$ receptors (rat or human) as well as different radioligands and concentrations.[42,52] The results also indicate that these phthalimide compounds interact with the same allosteric site where compound **14** binds to $mGlu_4$ because the quantitative specific binding curves of these compounds show complete replacement of the radioligand at a higher concentration.[51] In addition, the selectivity of **18** to $mGlu_1$, $mGlu_5$, and $mGlu_8$ was determined by the functional assays, which showed little activity against these mGlus.[51]

On the basis of their improved affinity, both compounds **17** and **18** could be better candidates for development of PET radioligands. Nonetheless,

Table 10.3 Selected SAR attributes of *N*-phenylpicolinamide derivatives.

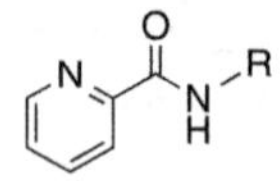

Compound	R	MW	CLogP	tPSA	Affinity IC_{50} (nM)	Log(IC_{50} ± SE)
14	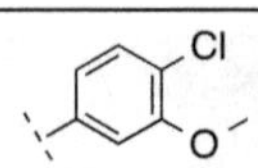	262.7	2.8	50.7	5.1	−8.29 ± 0.09
15	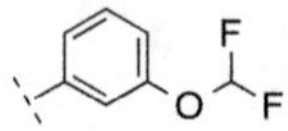	264.2	2.7	50.7	4.6	−8.34 ± 0.08
24	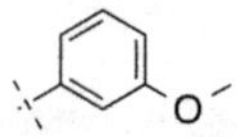	228.3	2.2	50.7	13.7	−7.86 ± 0.10
25	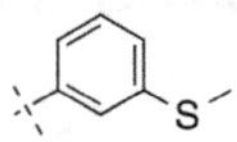	244.3	2.8	41.5	4.9	−8.31 ± 0.07
26	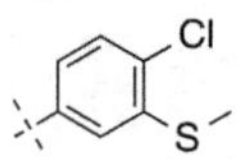	278.8	3.6	41.5	3.4	−8.46 ± 0.07
27	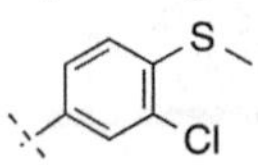	278.8	3.6	41.5	3.1	−8.50 ± 0.09
28	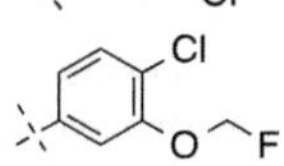	280.7	3.0	50.7	3.2	−8.49 ± 0.07
29	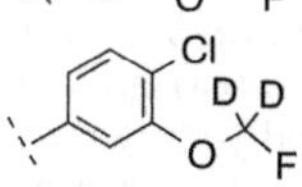	282.7	3.0	50.7	3.2	−8.49 ± 0.11

fluorine-18 is often the radionuclide of choice for both its physical and nuclear characteristics. Its half-life is long enough to carry out relatively extended imaging protocols when compared with what is possible with carbon-11, which facilitates kinetic studies and high-quality metabolic and plasma analyses. Therefore, we selected compound **18** for radiolabeling and *in vivo* PET imaging.

As shown in Scheme 10.3, [^{18}F]**18** was prepared from the nitro-precursor **23** with non-carrier-added (nca) [^{18}F]fluoride after careful optimization of reaction conditions. When finally formulated in 0.1 M citrate buffer (pH 4) with 10% ethanol, the specific activity of [^{18}F]**18** at the eEOS was 233.5 ± 177.8 GBq μmol^{-1} ($n = 4$). The average decay-corrected radiochemical yield (RCY) of [^{18}F]**18** was 16.4 ± 4.8% ($n = 4$). The radiochemical purity of [^{18}F]**18** was over 98%.

The characterization of [^{18}F]**18** was first conducted with male Sprague Dawley rats. The results of these studies demonstrated that [^{18}F]**18** crossed the BBB and occupied brain areas known to express mGlu$_4$. Time–activity curves (TACs) showed fast uptake and washout in different brain regions within 20 min. Pretreatment with 10 mg kg^{-1} of *N*-(3-(difluoromethoxy)phenyl)

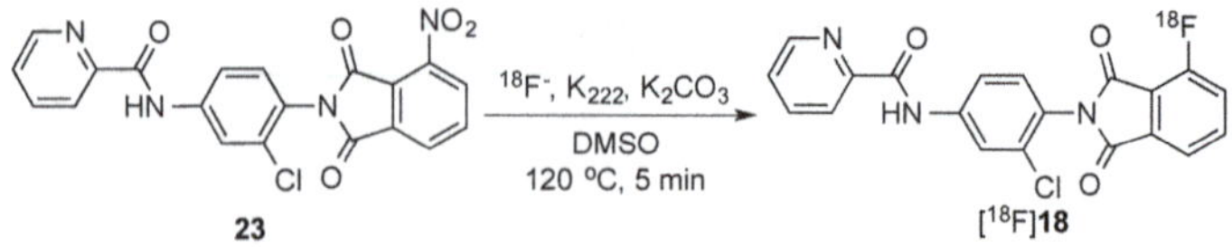

Scheme 10.3 Radiosynthesis of [^{18}F]**18**.

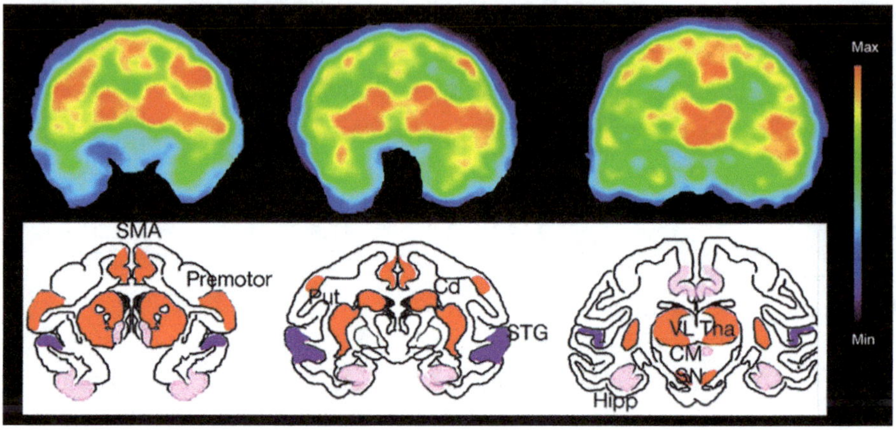

Figure 10.7 Distribution of [^{18}F]**18** in monkey brain. Color coded PET images show accumulation of [^{18}F]**18** in the selected three coronal slices of the monkey brain accumulated from the time interval 5–15 min after injection of the radioactivity and projection to mesostriatal, mesolimbic, and mesocortical areas, which are schematically shown over the atlas coronal sections in red (motor), pink (limbic), and purple (associative) areas. PET images show the highest accumulation of [^{18}F]**18** in the caudate (Cd), putamen (Put), thalamus (Tha), centromedian-parafascicularis thalamic nuclei (CM), substantia nigra (SN), hippocampus (Hipp) and several cortical areas including the supplementary motor area (SMA), premotor cortex, and supratemporal gyrus (STG).

picolinamide (DFMPP) (**15**) as an $mGlu_4$ blocking agent[35,63] or 10 mg kg^{-1} 3-((2-methyl-1,3-thiazole-4-yl)-ethynyl)pyridine (MTEP) as a selective $mGlu_5$ blocking agent[64] resulted in modest blocking effects in rat brains of 10–20% and 0–5% reduced uptake, respectively.

The results of the monkey studies showed in the brain high accumulation of [^{18}F]**18** in the striatum, thalamus, hippocampus, and cerebellum where previous literature has reported high expression of $mGlu_4$ (Figure 10.7).

TACs derived from the different monkey brain areas showed fast accumulation of [^{18}F]**18** with the highest accumulation about 3 min after the administration of the radioligand. About 50% of the activity stayed for 20 min followed by slower washout until 40 min, allowing suitable retention time for binding. The binding potential determined by the metabolite-corrected plasma input function and volume distribution model was highest in the thalamus (2.35) followed by the putamen (2.13), the caudate (1.91), the prefrontal cortex (1.47), and the cerebellum (1.19).

10.4.3 Re-Exploring the *N*-Phenylpicolinamide Derivatives

To improve the affinity and metabolic stability of mGlu$_4$ radiotracers, we synthesized a series of 20 modified *N*-phenylpicolinamide derivatives based on the previous SAR results.[35,40] In the structure–affinity relationship (SAR) study we modified compound **14** at three positions (3- or 4-phenyl, 6-pyrindyl). The modifications include the isosteric replacement of hydrogen by fluorine or deuterium, oxygen by sulfur, methoxy by cyano and fluoromethoxy groups, and changes to different halogen atoms. The radiolabeling strategy was also considered in lead optimization design to generate facile labeling positions for either ^{11}C or ^{18}F tracer. The SAR results showed that two modifications on the 3-phenyl position enhanced the affinity. The selected SAR results are given in Table 10.3.

The metabolic stability was an issue for **14**, in which the 3-methoxy group was identified as the soft group. Furthermore, the 3-phenyl position was also very sensitive to substitutions. It was reported that a simple change of the 3-difluoromethoxy group in compound **15** to a 3-trifluoromethoxy group imparted a more than 10-fold loss of mGlu$_4$ PAM functional activity.[35] We found that the 3-methylthio group was superior to the 3-methoxy group for mGlu$_4$ affinity by comparing compounds **24** (IC_{50} = 13.7 nM) and **25** (IC_{50} = 4.9 nM), showing a 2.8-fold enhancement in affinity. The 3-methylthio-4-chloro-phenyl derivative **26** (IC_{50} = 3.4 nM) was superior to the corresponding 3-methoxy-4-chloro-phenyl compound **14** (IC_{50} = 5.1 nM). On the other hand, as the 4-phenyl position of the *N*-phenylpicolinamide was tolerated with a relatively large substitution as demonstrated in compounds **16–19**, we synthesized compound **27**, which is the regioisomer of **26**. As shown in Table 10.3, the IC_{50} values of **26** and **27** are 3.4 and 3.1 nM, respectively, in which the affinity is slightly enhanced by 1.5–1.6 fold compared with **14** (IC_{50} = 5.1 nM). As the *N*-methylthiophenyl derivatives may have a different metabolic profile from the corresponding *N*-methoxyphenyl analogs, it was interesting to understand the effect on the metabolic stability of the picolinamide derivatives.

We then replaced 3-methoxy with 3-fluoromethoxy for two reasons: first, since both 3-methoxy- and 3-difluoromethoxy-analogs were active, we anticipated that fluoromethoxy analogs should also be active; second, it generates a position for fluorine-18 labeling, since fluorine-18 labeling is normally limited to chemical structures already containing a fluorine atom and the possible labeling strategies are limited for the preparation of radiotracers of high specific radioactivity. Indeed, the 3-fluoromethoxy compound **28** showed a slightly improved affinity (3.2 nM) compared with that (5.1 nM) of **14** (1.6-fold). However, 3-fluoromethoxy group may not be metabolically stable, since the 3-methoxy and 3-difluoromethoxy analogs **14** and **15** were metabolically unstable compounds. On the other hand, the 3-trifluoromethoxy analog of **15** was significantly more stable but lacked activity.[35] Hence, we had applied a 3-dideuteriumfluoromethoxy group to replace the 3-fluoromethoxy group as shown in compound **29**. Deuterium isotope effects have been used to reduce *in vivo* metabolic rates. For example, Zhang *et al.* reported that a deuterium-substituted analog (with $^{18}FD_2CO$) as a radioligand for peripheral benzodiazepine receptor (PBR) had remarkably prolonged the half-life ($T_{1/2}$) in mouse brain.[65] The deuterium substitution may reduce the rate of defluorination initiated by cleavage of the C–H bond without altering the binding affinity to $mGlu_4$. The result shows that the 3-dideuteriumfluoromethoxy modified compound **29** has excellent affinity.

On the basis of the affinity of these picolinamide derivatives, we subsequently determined the *in vitro* microsomal stability of selected compounds including **14, 15, 20** and **26–29** (Table 10.4). Compound **20** is one of a most potent $mGlu_4$ PAM to date and was shown to be orally active in a number of preclinical *in vivo* PD models.[33,45]

It is interesting to note that as the regioisomer of **26**, compound **27** displays a substantially decreased metabolic stability. These results show that compound **26** has enhanced affinity and/or increased *in vitro* microsomal stability compared with **14.** They also show that the dideuteriumfluoromethoxy-compound **29** ($T_{1/2}$ = 7.4 min) is somewhat more stable towards microsomal preparations than the corresponding fluoromethoxy-analog **28** ($T_{1/2}$ = 5.8 min) and the methoxy-analog **14** ($T_{1/2}$ = 4.9 min). It was reported that the cleaving rate of the C–H bond was about 6.7 times faster than that of the C–D bond at 25 °C.[65] On the other hand, the half life and the difference of the metabolic rates of the dideuteriumfluoromethoxy analog and the fluoromethoxy analog are dependent on the level of the enzyme. In developing the PET ligand for the PBR, Zhang *et al.* found that the $T_{1/2}$ in the plasma was 2.6 min for the deuterium-substituted analog (with $^{18}FD_2CO$) and 2.4 min for the non-deuterated analog. However, the $T_{1/2}$ of the deuterium-substituted analog in the brain was over 60 min, whereas that of for the non-deuterated analog was only 2.2 min.[65] We anticipate that the difference in the $T_{1/2}$ in the brain between compounds **29** and **28** as well as **14** could be more significant. Compared with **20**, compound **29** has the same affinity and a similar *in vitro* microsomal stability. It is clear that compound **29** has both enhanced affinity and improved *in vitro* microsomal stability compared with **14.**

The selectivities of **26** and **29** for other mGlu, including group I ($mGlu_1$ and $mGlu_5$), II ($mGlu_2$) and III ($mGlu_6$ and $mGlu_8$), were checked. The assays were

Table 10.4 *In vitro* properties of selected compounds.

Compound	Affinity IC_{50} (nM) (n = 3)	κ^a	SEM(κ) (n = 2)	$T_{1/2}$ (min)	Avg. P_e (10^{-6} cm s^{-1})
14	5.1	0.141	0.010	4.9, 6.0^b	256
15	4.6	0.132	0.008	5.2	214
20	3.2	0.099	0.005	7.0	257
26	3.4	0.106	0.005	6.5^b	
27	3.1	0.179	0.016	3.9^b	
28	3.2	0.120	0.010	5.8	272
29	3.2	0.093	0.005	7.4	

^{a}The decay constant that is slope of log concentration *vs.* time profile ($T_{1/2}$ = Ln 2/κ).
^{b}In a different run of microsomal stability assays (rat).

carried out at eight different concentrations between 0.1 nM and 30 μM of **26** and **29**. The results indicated that compound **26** did not have agonist and antagonist activities (>30 μM) toward $mGlu_1$, $mGlu_5$, $mGlu_2$, and $mGlu_6$, but displayed a weak agonist activity to $mGlu_8$ (EC_{50} = 14 μM).[66] Compound **29** showed little activity against these mGlus (>30 μM).

In addition, the permeability values of **14, 15, 20,** and **28** were measured using BBB parallel artificial membrane permeability assay (PAMPA) model at pH 7.4, which characterized the rate of crossing the BBB due to passive diffusion. The determined effective permeability (P_e) values are summarized in Table 10.4, in which the P_e results for internal high and low permeable standards are 160×10^{-6} cm s^{-1} for propranolol and $<2.8 \times 10^{-6}$ cm s^{-1} for atenolol, respectively. This result indicates that compounds **28** and **29** should have good BBB passive permeability. Although high BBB passive permeability does not necessary translate to sufficient unbound drug concentration in the brain, because of potential intrinsic clearance and efflux transport, it is beneficial for CNS drug candidates.

Our research further demonstrated that *N*-phenylpicolinamide is a good template for development of $mGlu_4$ ligands, which has generated extensive SAR results from us and other groups.[35,40,67] The SAR study has discovered a number of compounds with good affinities to $mGlu_4$. The dideuteriumfluoromethoxy-modified compound **29** is identified as a very promising $mGlu_4$ ligand, which has demonstrated enhanced affinity, improved *in vitro* microsomal stability, good selectivity and good permeability. Compound **29** is considered to be an attractive candidate for future labeling with fluorine-18 as an $mGlu_4$ PET tracer. Since a number of compounds have good affinity we are studying their PAM activity to $mGlu_4$ and potential therapeutic applications.

10.4.4 *N*-(3-([^{11}C]Methylthio)phenyl)picolinamide ([^{11}C]26)

Since compounds **26** and **27** displayed promising affinity, they were selected for *in vivo* evaluation as potential $mGlu_4$ radiotracers and were compared with [^{11}C]**14** to understand how their structure, affinity, and metabolic stability affect PET imaging.

Carbon-11 labeling was performed by methylation of the thiophenol precursors (**30** and **31**) using [^{11}C]CH_3I and K_2CO_3 in acetone at 50 °C for 3 min (Scheme 10.4). The complete radiosynthesis took 35–40 min from the end of bombardment (EOB) to the end of synthesis (EOS). The carbon-11 labeled compounds [^{11}C]**26** and [^{11}C]**27** were obtained in 19.0 ± 7.7% (n = 7) and 34.8 ± 8.0% (n = 4) RCYs from [^{11}C]CO_2, respectively. Their specific activities were 496 ± 138 GBq μmol^{-1} (n = 6) and 463 ± 263 GBq μmol^{-1} (n = 4) at EOS, respectively. The radiochemical purities of [^{11}C]**26** and [^{11}C]**27** determined by radio-TLC were 98.0 ± 0.5% (n = 7), and 97.5 ± 1.0% (n = 4), respectively. Similar results were also observed in the radio-high-performance liquid chromatography (HPLC) analyses.

PET imaging studies were conducted using male Sprague Dawley rats. The *in vivo* uptake and kinetics were examined using small-animal PET. The PET images (Figure 10.8) indicated that [^{11}C]**26** and [^{11}C]**27** crossed the BBB quickly and were mainly accumulated in the thalamus, hippocampus, cerebellum, and striatum, which were reported as the mGlu$_4$-rich regions of the rat brain.[5,68,69] We also acquired PET images of [^{11}C]**14** in the same time frames for comparison. To confirm the corresponding brain regions, anatomical templates of rat brains were used for the images with [^{11}C]**26** (Figure 10.8). The results suggested that [^{11}C]**26** gave the best contrast among the three radiotracers and displayed better images in the hippocampus,

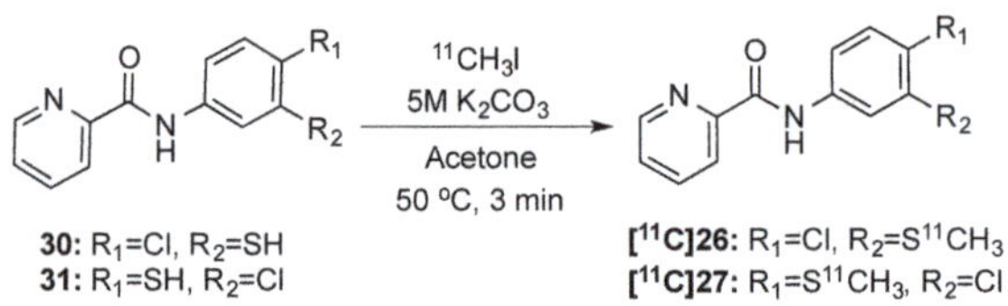

Scheme 10.4 Radiosyntheses of [^{11}C]**26** and [^{11}C]**27**.

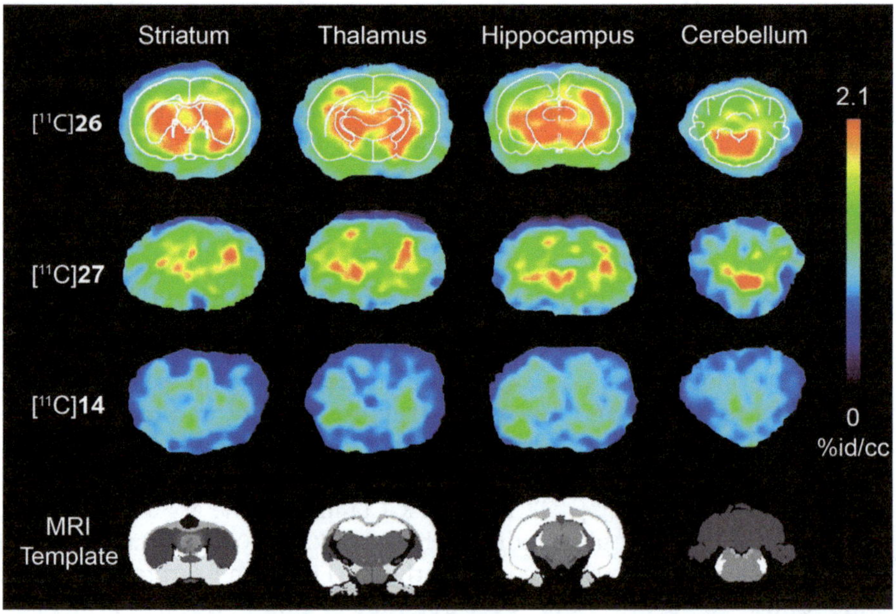

Figure 10.8 PET images of [^{11}C]**14**, [^{11}C]**26**, and [^{11}C]**27** in the striatum, thalamus, hippocampus and cerebellum. The images represent the distribution of radiotracer from 5 to 10 min after the injection. All images are normalized to the highest voxel value (%id/cc) of the data set. [^{11}C]**26** shows highest accumulation to the mGlu$_4$ expressing brain areas, which demonstrates the improved imaging quality compared with [^{11}C]**14**.

striatum, thalamus, and cerebellum. Using the mGlu$_4$-negligible region in the brain as a reference, we calculated the uptake ratio (contrast) of radioactivity between the mGlu$_4$-rich and the reference regions (areas immediately outside the cortical areas) from the reconstructed PET images. The maximum ratio was 1.67 ± 0.08 (n = 30; 10 images in 3 baseline studies) for [^{11}C]**14** and 2.10 ± 0.25 (n = 30) for [^{11}C]**26**. The results show that the signal to noise ratio of [^{11}C]**26** was enhanced up to 25% compared with [^{11}C]**14**.

The regional TACs of PET with [^{11}C]**14**, [^{11}C]**26**, and [^{11}C]**27** in the hippocampus, striatum, thalamus, and cerebellum of rat brain are given in Figure 10.9. The percent of the injected dose per milliliter (%id/cc) values at each time point obtained from the different PET images were averaged for each radiotracer and converted into TACs. As shown in the TACs, the radioactivity of [^{11}C]**14**, [^{11}C]**26**, and [^{11}C]**27** quickly increased in the brain within 2 min after a tail vein injection and then washed out within 20 min. The maximum radioactivity uptake depends on the region of the brain. The highest uptake for [^{11}C]**26** was seen in the thalamus, followed by the striatum, hippocampus, and cerebellum. The results indicate that [^{11}C]**26** has better brain uptake than [^{11}C]**14** and [^{11}C]**27** and is an improved PET radioligand for mGlu$_4$.

A blocking experiment for [^{11}C]**26** was carried out 2 h after the baseline scans to compare blocking images with baseline images. An mGlu$_4$ PAM **15** was used as an mGlu$_4$-blocking agent. The blocking agent was administered

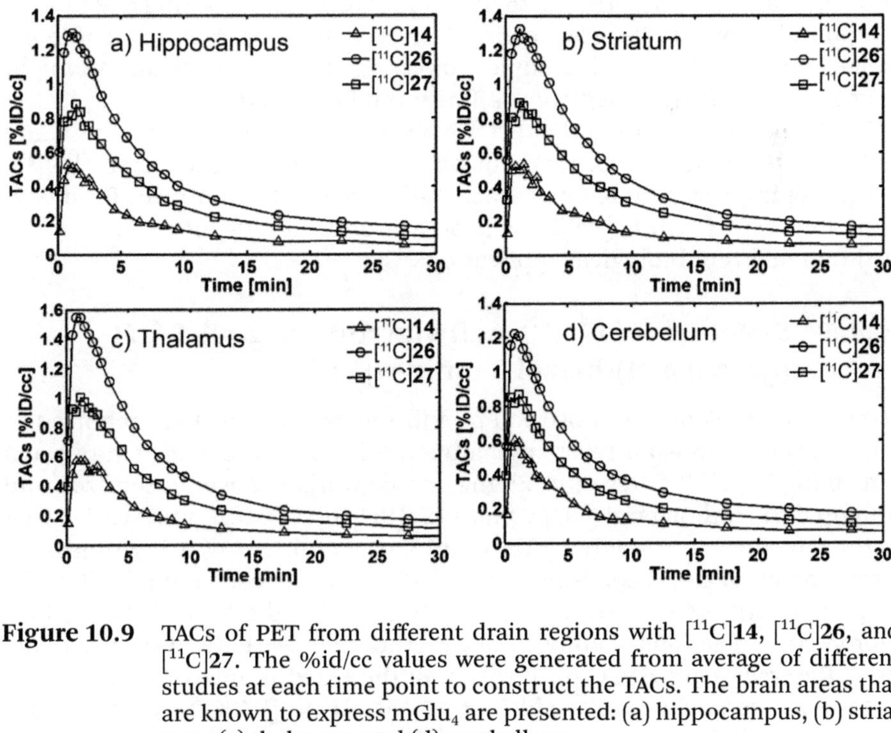

Figure 10.9 TACs of PET from different drain regions with [^{11}C]**14**, [^{11}C]**26**, and [^{11}C]**27**. The %id/cc values were generated from average of different studies at each time point to construct the TACs. The brain areas that are known to express mGlu$_4$ are presented: (a) hippocampus, (b) striatum, (c) thalamus, and (d) cerebellum.

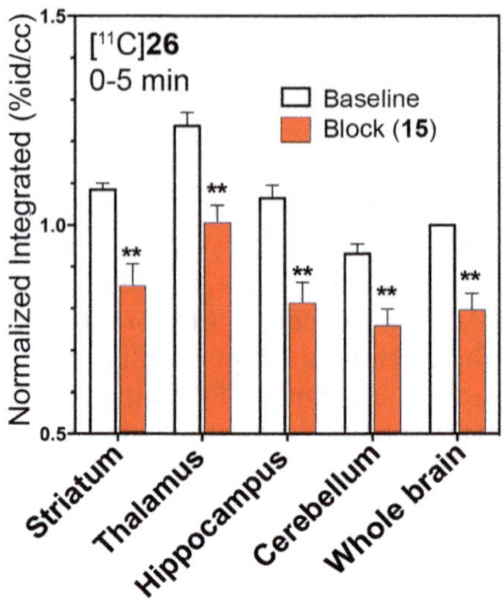

Figure 10.10 Blocking studies of $[^{11}C]$**26** with compound **15** during the first 5 min after the radiotracer injection. Rats were scanned for baseline and normalized to whole brain individually. The results ($n = 4$) show significant (**, $P < 0.01$) blocking effect in brain areas that are known to express $mGlu_4$.

intravenously 1 min before $[^{11}C]$**26** injection. The $mGlu_4$ blocking effects with compound **15** were 19–24% for first 5 min in selected regions of interest (ROIs) (Figure 10.10). The incomplete blocking effect by compound **15** might be attributed to its relatively weak affinity and fast kinetics.[35,67]

The PET studies showed that $[^{11}C]$**26** accumulated fast into the brain and had higher uptake, slower washout and 25% better contrast than $[^{11}C]$**14**, indicating improved imaging characteristics as a PET radiotracer for $mGlu_4$ compared with $[^{11}C]$**14**. $[^{11}C]$**26** will be a useful radioligand to investigate $mGlu_4$ in different biological applications.

10.4.5 5-Methyl-*N*-(4-[^{11}C] methylpyrimidin-2-yl)-4-(1H-pyrazol-4-yl)thiazol-2-amine ([^{11}C]20)

ADX88178 (**20**) is one of most potent and selective $mGlu_4$ PAM reported to date. It is efficacious for reversing haloperidol-induced catalepsy in rats at 3 and 10 mg kg^{-1}.[37,44,45] Recently, Zhang and co-workers have synthesized and evaluated the radioligand $[^{11}C]$ADX88178 ($[^{11}C]$**20**).[70] $[^{11}C]$**20** was synthesized by introducing a $[^{11}C]$methyl group into a pyrimidine ring using organotin precursor **32** *via* C–^{11}C coupling with tris(dibenzylideneacetone)dipalladium ($Pd_2(dba)_3$), tri(*o*-tolyl)phosphine (P(*o*-tol)$_3$), CuCl and a base, followed by a deprotection reaction to remove the 4-methoxybenzyl (PMB) group (Scheme 10.5). Reverse-phase HPLC purification of the reaction mixture gave $[^{11}C]$**20** in 16 ± 6% radiochemical yield (n = 10, based on $[^{11}C]CO_2$, corrected for decay). Starting from 22.2 to 24.0 GBq $[^{11}C]CO_2$, 0.54–1.10 GBq of $[^{11}C]$**20** was

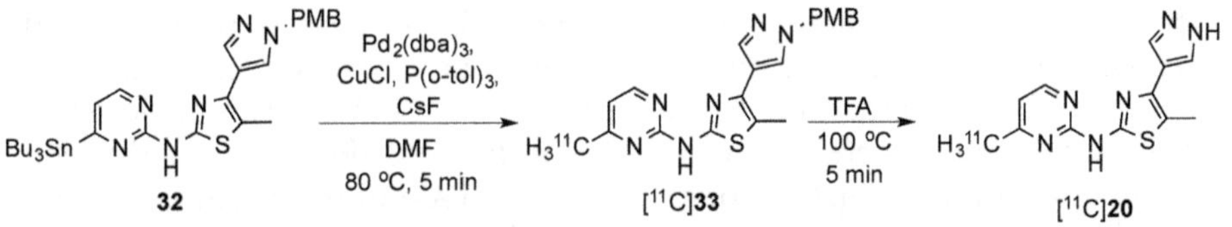

Scheme 10.5 Radiosynthesis of [^{11}C]**20**.

produced within 45 min (n = 10) of averaged synthesis time from the end of bombardment. The radiochemical purity of [^{11}C]**20** was greater than 98% and the specific activity was 90–120 GBq μmol^{-1} at the end of synthesis.

In vitro autoradiography of rat brain showed specific binding of [^{11}C]**20** in the cerebellum, striatum, thalamus, cerebral cortex, and medulla oblongata. The radioactivity was significantly decreased by treatment with unlabeled **20**. Specific binding of [^{11}C]**20** was further verified in the *ex vivo* biodistribution study in ddY mice brains. All radioactivity in brain regions peaked at 1 min after the injection of [^{11}C]**20** and then showed a quick decrease. The results showed that the distribution pattern of radioactive accumulation in the control section was similar to the reported biological distribution of mGlu$_4$ in the rat brain.[71] The radioactive accumulation was reduced dose-dependently by administration of multiple doses of unlabeled **20**. However, radioactive accumulation was also identified in some negligible regions, such as the cerebral cortex and midbrain, suggesting that [^{11}C]**20** may bind to other receptors. There has been no PET imaging using [^{11}C]**20** presented.

10.5 Functional Modulation of GPCRs During Parkinson-Disease-Like Neurodegeneration

A diverse set of allosteric modulators provide tools to investigate modulation of neural signaling and receptor functions. To obtain understanding of the target biology of mGlu$_4$ we have extensively investigated positive allosteric modulators as PET imaging ligands for mGlu$_4$. We investigated also biological activity of the developed compounds using pharmacological MRI (phMRI) approaches. These studies opened a new way to characterize compounds and explore their potential utility for therapeutic purposes.

We conducted experimental PET and phMRI studies using the rat 6-hydroxydopamine (6-OHDA) lesion model of PD, which provides a means to investigate the effects of mGlu stimulation on the intact control and the dopamine-denervated side of the brain simultaneously.

Dopamine receptors are GPCRs with seven membrane-spanning alpha-helical regions with extracellular and intracellular loops.[72] The glutamate and dopamine signaling are both coupled to the complex G-protein system, which could potentially involve cross-talk between receptors, transmitters, and signal modulations.[73–75]

By using [^{11}C]**14** as an $mGlu_4$ radiotracer for PET imaging we found enhanced $mGlu_4$ expression on the lesioned side in the striatum and hippocampus in this PD rat model. Similarly, using [^{18}F]FPEB as an $mGlu_5$ radioligand we found a decreased accumulation in the striatum and enhanced accumulation in the hippocampus. Using **15**, a PAM of $mGlu_4$, and MTEP, a NAM of $mGlu_5$, as stimulators for phMRI we showed in the same animals significant changes indicating enhanced blood volume on the lesioned side in different cortical areas.[75] Longitudinal PET imaging studies showed a decline in $mGlu_4$ expression and enhanced $mGlu_5$ expression in the striatum during progressive dopaminergic degeneration. In PD, the loss of dopamine-mediated inhibition in the striatum results in excess excitatory transmission and it is therefore possible that $mGlu_4$ activation improves PD symptoms directly and furthermore, inhibits the development of L-DOPA-induced dyskinesia, a frequent complication of dopamine replacement therapy that represents an abnormal form of synaptic plasticity. Finally, pharmacological manipulation $mGlu_4$ could open up new modalities of neuroprotection in PD and other degenerative diseases of the nervous system by decreasing excitotoxicity and modulating signaling pathways. It was shown in a recent study using the alpha-synuclein AT53T rat model of PD that there was a decline of the striatal $mGlu_1$ expression and no change in $mGlu_5$ expression, further demonstrating the important functional role of mGlus in Parkinson-disease-like neurodegeneration.[76]

10.6 Conclusions

Results of recent studies using PAMs of $mGlu_4$ as chemical tools consistently support their antiparkinsonian activity in animal models, thus opening a new avenue for developing non-dopaminergic treatments for PD and for identifying novel disease-modifying therapeutics. To better understand the role of $mGlu_4$ in health and disease conditions, we have focused on developing $mGlu_4$-selective PET tracers from $mGlu_4$ PAMs for *in vivo* study and will continue to optimize properties of $mGlu_4$ PET tracers. We have investigated target biology of this receptor *in vitro* using cells expressing $mGlu_4$, as well as *in vivo*, through the 6-OHDA-induced lesion rat model of Parkinson's disease. Results from cell studies showed the important role of the endogenous-glutamate-induced change on affinity of allosteric modulators of mGlus. Results of PET imaging studies showed interplay between presynaptic $mGlu_4$ and postsynaptic $mGlu_5$ expression in the lesioned side of the brain while the results of phMRI studies of the hemodynamic response showed enhanced signal changes in cortical areas in the lesioned side after challenge with $mGlu_4$ PAM or $mGlu_5$ NAM.

Acknowledgements

The research was supported by NIH grants NIBIB-R01EB012864 and NIMH-R01MH91684. The following supporting instrument grants (1S10RR029495-01, 1S10RR026666-01, and 1S10RR023452-01) are also appreciated. Finally, the authors would like to express their appreciation of the financial support for P.P. from The Orion Farmos Research Foundation, Saastamoinen Foundation, Sigrid Juselius Foundation, Osk Huttunen Foundation, and Kuopio University Foundation.

References

1. P. J. Conn, *Ann. N. Y. Acad. Sci.*, 2003, **1003**, 12–21.
2. J. C. J. Watkins and D. E. Jane, *Br. J. Pharmacol.*, 2006, **147**(S1), S100–S108.
3. P. J. P. Conn and J. P. Pin, *Annu. Rev. Pharmacol. Toxicol.*, 1997, **37**, 205–237.
4. G. P. Riedel, B. Platt and J. Micheau, *Behav. Brain Res.*, 2003, **140**, 1–47.
5. C. Corti, L. Aldegheri, P. Somogyi and F. Ferraguti, *Neuroscience (Oxford, U. K.)*, 2002, **110**, 403–420.
6. O. M. Valenti, G. Mannaioni, G. R. Seabrook, P. J. Conn and M. J. Marino, *J. Pharmacol. Exp. Ther.*, 2005, **313**, 1296–1304.
7. A. J. E. Robichaud, D. W. Engers, C. W. Lindsley and C. R. Hopkins, *ACS Chem. Neurosci.*, 2011, **17**, 433–449.
8. M. Amalric, S. Lopez, C. Goudet, G. Fisone, G. Battaglia, F. Nicoletti, J. P. Pin and F. C. Acher, *Neuropharmacology*, 2013, **66**, 53–64.
9. X. Huang, E. Dale, R. M. Brodbeck and D. Doller, *Curr. Top. Med. Chem. (Sharjah, United Arab Emirates)*, 2014, **14**, 1755–1770.
10. F. B.-G. Fettagutti, B. Balani-Guerra, M. Corsi, S. Nakanishi and C. Corti, *Eur J Neurosci.*, 1999, **11**, 2073–2082.
11. C. M. Niswender and P. J. Conn, *Annu. Rev. Pharmacol. Toxicol.*, 2010, **50**, 295–322.
12. W. Dauer and S. Przedborski, *Neuron*, 2003, **39**, 889–909.
13. J. Parkinson, *An essay on the shaking palsy*, Sherwood, Neely and Jones, London, UK, 1817.
14. G. L. Brownell and W. H. Sweet, *Nucleonics*, 1953, **11**, 6.
15. S. M. Pizer, D. A. Chesler and G. L. Brownell, in *Sharing of Computer Programs and technology in Nuclear medicine*, ed. F. H. Clark, B. F. Maskewitz and J. Gurney, U.S. Dept of Commerce, Springfield, VA, USAEC CONF-710425, 1971, p. 92.
16. D. A. Chesler, in *Tomographic Imaging in Nuclear Medicine*, ed. Gs. Freedman, Society of Nuclear Medicine, New York, 1972, pp. 176–183.
17. G. L. Brownell and C. A. Burnham, in *Tomographic imaging in nuclear medicine*, ed. G. S. Freedman, Society of Nuclear Medicine, New York, 1972, pp. 154–164.
18. J. Passchier, A. Gee, A. Willemsen, W. Vaalburg and A. v. Waarde, *Methods*, 2002, **27**, 278–286.
19. S. Sharma and C. Lindsley, *Curr. Top. Med. Chem.*, 2007, **7**, 1541–1542.
20. R. Xu, P. Zanotti-Fregonara, S. S. Zoghbi, R. L. Gladding, A. E. Woock, R. B. Innis and V. W. Pike, *J. Med. Chem.*, 2013, **56**, 9146–9155.
21. S. Patel, T. Hamill, B. Connolly, E. Jagoda, W. Li and R. Gibson, *Nucl. Med. Biol.*, 2007, **34**, 1009–1017.
22. J. Wang, W. Tueckmantel, A. Zhu, D. Pellegrino and A.-L. Brownell, *Synapse*, 2007, **61**, 951–961.

23. K. Brown Amira, Y. Kimura, S. Zoghbi Sami, G. Simeon Fabrice, J.-S. Liow, C. Kreisl William, A. Taku, M. Fujita, W. Pike Victor and B. Innis Robert, *J. Nucl. Med.*, 2008, **49**, 2042–2048.
24. S. M. Ametamey, V. Treyer, J. Streffer, M. T. Wyss, M. Schmidt, M. Blagoev, S. Hintermann, Y. Auberson, F. Gasparini, U. C. Fischer and A. Buck, *J. Nucl. Med.*, 2007, **48**, 247–252.
25. L. Chun, W.-h. Zhang and J.-f. Liu, *Acta Pharmacol. Sin.*, 2012, **33**, 312–323.
26. T. Muto, D. Tsuchiya, K. Morikawa and H. Jingami, *Proc. Natl. Acad. Sci. U. S. A.*, 2007, **104**, 3759–3764.
27. D. R. R. Hampson, E. M. Rose and J. E. Antflick, in *The glutamate receptors*, ed. R. W. Gereau and G. T. Swanson, Human Press, Totowa, 2008, pp. 363–386.
28. C. Goudet, B. Vilar, T. Courtiol, T. Deltheil, T. Bessiron, I. Brabet, N. Oueslati, D. Rigault, H.-O. Bertrand, H. McLean, H. Daniel, M. Amalric, F. Acher and J.-P. Pin, *FASEB J.*, 2012, **26**, 1682–1693.
29. M. Cajina, M. Nattini, D. Song, G. Smagin, E. B. Joergensen, G. Chandrasena, C. Bundgaard, D. B. Toft, X. Huang, F. Acher and D. Doller, *ACS Med. Chem. Lett.*, 2014, **5**, 119–123.
30. K. G. Liu and D. Doller, *Annu. Rep. Med. Chem.*, 2011, **46**, 3–17.
31. M. J. Marino, D. L. Williams, Jr., J. A. O'Brien, O. Valenti, T. P. McDonald, M. K. Clements, R. Wang, A. G. DiLella, J. F. Hess, G. G. Kinney and P. J. Conn, *Proc. Natl. Acad. Sci. U. S. A.*, 2003, **100**, 13668–13673.
32. M. Maj, V. Bruno, Z. Dragic, R. Yamamoto, G. Battaglia, W. Inderbitzin, N. Stoehr, T. Stein, F. Gasparini, I. Vranesic, R. Kuhn, F. Nicoletti and P. J. Flor, *Neuropharmacology*, 2003, **45**, 895–906.
33. C. W. Lindsley and C. R. Hopkins, *Expert Opin. Ther. Pat.*, 2012, **22**, 461–481.
34. Z. Zhang and A.-L. Brownell, in *Neuroimaging - Clinical applications*, ed. P. Bright, InTech - Open Access Publisher, Rijeka, Croatia, 2012, ch. 22, pp. 499–532.
35. D. W. N. Engers, C. M. Niswender, C. D. Weaver, S. Jadhav, U. N. Menon, R. Zamorano, P. J. Conn, C. W. Lindsley and C. R. Hopkins, *J. Med. Chem.*, 2009, **52**, 4115–4118.
36. C. K. Jones, D. W. Engers, A. D. Thompson, J. R. Field, A. L. Blobaum, S. R. Lindsley, Y. Zhou, R. D. Gogliotti, S. Jadhav, R. Zamorano, J. Bogenpohl, Y. Smith, R. Morrison, J. S. Daniels, C. D. Weaver, P. J. Conn, C. W. Lindsley, C. M. Niswender and C. R. Hopkins, *J. Med. Chem.*, 2011, **54**, 7639–7647.
37. M. Kalinichev, E. Le Poul, C. Bolea, F. Girard, B. Campo, M. Fonsi, I. Royer-Urios, S. E. Browne, J. M. Uslaner, M. J. Davis, J. Raber, R. Duvoisin, S. T. Bate, I. J. Reynolds, S. Poli and S. Celanire, *J. Pharmacol. Exp. Ther.*, 2014, **350**, 495–505.
38. S.-P. Hong, K. G. Liu, G. Ma, M. Sabio, M. A. Uberti, M. D. Bacolod, J. Peterson, Z. Z. Zou, A. J. Robichaud and D. Doller, *J. Med. Chem.*, 2011, **54**, 5070–5081.
39. C. W. Lindsley, C. M. Niswender, D. W. Engers and C. R. Hopkins, *Curr. Top. Med. Chem. (Sharjah, United Arab Emirates)*, 2009, **9**, 949–963.

40. C. Bolea, WO2009010454, 2009.
41. D. W. Engers, J. R. Field, U. Le, Y. Zhou, J. D. Bolinger, R. Zamorano, A. L. Blobaum, C. K. Jones, S. Jadhav, C. D. Weaver, P. J. Conn, C. W. Lindsley, C. M. Niswender and C. R. Hopkins, *J. Med. Chem.*, 2011, **54**, 1106–1110.
42. J. A. McCauley, J. W. Hess, N. J. Liverton, C. J. McIntyre, J. J. Romano and M. T. Rudd, WO2010033349 A1, 2010.
43. D. Engers, P. Gentry, R. Williams, J. Bolinger, D. Weaver, U. Menon, P. Conn, C. Lindsley, C. Niswender and C. Hopkins, *Bioorg. Med. Chem. Lett.*, 2010, **20**, 5175–5178.
44. C. Bolea, WO2010079239, 2010.
45. E. Le Poul, C. Bolea, F. Girard, S. Poli, D. Charvin, B. Campo, J. Bortoli, A. Bessif, B. Luo, A. J. Koser, L. M. Hodge, K. M. Smith, A. G. DiLella, N. Liverton, F. Hess, S. E. Browne and I. J. Reynolds, *J. Pharmacol. Exp. Ther.*, 2012, **343**, 167–177.
46. T. Utley, D. Haddenham, J. M. Salovich, R. Zamorano, P. N. Vinson, C. W. Lindsley, C. R. Hopkins and C. M. Niswender, *Bioorg. Med. Chem. Lett.*, 2011, **21**, 6955–6959.
47. O. M. Valenti, M. J. Marino, M. Wittmann, E. Lis, A. G. DiLella, G. G. Kinney and P. J. Conn, *J. Neurosci.*, 2003, **23**, 7218–7226.
48. K.-E. Bennouar, M. A. Uberti, C. Melon, M. D. Bacolod, H. N. Jimenez, M. Cajina, L. Kerkerian-Le Goff, D. Doller and P. Gubellini, *Neuropharmacology*, 2013, **66**, 158–169.
49. H. Iderberg, N. Maslava, A. D. Thompson, M. Bubser, C. M. Niswender, C. R. Hopkins, C. W. Lindsley, P. J. Conn, C. K. Jones and M. A. Cenci, *Neuropharmacology*, 2015, **95**, 121–129.
50. T. Kenakin and O. Onaran, *Trends Pharmacol. Sci.*, 2002, **23**, 275–280.
51. K.-E. Kil, P. Poutiainen, Z. Zhang, A. Zhu, J.-K. Choi, K. Jokivarsi and A.-L. Brownell, *J. Med. Chem.*, 2014, **57**, 9130–9138.
52. P. Poutiainen, K.-E. Kil, Z. Zhang, D. Kuruppu, B. Tannous and A.-L. Brownell, *Neuropharmacology*, 2015, **97**, 7.
53. C. R. Hopkins, C. M. Niswender, L. M. Lewis, C. D. Weaver and C. W. Lindsley, *Discovery of a potent, selective and in vivo active mGluR4 positive allosteric modulator*, Probe Reports from the NIH Molecular Libraries Program, National Center for Biotechnology Information (US), Bethesda (MD), 2010.
54. T. Kenakin and L. J. Miller, *Pharmacol. Rev.*, 2010, **62**, 265–304.
55. P. J. Conn, C. W. Lindsley, J. Meiler and C. M. Niswender, *Nat. Rev. Drug Discovery*, 2014, **13**, 692–708.
56. X. Rovira, F. Malhaire, P. Scholler, J.-P. Pin, C. Goudet, J. Rodrigo, P. Gonzalez-Bulnes, A. Llebaria and J. Giraldo, *FASEB J.*, 2015, **29**, 116–130.
57. C. R. Hopkins, C. W. Lindsley and C. M. Niswender, *Future Med. Chem.*, 2009, **1**, 501–513.
58. V. Katritch, G. Fenalti, E. Abola Enrique, V. Cherezov, C. Stevens Raymond and L. Roth Bryan, *Trends Biochem. Sci.*, 2014, **39**, 233–244.
59. D. J. Sheffler, K. J. Gregory, J. M. Rook and P. J. Conn, *Adv. Pharmacol. (San Diego, CA, U. S.)*, 2011, **62**, 37–77.

60. T. Branco and K. Staras, *Nat. Rev. Neurosci.*, 2009, **10**, 373–383.
61. H. Nickols Hilary and P. J. Conn, *Neurobiol. Dis.*, 2014, **61**, 55–71.
62. S. A. Hitchcock, *Curr. Opin. Chem. Biol.*, 2008, **12**, 318–323.
63. K.-E. Kil, Z. Zhang, K. Jokivarsi, C. Gong, J.-K. Choi, S. Kura and A.-L. Brownell, *Bioorg. Med. Chem.*, 2013, **21**, 5955–5962.
64. N. Cosford, L. Tehrani, J. Roppe, E. Schweiger, N. D. Smith, J. Andersen, L. Bristow, J. Brodkin, X. Jiang, I. McDonald, S. Rao, M. Washburn and M. A. Varney, *J. Med. Chem.*, 2003, **46**, 204–206.
65. M.-R. Zhang, J. Maeda, T. Ito, T. Okauchi, M. Ogawa, J. Noguchi, T. Suhara, C. Halldin and K. Suzuki, *Bioorg. Med. Chem.*, 2005, **13**, 1811–1818.
66. Q. Shi, J. E. Savage, S. J. Hufeisen, L. Rauser, E. Grajkowska, P. Ernsberger, J. T. Wroblewski, J. H. Nadeau and B. L. Roth, *J. Pharmacol. Exp. Ther.*, 2003, **305**, 131–142.
67. Z. Zhang, K.-E. Kil, P. Poutiainen, J.-K. Choi, H.-J. Kang, X.-P. Huang, B. L. Roth and A.-L. Brownell, *Bioorg. Med. Chem. Lett.*, 2015, **25**, 3956–3960.
68. S. R. Bradley, D. G. Standaert, K. J. Rhodes, H. D. Rees, C. M. Testa, A. I. Levey and P. J. Conn, *J. Comp. Neurol.*, 1999, **407**, 33–46.
69. A. O. Kinoshita, H. Ohishi, S. Nomura, R. Shigemoto, S. Nakanishi and N. Mizuno, *Neurosci. Lett.*, 1996, **207**, 199–202.
70. M. Fujinaga, T. Yamasaki, N. Nengaki, M. Ogawa, K. Kumata, Y. Shimoda, J. Yui, L. Xie, Y. Zhang, K. Kawamura and M.-R. Zhang, *Bioorg. Med. Chem. Lett.*, 2016, **26**, 370–374.
71. C. M. Testa, D. G. Standaert, A. B. Young and J. B. Penney, Jr., *J. Neurosci.*, 1994, **14**, 3005–3018.
72. D. P. Hurst, M. Schmeisser and P. H. Reggio, *Chem. Phys. Lipids*, 2013, **169**, 46–56.
73. D. Pellegrino, F. Cicchetti, X. Wang, A. Zhu, M. Yu, M. Saint-Pierre and A.-L. Brownell, *J. Nucl. Med.*, 2007, **48**, 1147–1153.
74. R. Sanchez-Pernaute, J. Q. Wang, D. Kuruppu, L. Cao, W. Tueckmantel, A. Kozikowski, O. Isacson and A. L. Brownell, *NeuroImage*, 2008, **42**, 248–251.
75. B. Jenkins, A. Zhu, P. Poutiainen, J.-K. Choi, K.-E. Kil, Z. Zhang, D. Kuruppu, N. Aytan, A. Dedeoglu and A.-L. Brownell, *Neuropharmacology*, 2015, **108**, 462–473.
76. T. Yamasaki, M. Fujinaga, K. Kawamura, Y. Shimoda, M. Takei, H. Hashimoto, J. Yui, H. Wakizaka, A. Hatori, L. Xie, K. Kumata, M.-R. Zhang, K. Furutsuka, N. Nengaki and S. Shiomi, *J. Neurosci.*, 2016, **36**, 375–384.

CHAPTER 11

Allosteric Modulators of Adenosine, P2Y and P2X Receptors

KENNETH A. JACOBSON*[a] AND ZHAN-GUO GAO[a]

[a]Molecular Recognition Section, Laboratory of Bioorganic Chemistry, National Institute of Diabetes and Digestive and Kidney Diseases, National Institutes of Health, Bethesda, MD 20892, USA
*E-mail: kajacobs@helix.nih.gov

11.1 Introduction

Extracellular purines and pyrimidines act at adenosine receptors (ARs) and P2Y receptors (P2YRs) for nucleotides, which are G-protein-coupled receptors (GPCRs), and at P2XRs, which are ligand-gated ion channels.[1–3] The endogenous agonists are released in response to cellular activity or stress, to induce receptor activation and intracellular signaling cascades. Adenosine acts as a protective and anti-inflammatory mediator, for example in the response to ischemia, through four receptor subtypes (A_1, A_{2A}, A_{2B}, and A_3 ARs). Adenosine is a local modulator rather than a circulating mediator, because it is rapidly degraded enzymatically and/or transported into cells in the vicinity of its release or production. P2YRs are activated by various mononucleotides and dinucleotides, especially phosphate or polyphosphate derivatives of adenosine and uridine. Major agonists of the P2YRs are ATP, ADP, UTP, UDP, and

RSC Drug Discovery Series No. 56
Allosterism in Drug Discovery
Edited by Dario Doller

Published by the Royal Society of Chemistry, www.rsc.org

UDP–glucose. Thus, the hydrolysis of some P2YR ligands, *i.e.* adenine nucleotides, to produce adenosine leads to activation of ARs. P2XRs are activated principally by extracellular ATP, which is rapidly hydrolyzed (deactivated) by ectonucleotidases.[4] Often the role of the P2XRs and P2YRs is to boost the immune response as a result of elevated nucleotides resulting from a challenge to an organ or tissue. The interplay of these signals from extracellular nucleotides and the nucleoside adenosine can be considered as a temporal sequence, in which the initial response, to enhance the immune defense against at physiological challenge through P2X and P2YRs, is followed by a prolonged suppression of the immune system through ARs.

There are numerous potential and actual applications for ligands of these purine and pyrimidine receptors in disease treatment and diagnosis. Many orthosteric ligands for these three classes of receptors, *i.e.* competitive agonists or competitive antagonists acting at the same site on the receptor as the endogenous ligand(s), have been reported and their structure–activity relationships (SARs) probed.[5] Also, modulators of orthosteric ligands (usually an endogenous agonist) that act at separate sites, *i.e.* allosteric modulators, have been reported for various members of these structurally distinct receptor families.[5–8] Usually, the structures of the allosteric modulators and their SARs do not resemble the structures of the endogenous ligands, as would be expected for compounds that bind to the same protein in a different manner. In addition to the indirect modulation of the action of extracellular purines and pyrimidines by allosteric receptor ligands, much effort has been devoted to altering the levels of these native agonists through the enzymes that produce and degrade them and through their channels and transporters.[9,10]

There are different ways in which positive allosteric modulators (PAMs) or negative allosteric modulators (NAMs) can manifest their action. The affinity, potency, efficacy, or signaling pathway of a given orthosteric ligand could be selectively affected, or a combination of these parameters could be affected.[11,12] Also, the effects of a given PAM or NAM may vary depending on which orthosteric ligand is being examined, *i.e.* the action of PAMs and NAMs are probe-dependent.[9] The structure-based design of both orthosteric and allosteric ligands is feasible for some of the ARs, P2YRs, and P2XRs, because X-ray crystallographic structures of members of each of these three receptor families are now available.[13–15] Many of the known orthosteric agonists and antagonists of the ARs, P2YRs, and P2XRs have the disadvantage of not being highly bioavailable, or, if action in the central nervous system (CNS) is desired, not able to readily cross the blood–brain barrier. Especially with respect to the nucleotide P2X and P2YRs, the known orthosteric ligands are often highly charged and prone to hydrolysis *in vivo*. The design of selective allosteric modulators of these receptors that are less highly charged and more drug-like would allow one to overcome some of the pharmacokinetic limitations of known agonists.

There are specific allosteric modulators that display high selectivity for a single target AR, P2YR, or P2XR, and other allosteric modulators operate throughout a class of receptors. For example, based on high-resolution X-ray structures, the sodium ion is a negative allosteric modulator of Family A GPCRs, including ARs.[16] It binds to a conserved site in the receptors that is involved in stabilizing the agonist-bound form of the receptor by an H-bond network and transmission of the agonist-induced conformational change to the intracellular loops that are in contact with the G-protein. Amiloride, a diuretic that inhibits epithelial Na^+ channels, and its analogues modulate various GPCRs, including ARs,[17,18] possibly through interaction with the conserved sodium binding site.[16]

Methods used to identify and characterize allosteric modulators of the purine receptors have been described.[7] Overall, the structural elucidation of the allosteric site with a bound allosteric modulator using crystallization and the ability to affect dissociation kinetics are the most accepted evidence of true allosteric modulation. If an agonist radioligand or fluorescent ligand is available, the dissociation rate in the absence and presence of a potential PAM (reduced rate) or NAM (elevated rate) is quantified.[19,20] The parameter allosteric enhancer (AE) score is determined from the ratio of levels of specifically bound agonist radioligand at a fixed time.[6] In functional assays, PAMs of purine receptors have been found to alter the concentration giving half the maximal effect (EC_{50}) or maximum efficacy of an orthosteric agonist or a combination thereof.[21] Cooperativity factors represent the degree of allosteric modulation for a given pair of allosteric or orthosteric ligands with respect to a given effector pathway.[12,22]

11.2 Adenosine Receptor (AR) Allosteric Modulation

Among the four ARs, well defined, selective allosteric modulators of the Gi-coupled subtypes (A_1 and A_3ARs) have been reported,[6,7] but the Gs-coupled subtypes (A_{2A} and A_{2B}ARs) have few such modulators.[23,24] SARs of PAMs acting at A_1 or A_3ARs have been probed, and NAMs have been reported for the ARs. These PAMs enhance the ability of endogenous adenosine to act as a protective substance in various organs and tissues. Thus, when adenosine is elevated due to stress to a system in the body, such as hypoxia, a PAM would serve to enhance its protective function in a site- and event-specific manner. This need-dependent action could have advantages over directly acting orthosteric agonists, which would stimulate the receptor action throughout the system, depending on the bioavailability of the agonist. With receptors as widespread in body as the ARs, this could be an important characteristic for avoiding some of the well-described side effects of AR agonists in the cardiovascular system, *e.g.* bradycardia of A_1AR agonists, and elsewhere.

When GPCRs form dimeric and multimeric assemblies, functional allosteric interactions exist between the protomers.[11,25–27] For example, agonist activation of the A_{2A}AR negatively modulates activity of a D_2 dopamine receptor to which it may be paired in the striatum.[27] The negative allostery is reciprocal, *i.e.* a dopamine agonist negatively modulates the binding of an adenosine A_{2A}AR agonist.[28] Dimeric pairs of the A_3AR with another A_3AR or a heterodimer with A_1AR display negative allostery.[25,26] In the case of tetramers and higher-order assemblies, this web of allostery can have complex pharmacological consequences. The G protein itself can be considered an allosteric modulator of binding at the GPCR, as demonstrated with fusion proteins between the A_1AR and various mutated $Gi_1\alpha$-subunits.[29] The remainder of this section will cover small-molecule modulators of GPCR action, rather than protein modulators engaged in a physical association. Nevertheless, one should keep in mind that even small-molecular allosteric modulators might conceivably act through a protein–protein interface, such as between two protomers of a GPCR heterodimer or homodimer.

11.2.1 Allosteric Modulators of the A_1AR

The classes of benzoylthiophenes and related heterocycles have provided numerous A_1AR PAMs, which are envisioned for the treatment of chronic pain.[6,22,30–36] The benzoylthiophene PD81,723 (**1**) and its congeners were among the first GPCR allosteric modulators reported (Figure 11.1).[30] It was discovered in a screen of potential A_1AR antagonist molecules, and an unexpected modest increase in radiolabeled agonist binding was observed. PD81,723 does not enhance agonist effects at $A_{2A}AR$ or A_3AR. PD81,723 increases the affinity of agonists binding at the A_1AR in a number of assay systems, and it behaves as an allosteric agonist in addition to acting as a PAM. Many analogues of PD81,723 have been reported leading to increased potency as a PAM and decreased inhibition of the binding of orthosteric radioligands. The mechanism of action of A_1AR PAMs has been explored pharmacologically and structurally. The site for binding of PD81,723 is on the A_1AR, and it stabilizes an active conformation of the receptor that enhances coupling to the G protein.[37] Compared with the agonist N^6-cyclopentyladenosine (CPA) (**33**), PD81,723 produced a relatively small degree of receptor internalization,[38] however it was found later that PD81,723 also enhanced the agonist-induced internalization of the A_1AR.[7,39]

The 2-amino-3-carbonyl thiophene moiety is common to most of the A_1AR PAMs in this family. The benzoyl group of PD81,723 can be substituted with other acyl and aroyl groups. Dialkyl substitution of the 4 and 5 positions of the thiophene ring can be greatly varied structurally, including with cycloalkyl groups, such as cyclohexyl (**2–4**)[40] and cycloheptyl (**5**) rings.[41] The 3-aroyl group may be substituted with a carboxylic acid (**5**), ester (**14**), amide or hydrazide group (**15, 16**).[42] Compounds **2** and **3** at 10 μM enhanced agonist radioligand binding at the human A_1AR by 151% and 123%, respectively, of the enhancement afforded by PD81,723.[6] Compound **5** enhanced A_1AR radioligand binding better than PD81,723 (AE score of 80% *vs.* 28%) and with less inhibition of orthosteric binding.[41] A novel urethane derivative, VCP333 (**13**), containing a heterocycloalkyl group was shown to act as a PAM at the A_1AR.[32] While compound **13** enhanced the maximum level of specific binding of A_1AR agonist radioligand (B_{max}) by ~10%, compound **9** increased it by ~50%.[43] 2-Amino-3-benzoyl thiophenes **11** and **12**, containing

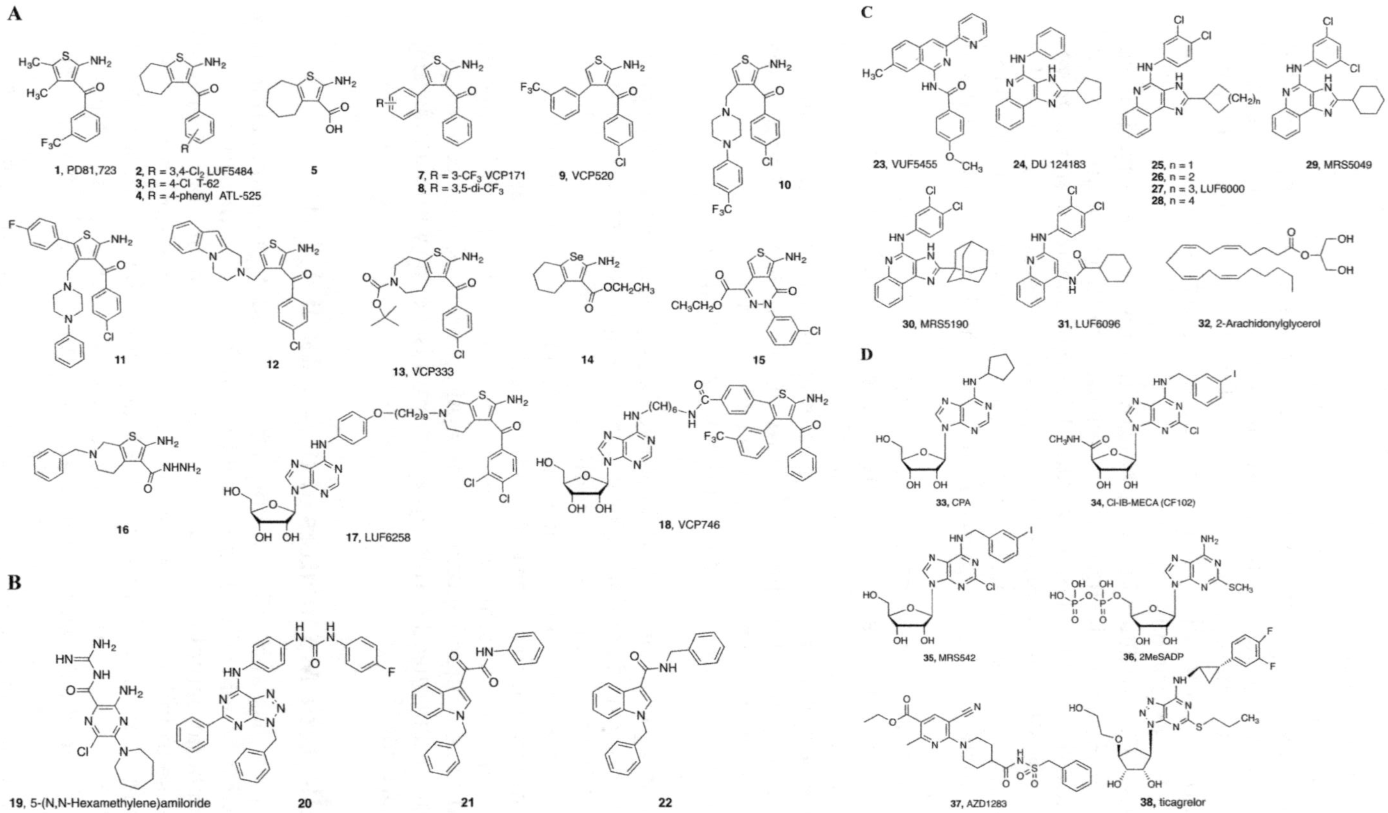

Figure 11.1 A. Structures of allosteric modulators of the A_1AR, which are principally benzoylthiophene derivatives (**1–4**, **7–13**) related to PD81,723 **1**. Compounds **17** and **18** are bitopic modulators of the A_1AR, containing both agonist and PAM (benzoylthiophene) moieties. Other 2-amino-3-substituted thiophene and selenophene derivatives are included. B. Structures of allosteric modulators of the A_{2A}AR, including a nonselective amiloride derivative **19** and a selective urea derivative **20** and a reported PAM **21** and NAM **22** of the A_{2B}AR. C. Structures of PAMs of the human A_3AR and a lipid NAM of A_3AR, **32**. D. Orthosteric nucleoside ligands **33–35** of ARs and agonist **36** and antagonists **37**, **38** of the P2Y_{12}R that are mentioned in the text.

a piperidinyl-*N*-methylene moiety at the 4 position, were especially potent in increasing the binding of radiolabeled agonist at the A_1AR, with B_{max} shifts of 14.1- and 8.1-fold, respectively.[44,45]

T-62 (**3**), a close analogue of PD81,723 that was reported by Baraldi *et al.*,[46] is active in the CNS when administered peripherally and has been evaluated clinically for pain treatment.[31] Its anti-nociceptive effect was demonstrated by a reduction of hypersensitivity in carrageenan-inflamed rats following administration intrathecally.[47] Tritiated T-62 was used as a radioligand to study the A_1AR allosteric binding site.[48]

Like T-62, a 3,5-di(trifluoromethyl) benzoylthiophene derivative (**8**) was shown to behave as an allo–agonist, *i.e.* having properties of an allosteric agonist, in addition to a PAM of the A_1AR.[49] Different effector systems downstream of the A_1AR were compared to reveal signaling bias. Compound **8** was more potent in activating the Gi protein, leading to increased [^{35}S]GTPγS binding, compared with phosphorylation of extracellular-signal-regulated kinase (ERK).

A benzoylthiophene PAM of the A_1AR, VCP171 (**7**), inhibited primary afferent synaptic transmission in a neuropathic pain model in which injury augments the adenosine tone in the dorsal horn of the spinal cord.[50] A_1AR agonists have been shown to have anti-ischemic properties in the brain and heart. The cardioprotective effects of PAMs of the A_1AR, under conditions in which endogenous adenosine is greatly elevated, have also been demonstrated.[51,52] Other activities of PAMs of the A_1AR in the heart and brain have been described.[7]

Bitopic PAM–agonists of the A_1AR have been designed, in which both pharmacophores, orthosteric and allosteric, are present on the same molecular conjugate.[22,51,53] This usually is accomplished by a tethering approach. The hybrid molecule **17** was found to have the optimal chain length within the series. The intention is to bridge two binding regions on the receptor protein, with enhancement of the affinity and allosteric efficacy. The hybrid molecule **18** appeared to be a biased agonist of the A_1AR, protecting a rat cardiomyoblast cell line and neonatal cardiomyocytes from ischemia without affecting rat atrial heart rate.[51]

11.2.2 Allosteric Modulators of the $A_{2A}AR$ and $A_{2B}AR$

$A_{2A}AR$ activation has anti-inflammatory effects,[54] and $A_{2B}AR$ modulation is associated with beneficial effects in certain mouse models of cardiovascular and metabolic pathologies.[55] One selective $A_{2A}AR$ agonist is approved for human use to induce dilation of the coronary artery during stress testing.[1] Compound **19**, a derivative of amiloride, was shown to be a NAM of the $A_{2A}AR$.[56] Several reports have described the allosteric enhancement at the $A_{2A}AR$[57,58] and the $A_{2B}AR$ by small molecules **20** and **21**,[24,59] respectively. Compound **20** increased the potency of an $A_{2A}AR$ agonist to relax rat aortic rings.[58] Compound **22** was described as a NAM of the $A_{2B}AR$. A PAM of the $A_{2A}AR$ AEA061 (structure not disclosed) derived from a natural product was found to attenuate inflammation induced by lipopolysaccharide (LPS) in the mouse, including inhibition of tumor necrosis factor (TNF) and other proinflammatory cytokines, and to enhance the affinity and B_{max} of orthosteric agonists.[23] However, the full medicinal chemistry characterization and SAR exploration of these chemical classes of allosteric modulators remain to be reported.

11.2.3 Allosteric Modulators of the A_3AR

Two selective A_3AR agonists are in advanced clinical trials for rheumatoid arthritis and psoriasis, and for liver cancer.[1] Most PAMs of the A_3AR come from three classes of substituted heterocycles: 3-(2-pyridinyl)isoquinolines (**23**), 1*H*-imidazo-[4,5-*c*]quinolin-4-amines (**24–30**) and 2,4-disubstituted quinolines (**31**).[21,60–63] These compounds do not enhance agonist effects at A_1AR or $A_{2A}AR$, but they may inhibit the binding of orthosteric radioligands at ARs to varying degrees. Various A_3AR PAMs were discovered among collections of heterocyclic antagonists of the same receptor.[21,60] There has been an effort to identify analogues that minimize the orthosteric binding inhibition and maximize the allosteric enhancing effects at the A_3AR. Other classes have been shown to modulate activity of the receptor; for example, the endocannabinoid 2-arachidonylglycerol (**32**) is a NAM of the human A_3AR.[64]

The imidazoquinolinamine PAM of the A_3AR, LUF6000 (**27**), was shown to modulate the receptor agonist efficacy in a functionally biased manner (Figure 11.2).[12] For example, LUF6000 has only a small effect on β-arrestin 2 translocation, and it has no effect on ERK1 or 2, induced by the protypical agonist 2-chloro-*N*^{6}-(3-iodobenzyl)-5′-*N*-methylcarboxamidoadenosine (Cl-IB-MECA) (**34**), but has pronounced effects on cAMP inhibition and cell hyperpolarization. Interestingly, LUF6000 has a larger enhancing effect on low-efficacy agonists than on high-efficacy agonists, indicating the fine-tuning modulatory property of this compound. The variation of the size of the cycloalkyl ring (**25–28**) altered the combination of enhancement of agonist potency and efficacy.[12,61] LUF6000, with a cyclohexyl ring, was the most effective in enhancement of A_3AR agonist efficacy and also increased the efficacy of an alternative native agonist of the A_3AR, inosine. Like adenosine, inosine is elevated in ischemia (an over 30-fold increase was found[65]), so this action of LUF6000 could have therapeutic relevance to benefit from an otherwise less potent metabolite of adenosine. LUF6000 itself displayed no agonism of the A_3AR. The cyclohexyl group may be substituted with a 1-adamantyl group, indicating tolerance of steric bulk in **30**.[62] Isomerization of the 3,4-dichlorophenyl substituents to 3,5-dichloro in **29** preserves allosteric action, but most other functionalization of this ring is not tolerated. Nucleoside antagonists of the A_3AR, *i.e.* adenosine derivatives that have been reduced in efficacy to the point of being selective antagonists, such as MRS542 (**35**), can be restored to full activation of the A_3AR in the presence of LUF6000. Thus, a residual, almost undetectable agonism of such derivatives can be amplified to restore full efficacy. However, non-nucleoside A_3AR antagonists were unaffected by this PAM.

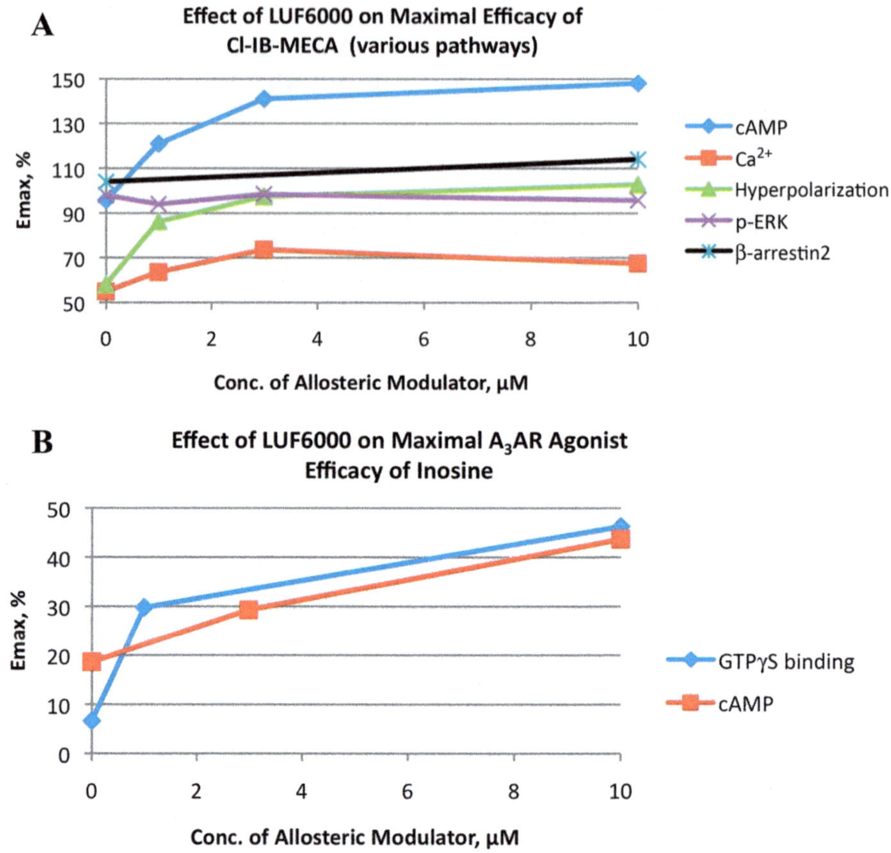

Figure 11.2 A. An A_3AR PAM LUF6000 **27** has anti-inflammatory effects and is being considered for clinical development. Enhancement by LUF6000 **27** of the maximal efficacy of A_3AR agonist 2-chloro-N^6-(3-iodobenzyl)-5′-*N*-methylcarboxamidoadenosine (Cl-IB-MECA, **34**) is dependent on the effector pathway. The effect of Cl-IB-MECA at 10 µM is normalized to 100% for each pathway. B. LUF6000 enhances the maximal efficacy of inosine (percent relative to Cl-IB-MECA, 10 µM), an alternate endogenous A_3AR agonist. Data taken from Gao *et al.*[12]

LUF6000 has anti-inflammatory effects, as was demonstrated in a model of rat adjuvant-induced arthritis, monoiodoacetate-induced osteoarthritis, and concanavalin-A-induced liver inflammation in mice.[66] However, species differences have been noted for the imidazoquinolinamine PAMs of the A_3AR. A report indicated no enhancement of the mouse or rat A_3AR by LUF6000 and LUF6096,[67] but positive allostery at this receptor is complex and requires further investigation.

A_3AR agonists are known to be cardioprotective in ischemia. A 2,4-disubstituted quinoline PAM of A_3AR, LUF6096 (**31**), was shown to enhance adenosine-dependent cardioprotection in an anesthetized canine model of myocardial ischemia (1 h)–reperfusion (3 h) injury.[68] Pretreatment with **31** prior to either ischemia or reperfusion greatly reduced infarct size.

11.2.4 Mutagenesis of ARs to Locate Residues Involved in Interaction with PAMs

The binding sites of A_1AR and A_3AR PAMs have been probed using site-directed mutagenesis of the ARs.[35,69–71]

The recognition of benzoylthiophene PAMs was found to involve a pocket formed by the second extracellular loop (EL2) of the A_1AR.[70] Two residues in that loop region, S150 and M162, were essential for the action of the prototypical PAM PD81,723 (**1**) and a biphenyl derivative ATL-525 (**4**). It was suggested that an outer vestibule of the A_1AR controls the association and dissociation of agonists from the deeper orthosteric binding site. Binding of agonist, but not antagonist, to the orthosteric site induces a conformation of EL2 that allows an allosteric enhancer to bind. This model provides a mechanism for the observations that allosteric enhancers slow the dissociation of orthosteric agonists but not antagonists.

Residues required for the binding and action of PAMs at the human A_3AR, 3-(2-pyridinyl)isoquinoline (**23**) and 1*H*-imidazo-[4,5-*c*]quinolin-4-amine (**24**), were probed using site-directed mutagenesis.[69] Allosteric enhancement of the A_3AR by these PAMs was dependent on two residues, Phe182(5.43) and Asn274(7.45) and two other residues were modulatory Trp243(6.48) and Aasn30(1.50), based on their Ala mutations. Asn274 was required for allosteric effect but not for binding of an orthosteric agonist. Compound **24** but not **23** required Asp107(3.49), based on its mutation to Asn. The binding of allosteric modulators of the ARs is thought to involve the EL regions of the receptor. Therefore, computational modeling of allosteric sites of these GPCRs is more challenging than for orthosteric sites, *i.e.* within the transmembrane (TM) region, because of the structural variation and conformational flexibility of the EL regions. Nevertheless, the binding of LUF6000 to a homology model of the human A_3AR has been simulated by molecular dynamics calculations.[72]

11.3 P2YR Allosteric Modulation

The subtypes of G protein-coupled P2YRs for nucleotides (and their endogenous agonists) are numbered $P2Y_1$ (ADP), $P2Y_2$ (ATP, UTP), $P2Y_4$ (UTP), $P2Y_6$ (UDP), $P2Y_{11}$ (ATP), $P2Y_{12}$ (ADP), $P2Y_{13}$ (ADP) and $P2Y_{14}$ (UDP–glucose).[2,14] Only a few PAMs and NAMs are known (Figure 11.3), and none are approved for human use. The only approved pharmaceuticals that act orthosterically at P2YRs are $P2Y_2$R agonist (for dry eye, in Japan only) and blockbuster $P2Y_{12}$R antagonists (antithrombotics). Nevertheless, there is much research activity focused on the P2YRs, which are widespread in the body.

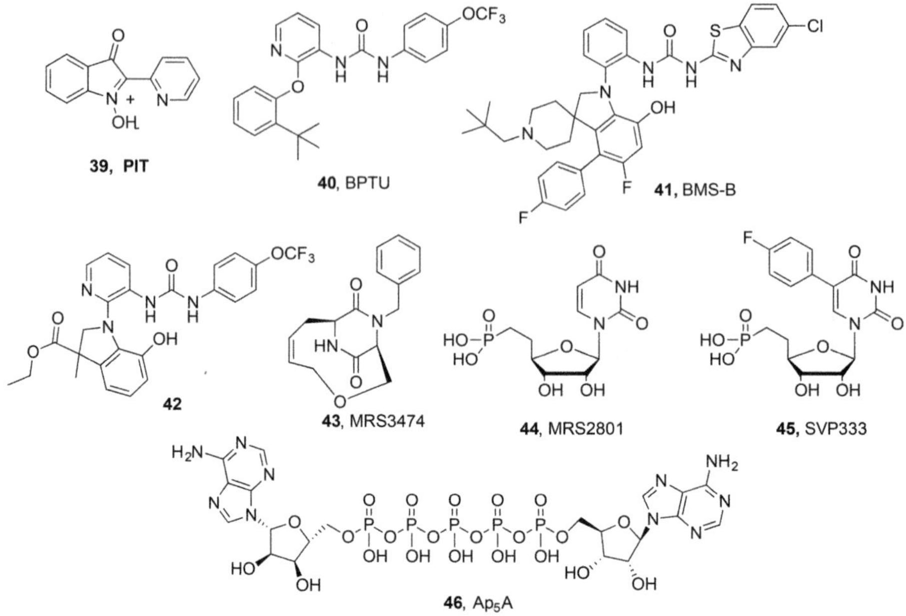

Figure 11.3 Structures of compounds found to act as allosteric modulators of P2YRs: NAMs of $P2Y_1R$ (**39–43**) and tentative PAMs of $P2Y_2R$ (**44, 45**) and $P2Y_4R$ (**46**).

11.3.1 NAMs of the $P2Y_1R$

The compound known as PIT (pyridyl isatogen tosylate) (**39**) has characteristics indicating that it may be NAM of the $P2Y_1R$, *i.e.* insurmountable antagonism, without effect on orthosteric radioligand binding. Its binding site on the $P2Y_1R$ has not been probed. However, a recent X-ray crystallographic structure of the $P2Y_1R$ identified the binding site of a potent and selective diarylurea antagonist 1-(2-(2-*tert*-butylphenoxy)pyridin-3-yl)-3-(4-(trifluoromethoxy)phenyl)urea (BPTU) (**40**) as completely distinct from the orthosteric nucleotide binding site.[73] The binding site of BPTU is located on the outer periphery of the receptor, in contact with the phospholipid bilayer. The urea of BPTU is doubly H-bonded to an available carbonyl group of a backbone amide in TM2, and the other interactions are mainly hydrophobic. This is a novel location for an allosteric modulator within the family of GPCRs. The allosteric property of BPTU was confirmed pharmacologically with a dissociation kinetics assay, and the residues in contact with BPTU were probed using mutagenesis.[73] The allosteric nature of the urea series of antithrombotic $P2Y_1R$ antagonists was not anticipated until the X-ray structure was determined. Earlier modeling proposed a binding site for the urea antagonists within the helical bundle, which has now been superseded.[74]

Analogue **41**, BMS-B (1-(5-chlorobenzo[d]thiazol-2-yl)-3-(2-(5-fluoro-4-(4-fluorophenyl)-7-hydroxy-1′-neopentylspiro[indoline-3,4′-piperidin]-1-yl)phenyl)urea), was tested in preclinical toxicity models, which required a spray-dried dispersion to improve oral absorption of this hydrophobic compound.[75] Urea analogue **42** was predicted to bind to the same site as BPTU using molecular modeling, and other diaryl urea analogues are evidently also NAMs. The hydroxyl group of the more potent **42** was predicted to interact with the $P2Y_1R$ through H bonding, thus further stabilizing the binding in this shallow cleft.

Bicyclic diketopiperazine **43** potently inhibited $P2Y_1R$ agonist activity on phospholipase C activation and calcium mobilization, but was weak in inhibiting activity of the $P2Y_6R$, another Gq-coupled receptor. It did not compete for binding of a radiolabeled nucleotide antagonist of the $P2Y_1R$, and therefore it is a putative NAM of the $P2Y_1R$.[76]

11.3.2 Allosteric Modulation of the $P2Y_2R$

The $P2Y_2R$ is activated by UTP, but not UDP or UMP. However, a phosphonate analogue of UMP **44** was found to act as a partial agonist of $P2Y_6R$ with EC_{50} = 1.6 μM.[77] In this analogue, the 5′-oxygen was replaced with a methylene bridge. A related 4-fluorophenyl phosphonate derivative **45** was even more potent agonist at the $P2Y_2R$ with EC_{50} = 0.4 μM, with 43% efficacy compared with the full agonist UTP.[78] Neither of these compounds had an effect at the $P2Y_4R$ or $P2Y_6R$. This compound series has characteristics of allosteric modulation, because a high concentration of **44** did not reduce the effect of a full agonist, as would be expected for a orthosteric partial agonist. However, more careful characterization is needed to define if it is a true allosteric agonist or not. The binding site for these phosphonates on the $P2Y_2R$ is unknown.

11.3.3 Allosteric Modulation of the $P2Y_4R$

Diadenosine polyphosphates, such as Ap_5A (**46**), potentiated the pharmacological response to UTP selectively at the $P2Y_4R$.[79] Dinucleotide **46** (100 μM) has no effect on intracellular Ca^{2+} mobilization in 1321N1 astrocytoma cells stably expressing the human $P2Y_4R$, but in the presence of an ED_{50} concentration of UTP (69 nM), it led to a 2.5-fold increase in the response. The authors stated, "These preliminary findings need to be investigated further in order to determine if the Ap_nAs are truly behaving as allosteric enhancers."

11.3.4 Modulation of the $P2Y_{12}R$

Three selective $P2Y_{12}R$ antagonists are approved for human use as antithrombotic agents.[2] X-ray crystallography has determined that the binding sites of agonist 2-methylthio-ADP (2MeSADP, **36**) and the nonnucleotide AZD1283 (ethyl 6-(4-((benzylsulfonyl)carbamoyl)piperidin-1-yl)-5-cyano-2-methylnicotinate) (**37**) are in different, but slightly overlapping regions of the helical bundle of the $P2Y_{12}R$, an important clinical target for antithrombotic drugs.[80,81] Several residues are involved in the recognition of both ligands according to the $P2Y_{12}R$ structures: Tyr105(3.33) forms π-stacking with the adenine or pyridine rings; positively charged R256(6.55) and K280(7.35) coordinate the phosphate or carbonyl and sulfonyl moieties;

Asn191(5.40) hydrogen bonds to the adenine or pyridine nitrogen atoms. Therefore, AZD1283 (**36**) is an orthosteric antagonist based on the structure, and, consistently, its binding was shown to be competitive pharmacologically. The available thiol groups for reaction with the irreversibly binding thienopyridine antagonists of the $P2Y_{12}R$, *i.e.* antithrombotic drugs clopidogrel and prasugrel (structures not shown) are also in the vicinity of the orthosteric nucleotide binding site. Van Giezen *et al.*[56] have found that ticagrelor (**38**) does not inhibit [^{3}H]ADP binding, but antagonizes ADP signaling. The authors suggested a possible non-competitive or an allosteric mechanism of action of ticagrelor at the $P2Y_{12}R$. However, ticagrelor was able to inhibit the binding of a more potent radioligand, [^{33}P]2MeSADP. In independent studies, Hoffmann *et al.*[82] and Zhang *et al.*[80] demonstrated competitive antagonism by ticagrelor. Therefore, allosteric modulators of the $P2Y_{12}R$ remain to be identified.

11.4 P2XR Allosteric Modulation

P2XRs are trimeric ion channels permeable to Na^+, K^+, and Ca^{2+} that have relevance for pain, CNS disorders and inflammation.[83,84] The seven P2X subunits can combine as homotrimers or heterotrimers, for example in the case of P2X2Rs and P2X3Rs.[85] Each P2XR subunit contains only two TMs and large extracellular and intracellular (with C- and N-termini) domains. An agonist-bound P2X4R structure features an orthosteric nucleotide binding site at each of the three subunit interfaces in the trimer, in proximity to cationic residues, and an open ion channel pore.[86] Some subtypes, *e.g.* the trimeric P2X7R ion channel, are not known to form heteromers. The medicinal chemistry of orthosteric and allosteric ligands of P2XRs was recently reviewed.[5,87,88] Most of the therapeutic interest in P2XRs involves blocking their effects with antagonists, although no selective compounds are approved for human use. Most of the translationally directed activity on P2XRs concerns the P2X3R (pain and bladder function), P2X4R (chronic pain), and P2X7R (inflammation and CNS disorders).[89] Some of the small, uncharged or charged molecules that inhibit action at the P2X ligand-gated ion channels are now thought to be allosteric modulators (NAMs), and PAMs of the P2XRs are also known (Figure 11.4). One of the first PAMs of a P2XR to be identified was MRS2219 (**47**), which enhances the action of ATP selectively at the P2X1R.

11.4.1 Allosteric Modulation of the P2X2R, P2X3R, and P2X2/3R

The activity of ATP at the P2X2R is modulated allosterically by cations, such as Ca^{2+} and Zn^{2+} and pH changes.[8,90–92] Key amino acid residues involved in these allosteric effects have been identified. Similar modulatory effects of cations have been noted for other subtypes, *i.e.* P2X1R, P2X3R, P2X4R,

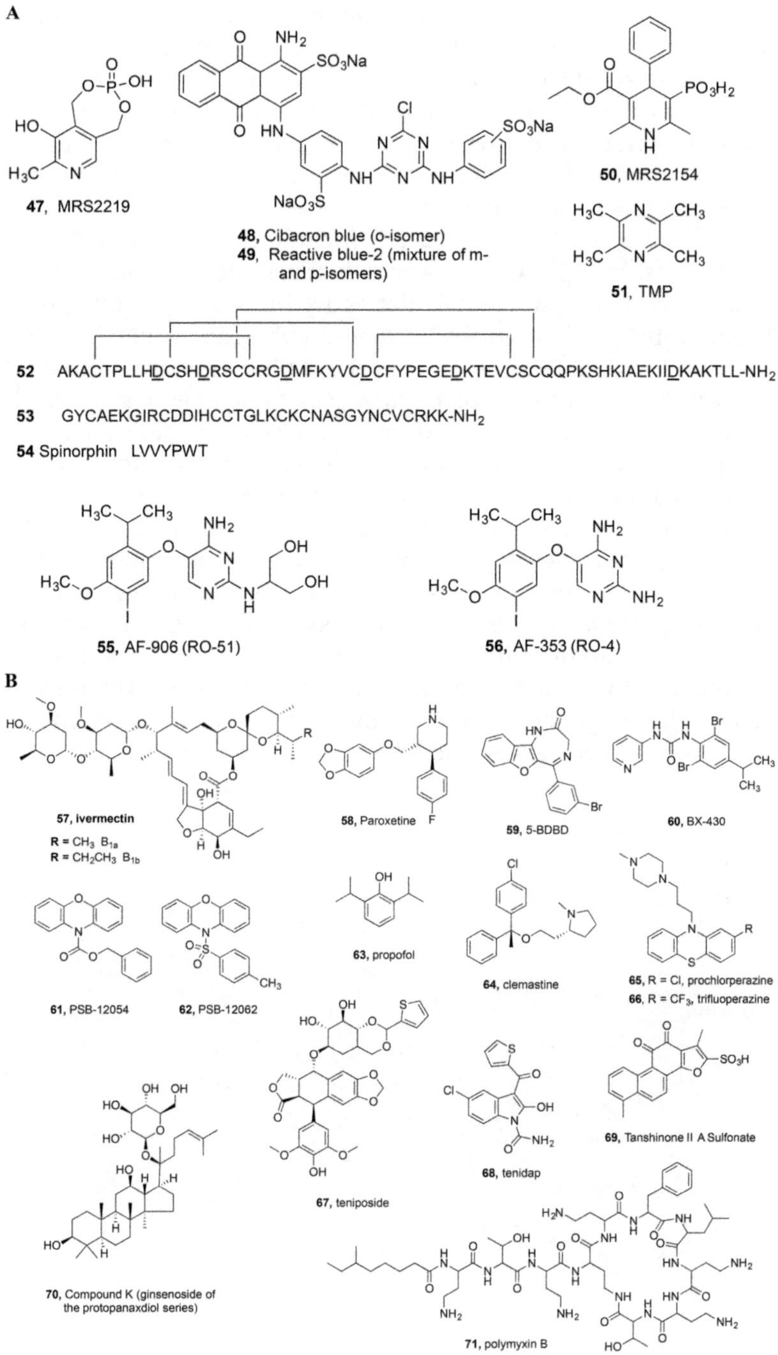

Figure 11.4 Structures of compounds found to act as allosteric modulators of: A. P2X1, P2X2/3, and P2X3Rs; and B. P2X4R and P2X7R.

and P2X7R.[93,94] Zn^{2+} increased the channel activity of P2X3R and P2X4R and decreased the activity of P2X1R and P2X7R. Acidification increases the potency of agonists at the rat P2X2R, and a His319 residue is a putative proton sensor, but other homotrimeric P2XRs are inhibited by acidification.[8] The P2X2R and P2X3R are sensitive to deviation from pH 7.1–7.2,[7] but allosteric modulation by H^+ and Zn^{2+} did not discriminate between the two possible heterotrimeric P2X2–3R variants.[95]

Small-molecule NAMs of the P2X2R include derivatives of suramin and anthraquinones.[96,97] The Ca^{+2} channel blocker nicardipine (structure not shown) and a related phosphonate derivative MRS2154 (**50**) antagonized the rat P2X2R, probably allosterically.[98] Cibacron blue (**48**), which is an isomer of the widely explored P2XR antagonist Reactive blue 2 (**49**), acts as a PAM of the rat P2X3R and even restores responsiveness to acutely desensitized receptors.[99] Other small-molecule NAMs of the P2X3R include tetramethylpyrazine **51** (from a Chinese traditional analgesic), two peptidic spider toxins (**52, 53**) and diaminopyrimidines RO-51 (**55**) and AF-353 (**56**).[100–103] The latter two compounds were discovered through library screening and lead optimization, in an effort to develop compounds for treating pain and pulmonary conditions. The peptide nonclassical opioid peptide spinorphin **54**, which inhibits enkephalinases, acts as a potent NAM of the P2X3R (concentration giving half maximum inhibition (IC_{50}) 8.3 pM).[104] P2X3R antagonist A-317491 (**72**) (Figure 11.5) was developed as a potential pain or overactive bladder treatment; it lacks oral bioavailability due to multiple negative charges, and its action proved to be orthosteric.[105]

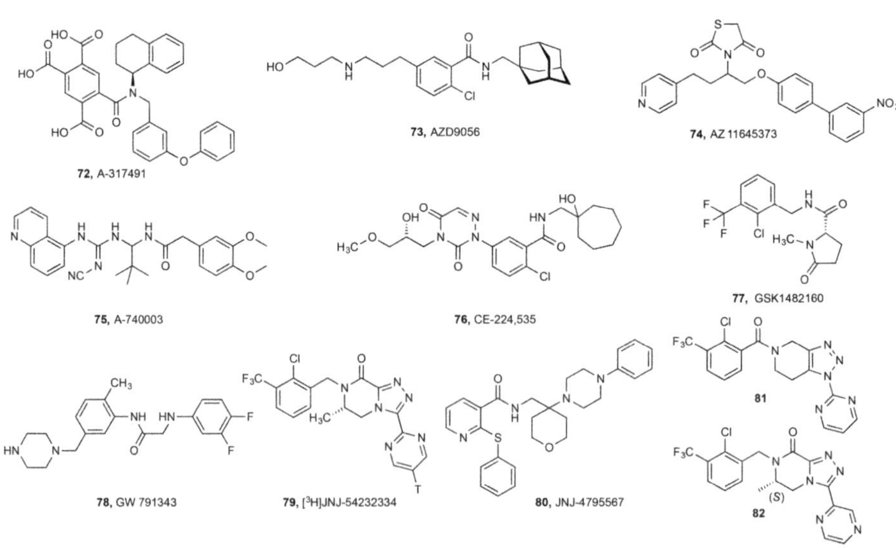

Figure 11.5 Additional antagonists of P2X3R, P2X4R, and P2X7R that either are orthosteric (**72**) or possibly allosteric (**73–82**).

11.4.2 Allosteric Modulation of the P2X4R

The P2X4R is present in microglial cells and regulates their activation and survival in pain states.[106] Metal cations regulate this receptor.[107–109] Ethanol also has a negative allosteric effect on P2X4R gating,[7] and therefore its modulation has been considered in the context of treatment of alcohol abuse.[110,111] The antiparasitic drug ivermectin **57** is a PAM of the P2X4R that has been probed for its site of interaction on the receptor.[8,112–114] Site-directed mutagenesis and molecular modeling based on the zebrafish P2X4R structure confirmed the role in both ethanol and ivermectin binding of an "action pocket" surrounded by Asp331 and Met336 in TM2 and Trp46 in TM1 segments.[115]

Cibacron blue (**48**) was the first NAM of the P2X4R to be reported.[116] The anaesthetic drug propofol (**63**) is a weak, but selective PAM of the rat P2X4R. Other small-molecule NAMs of the P2X4R include: the antidepressant drug paroxetine (**58**), a benzodiazepine derivative 5-(3-bromophenyl)-1,3-dihydro-2*H*-benzofuro[3,2-*e*]-1,4-diazepin-2-one (5-BDBD) (**59**), and the urea BX-430 (**60**), which was discovered by library screening[88,117–119] phenoxazine derivatives, including PSB-12054 (**61**) and PSB-12062 (**62**), were discovered to be NAMs of the P2X4R and their SAR probed.[120]

11.4.3 Allosteric Modulation of the P2X7R

P2X7R antagonists are sought for disease therapeutics.[121] Small-molecule NAMs of the human P2X7R include approved pharmaceutical molecules, such as the psychoactive drugs paroxetine (**58**), prochlorperazine (**65**) and trifluoperazine (**66**).[122,123] The experimental rheumatoid arthritis drug tenidap (**68**) and antibiotic polymyxin B (**71**) were discovered by Di Virgilio and colleagues to be NAMs of the P2X7R. Some P2XR NAMs show a species dependence; ivermectin (**57**) acts as a PAM of the human but not the rat P2X7R.[124,125] The chemotherapeutic agent teniposide (**67**) is a human P2X7R NAM but is inactive at the human P2X4R or rat P2X2R.[126] The sedating antihistamine clemastine (**64**) potentiates the human P2X7R by lowering the concentration threshold of activation by ATP.[127] Various non-pharmaceutical natural products have been found to modulate the P2X7R.[126] For example, a secondary metabolite from marine sponges, gelasine, and a phenolic tree resin component, garcinolic acid, (structures not shown) are PAMs of the human P2X7R.[126] Tanshinone II A sulfonate (**69**) was identified in library screening as a potent, selective NAM of P2X7R.[128] Steroid glycosides isolated from ginseng, including protopanaxdiols such as Compound K (**70**) and its congeners, are PAMs of the P2X7R.[129]

A selection of other chemically diverse antagonists of the P2X3R, P2X4R, and P2X7R, both pharmacological probe compounds and clinical candidates, are shown in Figure 11.5.[88,130] Because the orthosteric P2XR ligands tend to be negatively charged (to complement cationic residues in the receptor binding site), many of the uncharged P2XR antagonists are to be considered potential NAMs. P2X7R antagonists (**73–82**) include clinical compounds for inflammation, depression and other and CNS disorders,[131] but in general their site(s) of action on the receptor are not yet explored. AZD9056 **73** is an orally active P2X7R antagonist that was shown to be safe and efficacious in a Phase IIa clinical trial for treatment of Crohn's disease.[132] A-740003 (**75**) potently blocked P2X7R activation across mammalian species. Compound **79** is a useful high-affinity radioligand with a pK_i at the rat P2X7R of 9.3 ± 0.1.[133] Compound **81** was found to potently antagonize the P2X7R across species, with good physicochemical properties, oral bioavailability and partitioning into the CNS.[15,134] Orally administered compound **82** crossed the blood–brain barrier and dose-dependently engaged the P2X7R in the rat hippocampus.[135–137]

11.5 Conclusions

Allosteric modulators of purine and pyrimidine receptors provide a means of overcoming various pharmacokinetic and pharmacodynamic limitations of orthosteric ligands of these receptors. Many of the known orthosteric ligands, although they may be receptor-subtype-selective, display poor oral bioavailability or passage across the blood–brain barrier or are short-lived *in vivo*. In some cases, allosteric modulation is a means of achieving greater receptor subtype selectivity. Numerous allosteric modulators of A_1 and A_3ARs (mainly PAMs) and P2XRs (mainly NAMs), and one class of NAMs (diarylureas) of the P2Y_1R have been reported and their SAR explored. In some cases, X-ray crystallographic structures, molecular modeling, and site-directed mutagenesis provide knowledge of the receptor binding sites. Conditions that are envisioned for the allosteric modulators include chronic pain, ischemia, inflammatory diseases such as arthritis, cancer, neurodegeneration, and thrombosis.

Abbreviations

ADP	Adenosine diphosphate
AR	Adenosine receptor
ATL-525	[1,1′-Biphenyl]-4-yl(2-amino-4,5,6,7-tetrahydrobenzo[*b*]-thiophen-3-yl)methanone
ATP	Adenosine triphosphate
AZD1283	Ethyl 6-(4-((benzylsulfonyl)carbamoyl) piperidin-1-yl)-5-cyano-2-methylnicotinate
5-BDBD	5-(3-Bromophenyl)-1,3-dihydro-2*H*-benzofuro[3,2-*e*]-1,4-diazepin-2-one
BPTU	1-(2-(2-*tert*-Butylphenoxy)pyridin-3-yl)-3-(4-(trifluoromethoxy) phenyl)urea
Cl-IB-MECA	2-Chloro-*N*6-(3-iodobenzyl)-5′-*N*-methylcarboxamido-adenosine
CNS	Central nervous system
CPA	N^6-Cyclopentyladenosine

DU124183	*N*-Phenyl-2-cyclopentyl-1*H*-imidazo[4,5-*c*]quinolin-4-amine
EC_{50}	Concentration giving half the maximal effect
EL	Extracellular loop
ERK	Extracellular-signal-regulated kinase
Gi	Adenylyl cyclase-inhibitory guanine nucleotide-binding protein
GPCR	G protein-coupled receptor
Gs	Adenylyl cyclase-stimulatory guanine nucleotide-binding protein
HMA	5-(*N*,*N*-Hexamethylene)amiloride
IC_{50}	Concentration giving half the maximum inhibition
LUF5484	(2-Amino-4,5,6,7-tetrahydrobenzo[*b*]thiophen-3-yl)(3,4-dichlorophenyl)methanone
LUF6000	*N*-(3,4-Dichlorophenyl)-2-cyclohexyl-1*H*-imidazo[4,5-*c*] quinolin-4-amine
LUF6096	*N*-{2-[(3,4-Dichlorophenyl)amino]quinolin-4-yl} cyclohexanecarboxamide
LUF6258	*N*(6)-[2-Amino-3-(3,4-dichlorobenzoyl)-4,5,6,7-tetrahydrothieno-[2,3-*c*]pyridin-6-yl-9-nonyloxy-4-phenyl]-adenosine
PIT	Pyridyl isatogen tosylate
MRS542	2-Chloro-N^6-(3-iodobenzyl)-adenosine
MRS5049	2-Cyclohexyl-*N*-(3,5-dichlorophenyl)-1*H*-imidazo[4,5-*c*] quinolin-4-amine
MRS5190	2-(1-Adamantanyl)-*N*-(3,4-dichlorophenyl)-1*H*-imidazo[4,5-*c*] quinolin-4-amine
NAM	Negative allosteric modulator
NECA	5′-(*N*-Ethylcarboxamido)adenosine
PAM	Positive allosteric modulator
PD81,723	(2-Amino-4,5-dimethyl-3-thienyl)-[3-(trifluoromethyl)phenyl] methanone
SAR	Structure–activity relationship
T-62	2-Amino-3-(4-chlorobenzoyl)-5,6,7,8-tetrahydrobenzothiophene
TM	Transmembrane domain
UDP	Uridine diphosphate
UMP	Uridine monophosphate
UTP	Uridine triphosphate
VCP171	(2-Amino-4-(3-(trifluoromethyl)phenyl)thiophen-3-yl)(phenyl) methanone
VCP520	(2-Amino-4-(3-(trifluoromethyl)phenyl)thiophen-3-yl)(4-chlorophenyl)methanone
VCP333	*tert*-Butyl 2-amino-3-(4-chlorobenzoyl)-7,8-dihydro-4*H*-thieno-[2,3-*d*]azepine-6(5*H*)-carboxylate

Acknowledgements

The authors acknowledge support from the NIDDK Intramural Research Program, National Institutes of Health.

References

1. B. B. Fredholm, A. P. IJzerman, K. A. Jacobson, J. Linden and C. Müller, *Pharmacol. Rev.*, 2011, **63**, 1–34.
2. I. von Kügelgen and K. Hoffmann, *Neuropharmacology*, 2016, **104**, 50–61.
3. R. A. North and M. F. Jarvis, *Mol. Pharmacol.*, 2013, **83**, 759–769.
4. Y. Baqi, *Mini-Rev. Med. Chem.*, 2015, **15**, 21–33.
5. K. A. Jacobson and C. E. Müller, *Neuropharmacology*, 2016, **104**, 31–49.
6. R. Romagnoli, P. G. Baraldi, A. R. Moorman, P. A. Borea and K. Varani, *Future Med. Chem.*, 2015, **7**, 1247–1259.
7. K. A. Jacobson, Z. G. Gao, A. Göblyös and A. P. IJzerman, *Adv. Pharmacol.*, 2011, **61**, 187–221.
8. C. Coddou, Z. Yan, T. Obsil, J. P. Huidobro-Toro and S. S. Stojilkovic, *Pharmacol. Rev.*, 2011, **63**, 641–683.
9. L. Antonioli, P. Pacher, E. S. Vizi and G. Haskó, *Trends Mol. Med.*, 2013, **19**, 355–367.
10. G. Dahl, *Philos. Trans. R. Soc., B*, 2015, **370**, 20140191.
11. A. Christopoulos, J. P. Changeux, W. A. Catterall, D. Fabbro, T. P. Burris, J. A. Cidlowski, R. W. Olsen, J. A. Peters, R. R. Neubig, J. P. Pin and P. M. Sexton, *Pharmacol. Rev.*, 2014, **66**, 918–947.
12. Z. G. Gao, D. Verzijl, A. Zweemer, K. Ye, A. Göblyös, A. P. IJzerman and K. A. Jacobson, *Biochem. Pharmacol.*, 2011, **82**, 658–668.
13. M. Dias and F. H. Marshall, *J. Biol. Chem.*, 2015, **290**, 19469–19470.
14. K. A. Jacobson, S. Paoletta, V. Katritch, B. Wu, Z. G. Gao, Q. Zhao, R. C. Stevens and E. Kiselev, *Mol. Pharmacol.*, 2015, **88**, 220–230.
15. C. Habermacher, K. Dunning, T. Chataigneau and T. Grutter, *Neuropharmacology*, 2016, **104**, 18–30.
16. V. Katritch, G. Fenalti, E. E. Abola, B. L. Roth, V. Cherezov and R. C. Stevens, *Trends Biochem. Sci.*, 2014, **39**, 233–244.
17. Z. G. Gao and A. P. IJzerman, *Biochem. Pharmacol.*, 2000, **60**, 669–676.
18. Z. G. Gao, N. Melman, A. Erdmann, S. G. Kim, C. E. Müller, A. P. IJzerman and K. A. Jacobson, *Biochem. Pharmacol.*, 2003, **65**, 525–534.
19. A. Christopoulos and T. Kenakin, *Pharmacol. Rev.*, 2002, **54**, 323–374.
20. L. T. May, T. J. Self, S. J. Briddon and S. J. Hill, *Mol. Pharmacol.*, 2010, **78**, 511–523.
21. Z. G. Gao, S. G. Kim, K. A. Soltysiak, N. Melman, A. P. IJzerman and K. A. Jacobson, *Mol. Pharmacol.*, 2002, **62**, 81–89.
22. D. Guo, S. N. Venhorst, A. Massink, J. P. Veldhoven, G. Vauquelin, A. P. IJzerman and L. H. Heitman, *Br. J. Pharmacol.*, 2014, **171**, 5295–5312.
23. A. A. Welihinda and E. P. Amento, *J. Inflammation*, 2014, **11**(1), 37.
24. M. L. Trincavelli, C. Giacomelli, S. Daniele, S. Taliani, B. Cosimelli, S. Laneri, E. Severi, E. Barresi, I. Pugliesi, G. Greco and E. Novellino, *Biochim. Biophys. Acta, Gen. Subj.*, 2014, **1840**(3), 1194–1203.
25. S. J. Hill, L. T. May, B. Kellam and J. Woolard, *Br. J. Pharmacol.*, 2014, **171**, 1102–1113.

26. R. Corriden, L. E. Kilpatrick, B. Kellam, S. J. Briddon and S. J. Hill, *FASEB J.*, 2014, **28**, 4211–4222.
27. S. Ferré, J. Bonaventura, D. Tomasi, G. Navarro, E. Moreno, A. Cortés, C. Lluís, V. Casadó and N. D. Volkow, *Neuropharmacology*, 2016, **104**, 156–160.
28. V. Fernández-Dueñas, M. Gómez-Soler, K. A. Jacobson, T. S. Kumar, K. Fuxe, D. O. Borroto-Escuela and F. Ciruela, *J. Neurochem.*, 2012, **123**, 373–384.
29. E. Klaasse, R. A. de Ligt, S. F. Roerink, A. Lorenzen, G. Milligan, R. Leurs and A. P. IJzerman, *Eur. J. Pharmacol.*, 2004, **499**, 91–98.
30. R. F. Bruns and J. H. Fergus, *Mol. Pharmacol.*, 1990, **38**, 939–949.
31. I. Giorgi and P. Nieri, *Expert Opin. Ther. Pat.*, 2013, **23**, 1109–1121.
32. C. Valant, L. Aurelio, S. M. Devine, T. D. Ashton, J. M. White, P. M. Sexton, A. Christopoulos and P. J. Scammells, *J. Med. Chem.*, 2012, **55**, 2367–2375.
33. R. Romagnoli, P. G. Baraldi, A. P. IJzerman, A. Massink, O. Cruz-Lopez, L. C. Lopez-Cara, G. Saponaro, D. Preti, M. Aghazadeh Tabrizi, S. Baraldi and A. R. Moorman, *J. Med. Chem.*, 2014, **57**, 7673–7686.
34. F. Vincenzi, M. Targa, R. Romagnoli, S. Merighi, S. Gessi, P. G. Baraldi, P. A. Borea and K. Varani, *Neuropharmacology*, 2014, **81**, 6–14.
35. R. Romagnoli, P. G. Baraldi, C. Lopez-Cara, O. Cruz-Lopez, A. R. Moorman, A. Massink, A. P. IJzerman, F. Vincenzi, P. A. Borea and K. Varani, *Eur. J. Med. Chem.*, 2015, **101**, 185–204.
36. R. Romagnoli, P. G. Baraldi, A. P. IJzerman, A. Massink, O. Cruz-Lopez, L. C. Lopez-Cara, G. Saponaro, D. Preti, M. Aghazadeh Tabrizi, S. Baraldi and A. R. Moorman, *J. Med. Chem.*, 2014, **57**, 7673–7686.
37. S. Bhattacharya, R. L. Youkey, K. Ghartey, M. Leonard, J. Linden and A. L. Tucker, *Biochem. J.*, 2006, **396**, 139–146.
38. S. Bhattacharya and J. Linden, *Mol. Pharmacol.*, 1996, **50**, 104–111.
39. E. C. Klaasse, G. van den Hout, S. F. Roerink, W. J. de Grip, A. P. IJzerman and M. W. Beukers, *Eur. J. Pharmacol.*, 2005, **522**, 1–8.
40. P. A. M. van der Klein, A. P. Kourounakis and A. P. IJzerman, *J. Med. Chem.*, 1999, **42**, 3629–3635.
41. G. Nikolakopoulos, H. Figler, J. Linden and P. J. Scammels, *Bioorg. Med. Chem.*, 2006, **14**, 2358–2365.
42. L. Aurelio, A. Christopoulos, B. L. Flynn, P. J. Scammells, P. M. Sexton and C. Valant, *Bioorg. Med. Chem. Lett.*, 2011, **21**, 3704–3707.
43. C. Valant, L. Aurelio, V. B. Urmaliya, P. White, P. J. Scammells, P. M. Sexton and A. Christopoulos, *Mol. Pharmacol.*, 2010, **78**, 444–455.
44. R. Romagnoli, P. G. Baraldi, M. D. Carrion, C. L. Cara, M. K. Salvador, D. Preti, M. A. Tabrizi, A. R. Moorman, F. Vincenzi, P. A. Borea and K. Varani, *Eur. J. Med. Chem.*, 2013, **67**, 409–427.
45. R. Romagnoli, P. G. Baraldi, M. D. Carrion, C. L. Cara, O. Cruz-Lopez, M. K. Salvador, D. Preti, M. A. Tabrizi, J. Shryock and A. R. Moorman, *et al.*, *Bioorg. Med. Chem.*, 2012, **20**, 996–1007.
46. P. G. Baraldi, A. N. Zaid, I. Lampronti, F. Fruttarolo, M. G. Pavani, M. A. Tabrizi, J. C. Shryock, E. Leung and R. Romagnoli, *Bioorg. Med. Chem. Lett.*, 2000, **10**, 1953–1957.

47. X. Li, D. Conklin, H.-L. Pan and J. C. Eisenach, *J. Pharmacol. Exp. Ther.*, 2003, **305**, 950–955.
48. R. Romagnoli, *Bioorg. Med. Chem.*, 2006, **16**, 1402–1404.
49. L. Aurelio, C. Valant, H. Figler, B. L. Flynn, J. Linden, P. M. Sexton, A. Christopoulos and P. J. Scammells, *Bioorg. Med. Chem.*, 2009, **17**, 7353–7361.
50. W. L. Imlach, R. F. Bhola, L. T. May, A. Christopoulos and M. J. Christie, *Mol. Pharmacol.*, 2015, **88**, 460–468.
51. C. Valant, L. T. May, L. Aurelio, C. H. Chuo, P. J. White, J.-A. Baltos, P. M. Sexton, P. J. Scammells and A. Christopoulos, *Proc. Natl. Acad. Sci. U. S. A.*, 2014, **111**, 4614–4619.
52. A. Butcher, P. J. Scammels, P. J. White, S. M. Devine and R. B. B. Meyer, *Pharmaceuticals*, 2013, **6**, 546–556.
53. R. Narlawar, J. R. Lane, M. Doddareddy, J. Lin, J. Brussee and A. P. IJzerman, *J. Med. Chem.*, 2010, **53**, 3028–3037.
54. L. Antonioli, B. Csóka, M. Fornai, R. Colucci, E. Kókai, C. Blandizzi and G. Haskó, *Drug Discovery Today*, 2014, **19**, 1051–1068.
55. A. Eisenstein, S. Patterson and K. Ravid, *J. Cell. Physiol.*, 2015, **230**, 2891–2897.
56. J. J. Van Giezen, L. Nilsson, P. Berntsson, B. M. Wissing, F. Giordanetto, W. Tomlinson and P. J. Greasley, *J. Thromb. Haemostasis*, 2009, **7**, 1556–1565.
57. G. Yuan, N. G. Gedeon, T. C. Jankins and G. B. Jones, *Expert Opin. Drug Discovery*, 2015, **10**, 63–80.
58. I. Giorgi, G. Biagi, A. M. Bianucci, A. Borghini, O. Livi, M. Leonardi, D. Pietra, V. Calderone and A. Martelli, *Eur. J. Med. Chem.*, 2008, **43**, 1639–1647.
59. M. L. Trincavelli, S. Daniele, C. Giacomelli, S. Taliani, F. Da Settimo, B. Cosimelli, G. Greco, E. Novellino and C. Martini, *Biochim. Biophys. Acta, Mol. Cell Res.*, 2014, **1843**, 2957–2966.
60. Z. G. Gao, J. E. van Muijlwijk-Koezen, A. Chen, C. E. Müller, A. P. IJzerman and K. A. Jacobson, *Mol. Pharmacol.*, 2001, **60**, 1057–1063.
61. A. Göblyös, Z. G. Gao, J. Brussee, R. Connestari, S. Neves Santiago, K. Ye, A. P. IJzerman and K. A. Jacobson, *J. Med. Chem.*, 2006, **49**, 3354–3361.
62. Y. Kim, S. de Castro, Z. G. Gao, A. P. IJzerman and K. A. Jacobson, *J. Med. Chem.*, 2009, **52**, 2098–2108.
63. L. H. Heitman, A. Goblyos, A. M. Zweemer, R. Bakker, T. Mulder-Krieger, J. P. D. van Veldhoven, H. de Vries, J. Brussee and A. P. Ijzerman, *J. Med. Chem.*, 2009, **52**, 926–931.
64. J. R. Lane, M. W. Beukers, T. Mulder-Krieger and A. P. IJzerman, *Biochem. Pharmacol.*, 2010, **79**, 48–56.
65. T. Bäckström, M. Goiny, U. Lockowandt, J. Liska and A. Franco-Cereceda, *J. Appl. Physiol.*, 2003, **94**, 1122–1128.
66. S. Cohen, F. Barer, S. Bar-Yehuda, A. P. IJzerman, K. A. Jacobson and P. S. Fishman, *Mediators Inflammation*, 2014, 708746, DOI: 10.1155/2014/708746.

67. L. Du, Z. G. Gao, J. P. D. van Veldhoven, A. P. IJzerman, K. A. Jacobson and J. A. Auchampach, *FASEB J.*, 2012, **26**, 851.3.
68. L. Du, Z. G. Gao, K. Nithipatikom, A. P. IJzerman, J. P. van Veldhoven, K. A. Jacobson, G. J. Gross and J. A. Auchampach, *J. Pharmacol. Exp. Ther.*, 2012, **340**, 210–217.
69. Z. G. Gao, S. K. Kim, A. S. Gross, A. Chen, J. Blaustein and K. A. Jacobson, *Mol. Pharmacol.*, 2003, **63**, 1021–1031.
70. D. P. Kennedy, F. M. McRobb, S. A. Leonhardt, M. Purdy, H. Figler, M. A. Marshall, M. Chordia, R. Figler, J. Linden, R. Abagyan and M. Yeager, *Mol. Pharmacol.*, 2014, **85**, 301–309.
71. M. C. Peeters, L. E. Wisse, A. Dinaj, B. Vroling, G. Vriend and A. P. IJzerman, *Biochem. Pharmacol.*, 2012, **84**, 76–87.
72. G. Deganutti, A. Cuzzolin, A. Ciancetta and S. Moro, *Bioorg. Med. Chem.*, 2015, **23**, 4065–4071.
73. D. Zhang, Z. G. Gao, K. Zhang, E. Kiselev, S. Crane, J. Wang, S. Paoletta, C. Yi, L. Ma, W. Zhang, G. W. Han, H. Liu, V. Cherezov, V. Katritch, H. Jiang, R. C. Stevens, K. A. Jacobson, Q. Zhao and B. Wu, *Nature*, 2015, **520**, 317–321.
74. J. X. Qiao, T. C. Wang, R. Ruel, C. Thibeault, A. L'Heureux, W. A. Schumacher, S. A. Spronk, S. Hiebert, G. Bouthillier, J. Lloyd, Z. Pi, D. M. Schnur, L. M. Abell, J. Hua, L. A. Price, E. Liu, Q. Wu, T. E. Steinbacher, J. S. Bostwick, M. Chang, J. Zheng, Q. Gao, B. Ma, P. A. McDonnell, C. S. Huang, R. Rehfuss, R. R. Wexler and P. Y. S. Lam, *J. Med. Chem.*, 2013, **56**, 9275–9295.
75. X. Q. Chen, K. Stefanski, H. Shen, C. Huang, C. Caporuscio, W. Yang, P. Lam, C. Su, O. Gudmundsson and M. Hageman, *J. Pharm. Sci.*, 2014, **103**, 3924–3931.
76. P. Besada, L. Mamedova, C. J. Thomas, S. Costanzi and K. A. Jacobson, *Org. Biomol. Chem.*, 2005, **3**, 2016–2025.
77. L. Cosyn, S. Van Calenbergh, B. V. Joshi, H. Ko, R. L. Carter, T. K. Harden and K. A. Jacobson, *Bioorg. Med. Chem. Lett.*, 2009, **19**, 3002–3005.
78. S. Van Poecke, M. O. Barrett, T. S. Kumar, D. Sinnaeve, J. C. Martins, K. A. Jacobson, T. K. Harden and S. Van Calenbergh, *Bioorg. Med. Chem.*, 2012, **20**, 2304–2315.
79. K. Patel, A. Barnes, J. Camacho, C. Paterson, R. Boughtflower, D. Cousens and F. Marshall, *Eur. J. Pharmacol.*, 2001, **430**, 203–210.
80. K. Zhang, J. Zhang, Z. G. Gao, S. Paoletta, D. Zhang, G. W. Han, T. Li, L. Ma, W. Zhang, C. E. Müller, H. Yang, H. Jiang, V. Cherezov, V. Katritch, K. A. Jacobson, R. C. Stevens, B. Wu and Q. Zhao, *Nature*, 2014, **509**, 119–122.
81. S. Paoletta, D. Sabbadin, I. von Kügelgen, S. Hinz, V. Katritch, K. Hoffmann, A. Abdelrahman, J. Straßburger, Y. Baqi, Q. Zhao, R. C. Stevens, S. Moro, C. E. Müller and K. A. Jacobson, *J. Comput.-Aided Mol. Des.*, 2015, **29**, 737–756.
82. K. Hoffmann, D. A. Lutz, J. Straßburger, Y. Baqi, C. E. Müller and I. von Kügelgen, *J. Thromb. Haemostasis*, 2014, **12**, 1898–1905.

83. F. Rassendren and L. Ulmann, in *Pathologies of Calcium Channels*, ed. N. Weiss and A. Koschak, Springer, Berlin, 1st edn, 2014, pp. 615–633.
84. F. Di Virgilio, *Curr. Med. Chem.*, 2015, **22**, 866–877.
85. R. Hausmann, A. Kless and G. Schmalzing, *Curr. Med. Chem.*, 2015, **22**, 799–818.
86. M. Hattori and E. Gouaux, *Nature*, 2012, **485**, 207–212.
87. C. Lambertucci, D. Dal Ben, M. Buccioni, G. Marucci, A. Thomas and R. Volpini, *Curr. Med. Chem.*, 2015, **22**, 915–928.
88. C. E. Müller, *Curr. Med. Chem.*, 2015, **22**, 929–941.
89. F. Sáez-Orellana, P. A. Godoy, T. Silva-Grecchi, K. M. Barra and J. Fuentealba, *Pharmacol. Res.*, 2015, **101**, 109–115.
90. C. Coddou, Z. Yan and S. S. Stojilkovic, *Am. J. Physiol.: Cell Physiol.*, 2015, **308**, C729–C736.
91. B. F. King, L. E. Ziganshin, J. Pintor and G. Burnstock, *Br. J. Pharmacol.*, 1996, **117**, 1371–1373.
92. R. A. Lorca, C. Coddou, M. C. Gazitua, P. Bull, C. Arredondo and J. P. Huidobro-Toro, *J. Neurochem.*, 2005, **95**, 499–512.
93. S. S. Wildman, B. F. King and G. Burnstock, *Br. J. Pharmacol.*, 1999, **128**, 486–492.
94. K. Xiong, R. W. Peoples, J. P. Montgomery, Y. Chiang, R. R. Stewart, F. F. Weight and C. Li, *J. Neurophysiol.*, 1999, **81**, 2088–2094.
95. M. Kowalski, R. Hausmann, J. Schmid, A. Dopychai, G. Stephan, Y. Tang, G. Schmalzing, P. Illes and P. Rubini, *Neuropharmacology*, 2015, **99**, 115–130.
96. C. A. Trujillo, A. A. Nery, A. H. Martins, P. Majumder, F. A. Gonzalez and H. Ulrich, *Biochemistry*, 2006, **45**, 224–233.
97. Y. Baqi, R. Hausmann, C. Rosefort, J. Rettinger, G. Schmalzing and C. E. Müller, *J. Med. Chem.*, 2011, **4**, 817–830.
98. K. A. Jacobson, Y. C. Kim and B. F. King, *J. Auton. Nerv. Syst.*, 2000, **81**, 152–157.
99. K. Alexander, W. Niforatos, B. Bianchi, E. C. Burgard, K. J. Lynch, E. A. Kowaluk, M. F. Jarvis and T. van Biesen, *J. Pharmacol. Exp. Ther.*, 1999, **291**, 1135–1142.
100. Y. Gao, C. Xu, S. Liang, A. Zhang, S. Mu, Y. Wang and F. Wan, *Brain Res. Bull.*, 2008, **77**, 27–32.
101. N. V. Kabanova, A. A. Vassilevski, O. A. Rogachevskaja, M. F. Bystrova, Y. V. Korolkova, K. A. Pluzhnikov, R. A. Romanov, E. V. Grishin and S. S. Kolesnikov, *Biochim. Biophys. Acta, Biomembr.*, 2012, **1818**, 2868–2875.
102. A. Jahangir, M. Alam, D. S. Carter, M. P. Dillon, D. J. Bois, A. P. Ford, J. R. Gever, C. Lin, P. J. Wagner, Y. Zhai and J. Zira, *Bioorg. Med. Chem. Lett.*, 2009, **19**, 1632–1635.
103. J. R. Gever, R. Soto, R. A. Henningsen, R. S. Martin, D. H. Hackos, S. Panicker, W. Rubas, I. B. Oglesby, M. P. Dillon, M. E. Milla, G. Burnstock and A. P. Ford, *Br. J. Pharmacol.*, 2010, **160**, 1387–1398.
104. K. Y. Jung, H. D. Moon, G. E. Lee, H. H. Lim, C. S. Park and Y. C. Kim, *J. Med. Chem.*, 2007, **50**, 4543–4547.

105. M. F. Jarvis, E. C. Burgard, S. McGaraughty, P. Honore, K. Lynch, T. J. Brennan, A. Subieta, T. Van Biesen, J. Cartmell, B. Bianchi, W. Niforatos, K. Kage, H. Yu, J. Mikusa, C. T. Wismer, C. Z. Zhu, K. Chu, C. H. Lee, A. O. Steward, J. Polakowski, B. F. Cox, E. Kowaluk, M. Williams, J. Sullivan and C. Faltynek, *Proc. Natl. Acad. Sci. U. S. A.*, 2002, **99**, 17179–17184.
106. N. Vázquez-Villoldo, M. Domercq, A. Martín, J. Llop, V. Gómez-Vallejo and C. Matute, *Glia*, 2014, **62**, 171–184.
107. C. Coddou, C. Acuna-Castillo, P. Bull and J. P. Huidobro-Toro, *J. Biol. Chem.*, 2007, **282**, 36879–36886.
108. B. S. Khakh, W. R. Proctor, T. V. Dunwiddie, C. Labarca and H. A. Lester, *J. Neurosci.*, 1999, **19**, 7289–7299.
109. H. Zemkova, A. Khadra, M. B. Rokic, V. Tvrdonova, A. Sherman and S. S. Stojilkovic, *Pfluegers Arch.*, 2014, **467**, 713–726.
110. O. Ostrovskaya, L. Asatryan, L. Wyatt, M. Popova, K. Li, R. W. Peoples, R. L. Alkana and D. L. Davies, *J. Pharmacol. Exp. Ther.*, 2011, **337**, 171–179.
111. K. M. Franklin, L. Asatryan, M. W. Jakowec, J. R. Trudell, R. L. Bell and D. L. Davies, *Front. Neurosci.*, 2014, **8**, 176.
112. S. D. Silberberg, M. Li and K. J. Swartz, *Neuron*, 2007, **54**, 263–274.
113. E. Toulmé, F. Soto, M. Garret and E. Boué-Grabot, *Mol. Pharmacol.*, 2006, **69**, 576–587.
114. C. Coddou, S. S. Stojilkovic and J. P. Huidobro-Toro, *Rev. Neurosci.*, 2011, **22**, 335–354.
115. M. Popova, J. Trudell, K. Li, R. Alkana, D. Davies and L. Asatryan, *Purinergic Signalling*, 2013, **9**, 621–632.
116. K. J. Miller, A. D. Michel, I. P. Chessell and P. P. Humphrey, *Neuropharmacology*, 1998, **37**, 1579–1586.
117. K. Nagata, T. Imai, T. Yamashita, M. Tsuda, H. Tozaki-Saitoh and K. Inoue, *Mol. Pain*, 2009, **5**, 20–31.
118. B. Balázs, T. Dankó, G. Kovács, L. Köles, M. A. Hediger and A. Zsembery, *Cell. Physiol. Biochem.*, 2013, **32**, 11–24.
119. A. R. Ase, N. S. Honson, H. Zaghdane, T. A. Pfeifer and P. Séguéla, *Mol. Pharmacol.*, 2015, **87**, 606–616.
120. V. Hernandez-Olmos, A. Abdelrahman, A. El-Tayeb, D. Freudendahl, S. Weinhausen and C. E. Müller, *J. Med. Chem.*, 2012, **55**, 9576–9588.
121. R. Bartlett, L. Stokes and R. Sluyter, *Pharmacol. Rev.*, 2014, **66**, 638–675.
122. P. Dao-Ung, K. K. Skarratt, S. J. Fuller and L. Stokes, *Purinergic Signalling*, 2015, **11**, 481–490.
123. C. Hempel, W. Nörenberg, H. Sobottka, N. Urban, A. Nicke, W. Fischer and M. Schaefer, *et al.*, *Neuropharmacology*, 2013, **75**, 365–379.
124. J. M. Sanz, P. Chiozzi and F. Di Virgilio, *Eur. J. Pharmacol.*, 1998, **355**, 235–244.
125. D. Ferrari, C. Pizzirani, S. Gulinelli, G. Callegari, P. Chiozzi and M. Idzko, *et al.*, *Br. J. Pharmacol.*, 2007, **150**, 445–454.
126. W. Nörenberg, H. Sobottka, C. Hempel, T. Plötz, W. Fischer, G. Schmalzing and M. Schaefer, *Br. J. Pharmacol.*, 2012, **167**, 48–66.

127. D. Draganov, S. Gopalakrishna-Pillai, Y. R. Chen, N. Zuckerman, S. Moeller, C. Wang, D. Ann and P. P. Lee, *Sci. Rep.*, 2015, **5**, 16222.
128. W. Fischer, N. Urban, K. Immig, H. Franke and M. Schaefer, *Purinergic Signalling*, 2014, **10**, 313–326.
129. W. Nörenberg, C. Hempel, N. Urban, H. Sobottka, P. Illes and M. Schaefer, *J. Biol. Chem.*, 2011, **286**, 11067–11081.
130. M. Kaiser, H. Sobottka, W. Fischer, M. Schaefer and W. Nörenberg, *J. Pharmacol. Exp. Ther.*, 2014, **350**, 531–542.
131. R. M. Helliwell, C. O. ShioukHuey, K. Dhuna, J. C. Molero, J. M. Ye, C. C. Xue and L. Stokes, *Br. J. Pharmacol.*, 2015, **172**, 3326–3340.
132. C. C. Chrovian, J. C. Rech, A. Bhattacharya and M. A. Letavic, *Prog. Med. Chem.*, 2014, **53**, 65–100.
133. N. Mehta, M. Kaur, M. Singh, S. Chand, B. Vyas, P. Silakari, M. S. Bahia and O. Silakari, *Bioorg. Med. Chem. Lett.*, 2014, **22**, 54–88.
134. A. Eser, J. F. Colombel, P. Rutgeerts, S. Vermeire, H. Vogelsang, M. Braddock, T. Persson and W. Reinisch, *Inflammatory Bowel Dis.*, 2015, **21**, 2247–2253.
135. B. Lord, M. K. Ameriks, Q. Wang, L. Fourgeaud, M. Vliegen, W. Verluyten, P. Haspeslagh, N. I. Carruthers, T. W. Lovenberg, P. Bonaventure, M. A. Letavic and A. Bhattacharya, *Eur. J. Pharmacol.*, 2015, **765**, 551–559.
136. B. M. Savall, D. Wu, M. De Angelis, N. I. Carruthers, H. Ao, Q. Wang, B. Lord, A. Bhattacharya and M. A. Letavic, *ACS Med. Chem. Lett.*, 2015, **6**, 671–676.
137. M. K. Ameriks, H. Ao, N. I. Carruthers, B. Lord, S. Ravula, J. C. Rech, B. M. Savall, J. L. Wall, Q. Wang, A. Bhattacharya and M. A. Letavic, *Bioorg. Med. Chem. Lett.*, 2016, **26**, 257–261.

CHAPTER 12

Positive Allosteric Modulators of G-Protein-Coupled Receptors that Act via *Covalent Mechanisms of Action*†

WHITNEY M. NOLTE*[a] AND PHILIP A. CARPINO[a]

[a]Worldwide Medicinal Chemistry, Worldwide Research & Development, Pfizer Inc., 610 Main Street, Cambridge, Massachusetts, 02139, USA
*E-mail: Whitney.Nolte@pfizer.com

12.1 Introduction

G-protein-coupled receptors (GPCRs) belong to a family of seven-transmembrane-domain receptors that promote intracellular signal transduction by endogenous extracellular ligands (*e.g.*, peptide hormones, neurotransmitters, or metal ions) or external stimuli (*e.g.*, light) *via* recruitment of G proteins. GPCRs are organized into three classes—class A (rhodopsin-like), class B (secretin-like), and class C (glutamate receptor-like)—based on sequence homology and structure. Class B and C GPCRs are characterized by large extracellular N-terminal domains, in contrast to class A GPCRs. Because dysregulation of GPCR signaling can contribute to the pathophysiology of

†The authors are employees of Pfizer, Inc.

RSC Drug Discovery Series No. 56
Allosterism in Drug Discovery
Edited by Dario Doller

Published by the Royal Society of Chemistry, www.rsc.org

many diseases, GPCRs have attracted considerable interest as drug targets for therapeutic intervention.[1] Small-molecule and peptide drugs have been identified that modulate over a third of the non-olfactory GPCRs by varied mechanisms ranging from agonist stimulation of receptors to antagonist inhibition of the actions of endogenous ligands at receptors.

The pharmacological activities of endogenous ligands of GPCRs can be modified by drugs called allosteric modulators that bind at sites on the receptors which are distinct from the orthosteric binding sites. Allosteric modulators alter the affinity and/or efficacy of endogenous ligands by inducing specific conformational changes in GPCRs. Positive allosteric modulations (PAMs) enhance the activity of endogenous ligands, in contrast to negative allosteric modulators (NAMs) which decrease activity. GPCR PAMs and NAMs can exhibit several advantages over drugs that act at orthosteric binding sites. Allosteric modulators are efficacious only when endogenous ligands are present, allowing regulation of physiological systems in a spatial–temporal manner which is not generally possible with orthosteric ligands.[2] As allosteric binding sites tend to be less conserved compared with orthosteric binding sites, allosteric modulators can demonstrate greater selectivity for one GPCR subtype over closely related subtypes, potentially resulting in enhanced safety profiles.[3] For some class B GPCRs for which the endogenous ligands are large peptides, small-molecule allosteric modulators can be identified when it has proven challenging to identify small-molecule ligands that bind at the orthosteric sites. This is because the allosteric binding sites on these class B GPCRs tend to be more chemically tractable than the orthosteric binding sites.[4] Several recent reviews describe in greater detail the identification and pharmacological characterization of PAMs and NAMs of GPCRs.[2,5–7]

A class B GPCR for which several small-molecule PAMs but few low-molecular-weight agonists have been identified is the glucagon-like peptide-1 (GLP-1) receptor (GLP-1R), a pharmacological target involved in the potentiation of glucose-dependent insulin secretion for the treatment of type 2 diabetes.[8] The discovery of PAMs of GLP-1R circumvents the challenges of finding small-molecule agonists that can mimic the complex two-step process by which the major endogenous ligands of GLP-1R, *e.g.* GLP-1(7–37) and GLP-1(7–36)NH_2 (hereafter referred to as GLP-1), activate the receptor. Examples of PAMs of GLP-1R include 4-(3-benzyloxyphenyl)-2-ethylsulfinyl-6-(trifluoromethyl) pyrimidine (BETP, **1**), 6,7-dichloro-2-methylsulfonyl-3-*tert*-butylaminoquinoxaline (**2**, historically known as "compound 2"), quercetin (**3**), VU0453379 (**4**), and HIT-465 (**5**; Figure 12.1).[9–14] Several of these PAMs have been shown to increase insulin secretion in islet preparations and glucose-dependent insulin secretion in animal models of diabetes.[11,12]

Studies to identify binding sites of two PAMs of GLP-1R, compound 2 and BETP, revealed an unexpected mechanism of action. In contrast to the majority of PAMs of GPCRs, which bind reversibly to their target proteins to elicit pharmacological effects, compound 2 and BETP were shown to potentiate the activities of several peptides at GLP-1R by a covalent, irreversible

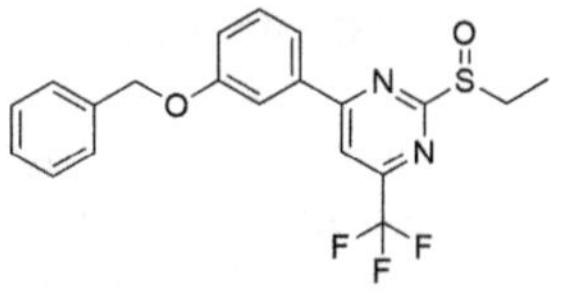

1, BETP

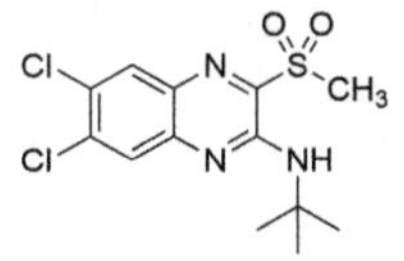

2, "Compound 2"

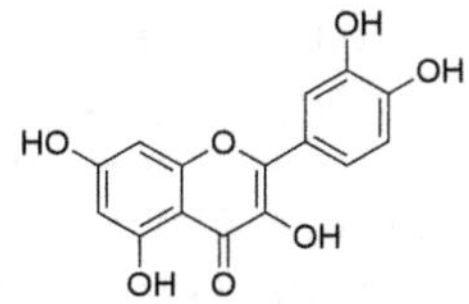

3, Quercetin

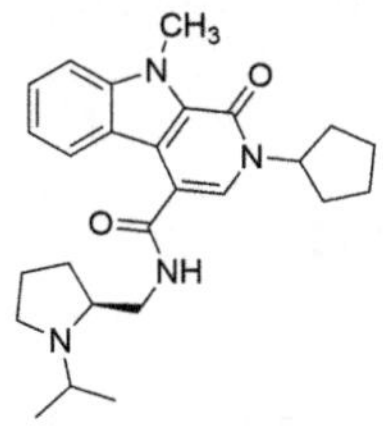

4, VU0453379

Structure not disclosed

5, HIT-465

Figure 12.1 Structures of PAMs of GLP-1R disclosed in the literature.

mechanism. Compound 2 and BETP represent the first known examples of a new class of positive allosteric modulators termed covalent PAMs (cPAMs) of GPCRs. cPAMs of GPCRs may have several unique advantages compared with reversible PAMs of GPCRs, including more complete and prolonged target engagement, for which termination of activity requires protein degradation. In this chapter, we will provide an overview of the pharmacological activity of compound 2 and BETP, which behave as cPAMs at GLP-1R and a glucagon receptor (GCGR) variant (Phe345Cys GCGR), and describe studies that led to the identification of the covalent mechanism of action of these compounds.

12.2 Pharmacology of Compound 2 and BETP

Compound 2 was initially disclosed in the literature as a small-molecule agonist and allosteric modulator (*i.e.*, ago-allosteric modulator) of GLP-1R.[15] In GLP-1R agonist assays, compound 2 promoted cAMP accumulation, extracellular-signal-regulated kinase 1/2 (ERK1/2) phosphorylation, and β-arrestin recruitment.[16] In PAM assays evaluating cAMP accumulation and β-arrestin recruitment, compound 2 potentiated the activity of several different peptides including GLP-1, GLP-1(9–36)NH_2 (a primary metabolite of GLP-1), and oxyntomodulin (a combined GLP-1R–GCGR co-agonist). GLP-1(9–36)NH_2 is a weak stimulator of cAMP accumulation at GLP-1R, unlike oxyntomodulin which is a more potent agonist at the receptor. PAM activities of compound 2 were not blocked by a GLP-1R orthosteric antagonist, exendin(9–39), confirming an allosteric mode of action. The PAM activity of compound 2 was found to be probe-dependent, exhibiting the strongest cooperativity for GLP-1(9–36)NH_2 and oxyntomodulin.[17–19]

BETP was also initially identified as an ago-allosteric modulator of GLP-1R.[12] BETP increased GLP-1R signaling in an additive manner with GLP-1. In agonist assays, BETP was shown to weakly promote cAMP accumulation, calcium mobilization, and β-arrestin recruitment.[16] In PAM assays examining promotion of cAMP accumulation, BETP potentiated the activity of several different orthosteric ligands, exhibiting robust cooperativity with GLP-1(9–36)NH_2 and oxyntomodulin, similar to compound 2.[17,20]

12.3 Mechanistic Studies with BETP and Compound 2

Ago-allosteric modulators and orthosteric peptide agonists of GLP-1R have been shown to require different regions of GLP-1R for pharmacological activity. BETP retained activity at GLP-1R mutants that lacked the extracellular domain, unlike GLP-1.[12] Alanine replacement of residues in extracellular loop 2 affected neither agonist signaling nor bias of compound 2 despite substantial modification of the activity of orthosteric peptide ligands.[21] In mutagenesis studies in which single-nucleotide polymorphisms (SNPs) were introduced into GLP-1R, introduction of a Ser333Cys SNP into intracellular loop 3 of GLP-1R was found to attenuate the activity of compound 2 as both an agonist and a PAM potentiating the effects of oxyntomodulin.[22,23]

An analog of compound 2, called "compound 3" in the literature (**6**, see Figure 12.2), was evaluated in GLP-1R point mutants and GLP-1R–GCGR chimeras.[24] Compound 3 behaves as ago-allosteric modulator of GLP-1R, but not GCGR, a receptor with high homology to GLP-1R (~44%). A chimeric GLP-1R–GCGR protein which contained the least GLP-1R sequence relative to GCGR and that retained sensitivity to compound 3 consisted of residues 173–347 of GLP-1R with residues 1–169 and 346–477 of GCGR.[24] Residues 173–347 of GLP-1R, which include transmembrane domain 2 through intracellular loop 3, were necessary for the activity of compound 3. Additional chimeras and point mutants indicated the importance of transmembrane domains 2 and 7 in GLP-1R, in particular residues Phe195 and Thr391, to the activity of compound 3, but not to the activity of the orthosteric peptide agonist, GLP-1.

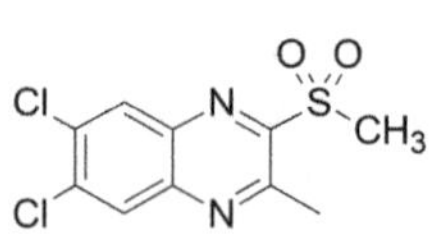

6, "Compound 3"

Figure 12.2 Structure of compound 3.

The agonist activity of compound 2 was found to be dependent on the oligomerization status of GLP-1R. The cAMP agonist activity of compound 2 was eliminated by impairing GLP-1R dimerization *via* mutagenesis within transmembrane domain 4, but ERK1 and ERK2 signaling and the PAM activity of either compound 2 or BETP with oxyntomodulin were not affected.[25] PAM activity of BETP and compound 2 was retained in the dimeric and monomeric forms of GLP-1R, in contrast to the cAMP agonist activity of compound 2, which was dependent on the dimeric form of the receptor.

Results of mutagenesis and pharmacology studies demonstrated that BETP and compounds 2 and 3 could act at allosteric site(s) within GLP-1R. The key residues for activity were dependent on the downstream endpoints that were measured.

12.4 Covalent Mechanism for BETP and Compound 2

BETP was shown to increase glucose-dependent insulin secretion in a rat model following intravenous administration.[12] An analysis of the plasma concentrations of BETP in efficacy studies revealed the presence of minimal intact parent drug, but modest levels of a major metabolite.[26] Results of subsequent studies demonstrated that BETP could be rapidly degraded in the presence of plasma, liver microsomes, and bovine serum albumin (BSA). Previous work with BETP overlooked the electrophilic nature of the compound and the potential reactivity of the compound in the presence of thiol nucleophiles. Results of conjugate-addition studies showed that BETP could readily react with glutathione to give a product in which the ethylsulfinyl group at the C2 position on the 6-(trifluoromethyl)pyrimidine heterocycle was displaced (Figure 12.1).[26] The major metabolite observed in the *in vivo* efficacy studies with BETP was identified as the glutathione addition product.[26] Compound 2 and related analogs were shown to be similarly unstable in the presence of nucleophiles, such as 1,4-dithiothreitol, which can displace the methylsulfone group in these analogs.[13]

The reactivity of BETP and compound 2 in the presence of glutathione or 1,4-dithiothreitol prompted studies to probe a potential covalent mechanism of action for these compounds.[27] Results from washout studies in which GLP-1R-expressing cells were pre-treated with BETP or compound 2 prior to washout and treatment with GLP-1(9–36)NH_2 indicated that the PAM activity of both compounds and the agonist activity of compound 2 appeared to be irreversible. A clickable analog of BETP called PETP (4-(3-(prop-2-ynyloxy) phenyl)-2-ethylsulfinyl-6-(trifluoromethyl)pyrimidine, 7, Figure 12.3) was used to demonstrate a covalent mechanism of action. Interestingly, these clickable analogs revealed that BETP could alkylate a broad range of other proteins, suggesting potential issues with polypharmacology. Mass-spectrometry-based proteomics studies of GLP-1R, purified from cells treated with BETP, showed that two cysteine residues in GLP-1R (*i.e.*, Cys347 and Cys438) were covalently modified by BETP. In subsequent mutagenesis studies, these

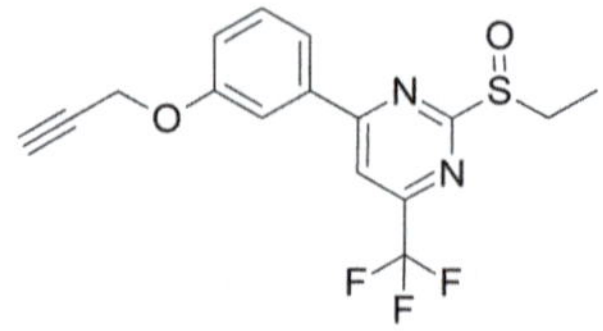

7, PETP

Figure 12.3 Structure of a clickable analog of BETP.

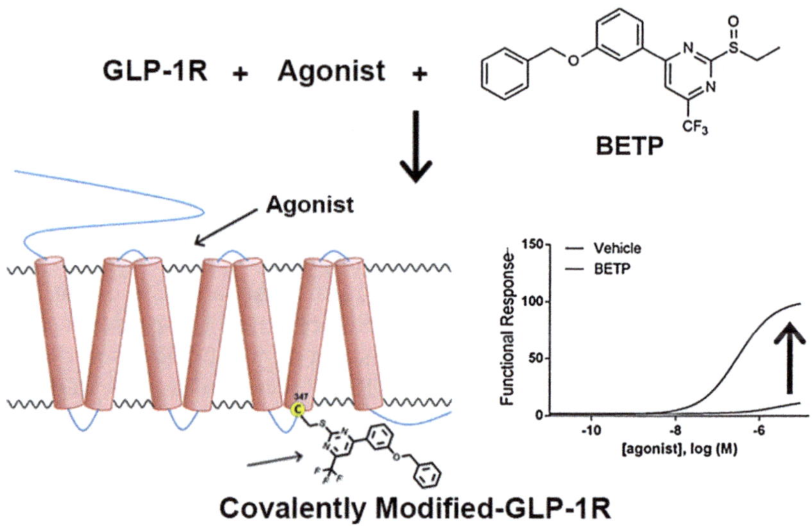

Figure 12.4 Illustration outlining the covalent mechanism of action of BETP at wild-type GLP-1R. BETP alkylates a cysteine residue at the juxtaposition of ICL3 and TM6 of each receptor and potentiates weak agonist signaling.

two cysteine residues and other transmembrane-domain cysteine residues were individually mutated to alanine residues. Only the Cys347Ala mutant abrogated BETP potentiation of GLP-1(9–36)NH_2 at GLP-1R without substantially affecting GLP-1 signaling. The PAM and agonist activities of compound 2 were also absent at the Cys347Ala mutant. Taken together, these data indicated that BETP and compound 2 both acted by covalently modifying Cys347 of GLP-1R (Figure 12.4).

The site of the covalent modification of GLP-1R by BETP and compound 2, Cys347, was shown to be located at the juxtaposition of intracellular loop 3 (ICL3) and transmembrane domain 6 (TM6). While most GPCR PAMs bind near the extracellular face of a GPCR, only a few examples of molecules binding reversibly to allosteric sites on the intracellular face of the receptor have been disclosed.[28] An intracellular binding site at the edge of ICL3 is consistent with the results of prior mutagenesis studies of GLP-1R signaling and other mutagenesis studies probing the mechanism of these ago-allosteric modulators. ICL3 of GLP-1R has previously been shown to be important for cAMP signaling of orthosteric agonists through mutagenesis studies.[29] These findings are further consistent with the results of studies showing that chimeras of GCGR which include residues 173–347 of GLP-1R were sufficient to confer sensitivity to compound 3. These chimeras contain the Cys347 residue, which is necessary for the function of BETP and compound 2. The reduced agonist activity of compound 2 in the Ser333Cys GLP-1R variant could be due to an alternative alkylation site since Cys333 is located near Cys347 in ICL3.[22]

Sensitivity to BETP was found to be transferable to a closely related class B GPCR, GCGR, through mutagenesis of a single residue. Studies showed that BETP was devoid of PAM activity at the wild-type GCGR.[20] Sequence alignment between GLP-1R and GCGR revealed that GCGR lacks an analogous cysteine to Cys347 of GLP-1R. Based upon this sequence alignment, a BETP-sensitive GCGR mutant, Phe345Cys GCGR, was generated. Treatment of Phe345Cys GCGR with BETP led to the potentiation of orthosteric agonists including glucagon(1–29) and glucagon(3–29), a DPP4 metabolite.[27,30] Introduction of a single cysteine residue at the juxtaposition of TM6 and ICL3 of GCGR was able to confer sensitivity to BETP (Figure 12.5).

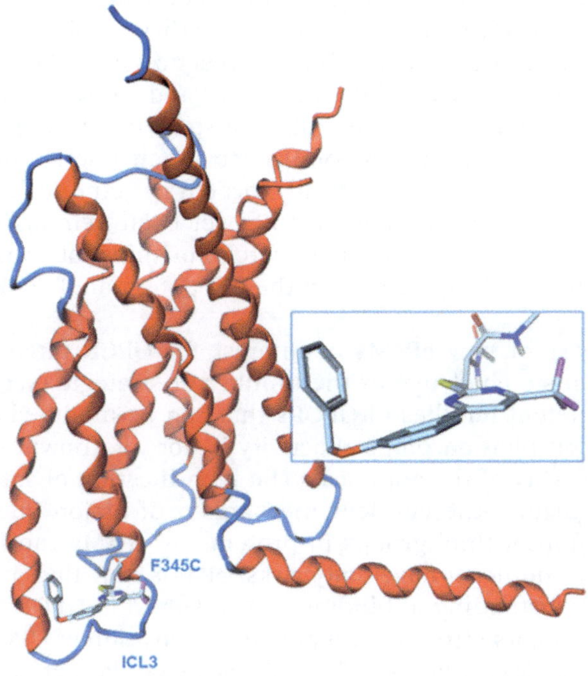

Figure 12.5 Model of the BETP adduct of the F345C GCGR mutant, based upon the published GCGR crystal structure.[35] The inset is a magnification of the putative 4-(3-(benzyloxy)phenyl)-6-(trifluoromethyl)pyrimidin-2-yl adduct at Cys345 of the Phe345Cys mutant of GCGR (Reproduced with permission from ref. 20).

The sensitization of GCGR to BETP *via* a single amino acid change indicates common mechanisms of activation between GLP-1R and GCGR. These mechanisms may rely upon the introduction of steric bulk at the base of TM6 to disrupt interactions that stabilize inactive receptor conformations. A similar mechanism involving disruption of inactive receptor conformations in class A receptors has been proposed. Inactive conformations in class A receptors are thought to be stabilized *via* an ionic lock consisting of a salt bridge connecting the base of TM3 to the base of TM6. Mutagenesis of ionic lock residues has been shown to increase basal receptor activity and enhance functional activity of weak agonists, paralleling the potentiation of weak GLP-1R agonists such as GLP-1(9–36)NH_2 by BETP and compound 2 at GLP-1R.[31]

12.5 Conclusions

The majority of PAMs of GPCRs bind reversibly at allosteric sites to potentiate the activity of orthosteric ligands. In this chapter, we describe the first examples of PAMs that act *via* covalent, irreversible mechanisms, representing a new class of GPCR pharmacological agents. These cPAMs contain electrophilic groups which can react with free thiol groups in strategically located cysteine residues, yielding covalent adducts of GPCRs. One cPAM of the GLP-1R, BETP, was shown to react with several different cysteine residues present in the receptor, but the reaction of BETP with a specific cysteine residue (*i.e.*, Cys347) was necessary and sufficient to confer PAM responsiveness of GLP-1R to different orthosteric peptide ligands. Modulation of the electrophilic character of BETP to enhance selectivity of the reaction of BETP with the key strategically located cysteine residue in GLP-1R over the other cysteine residues has not yet been demonstrated. Future work to delineate the structure–activity relationships for compound 2 and BETP will be required to understand the full potential of cPAMs that act at GLP-1R.

The discovery of new cPAMs of interest for GPCR targets other than GLP-1R presents a challenge as the number of strategic cysteine residues across the different families of GPCRs that can be modified in a covalent manner to elicit pharmacological activity is not yet known, probably limiting the generality of the approach. The identification of safe cPAMs may also prove difficult—the covalent mechanism of action and the non-selective alkylation of thiol groups in proteins by cPAMs can result in substantial safety and immunogenicity risks. cPAMs may find applications in improving understanding of binding sites involved in GPCR signaling and provide new insights into the mechanism of activation of GPCRs.[32–34] Additional opportunities for cPAMs of GPCRs as a new class of drugs remain to be determined.

References

1. B. C. Heng, D. Aubel and M. Fussenegger, *Biotechnol. Adv.*, 2013, **31**, 1676.
2. N. T. Burford, J. Watson, R. Bertekap and A. Alt, *Biochem. Pharmacol.*, 2011, **81**, 691.
3. H. H. Nickols and P. J. Conn, *Neurobiol. Dis.*, 2014, **61**, 55.
4. J. A. Lewis, E. P. Lebois and C. W. Lindsley, *Curr. Opin. Chem. Biol.*, 2008, **12**, 269.
5. P. J. Conn, A. Christopoulos and C. W. Lindsley, *Nat. Rev. Drug Discovery*, 2009, **8**, 41.
6. P. J. Conn, C. W. Lindsley, J. Meiler and C. M. Niswender, *Nat. Rev. Drug Discovery*, 2014, **13**, 692.
7. R. Raddatz, H. Schaffhauser and M. J. Marino, *Biochem. Pharmacol.*, 2007, **74**, 383.
8. C. Koole, K. Pabreja, E. E. Savage, D. Wootten, S. G. B. Furness, L. J. Miller, A. Christopoulos and P. M. Sexton, *Biochem. Soc. Trans.*, 2013, **41**, 172.
9. A. Nakane, Y. Gotoh, J. Ichihara and H. Nagata, *Anal. Biochem.*, 2015, **491**, 23.
10. L. C. Morris, E. L. Days, M. Turney, D. Mi, C. W. Lindsley, C. D. Weaver and K. D. Niswender, *J. Biomol. Screening*, 2014, **19**, 847.
11. L. C. Morris, K. D. Nance, P. R. Gentry, E. L. Days, C. D. Weaver, C. M. Niswender, A. D. Thompson, C. K. Jones, C. W. Locuson, R. D. Morrison, J. S. Daniels, K. D. Niswender and C. W. Lindsley, *J. Med. Chem.*, 2014, **57**, 10192.
12. K. W. Sloop, F. S. Willard, M. B. Brenner, J. Ficorilli, K. Valasek, A. D. Showalter, T. B. Farb, J. X. Cao, A. L. Cox, M. D. Michael, S. M. Gutierrez Sanfeliciano, M. J. Tebbe and M. J. Coghlan, *Diabetes*, 2010, **59**, 3099.
13. M. Teng, M. D. Johnson, C. Thomas, D. Kiel, J. N. Lakis, T. Kercher, S. Aytes, J. Kostrowicki, D. Bhumralkar, L. Truesdale, J. May, U. Sidelman, J. T. Kodra, A. S. Jorgensen, P. H. Olesen, J. C. de Jong, P. Madsen, C. Behrens, I. Pettersson, L. B. Knudsen, J. J. Holst and J. Lau, *Bioorg. Med. Chem. Lett.*, 2007, **17**, 5472.
14. D. Wootten, J. Simms, C. Koole, O. L. Woodman, R. J. Summers, A. Christopoulos and P. M. Sexton, *J. Pharmacol. Exp. Ther.*, 2011, **336**, 540.
15. L. B. Knudsen, D. Kiel, M. Teng, C. Behrens, D. Bhumralkar, J. T. Kodra, J. J. Holst, C. B. Jeppesen, M. D. Johnson, J. C. de Jong, A. S. Jorgensen, T. Kercher, J. Kostrowicki, P. Madsen, P. H. Olesen, J. S. Petersen, F. Poulsen, U. G. Sidelmann, J. Sturis, L. Truesdale, J. May and J. Lau, *Proc. Natl. Acad. Sci. U. S. A.*, 2007, **104**, 937.
16. D. Wootten, E. E. Savage, F. S. Willard, A. B. Bueno, K. W. Sloop, A. Christopoulos and P. M. Sexton, *Mol. Pharmacol.*, 2013, **83**, 822.
17. D. Wootten, E. E. Savage, C. Valant, L. T. May, K. W. Sloop, J. Ficorilli, A. D. Showalter, F. S. Willard, A. Christopoulos and P. M. Sexton, *Mol. Pharmacol.*, 2012, **82**, 281.

18. C. Koole, D. Wootten, J. Simms, C. Valant, R. Sridhar, O. L. Woodman, L. J. Miller, R. J. Summers, A. Christopoulos and P. M. Sexton, *Mol. Pharmacol.*, 2010, **78**, 456.
19. K. Coopman, Y. Huang, N. Johnston, S. J. Bradley, G. F. Wilkinson and G. B. Willars, *J. Pharmacol. Exp. Ther.*, 2010, **334**, 795.
20. F. S. Willard, D. Wootten, A. D. Showalter, E. E. Savage, J. Ficorilli, T. B. Farb, K. Bokvist, J. Alsina-Fernandez, S. G. Furness, A. Christopoulos, P. M. Sexton and K. W. Sloop, *Mol. Pharmacol.*, 2012, **82**, 1066.
21. C. Koole, D. Wootten, J. Simms, E. E. Savage, L. J. Miller, A. Christopoulos and P. M. Sexton, *J. Biol. Chem.*, 2012, **287**, 3659.
22. C. Koole, D. Wootten, J. Simms, L. J. Miller, A. Christopoulos and P. M. Sexton, *J. Pharmacol. Exp. Ther.*, 2015, **353**, 52.
23. C. Koole, D. Wootten, J. Simms, C. Valant, L. J. Miller, A. Christopoulos and P. M. Sexton, *Mol. Pharmacol.*, 2011, **80**, 486.
24. C. Rye Underwood, S. Moller Knudsen, B. Schjellerup Wulff, H. Brauner-Osborne, J. Lau, L. B. Knudsen, G. H. Peters and S. Reedtz-Runge, *Pharmacology*, 2011, **88**, 340.
25. K. G. Harikumar, D. Wootten, D. I. Pinon, C. Koole, A. M. Ball, S. G. Furness, B. Graham, M. Dong, A. Christopoulos, L. J. Miller and P. M. Sexton, *Proc. Natl. Acad. Sci. U. S. A.*, 2012, **109**, 18607.
26. H. Eng, R. Sharma, T. S. McDonald, D. J. Edmonds, J. P. Fortin, X. Li, B. D. Stevens, D. A. Griffith, C. Limberakis, W. M. Nolte, D. A. Price, M. Jackson and A. S. Kalgutkar, *Drug Metab. Dispos.*, 2013, **41**, 1470.
27. W. M. Nolte, J. P. Fortin, B. D. Stevens, G. E. Aspnes, D. A. Griffith, L. R. Hoth, R. B. Ruggeri, A. M. Mathiowetz, C. Limberakis, D. Hepworth and P. A. Carpino, *Nat. Chem. Biol.*, 2014, **10**, 629.
28. L. Dowal, D. S. Sim, J. R. Dilks, P. Blair, S. Beaudry, B. M. Denker, G. Koukos, A. Kuliopulos and R. Flaumenhaft, *Proc. Natl. Acad. Sci. U. S. A.*, 2011, **108**, 2951.
29. S. K. Mathi, Y. Chan, X. Li and M. B. Wheeler, *Mol. Endocrinol.*, 1997, **11**, 424.
30. S. A. Hinke, J. A. Pospisilik, H. U. Demuth, S. Mannhart, K. Kuhn-Wache, T. Hoffmann, E. Nishimura, R. A. Pederson and C. H. McIntosh, *J. Biol. Chem.*, 2000, **275**, 3827.
31. J. Leschner, G. Wennerberg, J. Feierler, M. Bermudez, B. Welte, I. Kalatskaya, G. Wolber and A. Faussner, *J. Pharmacol. Exp. Ther.*, 2013, **344**, 85.
32. D. Weichert, A. C. Kruse, A. Manglik, C. Hiller, C. Zhang, H. Hubner, B. K. Kobilka and P. Gmeiner, *Proc. Natl. Acad. Sci. U. S. A.*, 2014, **111**, 10744.
33. D. Weichert and P. Gmeiner, *ACS Chem. Biol.*, 2015, **10**, 1376.
34. P. M. Kulkarni, A. R. Kulkarni, A. Korde, R. B. Tichkule, R. B. Laprairie, E. M. Denovan-Wright, H. Zhou, D. R. Janero, N. Zvonok, A. Makriyannis, M. G. Cascio, R. G. Pertwee and G. A. Thakur, *J. Med. Chem.*, 2016, **59**, 44.
35. F. Y. Siu, M. He, C. de Graaf, G. W. Han, D. Yang, Z. Zhang, C. Zhou, Q. Xu, D. Wacker, J. S. Joseph, W. Liu, J. Lau, V. Cherezov, V. Katritch, M. W. Wang and R. C. Stevens, *Nature*, 2013, **499**, 444.

CHAPTER 13

Mechanism of Action of a GluN2C- and GluN2D-Selective NMDA Receptor Positive Allosteric Modulator

KATIE L. STRONG[a], MATTHEW P. EPPLIN[b], YAO JING[b], STEPHEN F. TRAYNELIS[a] AND DENNIS C. LIOTTA*[b]

[a]Emory University, Department of Pharmacology, 1510 Clifton Road, Atlanta, GA, 30322, USA; [b]Emory University, Department of Chemistry, 1521 Dickey Drive, Atlanta, GA, 30322, USA
*E-mail: dliotta@emory.edu

13.1 Introduction

The *N*-methyl-D-aspartate (NMDA) receptor is one of four known mammalian ionotropic glutamate receptors (iGluRs) along with the 2-amino-3-(5-methyl-3-oxo-1,2-oxazol-4-yl)propanoic acid (AMPA), kainate, and δ receptors.[1] The receptor plays a role in mediating neuroplasticity within the brain and is important for normal neuronal development as well as higher cognitive function. Both hypo- and hyperfunction of the receptor have been implicated in neuropathological conditions including schizophrenia,[2–4] Alzheimer's disease,[5] Parkinson's disease,[6] epilepsy,[7] and depression.[8,9] The NMDA receptor is unique among receptors in its class in terms of pharmacological and

RSC Drug Discovery Series No. 56
Allosterism in Drug Discovery
Edited by Dario Doller

Published by the Royal Society of Chemistry, www.rsc.org

functional properties and this, along with its involvement in neurological conditions, creates a therapeutic rationale for developing both positive and negative modulators of the receptor.

NMDA receptors exhibit a high permeability to Ca^{2+} ions, and thus mediate a slow Ca^{2+}-permeable component of synaptic transmission[10] upon synaptic release of glutamate. Receptor activation requires the binding of both co-agonists glutamate and glycine.[11] The ion channel is blocked in a voltage-dependent manner by extracellular Mg^{2+}, and thus membrane depolarization that reduces the Mg^{2+} block of the channel pore can increase the response to agonist activation.[12] The opening of the pore, inferred by the time required for maximally effective concentrations of co-agonists to induce a peak response, takes roughly 5–15 milliseconds at room temperature. This is significantly slower than the AMPA receptors, which open in the microsecond range.[13] The NMDA receptor exists as a tetrameric assembly of two obligate heterodimers, each consisting of a glycine-binding GluN1 subunit and a glutamate-binding GluN2 subunit.[14,15] This is in contrast to both AMPA and kainate receptors, which can exist as homotetramers. While only one gene (*GRIN1*) encodes the GluN1 subunit, four genes (*GRIN2A–D*) encode the four GluN2 subunits, designated GluN2A–GluN2D.[1] These subunits impart unique properties to the receptor, creating differences in deactivation time course following removal of glutamate, open probability, channel open duration, and agonist sensitivity.[11,16–18] For example, receptors containing the GluN2A subunit show faster deactivation time course, higher agonist sensitivity, and an open probability that is nearly tenfold higher than those of other receptor subtypes.[19,20] Receptors containing the GluN2C or D subunit have reduced Ca^{2+} permeability, Mg^{2+} blockage, and single-channel conductance compared with receptors containing GluN2A or B, and a single residue contained in the M3 transmembrane helix is responsible for these differences in pore properties.[21] There are also pharmacological implications due to differential subunit expression in the brain during its development. For example, pre-natal rats primarily express the GluN2B and GluN2D subunits, with increased expression following birth. In adulthood, all subunits are expressed, but GluN2B expression is significantly decreased in the frontal lobe compared with previous levels.[22–24]

Although GluN1–GluN2 heterodimers are the best understood, two glycine-binding GluN3 subunits, GluN3A and GluN3B, also exist. The obligatory GluN1 subunit can associate with a combination of the GluN3 subunits forming GluN1–GluN3 heterodimers in some heterologous expression systems. Native NMDA receptors thought to contain GluN3 subunits show a reduced sensitivity to Mg^{2+} blockage and lower permeability to Ca^{2+} in comparison with GluN1–GluN2 receptors.[25] Receptors can also contain two different GluN2 subunits along with the two GluN1 subunits, and these triheteromeric NMDA receptors show unique receptor properties compared with diheteromeric receptors.[26,27] The GluN1–GluN2A–GluN2B triheteromers have been suggested to constitute the majority of receptors in both the hippocampus and frontal cortex in adult rat brains,[26,28,29] but more work is needed to confirm these suggestions.

Sequence homology is well conserved among the iGluRs, suggesting that they share a similar architecture. Each receptor in the family consists of four semiautonomous domains: the extracellular amino-terminal domain (ATD), the extracellular ligand-binding domain (LBD), the transmembrane domain (TMD), and the intracellular carboxyl-terminal domain (CTD). Two crystal structures of the ATD–LBD–TMD complex within the GluN1–GluN2B receptor were recently solved, providing deeper insight into the architecture of the receptor as well as into structural differences between the AMPA and NMDA receptors.[14,15] In general, the receptor was shown to exist as a dimer of dimers, with the ATD and LBD tightly packed together as an apparent single unit, while in the AMPA receptor the two segments are separated by linkers, suggesting that they may exist as two functionally distinct units. This may help explain why the ATD of the NMDA receptor plays a larger role in regulating ion–channel activity than the ATD of the AMPA receptor.[30,31]

The diversity of function that stems from the GluN2 subunit composition has helped make the NMDA receptor an important drug target. Variations in subunit expression suggest that pharmacological agents that distinguish between the four GluN2 subunits could have distinct effects on different brain structures and functions. A number of modulators of the NMDA receptor have been studied for potential therapeutic relevance, including voltage-dependent channel blockers, subunit-selective inhibitors, and subunit-selective positive allosteric modulators.[32–41]

Channel blockers require receptor activation, which provides access of the blocker to its binding site deep within the transmembrane ion channel pore.[42] Well-studied NMDA receptor channel blockers include amantadine, memantine, ketamine, MK-801, and phencyclidine (PCP). Amantadine and memantine, FDA-approved for the symptomatic treatment of Alzheimer's disease, exhibit fast unblocking rates and lower affinity for the receptor compared with other channel blockers.[32] This property allows for the preferential targeting of receptors that have been opened for prolonged periods of time compared with transiently opened receptors, which has been argued to be one reason they are clinically tolerated.[43–46] Ketamine, typically used as an anesthetic for children[47] and in veterinary medicine,[48] has also seen increased clinical interest due to success in multiple trials for treatment-resistant depression,[33] preliminary clinical trials for obsessive–compulsive disorder (OCD),[49] and post-traumatic stress disorder (PTSD).[50] All reported that channel blockers show less than tenfold GluN2 subunit-selectivity,[51] and this lack of selectivity among subunits may be associated with negative side effects, including decreased motor function, altered cardiovascular function, hallucinations, and delusions.[45]

The first subunit-selective modulator of the NMDA receptor identified was the GluN2B-containing NMDA receptor antagonist, ifenprodil, which is 200- to 400-fold more active at GluN1–GluN2B subunits than other GluN1–GluN2 subunits.[34,52] Crystallographic studies revealed that ifenprodil binds between the ATD GluN1–GluN2B heterodimer interface at a site that is distinct from the Zn^{2+} binding site.[53] Ifenprodil and related analogs such as traxoprodil and Ro 25-6981 have been tested in clinical trials for neuropathic pain, traumatic brain injury, and depression,[54–59] although these compounds have been discontinued due to low efficacy and adverse central nervous system side effects.[60–63] Recently, a negative allosteric modulator of glycine binding, TCN-201, was discovered that was highly selective for GluN2A-containing receptors.[35,64,65] Results from chimeric studies utilizing the GluN2A- and GluN2D-containing NMDA receptors suggest that TCN-201 binds at the interface of GluN1–GluN2A LBD.[66]

Additionally, two classes of noncompetitive GluN2C- and GluN2D- selective antagonists have also been identified: QNZ-46[36] and DQP-1105.[37] QNZ-46 is voltage-independent and exhibits a mechanism of action that depends on glutamate, but not glycine, binding.[67] Site-directed mutagenesis identified multiple amino acid residues in the S2 region of the LBD that appear to be structural determinants of activity of these inhibitors.[67] However, the binding site for this class of compounds remains to be elucidated. Results from site directed mutagenesis suggest that DQP-1105 is a negative allosteric modulator that shares the same structural determinants in the S2 region of the LBD as QNZ-46.[68] A novel class of iminothiazolidinone-ring-containing negative allosteric modulators has also recently been discovered. Saturating concentrations of these compounds exhibit submaximal inhibition, and the best-in-class modulators show a modest preference for GluN2C- and GluN2D-containing receptors over ones containing GluN2A and GluN2B subunits.[38]

A series of phenanthrene- or naphthalene-containing modulators referred to as UBP compounds shows complex pharmacology properties, acting as positive and negative modulators at the NMDA receptor. For instance, the allosteric modulator naphthalene UBP618 inhibits all four GluN2 subunits with concentration giving half maximal inhibition (IC_{50}) values around 2 μM.[69] The antagonist UBP141 competitively inhibits GluN2D-containing receptors with 11-fold and 9-fold higher potency than GluN2A or GluN2B, respectively; an inhibition constant (K_i) value of 2.36 μM was reported at GluN2D compared with a K_i value of 22.0 μM at GluN2A and 17.0 μM at GluN2B.[39] Some compounds do not exclusively act as potentiators or inhibitors, but show a complex mixed mode of action. For example, allosteric modulator UPB710 potentiates GluN2A- and GluN2B-containing receptors (concentration giving half maximal effect (EC_{50}) value reported greater than 30 μM), but partially inhibits GluN2C- and GluN2D-containing receptors with an IC_{50} value of 100 μM or greater.[69] Structure–activity relationship (SAR) studies suggest these compounds bind in the LBD region.[70] A number of coumarin compounds also act as negative and positive allosteric modulators, including UBP608 which exhibits modest selectivity for inhibition at GluN2A-containing receptors,[69,71] and UBP714, a positive modulator at GluN2A, GluN2B, and GluN2D subunits.[71]

Two classes of subunit-selective allosteric potentiators, (3-chlorophenyl)(6,7-dimethoxy-1-((4-methoxyphenoxy)methyl)-3,4-dihydroisoquinolin-2(1*H*)-yl)methanone (CIQ) and pyrrolidinone (PYD), have been reported.

Analogs of CIQ exhibit selective potentiation at GluN2C- and GluN2D-containing receptors[40,72] (discussed below) and PYD exhibits 50-fold selectivity at the GluN2C subunit over GluN2A, GluN2B, and GluN2D subunits.[41] A binding site of the PYD compounds has been proposed to exist between the LBD and ATD interface on the basis of crystallographic data, docking, and mutagenesis.[73]

13.2 Therapeutic Rationale for NMDA Receptor Positive Allosteric Modulators

13.2.1 Schizophrenia

Schizophrenia, reported to affect 1% of the world's population,[74] is accompanied by symptoms that are characterized as positive and negative. Positive symptoms are behaviors that are usually not present in a healthy individual, such as hallucinations, delusions, and agitated body movements. Negative symptoms, on the other hand, can be thought of as behavioral deficits and include withdrawal, lack of motivation, and lack of affective response. Schizophrenia is commonly accompanied by cognitive impairment, which is typically subtle, and can include problems with working memory, attention, and decision-making.[75] While many people are treated with antipsychotic medication, this treatment only reduces symptomology without being curative, and often these medications affect positive symptoms, but do very little to alleviate the debilitating negative symptoms.[76,77]

Genetic linkage studies have identified genes implicated in schizophrenia that encode or regulate a number of proteins that influence NMDA receptor activity. Neuregulin-1 (*NRG1*) regulates expression of glutamate subtypes and has specifically been linked to GluN2C expression,[78] and post-mortem schizophrenia patients and control patients with a specific *NRG1* polymorphism were reported to have lower levels of GluN2C expression in the cerebellum.[79] It was also shown that mice hypomorphic for *nrg1* exhibited behavioral abnormalities that could be reversed with the antipsychotic clozapine and that these hypomorphic mice expressed fewer functional NMDA receptors than control mice.[80] A second gene related to schizophrenia is *G72*, which interacts with D-amino acid oxidase (DAAO). DAAO is an enzyme that catabolizes through oxidation D-serine, an NMDA receptor co-agonist, suggesting that schizophrenia is implicated in the NMDA receptor pathway.[81] Other schizophrenia risk genes related to the NMDA receptor activity include *RGS4*,[82] dysbindin,[83] and *PPP3CC*.[84] Studies have also demonstrated altered NMDA expression levels in the brains of those diagnosed with schizophrenia,[79,85–87] and for these reasons the NMDA receptor is one pathway that has garnered considerable attention for its potential role in schizophrenia.

The glutamate hypofunction hypothesis, which links NMDA receptor activity and schizophrenia, originated in the 1950's when it was reported that the effects of using PCP, an NMDA receptor channel blocker, were virtually indistinguishable from schizophrenic symptomology.[88,89] NMDA receptor channel blockers MK-801 [90] and ketamine[91–94] also induce schizophrenia-like symptoms, and for this reason, the administration of NMDA receptor channel blockers has become a widely used model of schizophrenia in humans and animals.[95,96] Moreover, if blocking NMDA receptors can reproduce symptoms of schizophrenia, it stands to reason that NMDA receptor deficits may underlie a subset of symptoms in some patients with this disorder.

Schizophrenia is most probably a result of dysfunction of a number of neurotransmitter systems, including gamma-aminobutyric acid (GABA). Post-mortem analysis of schizophrenia patients revealed a decrease in GABA in the nucleus accumbens and the thalamus.[97] The hypofunction of the GABAergic system secondary to a reduction of GABA and glutamate decarboxylase (GAD) in fast-spiking, parvalbumin-containing interneurons has been proposed to contribute to the disease. In this model, NMDA receptors in GABAergic neurons serve as homeostatic sensors for principal cell activity. Hypofunction of the NMDA receptor system results in interneuron adaptation, which triggers reduced GABAergic function causing increased principal cell activity to restore normal activity. This leads to enhanced principal cell activity and overactivation of the dopaminergic system.[98]

13.2.2 Cognitive Enhancement

Activation of NMDA receptors can lead to long-term potentiation (LTP) in some cells, a mechanism proposed to underlie synaptic strength,[99] and changes in synaptic strength are believed to play a role in forming memories and learning.[100] Additionally, the deletion of the GluN1 NMDA gene in the CA1 pyramidal cells in the hippocampus leads to an absence of LTP in the CA1 cells, and hippocampal GluN1-knockout mice show deficiencies in spatial learning,[101] object recognition, and contextual fear memory.[102] Several lines of evidence suggest that enhancement of the function of GluN2B-containing NMDA receptors could improve memory formation. Overexpression of GluN2B in the prefrontal cortex (PFC) of rats resulted in enhanced LTP, but not long-term depression (LTD), and the GluN2B-selective antagonist Ro25-6981 abolished this improvement, suggesting that overexpression of GluN2B was responsible for the effect.[103] Similarly, GluN2B receptor overexpression in the forebrain of mice resulted in enhanced activation of NMDA receptors in response to stimulation,[104] which was shown to improve learning and long-term memory, spatial performance, and cued and contextual fear conditioning of the transgenic mice compared with wild-type mice. Additionally, mice with this enhanced activation exhibit faster fear extinction.[103,105–107] Both GluN2A- and GluN2B-containing receptors are heavily expressed in the PFC, but compared with GluN2A subunits the GluN2B has a slower deactivation time course,[16] which prolongs excitatory synaptic signals and the resulting depolarization that is required for working memory.[108–110] This is one hypothesis to explain the integral role of GluN2B-containing receptors in the process of higher cognitive thought.

13.2.3 Anxiety Disorders

While fear is a normal human emotion, continued fear when the threat of danger has passed can lead to a number of debilitating anxiety disorders, the main treatments for which include benzodiazepines and anti-depressants, such as selective serotonin-reuptake inhibitors (SSRIs).[111] Cognitive behavioral therapy (CBT), a form of psychotherapy has also proven successful in treating many types of anxiety disorders,[112] but not in conjunction with medication.[113,114] Ketamine, the NMDA receptor channel blocker, has seen a reemergence in the clinical world as its experimental use has been accompanied with potentially useful effects for the treatment of OCD[49] and PTSD,[50] despite an incomplete understanding of the mechanism of action.

NMDA receptors play a fundamental role in learning and memory, and it is thought that the molecular mechanisms in the amygdala behind the acquisition of fear have much in common with learning and memory.[115] NMDA antagonists that weaken LTP or LTD in the amygdala have been shown to impair fear acquisition and fear extinction.[116,117] Results of studies with a number of NMDA antagonists have demonstrated blocked fear acquisition[118–120] and reduced fear extinction[121–123] in animal models of fear conditioning, while amygdala infusions of the GluN2B-selective inhibitors ifenprodil[124] and traxoprodil[125] impair fear conditioning. D-Cycloserine (DSC), a co-agonist for the glycine site, is a partial agonist at GluN2A, GluN2B, and GluN2D, but causes greater potentiation at GluN2C than saturating concentrations of glycine.[126,127] Interestingly, DSC enhanced fear extinction when infused into the amygdala, suggesting that modulation of the GluN2C receptor might be a means to heighten fear extinction.[128] GluN2C-knockout mice are impaired during tests of fear acquisition[129] and direct amygdala injections of CIQ, a positive allosteric modulator of GluN2C- and GluN2D-containing receptors, have resulted in mice that have enhanced fear acquisition and improved fear extinction[130] (discussed below). The combined results of these studies highlight the potential for a positive allosteric modulator in treating anxiety disorders.

13.3 Mechanism of Action and Structural Determinants of CIQ

13.3.1 Mechanism of Action

An analogue of CIQ (**1**) was initially discovered in a Ca^{2+} imaging high-throughput screen using 96-well plates to identify novel GluN2C- and GluN2D-containing NMDA receptor modulators. A baby hamster kidney (BHK) cell line expressing either GluN2C or GluN2D was used in the primary screen, which measured an increase or decrease in Ca^{2+} flux, to test approximately 100 000 compounds.[131] Of this initial primary screening library, 839 compounds (10 μM concentration) altered the response by more than 40% of the control response (hit rate of 1.4%), and 516 compounds were taken forward in the secondary screen, a two-electrode voltage clamp assay of *Xenopus* oocytes expressing rat recombinant GluN1–GluN2D. At 10 μM, only 54 compounds produced more than 25% inhibition or potentiation and warranted the analysis of full dose–response curves for GluN2A-, GluN2B-, GluN2C-, GluN2D-containing receptors. From the tertiary screen of 54 compounds, six classes of potentiators (less than 0.01% of the compounds tested in the primary screen) were discovered, and one of these potentiators was a tetrahydroisoquinoline compound, 1180. Co-application of compound 1180 with saturating glutamate and glycine enhanced the current response to 160–180% of those for control with an EC_{50} value at GluN2C-containing receptors and GluN2D-containing receptors of 11 μM and 13 μM, respectively. Early SAR studies identified a chloro-derivative CIQ (**1**), which had improved activity at both GluN2C- and GluN2D-containing receptors over the initial screening hit, 1180 (Figure 13.1).[40]

The CIQ class of compounds, the first reported positive allosteric modulators selective for GluN2C- and GluN2D-containing NMDA receptors, provided a unique tool to study the role of GluN2C and GluN2D subunits in brain function and dysfunction. While there are a number of subunit-selective antagonists for NMDA receptors, there was until recently a lack of subunit-selective potentiators, other than the aminoglycosides and spermine, which act selectively on GluN2B-containing receptors.[1] There are now several reports on the mechanism of action and binding site of this class of tetrahydroisoquinoline compounds on GluN2C- and GluN2D-containing receptors.[40,132]

In a two-electrode voltage clamp assay in rat recombinant NR2C- and NR2D-containing receptors expressed in *Xenopus laevis* oocytes, CIQ was reported to have an EC_{50} value of ~3 μM, and increased the maximal current response during co-application with glutamate and glycine approximately twofold compared with controls.[40] Similar EC_{50} values were reported irrespective of the GluN1 splice variant paired with GluN2C and GluN2D. One important feature of CIQ and its analogues is their relatively high calculated log of the partition coefficient (cLogP) and low maximum solubility, which generally is about 20 μM.[40] While this minimally affects results for freshly made CIQ at concentrations below 20 μM, higher concentrations require inclusion

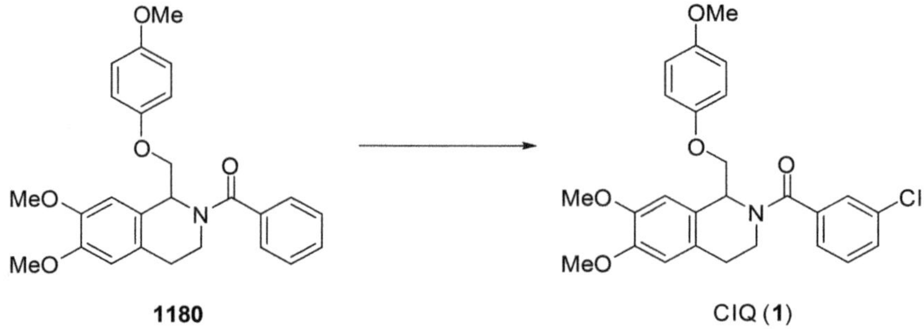

Figure 13.1 Scaffold of initial screening hit (1180) and CIQ (**1**).

of an agent such as 2-hydroxypropyl-β-cyclodextrin or dimethyl sulfoxide (DMSO) to allow CIQ to remain in solution. In addition, we have observed that CIQ can slowly precipitate over 6 hours (Ogden, Traynelis, unpublished data), so it is best to make solutions fresh. Limited solubility can also complicate accurate determination of the EC_{50} value, since either the response to higher concentrations cannot be determined or the actual free solution level above the solubility limit is lower than the nominal concentration. As a result, the reported EC_{50} values for CIQ vary slightly depending how the experiment was performed and whether 2-hydroxypropyl-β-cyclodextrin was included in the experiments (compare EC_{50} values between Mullasseril *et al.*, Santangelo *et al.*, and Ogden and Traynelis).[72,132,133]

Potentiation of rat and human GluN2C-containing NMDA receptors was similar, but CIQ may be slightly more potent in human GluN2D-containing receptors compared with rat GluN2D-containing receptors (EC_{50} value of 3 μM and 6 μM for human and rat, respectively).[134] CIQ had no detectable effects on GluN2A-containing NMDA receptors (Figure 13.2), AMPA, or kainate receptors.[40] CIQ at the highest concentrations may cause modest (5–10%) inhibition of GluN2B-containing receptors at 10 μM.[40]

Tetrameric NMDA receptors are comprised of two GluN1 subunits and two GluN2 subunits. NMDA receptors that can contain two different GluN2 subunits, are often referred to as triheteromeric receptors, and have been recently studied using a variety of techniques.[26,27] The function of triheteromeric NMDA receptors in oocytes can be isolated using mutagenesis to reduce responses of diheteromeric receptors,[27] and this approach has been

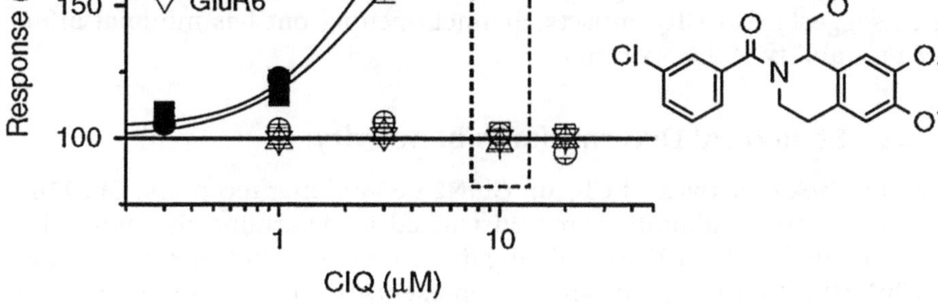

Figure 13.2 Concentration–response curve of CIQ at GluN2A-, GluN2B-, GluN2C-, and GluN2D-containing receptors, along with GluR1 and GluR6 receptors. Reprinted by permission from Macmillan Publishers Ltd: [Nature Communications] (Mullasseril, P. *et al. Nat. Commun.* **2010**, 1, 90), copyright 2010.

used to evaluate the actions of CIQ on triheteromeric receptors containing GluN1–GluN2A–GluN2B, GluN1–GluN2A–GluN2C, and GluN1–GluN2A–GluN2D. Data obtained for receptors comprised of GluN1, GluN2C, and a magnesium-insensitive, low-glutamate-potency GluN2A mutant (GluN2A*)[27] suggested that CIQ could potentiate the response of triheteromeric receptors to maximally effective concentrations of glutamate and glycine. Similar results were found when GluN2D was used instead of GluN2C.[40] Receptors that contained GluN1–GluN2A–GluN2B were insensitive to CIQ. These data suggest that only one copy of GluN2C and GluN2D is needed for CIQ potentiation of NMDA receptors' responses.[40]

The binding of CIQ to GluN2C- and GluN2D-containing NMDA receptors had no detectable effect on the glutamate or glycine EC_{50} value, and it did not have an impact on the IC_{50} of Mg^{2+} inhibition. Equal levels of potentiation were observed in the presence of glutamate and the partial agonist NMDA, but no potentiation was recorded when only glutamate or glycine was co-applied with CIQ. CIQ potentiation was voltage-independent, and was unaffected by acidic pH, suggesting that CIQ plays no role in relieving tonic proton inhibition.[40] CIQ potentiation was enhanced at low concentrations of agonist, an observation that requires further study.[72]

CIQ can potentiate the response of native NMDA receptors in neurons in the subthalamic nucleus,[40,135] which express GluN2D.[135,136] CIQ is also active on hippocampal interneurons, which express GluN2D.[137] In contrast, no effect of CIQ could be detected on NMDA receptor responses in CA1 hippocampal pyramidal cells,[40] which are thought to express mostly the GluN2A and GluN2B subunits.[24] These data suggest that CIQ can serve as a tool compound for studying the subunit composition of native neurons.[40]

To further study the mechanism by which CIQ potentiates the NMDA receptor, a single-channel analysis was conducted on recombinant GluN1–GluN2D receptors in excised outside-out patches.[40] Co-application of 10 μM CIQ with maximally effective concentrations of glutamate and glycine increased open probability to 190% of the control value and decreased the mean shut time. Interestingly, CIQ had no effect on the mean open time, which suggests that CIQ impacts channel opening but has minimal influence on stability of the open state.

13.3.2 Structural Determinants of Activity

Given the lack of activity of CIQ on GluN2A-containing receptors, GluN2A–GluN2D chimeric subunits were constructed to determine the molecular determinants of CIQ activity. Initial gain-of-function studies revealed that the GluN2D L–S1–M1 region, where L represents the linker between the ATD and LBD, was required for CIQ potentiation. Chimeric receptors that contained varying and smaller portions of these regions from GluN2D inserted into GluN2A were constructed and tested, and the regions that elicited activity in the GluN2D-containing GluN2A chimeric receptors in the presence of CIQ were narrowed down to the linker between ATD and S1 and one residue in the M1 transmembrane helix that is shared between GluN2C and D but divergent from GluN2A and B (GluN2D Thr592). GluN2A-containing GluN2D chimeric receptors containing a 16-amino-acid portion of the linker between the ATD and S1 region of the LBD, along with residue T592 in the M1 region were potentiated by CIQ, suggesting that these are key structural determinants of CIQ activity.[40]

This chimeric strategy was further expanded to refine the structural determinants of CIQ activity. Further chimeric receptors that contained the linker preceding the M1 transmembrane helix in addition to five residues (residues 590–594) within the M1 transmembrane helix from GluN2D was sufficient to transfer potentiating activity to GluN2A, suggesting that the ATD–S1 linker, while important, was not necessary for CIQ function. Furthermore, it is unlikely that CIQ can interact with both the ATD–S1 linker and the M1 region, as crystallographic data for the NMDA receptor places these two sites too far apart.[14,15] Although CIQ potentiated responses for GluN2A chimeric receptors when both pre-M1 and M1 regions were transferred from GluN2D to GluN2A, CIQ did not potentiate chimeric receptors when only the pre-M1 linker or M1 residues were transferred. These data indicate that the pre-M1 and M1 regions of the GluN2D subunit contain essential determinants of CIQ activity.

Loss-of-function studies that introduced parts of GluN2A into GluN2D confirmed the important roles of the pre-M1 linker region and the M1 region of the transmembrane helix. Insertion of the GluN2A linker between the ATD and S1 region of the LBD into GluN2D had no detrimental effects on CIQ activity, consistent with the idea that this region was not required for CIQ activity. It should be noted however, that the insertion of the linker plus the ATD did reduce potentiation by CIQ. In contrast, site-directed mutagenesis of a single residue that is conserved in the M1 region of the transmembrane domain in GluN2C- and GluN2D-containing receptors, but not in GluN2A- and GluN2B-containing receptors (GluN2D T592I), completely abolished CIQ activity.[40] To elucidate details of the structural determinants in the pre-M1–M1 region, an alanine screen of 23 residues in the GluN2D M1 helix was conducted.[40] From this screen, six residues reduced potentiation by CIQ, while two residues increased potentiation.

Additional site-directed mutagenesis of residues 5 Å away from the GluN2 M1 helix revealed a cluster of residues, including four residues in the GluN2D pre-M1 linker in addition to two residues in the GluN1 M4 helix which perturbed CIQ potentiation. Interestingly, the GluN2D pre-M1 region of the receptor has been implicated in the gating mechanism of the receptor. In the AMPA and NMDA crystal structures, these amino acids compose a pre-M1 cuff helix that make van der Waals contacts with the M3 helix, and thus may stabilize the closed conformation of the receptor.[14,15,138]

CIQ modulates the NMDA receptor by increasing the frequency of channel opening,[40] so it stands to reason that mutations in the pre-M1 helix, a region implicated in gating, might disrupt the enhancement of channel opening caused by CIQ.[132] These data raise the possibility that CIQ interacts with the pre-M1 and M1 helices of the GluN2D subunit, and that this interaction leads to potentiation of the receptor response by promoting channel opening of the receptor (Figure 13.3). A binding site in the TMD region of the receptor could mean that CIQ is required to partition into the plasma membrane, but whole-cell voltage clamp recording suggests that CIQ is accessing its binding site extracellularly and not intracellularly or through diffusion into the plasma membrane.[132]

While the chimeric data point towards the pre-M1–M1 regions as primary structural determinants of activity, several other candidate binding sites exist on the NMDA receptor.[132] One well-established binding site for subunit-selective allosteric modulators that has been identified on the NMDA receptor is the dimer interface between GluN1 and GluN2 ATD, a site where the GluN2B-selective inhibitor ifenprodil binds.[53] Even though the linker between the ATD and LBD appears able to influence CIQ activity, CIQ does not share a binding site with ifenprodil on the ATD interface nor is it binding to the linker region. Potentiation of the NMDA receptor by CIQ was still observed in receptors where both the GluN1 ATD and GluN2 ATD, along with the ATD–LBD linker from both subunits were deleted, suggesting that the binding site resides elsewhere.[132] Further studies also ruled out other well-established binding sites.[132] CIQ has no impact on NMDA receptor (NMDAR) activity (degree of potentiation or potency) of channel blockers Mg^{2+}, or ketamine, suggesting that CIQ is most likely not interacting with the ion channel pore. In addition, point mutations made to the S2 region of the LBD for GluN2D that reduce the potency of GluN2C- and GluN2D-selective inhibitors DQP[68] and QNZ[67]

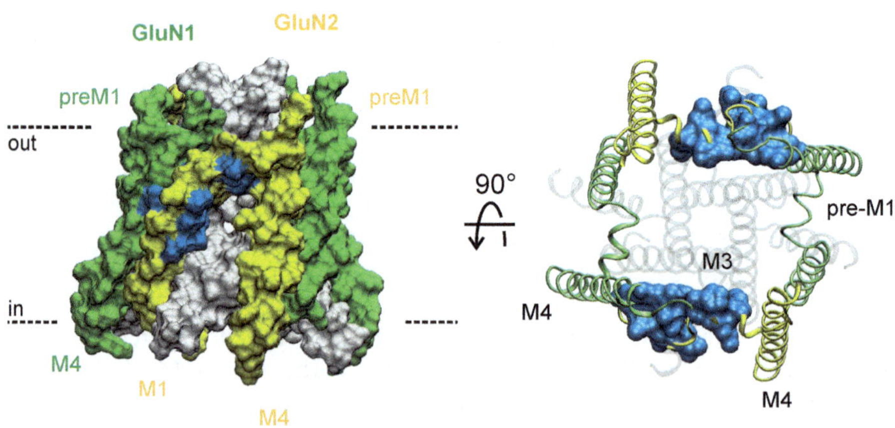

Figure 13.3 Homology model of transmembrane domain region of GluN1–GluN2D NMDA receptor where amino acid mutations on the GluN2D pre M1 and M1 helix that positively affect CIQ are depicted in blue. The outside of the NMDA receptor is shown on the left and a view down the pore axis is shown on the right. Reprinted with permission from Aspet Journals (Ogden, K. K. and Traynelis, S. F, Contribution of the M1 Transmembrane Helix and Pre-M1 Region to Positive Allosteric Modulation and Gating of *N*-Methyl-D-Aspartate Receptors, *Mol Pharmacol* May 2013 83:1045–1056).

activity have no effect on CIQ activity, suggesting that CIQ does not share structural determinants with these negative allosteric modulators.[132] Site directed mutagenesis of residues at the GluN1–GluN2 LBD dimer interface that correspond to those that alter the potency of the GluN2A-selective modulator TCN-201[66] had no effect on the potency of CIQ. While these data indicate that the binding site of TCN-201 is not identical to that of CIQ, this site has considerable space along the dimer interface, and CIQ could still reside at a site nearby or partially occupying the corresponding site for TCN-201 in GluN2D. Thus, while we have identified key structural determinants for CIQ's actions in the pre-M1 and M1 regions,[132] more work will be needed to definitively identify the binding site.

13.4 Off-Target Testing and the Selectivity of CIQ for the NMDA Receptor

CIQ has an EC_{50} value at GluN2C- and GluN2D-containing receptors of approximately 3–6 μM, and has no effect on AMPA or kainate receptors at a concentration of 20 μM.[40] Additional two-electrode voltage clamp assays of *Xenopus* oocytes expressing 11 other ion channels revealed that 10 μM of CIQ had a modest inhibitory effect at nicotinic acetylcholine receptors, but had no influence on $GABA_A$, $GABA_C$, glycine, 5-HT_{3A}, or P2X2 receptors. In radioligand displacement assays that screened CIQ against 62 receptors, transporters, and ion channels in the CNS, CIQ only displaced ligands at the κ opioid receptor, and the 5-hydroxytryptamine $(5\text{-HT})_{2B}$ and 5-HT_6 receptors at relevant concentrations that might be encountered during an experiment (K_i < 10 μM).[130]

Additional off-target testing was conducted on benzyl-containing analog of CIQ (compound 83 from Santangelo *et al.*[72]), which is tenfold more potent than CIQ with an EC_{50} at GluN2C- and GluN2D-containing receptors of 0.3 μM and 0.3 μM. This analog significantly affected seven of the 21 receptors tested, but in all cases the inhibition was less than 10% and the potentiation was less than 5%.[72] In the National Institute of Mental Health Psychoactive Drug Screening Program (PDSP) assay of 42 targets, compound 2 only significantly inhibited the κ opioid receptor with a K_i greater than 15 μM, which is approximately fivefold higher than the EC_{50} value at NMDA receptors.[72] Thus, these initial members of the tetrahydroisoquinoline potentiators appear relatively selective for GluN2C-and GluN2D-containing NMDARs, suggesting that optimization of this scaffold could further refine the properties of this class of compounds.

13.5 Pharmacokinetics of CIQ

The half-life, blood–brain barrier penetration, and the bioavailability of CIQ has been determined.[130] CIQ exhibited a half-life of 46 minutes following intravenous (i.v.) injection (5 mg kg^{-1}) and 81 minutes following intraperitoneal (i.p.) injection (20 mg kg^{-1}) in mice. Under the same i.v. and i.p.

injection conditions, CIQ peaked in the brain at 30 minutes and decreased with first-order kinetics. The bioavailability of CIQ was reported as 0.42 when injected *via* an i.p. route and the brain-to-plasma concentration ratio was 5.4.[130] Although CIQ is highly brain-penetrant, results from recent work suggest that CIQ is highly bound to CNS tissue and that the free fraction may be quite low, requiring more work to understand its availability (Jensen, Egebjerg, Bundgaard, Traynelis, unpublished data).

13.6 Utility of CIQ as a Pharmacological Probe

13.6.1 CIQ as a Pharmacological Probe for Fear Acquisition and Fear Extinction

Based on evidence that GluN2C-containing NMDA receptors are implicated in fear acquisition and fear extinction[128,129] (discussed in Section 13.2.2), racemic CIQ was administered to mice and utilized in fear conditioning and fear extinction tests where a tone acted as the conditioned stimulus (CS) and foot shocks as the unconditioned stimulus (US).[130] CIQ (3 μg, 10 μg, and 30 μg) was injected directly into the amygdala immediately following fear conditioning training, and 24 hours later, the effect of this CIQ treatment on fear retention was studied by measuring freezing levels of the mice in the presence of the CS alone. Compared with the vehicle and 3 μg injections, injections of 10 μg and 30 μg of CIQ elicited higher freezing levels in the test for fear retention, suggesting that CIQ can enhance fear retention in a dose-dependent manner. Additionally, CIQ was also studied in tests of fear extinction. Immediately following fear extinction training, direct amygdala CIQ infusions (3 μg, 10 μg, and 30 μg) were administered, and 24 hours later, the percentage freezing of the treatment groups was measured in an extinction session that consisted of only the CS. In this test, mice treated with 10 μg of CIQ exhibited lower freezing percentages compared with those receiving the vehicle. Injections of 3 μg and 30 μg gave results similar to the vehicle, and the authors suggest that at 30 μg, CIQ infusions may exist outside of the basolateral amygdala and interaction with either the NMDA receptor or an off-target receptor may neutralize the fear-extinction effects of CIQ. CIQ did not have a significant effect on mice in the Morris Water Maze, suggesting that CIQ treatment is specific to fear learning. CIQ also did not influence locomotor activity or coordinated movement, which could be confounding variables in tests that measure freezing activity.[130]

The results were consistent with those from previous work showing that DSC enhances fear extinction[128] and GluN2C-knockout mice have impaired abilities in fear acquisition,[129] suggesting that CIQ is influencing fear learning by potentiating the NMDA receptor. Whereas the modulation of fear learning may be due to CIQ's off-target interaction with 5-HT_2, 5-HT_6, or the κ-opioid receptor (discussed in Section 13.4), these effects from direct amygdala infusions have been interpreted as a result of CIQ's interaction with the NMDA receptor. Future studies may also include determination of downstream signaling cascades of CIQ allosteric modulation of the NMDA receptor, as well as identifying the neuronal subtypes in the amygdala that CIQ influences. Modulation of the GluN2C-containing NMDA receptor may represent a strategy to alleviate symptoms associated with fear acquisition and fear learning, including anxiety, OCD, and PTSD, and CIQ is one such small molecule that could aid in the development of therapeutics for these disorders.

13.6.2 CIQ as a Pharmacological Probe for Schizophrenia

A number of lines of evidence that support the glutamate hypofunction hypothesis,[98] including genetic linkage studies,[139] and postmortem brain analysis of schizophrenia patients[79,85–87] suggest that the dysfunction of the NMDA receptor may be a contributing factor to the pathophysiology of schizophrenia. If NMDA receptor hypofunction is indeed responsible for some aspects of schizophrenia, one therapeutic strategy to alleviate the negative symptoms of schizophrenia may be to use positive allosteric modulators for NMDA receptors. Furthermore, if an understanding of which NMDAR subtypes contribute to the hypofunction is achieved, a rational and potentially well-tolerated treatment could be devised using subunit-selective positive allosteric modulators. One intriguing hypothesis is that hypofunction of GluN2C- and/or GluN2D-containing NMDA receptors plays an important role in schizophrenic symptoms.[98] If this is the case, one would predict that GluN2C–D-selective positive allosteric modulators could be of utility in schizophrenia. For this reason, the effects of racemic CIQ in animal models of schizophrenia were recently reported.[140] The effect of systemic CIQ on schizophrenia-like behaviors (*i.e.*, prepulse inhibition (PPI), hyperlocomotion, and stereotypy) induced in mice using the NMDA receptor channel blocker MK-801 (0.15 mg kg^{-1}) and the dopamine agonist methamphetamine (1 mg kg^{-1} and 3 mg kg^{-1}) was described. In these behavioral tests, i.p. injection of CIQ was followed (15 minutes later) by i.p. injection of either MK-801 or methamphetamine to test if CIQ treatment could reverse schizophrenia-like behaviors. A dose of 20 mg kg^{-1} CIQ rescued MK-801 (0.15 mg kg^{-1})-induced PPI deficits, but had no effect on the startle amplitude, which had been increased with MK-801 treatment. CIQ alone had no effect on PPI or startle amplitude compared with the vehicle. In contrast, CIQ had no effect on methamphetamine (3 mg kg^{-1})-induced PPI reduction, but CIQ treatment following administration of methamphetamine did influence the startle amplitude. At a dose of 3 mg kg^{-1}, methamphetamine significantly decreased the startle amplitude compared with the vehicle, and CIQ (20 mg kg^{-1}) was able to reverse this effect. It has previously been reported that MK-801, a NMDA receptor antagonist, and methamphetamine, which elevates dopamine levels, elicit different effects on the startle response, suggesting that glutamate signaling and dopaminergic signaling have different effects on the neuronal pathways of the startle response. CIQ has contrasting effects on the behaviors induced by MK-801 and methamphetamine, as well.

This would also imply that the molecular target of CIQ, whether this is the NMDA receptor or an off-target receptor, interacts differently with the neurocircuitry of MK-801-induced and methamphetamine-induced schizophrenia-like behaviors.

It was also reported that CIQ had no significant effect on MK-801-induced hyperlocomotion, but CIQ at a dose of 20 mg kg^{-1} did reduce methamphetamine (3 mg kg^{-1})-induced hyperlocomotion. In terms of stereotypic behaviors, CIQ decreased the stereotypical behaviors brought on by MK-801 (0.15 mg kg^{-1}) and methamphetamine (1 mg kg^{-1}) treatment. CIQ was also tested for its effect on MK-801 memory impairment in a Y-maze spontaneous alternation test, and it was found that CIQ at a dose of 20 mg kg^{-1} significantly attenuated the impairment by MK-801, but the effect of CIQ alone was not significantly different from that of the vehicle. Subtle cognitive decline is a hallmark of schizophrenia, and this data indicates that the same molecular target of CIQ may serve as a target for cognitive decline.[140] While these results are intriguing, more data is needed to elucidate the pharmacokinetics of CIQ in nervous tissue, and the role of potentiation of GluN2C and GluN2D receptors in models of schizophrenia. Furthermore, given data showing off-target actions of CIQ at the κ opioid receptor, the 5-HT_{2A}, the 5-HT_{2B}, and 5-HT_6 receptors, and the norepinephrine receptor (NET),[130] it will be important to confirm if actions of CIQ observed *in vivo* reflect actions at the NMDA receptor. It cannot be ruled out that the reduction of schizophrenia-like behavior could be a result of an interaction between CIQ and an off-target receptor, such as 5-HT_6 at which CIQ can bind with reasonable potency (K_i of 0.51 μM[130]). Studies utilizing GluN2C- and GluN2D-knockout mice or genetic approaches to acutely reduce GluN2C or GluN2D expression may help to elucidate this.

Additional evidence for a potential role of GluN2D can be derived from behavioral studies of GluN2D knockout mice. For example, the dissociative anesthetic ketamine can induce symptoms reminiscent of schizophrenia.[91–94] Ketamine induces increased locomotor activity, 2-deoxy-glucose uptake in prefrontal cortex, and increased power of gamma-band oscillations in wild-type mice.[141–143] Interestingly, the actions of ketamine on locomotor activity[144] and 2-deoxy-glucose uptake[143] were eliminated in GluN2D-knockout mice, and the effects on gamma-band oscillations significantly reduced,[143] raising the possibility that GluN2D hypofyunction may contribute to some of the actions of ketamine that mimic schizophrenia symptoms. In addition, GluN2D knockout mice show decreased motor impairment compared with wild-type mice following systematic PCP treatment (1 mg kg^{-1} and 3 mg kg^{-1}), indicative of GluN2D-containing NMDA receptors being involved in PCP-induced motor impairment.[145] In agreement with results from previous studies, CIQ (20 mg kg^{-1}, i.p. injection) did not alter motor performance in wild-type mice,[140] but CIQ did negatively affect motor performance in GluN2D-knockout mice, suggesting that CIQ is interacting with a receptor other than GluN2D to influence PCP-influenced motor skills.[145]

13.6.3 CIQ as a Pharmacological Probe to Study Parkinson's Disease

One of the hallmark features of Parkinson's disease (PD) is a loss of dopaminergic neurons in the substantia nigra,[146] which leads to the dysfunction of other neurotransmitter systems, including glutamatergic neurotransmission.[63] Interactions between glutamate and dopamine are in part modulated by the NMDA receptor, and GluN2D-containing NMDA receptors are found in the cholinergic interneurons of the striatum.[147,148] These large cholinergic interneurons, despite constituting only 2% of the total neuronal population in the striatum, are implicated in the pathophysiology of PD due to their large axonal branching.[149–151] For this reason, allosteric modulators of the NMDA receptor have been discussed as potential therapeutics for PD,[6] and several reports have examined the effect of CIQ on cholinergic interneurons and dopamine release in the striatum.[152,153] Since GluN2C-containing NMDA receptors are not found in the striatum,[154] these studies have helped to further elucidate the roles of the GluN2D subunit in PD.

The effect of CIQ (20 μM) on both tonic and phasic dopamine release in corticostriatal brain slices from control mice and mice with partial degeneration of dopaminergic terminals in the striatum was examined.[152] During single stimulation pulses meant to mimic tonic activity, CIQ reduced dopamine release, which could be reversed by the GluN2C–GluN2D-preferring competitive antagonist UBP141,[39] suggesting that CIQ's ability to inhibit dopamine release is modulated by the NMDA receptor. The inhibition was also blocked by pirenzepine, a muscarinic receptor antagonist. Dopamine release is affected by acetylcholine (ACh) acting on muscarinic receptors, and the reversal of CIQ's effects in the presence of a muscarinic receptor antagonist implicates a reduction of acetylcholine, presumably released from cholinergic interneurons. To further explore this hypothesis, the effect of CIQ on the firing rate of cholinergic interneurons was determined. Cholinergic neurons exhibited a significant increase in firing following CIQ application, and this increase in firing was not observed in the presence of the NMDA receptor antagonists (2R)-amino-5-phosphonovaleric acid (APV) or UBP141,[153] demonstrating that CIQ activity at cholinergic interneurons is mediated through GluN2D-containing NMDA receptors.

CIQ also inhibited dopamine release during phasic stimulation, but this initial inhibition was followed by long-lasting potentiation. While the inhibition could be reversed by pirenzepine, the potentiation could not be; CIQ may inhibit dopamine release *via* a mechanism that involves muscarinic receptors, but the CIQ-induced potentiation is most probably independent of the muscarinic receptors. Since one off-target effect of CIQ was modest inhibition at nicotinic ACh receptors,[130] CIQ modulation of dopamine release may be dependent on nicotinic ACh receptors. However, the nicotinic receptor antagonist mecamylaime had no effect on the CIQ-induced dopamine depression or potentiation, ruling out the possibility that dopamine modulation is reliant on an interaction between CIQ and nicotinic receptors.

To determine if the effects of CIQ were altered in a mouse model of presymptomatic PD, 6-hydroxydopamine (6-OHDA) was injected into the substantia nigra pars compacta to produce a partially dopamine-depleted striatum. In both stimulations of tonic and phasic release of dopamine, CIQ induced a short-lasting depression in dopamine release followed by a longer-lasting potentiation. Unlike in the control mice, CIQ had no effect on the firing rate of cholinergic interneurons in the lesioned striatum.[152,153] This result suggests that exposure to 6-OHDA in the striatum may cause a loss of expression of GluN2D-containing NMDA receptors in cholinergic interneurons, which is consistent with the results of other studies on the dopamine-depleted striatum.[153,155]

The effect of CIQ on medium spiny projection neurons (MSNs) in the striatum of control mice and 6-OHDA lesioned mice has also been studied.[153] When the spontaneous inhibitory (sIPSCs) and excitatory (sEPSCs) postsynaptic currents in MSNs were measured, CIQ decreased the frequency of sIPSCs, but had no effect on the sIPSC amplitude. The ability of CIQ to depress GABAergic neurotransmission was abolished in the presence of the NMDA receptor antagonist APV, suggesting that this effect is mediated through the NMDA receptor. Scopolamine, a muscarinic receptor antagonist, was able to inhibit the CIQ-induced decrease of sIPSC frequency, implicating that the increase in cholinergic interneuron firing and subsequent release of ACh following application of CIQ plays a role in CIQ-mediated GABAergic synaptic transmission depression.[153] In control mice, CIQ had no effect on the frequency or amplitude of sEPSCs in MSNs. In a mouse model of presymptomatic PD where the striatum was lesioned due to 6-OHDA exposure, the decrease in sIPSC frequency was not observed following CIQ application, but in contrast to control mice, an increase in sEPSCs frequency was recorded in the presence of CIQ. Similarly to control mice, this effect could be blocked by APV application, suggesting that the increase in sEPSC frequency is mediated through an interaction between CIQ and NMDA receptors. Taken together, the results of these studies show that in a dopamine-depleted striatum, GluN2D-containing NMDA receptors are unable to decrease inhibitory transmission to the same degree as in the control animals. GluN2D-containing striatal NMDA receptors in a presymptomatic model of PD also differ from those of the controls by exhibiting a potentiating effect on excitatory inputs.[153]

CIQ has also been used as a tool compound to study EPSCs and spike firing rate of neurons in the subthalamic nucleus (STN),[135] a brain region whose dysfunction has been indicated in PD.[156–160] Symptoms associated with Parkinson's disease, including motor dysfunction, have been correlated with modified firing patterns in the STN caused by loss of substantia nigra pars compacta (SNc) neurons.[161,162] Previously it was shown that NMDA receptors in SNc neurons contain GluN2B and GluN2D subunits, but not GluN2A.[163] Western blotting in rat STN at various time points during development (P7 up to P120) demonstrated the expression of GluN2D-containing NMDA receptors in the STN as well.[135] CIQ, along with the GluN2B-selective inhibitor ifenprodil,[34,52] the GluN2C–D-selective inhibitor DQP-1105,[37] and TCN-201,[35] a GluN2A-selective inhibitor, were used to elucidate which GluN2 subunits were responsible for NMDA receptor currents in the STN. While CIQ potentiated current amplitude of NMDA- and glycine-evoked currents and DQP-1105 and ifenprodil reduced the current, TCN-201 had no effect on the peak amplitude of the current, suggesting that GluN2B- and GluN2D-containing receptors are the primary receptors expressed in the STN and that these subunits contribute to the NMDA-mediated currents.[135] These results are in agreement with those of previous *in situ* hybridization studies.[24,136] CIQ was also used as a pharmacological probe to show that in the STN, GluN2D subunits, along with GluN2B subunits, mediate NMDA receptor synaptic activity. Compared with the vehicle, CIQ increased the peak amplitude of evoked ESPCs. In GluN2D-knockout mice, CIQ had no effect on the evoked currents, suggesting that an interaction between CIQ and the GluN2D subunit is responsible for the observed effect. Direct injections of CIQ into the STN of rats also led to a significant increase in the firing rate of subthalamic neurons, and the magnitude of potentiation was related to baseline firing; the effect of CIQ was greater when baseline firing was low compared with when baseline firing was already high. The effect was reversible and significantly weakened in the presence of MK-801, suggesting that CIQ is again interacting with the NMDA receptor.[135] The results of this *in vivo* study, showing that NMDA receptor modulators can influence the spike firing rate of STN neurons, reveal that the GluN2D subunit is a potential molecular target for STN modulation, and therefore alleviation of PD symptoms.

13.7 Overview of SAR for the Tetrahydroisoquinoline Class of Compounds Selective for the GluN2C- and GluN2D-Containing NMDA Receptors

For SAR modification, CIQ was divided into three rings. Extensive work on the A-ring revealed that a bromo, iodo, trifluromethyl, and phenyl group in the *meta*-position (R_2 in Figure 13.4) yielded the most potent compounds. The amide linker appears to be required for activity as well; replacement of the amide with a urea, thiourea, sulfonamide, or tertiary amine linker led to a virtually complete loss in activity.[72] Modifications to the CIQ scaffold on the B-ring indicate that this ring has tight substituent requirements for potentiation at the GluN2C- and GluN2D-containing receptors; activity was observed only when a methoxy and thiomethyl group existed at the *para* position (R_4), and a thiomethyl group was only active if a chloro-group was at the R_2 position on the A-ring. In terms of the linker between the core and B-ring, the ether linkage was optimal for activity and solubility. A carbon–carbon double bond linker was also active, however the low solubility of these analogues made them difficult to work with.[72] Only a few modifications of the C-ring have been reported, but it is notable that a variant with a single methoxy group at the R_8 position of the ring had similar activity to CIQ at GluN2C- and

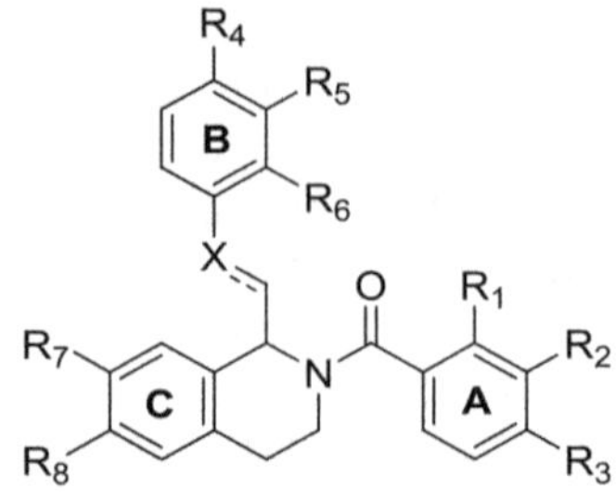

Figure 13.4 Generic structure of CIQ depicting how rings were divided and labeled for SAR development.

GluN2D-containing receptors, suggesting that the methoxy group at the R_7 position is not absolutely essential for potentiation. While a methoxy group in the R_8 position was active, R_8 benzyloxy functionality was more potent, with EC_{50} values at GluN2C and GluN2D ranging from 0.3 μM to 0.4 μM when the A-ring contained a chloro or bromo group in the R_1 position.[72]

The enantiomers of CIQ were synthesized utilizing chiral synthesis, and biological testing revealed that the (+)-enantiomer was active at GluN2C- and GluN2D-containing NMDA receptors, while the (−)-enantiomer was inactive. Although the absolute stereochemistry of the two enantiomers remains undetermined, a stereoselective model[72] for the reduction of the imine utilized to synthesize the enantiomers predicts that the active enantiomer with a (+) optical rotation value is the (*R*)-enantiomer.[164] The stereoselective nature of the CIQ was shared with several related analogues.[165]

13.8 Conclusions

NMDA receptors are ionotropic glutamate receptors that mediate a slow, Ca^{2+}-permeable component of excitatory synaptic transmission. These receptors are fundamentally important for neurological processes such as learning and memory, and their dysfunction has been implicated in a variety of neurological disorders. For this reason, identification of new pharmacological and therapeutic probes that selectively potentiate receptors of specific subunit composition could be of therapeutic relevance, especially in the areas of schizophrenia, anxiety, and cognitive enhancement. One such subunit-selective potentiator is CIQ, the first reported positive allosteric modulator for GluN2C- and GluN2D-containing NMDA receptors. CIQ is a voltage-independent allosteric modulator that causes potentiation of responses to maximally active concentrations of glutamate and glycine by increasing channel opening frequency.[40] The structural determinants of action reside in the pre-M1 linker and the M1 transmembrane domain, and residues residing in the linker between the GluN2 ATD and LBD can also influence the effects of this class of modulators.[40,132]

CIQ has been tested in animal models of fear acquisition and extinction and schizophrenia. Racemic CIQ was directly injected into the amygdala immediately following either fear conditioning or fear extinction training, and CIQ was able to enhance the retention of fear and the extinction of this learned fear.[130] Racemic CIQ was also tested in PPI impairment studies, an animal model for schizophrenia. Systematic administration of CIQ was able to rescue MK-801-induced deficits in PPI, but not when it was methamphetamine-induced. These results suggest that GluN2C- and GluN2D-containing receptors may be a relevant target for schizophrenia, although more data is needed to fully understand the role actions of CIQ and the role of GluN2D in these behaviors.[140] Additional studies have also utilized CIQ as a tool compound to study the dopamine-depleted brain in animal models of Parkinson's disease.[152,153]

Although CIQ and the class of compounds are highly selective for GluN2C- and GluN2D-containing NMDA receptors, they retain important off-target activity, including actions at serotonin and nicotinic acetylcholine receptors. Thus, the effects of direct amygdala infusions of CIQ and systematic administration of CIQ could be influenced by off-target effects, and future studies will be needed to fully understand the role of GluN2D NMDA receptors in these actions. Future work is also needed to better understand the enantioselectivity of this class of compounds, as only the (+)-enantiomer of CIQ is active *in vitro* on GluN2C- and GluN2D-containing NMDA receptors.[72,164] Despite the stereoselectivity of one enantiomer *in vitro*, all literature reports of *in vivo* studies have been conducted with the racemic mixture. Although additional studies are required, this class of positive tetrahydroisoquinoline allosteric modulators appears to be a useful and reliable tool for studying GluN2C- and GluN2D-containing NMDA receptors.

Acknowledgements

This work was supported by NS 036654 (SFT). We thank Dr Daniel Monaghan and Dr Alasdair Gibb for their critical comments on this manuscript.

References

1. S. F. Traynelis, L. P. Wollmuth, C. J. McBain, F. S. Menniti, K. M. Vance, K. K. Ogden, K. B. Hansen, H. Yuan, S. J. Myers and R. Dingledine, *Pharmacol. Rev.*, 2010, **62**, 405–496.
2. J. T. Coyle, *Schizophr. Bull.*, 2012, **38**, 920–926.
3. D. T. Balu, Y. Li, M. D. Puhl, M. A. Benneyworth, A. C. Basu, S. Takagi, V. Y. Bolshakov and J. T. Coyle, *Proc. Natl. Acad. Sci.*, 2013, **110**, E2400–E2409.
4. J. W. Olney, J. W. Newcomer and N. B. Farber, *J. Psychiatr. Res.*, 1999, **33**, 523–533.
5. B. Reisberg, R. Doody, A. Stöffler, F. Schmitt, S. Ferris and H. J. Möbius, *N. Engl. J. Med.*, 2003, **348**, 1333–1341.
6. P. J. Hallett and D. G. Standaert, *Pharmacol. Ther.*, 2004, **102**, 155–174.
7. H. Yuan, C.-M. Low, O. A. Moody, A. Jenkins and S. F. Traynelis, *Mol. Pharmacol.*, 2015, **88**, 203–217.

8. P. Skolnick, P. Popik and R. Trullas, *Trends Pharmacol. Sci.*, 2009, **30**, 563–569.
9. J. W. Murrough, *Clin. Pharmacol. Ther.*, 2012, **91**, 303–309.
10. A. B. MacDermott, M. L. Mayer, G. L. Westbrook, S. J. Smith and J. L. Barker, *Nature*, 1986, **321**, 519–522.
11. H. Furukawa, S. K. Singh, R. Mancusso and E. Gouaux, *Nature*, 2005, **438**, 185–192.
12. M. L. Mayer, G. L. Westbrook and P. B. Guthrie, *Nature*, 1984, **309**, 261–263.
13. I. D. Forsythe and G. L. Westbrook, *J. Physiol.*, 1988, **396**, 515–533.
14. E. Karakas and H. Furukawa, *Science*, 2014, **344**, 992–997.
15. C.-H. Lee, W. Lu, J. C. Michel, A. Goehring, J. Du, X. Song and E. Gouaux, *Nature*, 2014, **511**, 191–197.
16. P. Paoletti, C. Bellone and Q. Zhou, *Nat. Rev. Neurosci.*, 2013, **14**, 383–400.
17. S. G. Cull-Candy and D. N. Leszkiewicz, *Sci. Signaling*, 2004, **2004**, re16–re16.
18. P. Paoletti, *Eur. J. Neurosci.*, 2011, **33**, 1351–1365.
19. S. Vicini, J. F. Wang, J. H. Li, W. J. Zhu, Y. H. Wang, J. H. Luo, B. B. Wolfe and D. R. Grayson, *J. Neurophysiol.*, 1998, **79**, 555–566.
20. K. Erreger, S. M. Dravid, T. G. Banke, D. J. A. Wyllie and S. F. Traynelis, *J. Physiol.*, 2005, **563**, 345–358.
21. B. Siegler Retchless, W. Gao and J. W. Johnson, *Nat. Neurosci.*, 2012, **15**, 406–413, S1–2.
22. M. Watanabe, Y. Inoue, K. Sakimura and M. Mishina, *NeuroReport*, 1992, **3**, 1138–1140.
23. C. Akazawa, R. Shigemoto, Y. Bessho, S. Nakanishi and N. Mizuno, *J. Comp. Neurol.*, 1994, **347**, 150–160.
24. H. Monyer, N. Burnashev, D. J. Laurie, B. Sakmann and P. H. Seeburg, *Neuron*, 1994, **12**, 529–540.
25. L. A. Kehoe, Y. Bernardinelli, D. Muller, L. A. Kehoe, Y. Bernardinelli and D. Muller, *Neural Plast.*, 2013, **2013**, e145387.
26. K. B. Hansen, K. K. Ogden, H. Yuan and S. F. Traynelis, *Neuron*, 2014, **81**, 1084–1096.
27. C. J. Hatton and P. Paoletti, *Neuron*, 2005, **46**, 261–274.
28. C. Rauner and G. Köhr, *J. Biol. Chem.*, 2011, **286**, 7558–7566.
29. J. Luo, Y. Wang, R. P. Yasuda, A. W. Dunah and B. B. Wolfe, *Mol. Pharmacol.*, 1997, **51**, 79–86.
30. M. Gielen, B. S. Retchless, L. Mony, J. W. Johnson and P. Paoletti, *Nature*, 2009, **459**, 703–707.
31. H. Yuan, K. B. Hansen, K. M. Vance, K. K. Ogden and S. F. Traynelis, *J. Neurosci.*, 2009, **29**, 12045–12058.
32. H. S. Chen, J. W. Pellegrini, S. K. Aggarwal, S. Z. Lei, S. Warach, F. E. Jensen and S. A. Lipton, *J. Neurosci.*, 1992, **12**, 4427–4436.
33. H. Hasselmann, *Curr. Neuropharmacol.*, 2014, **12**, 57–70.
34. K. Williams, *Mol. Pharmacol.*, 1993, **44**, 851–859.

35. E. Bettini, A. Sava, C. Griffante, C. Carignani, A. Buson, A. M. Capelli, M. Negri, F. Andreetta, S. A. Senar-Sancho, L. Guiral and F. Cardullo, *J. Pharmacol. Exp. Ther.*, 2010, **335**, 636–644.
36. C. A. Mosley, T. M. Acker, K. B. Hansen, P. Mullasseril, K. T. Andersen, P. Le, K. M. Vellano, H. Bräuner-Osborne, D. C. Liotta and S. F. Traynelis, *J. Med. Chem.*, 2010, **53**, 5476–5490.
37. T. M. Acker, A. Khatri, K. M. Vance, C. Slabber, J. Bacsa, J. P. Snyder, S. F. Traynelis and D. C. Liotta, *J. Med. Chem.*, 2013, **56**, 6434–6456.
38. B. M. Katzman, R. E. Perszyk, H. Yuan, Y. A. Tahirovic, A. E. Sotimehin, S. F. Traynelis and D. C. Liotta, *Bioorg. Med. Chem. Lett.*, 2015, **25**, 5583–5588.
39. B. M. Costa, B. Feng, T. S. Tsintsadze, R. M. Morley, M. W. Irvine, V. Tsintsadze, N. A. Lozovaya, D. E. Jane and D. T. Monaghan, *J. Pharmacol. Exp. Ther.*, 2009, **331**, 618–626.
40. P. Mullasseril, K. B. Hansen, K. M. Vance, K. K. Ogden, H. Yuan, N. L. Kurtkaya, R. Santangelo, A. G. Orr, P. Le, K. M. Vellano, D. C. Liotta and S. F. Traynelis, *Nat. Commun.*, 2010, **1**, 90.
41. S. S. Zimmerman, A. Khatri, E. C. Garnier-Amblard, P. Mullasseril, N. L. Kurtkaya, S. Gyoneva, K. B. Hansen, S. F. Traynelis and D. C. Liotta, *J. Med. Chem.*, 2014, **57**, 2334–2356.
42. J. E. Huettner and B. P. Bean, *Proc. Natl. Acad. Sci.*, 1988, **85**, 1307–1311.
43. H.-S. V. Chen and S. A. Lipton, *J. Neurochem.*, 2006, **97**, 1611–1626.
44. H. S. Chen and S. A. Lipton, *J. Physiol.*, 1997, **499**, 27–46.
45. S. A. Lipton, *NeuroRx*, 2004, **1**, 101–110.
46. S. A. Lipton, *Trends Neurosci.*, 1993, **16**, 527–532.
47. R. D. Mellon, A. F. Simone and B. A. Rappaport, *Anesth. Analg.*, 2007, **104**, 509–520.
48. M. J. Wright, *J. Am. Vet. Med. Assoc.*, 1982, **180**, 1462–1471.
49. C. I. Rodriguez, L. S. Kegeles, A. Levinson, T. Feng, S. M. Marcus, D. Vermes, P. Flood and H. B. Simpson, *Neuropsychopharmacology*, 2013, **38**, 2475–2483.
50. A. Feder, M. K. Parides and J. W. Murrough, *et al.*, *JAMA Psychiatry*, 2014, **71**, 681–688.
51. S. M. Dravid, K. Erreger, H. Yuan, K. Nicholson, P. Le, P. Lyuboslavsky, A. Almonte, E. Murray, C. Mosley, J. Barber, A. French, R. Balster, T. F. Murray and S. F. Traynelis, *J. Physiol.*, 2007, **581**, 107–128.
52. B. L. Chenard, J. Bordner, T. W. Butler, L. K. Chambers, M. A. Collins, D. L. De Costa, M. F. Ducat, M. L. Dumont and C. B. Fox, *J. Med. Chem.*, 1995, **38**, 3138–3145.
53. E. Karakas, N. Simorowski and H. Furukawa, *Nature*, 2011, **475**, 249–253.
54. B. Gotti, D. Duverger, J. Bertin, C. Carter, R. Dupont, J. Frost, B. Gaudilliere, E. T. MacKenzie, J. Rousseau and B. Scatton, *J. Pharmacol. Exp. Ther.*, 1988, **247**, 1211–1221.
55. K. Taniguchi, K. Shinjo, M. Mizutani, K. Shimada, T. Ishikawa, F. S. Menniti and A. Nagahisa, *Br. J. Pharmacol.*, 1997, **122**, 809–812.

56. B. A. Chizh, P. M. Headley and T. M. Tzschentke, *Trends Pharmacol. Sci.*, 2001, **22**, 636–642.
57. L. Yurkewicz, J. Weaver, M. R. Bullock and L. F. Marshall, *J. Neurotrauma*, 2005, **22**, 1428–1443.
58. S. H. Preskorn, B. Baker, S. Kolluri, F. S. Menniti, M. Krams and J. W. Landen, *J. Clin. Psychopharmacol.*, 2008, **28**, 631–637.
59. S. Boyce, A. Wyatt, J. K. Webb, R. O'Donnell, G. Mason, M. Rigby, D. Sirinathsinghji, R. G. Hill and N. M. J. Rupniak, *Neuropharmacology*, 1999, **38**, 611–623.
60. K. L. Nicholson, R. S. Mansbach, F. S. Menniti and R. L. Balster, *Behav. Pharmacol.*, 2007, **18**, 731–743.
61. L. Hoyte, P. Barber, A. Buchan and M. Hill, *Curr. Mol. Med.*, 2004, **4**, 131–136.
62. D. A. Dawson, G. Wadsworth and A. M. Palmer, *Brain Res.*, 2001, **892**, 344–350.
63. K. R. Gogas, *Curr. Opin. Pharmacol.*, 2006, **6**, 68–74.
64. S. Edman, S. McKay, L. J. MacDonald, M. Samadi, M. R. Livesey, G. E. Hardingham and D. J. A. Wyllie, *Neuropharmacology*, 2012, **63**, 441–449.
65. S. McKay, N. Griffiths, P. Butters, E. Thubron, G. Hardingham and D. Wyllie, *Br. J. Pharmacol.*, 2012, **166**, 924–937.
66. K. B. Hansen, K. K. Ogden and S. F. Traynelis, *J. Neurosci.*, 2012, **32**, 6197–6208.
67. K. B. Hansen and S. F. Traynelis, *J. Neurosci.*, 2011, **31**, 3650–3661.
68. T. M. Acker, H. Yuan, K. B. Hansen, K. M. Vance, K. K. Ogden, H. S. Jensen, P. B. Burger, P. Mullasseril, J. P. Snyder, D. C. Liotta and S. F. Traynelis, *Mol. Pharmacol.*, 2011, **80**, 782–795.
69. B. M. Costa, M. W. Irvine, G. Fang, R. J. Eaves, M. B. Mayo-Martin, D. A. Skifter, D. E. Jane and D. T. Monaghan, *J. Pharmacol. Exp. Ther.*, 2010, **335**, 614–621.
70. B. M. Costa, M. W. Irvine, G. Fang, R. J. Eaves, M. B. Mayo-Martin, B. Laube, D. E. Jane and D. T. Monaghan, *Neuropharmacology*, 2012, **62**, 1730–1736.
71. M. W. Irvine, B. M. Costa, A. Volianskis, G. Fang, L. Ceolin, G. L. Collingridge, D. T. Monaghan and D. E. Jane, *Neurochem. Int.*, 2012, **61**, 593–600.
72. R. M. Santangelo Freel, K. K. Ogden, K. L. Strong, A. Khatri, K. M. Chepiga, H. S. Jensen, S. F. Traynelis and D. C. Liotta, *J. Med. Chem.*, 2013, **56**, 5351–5381.
73. A. Khatri, P. B. Burger, S. A. Swanger, K. B. Hansen, S. Zimmerman, E. Karakas, D. C. Liotta, H. Furukawa, J. P. Snyder and S. F. Traynelis, *Mol. Pharmacol.*, 2014, **86**, 548–560.
74. J. Perälä, J. Suvisaari, S. I. Saarni, K. Kuoppasalmi, E. Isometsä, S. Pirkola, T. Partonen, A. Tuulio-Henriksson, J. Hintikka, T. Kieseppä, T. Härkänen, S. Koskinen and J. Lönnqvist, *Arch. Gen. Psychiatry*, 2007, **64**, 19–28.

75. National Institute of Mental Health, *Schizophrenia*, 2016, retrieved on Jan. 25, 2016 from http://www.nimh.nih.gov/health/topics/schizophrenia/index.shtml.
76. A. Kopelowicz, R. Zarate, K. Tripodis, V. Gonzalez and J. Mintz, *Am. J. Psychiatry*, 2000, **157**, 987–993.
77. S. Leucht, G. Pitschel-Walz, D. Abraham and W. Kissling, *Schizophr. Res.*, 1999, **35**, 51–68.
78. M. Ozaki, M. Sasner, R. Yano, H. S. Lu and A. Buonanno, *Nature*, 1997, **390**, 691–694.
79. A. Schmitt, J. Koschel, M. Zink, M. Bauer, C. Sommer, J. Frank, J. Treutlein, T. Schulze, T. Schneider-Axmann, E. Parlapani, M. Rietschel, P. Falkai and F. A. Henn, *Eur. Arch. Psychiatry Clin. Neurosci.*, 2010, **260**, 101–111.
80. H. Stefansson, H. Petursson, E. Sigurdsson, V. Steinthorsdottir, S. Bjornsdottir, T. Sigmundsson, S. Ghosh, J. Brynjolfsson, S. Gunnarsdottir, O. Ivarsson, T. T. Chou, O. Hjaltason, B. Birgisdottir, H. Jonsson, V. G. Gudnadottir, E. Gudmundsdottir, A. Bjornsson, B. Ingvarsson, A. Ingason, S. Sigfusson, H. Hardardottir, R. P. Harvey, D. Lai, M. Zhou, D. Brunner, V. Mutel, A. Gonzalo, G. Lemke, J. Sainz, G. Johannesson, T. Andresson, D. Gudbjartsson, A. Manolescu, M. L. Frigge, M. E. Gurney, A. Kong, J. R. Gulcher and K. Stefansson, *Am. J. Hum. Genet.*, 2002, **71**, 877–892.
81. I. Chumakov, M. Blumenfeld, O. Guerassimenko, L. Cavarec, M. Palicio, H. Abderrahim, L. Bougueleret, C. Barry, H. Tanaka, P. L. Rosa, A. Puech, N. Tahri, A. Cohen-Akenine, S. Delabrosse, S. Lissarrague, F.-P. Picard, K. Maurice, L. Essioux, P. Millasseau, P. Grel, V. Debailleul, A.-M. Simon, D. Caterina, I. Dufaure, K. Malekzadeh, M. Belova, J.-J. Luan, M. Bouillot, J.-L. Sambucy, G. Primas, M. Saumier, N. Boubkiri, S. Martin-Saumier, M. Nasroune, H. Peixoto, A. Delaye, V. Pinchot, M. Bastucci, S. Guillou, M. Chevillon, R. Sainz-Fuertes, S. Meguenni, J. Aurich-Costa, D. Cherif, A. Gimalac, C. V. Duijn, D. Gauvreau, G. Ouelette, I. Fortier, J. Realson, T. Sherbatich, N. Riazanskaia, E. Rogaev, P. Raeymaekers, J. Aerssens, F. Konings, W. Luyten, F. Macciardi, P. C. Sham, R. E. Straub, D. R. Weinberger, N. Cohen and D. Cohen, *Proc. Natl. Acad. Sci.*, 2002, **99**, 13675–13680.
82. D. W. Morris, A. Rodgers, K. A. McGhee, S. Schwaiger, P. Scully, J. Quinn, D. Meagher, J. L. Waddington, M. Gill and A. P. Corvin, *Am. J. Med. Genet., Part B*, 2004, **125B**, 50–53.
83. R. E. Straub, Y. Jiang, C. J. MacLean, Y. Ma, B. T. Webb, M. V. Myakishev, C. Harris-Kerr, B. Wormley, H. Sadek, B. Kadambi, A. J. Cesare, A. Gibberman, X. Wang, F. A. O'Neill, D. Walsh and K. S. Kendler, *Am. J. Hum. Genet.*, 2002, **71**, 337–348.
84. D. J. Gerber, D. Hall, T. Miyakawa, S. Demars, J. A. Gogos, M. Karayiorgou and S. Tonegawa, *Proc. Natl. Acad. Sci.*, 2003, **100**, 8993–8998.
85. M. Beneyto and J. H. Meador-Woodruff, *Neuropsychopharmacology*, 2007, **33**, 2175–2186.

86. S. Akbarian, N. J. Sucher, D. Bradley, A. Tafazzoli, D. Trinh, W. P. Hetrick, S. G. Potkin, C. A. Sandman, W. E. Bunney and E. G. Jones, *J. Neurosci.*, 1996, **16**, 19–30.
87. C. S. Weickert, S. J. Fung, V. S. Catts, P. R. Schofield, K. M. Allen, L. T. Moore, K. A. Newell, D. Pellen, X.-F. Huang, S. V. Catts and T. W. Weickert, *Mol. Psychiatry*, 2013, **18**, 1185–1192.
88. E. D. Luby, B. D. Cohen, G. Rosenbaum, J. S. Gottlieb and R. Kelley, *AMA Arch. Neurol. Psychiatry*, 1959, **81**, 363–369.
89. E. F. Domino, *Int. Rev. Neurobiol.*, 1964, **6**, 303–347.
90. J.-H. Wu, H. Zou, J. Yu, X.-D. Zhou, Q.-L. Xie and M.-L. Jin, *Shengli Xuebao*, 2003, **55**, 381–387.
91. J. H. Krystal, L. P. Karper, J. P. Seibyl, G. K. Freeman, R. Delaney, J. D. Bremner, G. R. Heninger, M. B. Bowers and D. S. Charney, *Arch. Gen. Psychiatry*, 1994, **51**, 199–214.
92. D. Umbricht, L. Schmid, R. Koller, F. X. Vollenweider, D. Hell and D. C. Javitt, *Arch. Gen. Psychiatry*, 2000, **57**, 1139–1147.
93. A. Parwani, M. A. Weiler, T. A. Blaxton, D. Warfel, M. Hardin, K. Frey and A. C. Lahti, *Psychopharmacology (Berlin, Ger.)*, 2005, **183**, 265–274.
94. A. J. Saykin, R. C. Gur, R. E. Gur, P. D. Mozley, L. H. Mozley, S. M. Resnick, D. B. Kester and P. Stafiniak, *Arch. Gen. Psychiatry*, 1991, **48**, 618–624.
95. D. Lodge and M. S. Mercier, *Br. J. Pharmacol.*, 2015, **172**, 4254–4276.
96. J. D. Jentsch and R. H. Roth, *Neuropsychopharmacology*, 1999, **20**, 201–225.
97. T. L. Perry, S. J. Kish, J. Buchanan and S. Hansen, *Lancet London Engl.*, 1979, **1**, 237–239.
98. J. E. Lisman, J. T. Coyle, R. W. Green, D. C. Javitt, F. M. Benes, S. Heckers and A. A. Grace, *Trends Neurosci.*, 2008, **31**, 234–242.
99. G. L. Collingridge, S. J. Kehl and H. McLennan, *J. Physiol.*, 1983, **334**, 33–46.
100. J. Lisman, *Philos. Trans. R. Soc., B*, 2003, **358**, 829–842.
101. J. Z. Tsien, P. T. Huerta and S. Tonegawa, *Cell*, 1996, **87**, 1327–1338.
102. C. Rampon, Y.-P. Tang, J. Goodhouse, E. Shimizu, M. Kyin and J. Z. Tsien, *Nat. Neurosci.*, 2000, **3**, 238–244.
103. Y. Cui, J. Jin, X. Zhang, H. Xu, L. Yang, D. Du, Q. Zeng, J. Z. Tsien, H. Yu and X. Cao, *PLoS One*, 2011, **6**, 1–10.
104. Y.-P. Tang, E. Shimizu, G. R. Dube, C. Rampon, G. A. Kerchner, M. Zhuo, G. Liu and J. Z. Tsien, *Nature*, 1999, **401**, 63–69.
105. X. Cao, Z. Cui, R. Feng, Y.-P. Tang, Z. Qin, B. Mei and J. Z. Tsien, *Eur. J. Neurosci.*, 2007, **25**, 1815–1822.
106. S. A. Jacobs and J. Z. Tsien, *PLoS One*, 2012, **7**, e36387.
107. T. L. White and S. L. Youngentob, *Brain Res.*, 2004, **1021**, 1–7.
108. X.-J. Wang, *J. Neurosci.*, 1999, **19**, 9587–9603.
109. X. J. Wang, *Trends Neurosci.*, 2001, **24**, 455–463.
110. S. A. Monaco, Y. Gulchina and W.-J. Gao, *Neurosci. Biobehav. Rev.*, 2015, **56**, 127–138.
111. C. Allgulander, B. Bandelow, E. Hollander, S. A. Montgomery, D. J. Nutt, A. Okasha, M. H. Pollack, D. J. Stein and R. P. Swinson, World Council of Anxiety, *CNS Spectrums*, 2003, **8**, 53–61.

112. R. E. Stewart and D. L. Chambless, *J. Consult. Clin. Psychol.*, 2009, **77**, 595–606.
113. H. Rodrigues, I. Figueira, R. Gonçalves, M. Mendlowicz, T. Macedo and P. Ventura, *J. Affective Disord.*, 2011, **129**, 219–228.
114. D. H. Barlow, J. M. Gorman, M. K. Shear and S. W. Woods, *JAMA*, 2000, **283**, 2529–2536.
115. J. P. Johansen, C. K. Cain, L. E. Ostroff and J. E. LeDoux, *Cell*, 2011, **147**, 509–524.
116. G. L. Dalton, D. C. Wu, Y. T. Wang, S. B. Floresco and A. G. Phillips, *Neuropharmacology*, 2012, **62**, 797–806.
117. E. P. Bauer, G. E. Schafe and J. E. LeDoux, *J. Neurosci.*, 2002, **22**, 5239–5249.
118. M. J. Miserendino, C. B. Sananes, K. R. Melia and M. Davis, *Nature*, 1990, **345**, 716–718.
119. H. Lee and J. J. Kim, *J. Neurosci.*, 1998, **18**, 8444–8454.
120. C.-H. Lin, S.-H. Yeh, H.-Y. Lu and P.-W. Gean, *J. Neurosci.*, 2003, **23**, 8310–8317.
121. J. D. Baker and J. L. Azorlosa, *Behav. Neurosci.*, 1996, **110**, 618–620.
122. W. A. Falls, M. J. Miserendino and M. Davis, *J. Neurosci.*, 1992, **12**, 854–863.
123. E. Santini, R. U. Muller and G. J. Quirk, *J. Neurosci.*, 2001, **21**, 9009–9017.
124. S. M. Rodrigues, G. E. Schafe and J. E. LeDoux, *J. Neurosci.*, 2001, **21**, 6889–6896.
125. D. L. Walker and M. Davis, *Learn. Mem. Cold Spring Harbor N*, 2008, **15**, 67–74.
126. S. M. Dravid, P. B. Burger, A. Prakash, M. T. Geballe, R. Yadav, P. Le, K. Vellano, J. P. Snyder and S. F. Traynelis, *J. Neurosci.*, 2010, **30**, 2741–2754.
127. A. Sheinin, S. Shavit and M. Benveniste, *Neuropharmacology*, 2001, **41**, 151–158.
128. D. L. Walker, K. J. Ressler, K.-T. Lu and M. Davis, *J. Neurosci.*, 2002, **22**, 2343–2351.
129. B. G. Hillman, S. C. Gupta, D. J. Stairs, A. Buonanno and S. M. Dravid, *Neurobiol. Learn. Mem.*, 2011, **95**, 404–414.
130. K. K. Ogden, A. Khatri, S. F. Traynelis and S. A. Heldt, *Neuropsychopharmacology*, 2014, **39**, 625–637.
131. K. B. Hansen, P. Mullasseril, S. Dawit, N. L. Kurtkaya, H. Yuan, K. M. Vance, A. G. Orr, T. Kvist, K. K. Ogden, P. Le, K. M. Vellano, I. Lewis, S. Kurtkaya, Y. Du, M. Qui, T. J. Murphy, J. P. Snyder, H. Bräuner-Osborne and S. F. Traynelis, *J. Pharmacol. Exp. Ther.*, 2010, **333**, 650–662.
132. K. K. Ogden and S. F. Traynelis, *Mol. Pharmacol.*, 2013, **83**, 1045–1056.
133. R. M. Santangelo, T. M. Acker, S. S. Zimmerman, B. M. Katzman, K. L. Strong, S. F. Traynelis and D. C. Liotta, *Expert Opin. Ther. Pat.*, 2012, **22**, 1337–1352.
134. M. Hedegaard, K. B. Hansen, K. T. Andersen, H. Bräuner-Osborne and S. F. Traynelis, *Neurochem. Int.*, 2012, **61**, 601–609.
135. S. A. Swanger, K. M. Vance, J.-F. Pare, F. Sotty, K. Fog, Y. Smith and S. F. Traynelis, *J. Neurosci.*, 2015, **35**, 15971–15983.

136. D. G. Standaert, C. M. Testa, A. B. Young and J. B. Penney, *J. Comp. Neurol.*, 1994, **343**, 1–16.
137. J. von Engelhardt, C. Bocklisch, L. Tönges, A. Herb, M. Mishina and H. Monyer, *Front. Cell. Neurosci.*, 2015, **9**, 95.
138. A. I. Sobolevsky, M. P. Rosconi and E. Gouaux, *Nature*, 2009, **462**, 745–756.
139. B. Moghaddam, *Neuron*, 2003, **40**, 881–884.
140. P. S. Suryavanshi, R. R. Ugale, D. Yilmazer-Hanke, D. J. Stairs and S. M. Dravid, *Br. J. Pharmacol.*, 2014, **171**, 799–809.
141. G. E. Duncan, S. Miyamoto, J. N. Leipzig and J. A. Lieberman, *Brain Res.*, 1999, **843**, 171–183.
142. S. Miyamoto, J. N. Leipzig, J. A. Lieberman and G. E. Duncan, *Neuropsychopharmacology*, 2000, **22**, 400–412.
143. K. Sapkota, Z. Mao, P. Synowicki, D. Lieber, M. Liu, T. Ikezu, V. Gautam and D. T. Monaghan, *J. Pharmacol. Exp. Ther.*, 2016, **356**, 702–711.
144. Y. Hagino, S. Kasai, W. Han, H. Yamamoto, T. Nabeshima, M. Mishina and K. Ikeda, *PLoS One*, 2010, **5**, e13722.
145. H. Yamamoto, E. Kamegaya, W. Sawada, R. Hasegawa, T. Yamamoto, Y. Hagino, Y. Takamatsu, K. Imai, H. Koga, M. Mishina and K. Ikeda, *Mol. Brain*, 2013, **6**, 56.
146. W. R. Gibb, *Neurol. Clin.*, 1992, **10**, 361–376.
147. S. L. C. Brothwell, J. L. Barber, D. T. Monaghan, D. E. Jane, A. J. Gibb and S. Jones, *J. Physiol.*, 2008, **586**, 739–750.
148. G. B. Landwehrmeyer, D. G. Standaert, C. M. Testa, J. B. Penney and A. B. Young, *J. Neurosci.*, 1995, **15**, 5297–5307.
149. A. Pisani, G. Bernardi, J. Ding and D. J. Surmeier, *Trends Neurosci.*, 2007, **30**, 545–553.
150. P. Bonsi, D. Cuomo, G. Martella, G. Madeo, T. Schirinzi, F. Puglisi, G. Ponterio and A. Pisani, *Front. Neuroanat.*, 2011, **5**, 6.
151. J. A. Goldberg, J. B. Ding and D. J. Surmeier, in *Muscarinic Receptors*, ed. A. D. Fryer, A. Christopoulos and N. M. Nathanson, Springer, Berlin, Heidelberg, 2012, pp. 223–241.
152. X. Zhang, Z.-J. Feng and K. Chergui, *Br. J. Pharmacol.*, 2014, **171**, 3938–3945.
153. Z.-J. Feng, X. Zhang and K. Chergui, *Exp. Neurol.*, 2014, **255**, 154–160.
154. C. Bloomfield, P. O'Donnell, S. J. French and S. Totterdell, *Neuroscience*, 2007, **150**, 639–646.
155. X. Zhang, Z.-J. Feng and K. Chergui, *J. Neurochem.*, 2014, **129**, 581–590.
156. A. A. Kühn, L. Doyle, A. Pogosyan, K. Yarrow, A. Kupsch, G.-H. Schneider, M. I. Hariz, T. Trottenberg and P. Brown, *Brain*, 2006, **129**, 695–706.
157. V. Litvak, A. Jha, A. Eusebio, R. Oostenveld, T. Foltynie, P. Limousin, L. Zrinzo, M. I. Hariz, K. Friston and P. Brown, *Brain*, 2011, **134**, 359–374.
158. P. J. Magill, J. P. Bolam and M. D. Bevan, *Neuroscience*, 2001, **106**, 313–330.
159. P. J. Magill, A. Sharott, M. D. Bevan, P. Brown and J. P. Bolam, *J. Neurophysiol.*, 2004, **92**, 700–714.

160. M.-K. Pan, C.-H. Tai, W.-C. Liu, J.-C. Pei, W.-S. Lai and C.-C. Kuo, *J. Clin. Invest.*, 2014, **124**, 4629–4641.
161. C. Hamani, J. A. Saint-Cyr, J. Fraser, M. Kaplitt and A. M. Lozano, *Brain*, 2004, **127**, 4–20.
162. C. J. Wilson and M. D. Bevan, *Neuroscience*, 2011, **198**, 54–68.
163. S. Jones and A. J. Gibb, *J. Physiol.*, 2005, **569**, 209–221.
164. R. M. Santangelo Freel, K. K. Ogden, K. L. Strong, A. Khatri, K. M. Chepiga, H. S. Jensen, S. F. Traynelis and D. C. Liotta, *J. Med. Chem.*, 2014, **57**, 4975.
165. R. E. Perszyk, J. O. DiRaddo, K. L. Strong, C.-M. Low, K. K. Ogden, A. Khatri, G. A. Vargish, K. A. Pelkey, L. Tricoire, D. C. Liotta, Y. Smith, C. J. McBain and S. F. Traynelis, *Mol. Pharm.*, 2016, in press.

CHAPTER 14

Development of AMPA Receptor Modulators as Cognition Enhancers

CRAIG JAMIESON*[a] AND EMMA L. DUFFY[a]

[a]Department of Pure and Applied Chemistry, University of Strathclyde, 295 Cathedral Street, Glasgow G1 1XL, UK
*E-mail: craig.jamieson@strath.ac.uk

14.1 Introduction

In the mammalian central nervous system (CNS), L-glutamate is the most abundant excitatory neurotransmitter and is the cognate ligand for two receptor families that have distinct functionalities. The first of these families are the metabotropic (G-protein-coupled) receptors which are believed to control slow metabolic responses, whereas the ionotropic (ligand-gated ion channels) are responsible for mediating the fast synaptic response to L-glutamate. From consideration of the endogenous ligands found to activate them, and the pharmacological responses evoked, the ionotropic glutamate receptors can be further stratified into three subclasses, namely the α-amino-3-hydroxy-5-methyl-4-isoxazole-propionic acid (AMPA) receptors, *N*-methyl-D-aspartate (NMDA) receptors, and 2-carboxy-3-carboxymethyl-4-isopropenylpyrrolidine (kainate) receptors.[1,2] The majority of fast excitatory responses evoked in the CNS can be attributed to modulation of the

RSC Drug Discovery Series No. 56
Allosterism in Drug Discovery
Edited by Dario Doller

Published by the Royal Society of Chemistry, www.rsc.org

AMPA receptor, and this protein class is also believed to participate in forms of synaptic plasticity which may encode both learning and memory. Despite the fact that the physiological significance of AMPA receptors in the brain has been widely recognised, there are currently no medicines that have been specifically designed as AMPA receptor modulators employed in the clinical setting. Having stated this, in the last decade there has been an increasing interest in the application of positive allosteric modulators of the AMPA receptor to enhance glutamatergic function whilst potentially avoiding excitotoxic side-effects often associated with direct agonists of ionotropic glutamate receptors. To date, numerous classes of relatively potent and selective AMPA receptor positive modulators have now been described, many of which have appreciable CNS exposure. Compounds from various different chemotypes have demonstrated enhancements in synaptic transmission, and long-term potentiation (LTP), and have been associated with elevation in expression of neurotrophic factors. Based on all of the above, the therapeutic potential of numerous AMPA receptor positive allosteric modulators has been actively examined for a range of conditions such as schizophrenia, depression, Alzheimer's disease, and Parkinson's disease. However, despite considerable activity in the area, only a paucity of compounds (CX516 (Ampalex), CX717, CX691 (farampator) and LY451395) have been progressed as far as phase II clinical trials.

In relation to schizophrenia, treatment with molecules whose primary mechanism of action is dopamine (D_2) antagonism has proved to be effective at ameliorating the positive symptoms of the disease. Although significant advances have been made in the development of atypical antipsychotic agents, a substantial proportion of this patient population still suffer from cognitive deficits associated with the disease. Indeed, it has been posited that the cognitive domain is a core facet of schizophrenia and the lack of appropriate treatment units therefore represents a significant unmet medical need.[3,4] A major additional up-side of any positive allosteric modulator of the AMPA receptor is that it could also be used to treat the cognitive impairment associated with Alzheimer's and other neurodegenerative disorders, highlighting the significant therapeutic potential of this class of agent.

The major focus of this chapter will be to provide an overview of the chemical landscape associated with AMPA receptor positive allosteric modulators, to survey the clinical efforts made with a variety of compounds, and highlight future opportunities in the area. In addition, we aim to connect the chemotypes emerging from this area to the available body of biostructural data being used to inform compound design and understand function.

14.2 Structure and Function of the AMPA Receptors

There are four distinct genes, termed GluA1 to GluA4, which code for the AMPA receptors, and these are known synonymously as GluR1 to GluR4.[5] For each of these subunits, there are two known splice variants labeled 'flip' (i) or 'flop' (o) which differ by less than 10 amino acids in their extracellular domain. Of particular note, is that the asparagine at position 754 is mutated to serine in the flip isoform. In addition, further heterogeneity in the sequence of the receptors is derived from RNA editing, with the most well characterized splice variant being a Gln → Arg mutation within the pore domain of the GluA2 isoform, which confers reduced permeability to calcium.

The functionally active AMPA receptor is a tetramer, and this is usually comprised of heteromeric subunits from GluA1 to GluA4, and through this assembly, calcium ions enter the cell upon receptor activation. The exact composition of the functional heterotetramer assembly varies, however, it can be asserted that each heterotetramer most often contains one or more GluA2 subunits, and this is reasoned to be significant in modulation of permeability to Ca^{2+} cations. Taken as a whole, the differences arising through heteromeric subunit combinations, allied with the heterogeneous nature of AMPA receptor subunit expression in the CNS, RNA editing and splice variant composition undoubtedly gives rise to considerable variability in both pharmacological and biophysical properties. Consequently, this also results in heterogeneous synaptic responses throughout the brain *via* AMPA receptor activation.

Each of the constituent monomers of the receptor assembly is modular in nature (Figure 14.1) consisting of two principal extracellular domains: the N-terminal domain (NTD) and the ligand-binding domain (LBD) which has two sub-domains which are denoted S1 and S2; a transmembrane (TM) domain which is somewhat related to bacterial potassium channels such as KcsA, and a C-terminal domain. Pioneering work from Gouaux and co-workers has enabled the determination of the structure of the intact rat GluA2,[6,7] and illuminates both the overall topology and activity of this class of receptor. The combined structure displays varying levels of symmetry, with four-fold

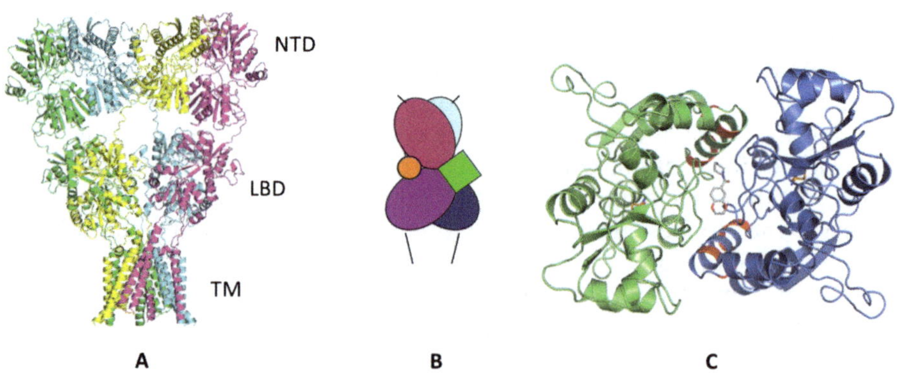

Figure 14.1 A. Full length X-ray crystal structure of the tetrameric GluA2 receptor. B. Schematic view of clam shell arrangement of the ligand binding domain indicating binding sites: yellow, glutamate orthosteric site; green, allosteric site. C. X-ray crystal structure of isolated GluA2 S1S2J LBD showing binding of compound **3** at the interface between the protein dimers.

symmetry observed in the transmembrane regions but a two-fold, dimer-of-dimers configuration in the extracellular domains, giving rise to C_2-symmetry in the LBD itself.

X-ray structures of individual GluA2–4 receptor domains have also been obtained,[8–10] and additionally, there have been a series of studies of the GluA2 LBD, which include determination of interactions between this isoform and compounds with varying mechanisms of action,[8,11–13] an example of which is shown in Figure 14.1C. The specific topology of the LBD and interactions with positive allosteric modulators is further elaborated on in Section 14.4. The AMPA receptor LBD is crucial in controlling pharmacological specificity in addition to tuning the important biophysical parameters of desensitisation and deactivation (*vide infra*). This fact became apparent from studies involving exchange of LBD moieties between the AMPA GluA3 and kainate GluA6 where the respective receptors also exchanged their associated pharmacological characteristics.[14,15]

As stated above, two key functional properties of AMPA receptors which control both the duration and amplitude of synaptic responses to glutamate are desensitisation and deactivation. Desensitisation is a phenomenon whereby an ion channel closes despite agonist remaining bound to the receptor. In contrast to NMDA receptors, desensitisation takes place within approximately 10 ms of receptor activation by agonist. A greater understanding of the precise biophysics controlling desensitisation of the AMPA receptor has been enabled through X-ray crystallographic studies with a soluble S1S2 ligand binding domain construct,[8] and more recently on the full length GluA2 receptor.[16] Data from this study indicates that rearrangements in receptor subunit interactions uncouple the ion channel gate from the ligand binding domain following binding of agonist.

In contrast, deactivation results in inactivation of the channel following dissociation of orthosteric agonist (glutamate), and is more usually represented as the decay of agonist-induced current following removal of agonist. Again, in contrast to NMDA receptors, this takes place in a very rapid timeframe (approximately 1–2 ms), and most probably is attributable to the comparatively low affinity that glutamate has for the receptor. Both of these processes can be tuned through action of allosteric modulators of the AMPA receptor through either affecting agonist binding, or the conformational changes associated with receptor gating and/or desensitisation.[17]

Crystallographic evidence suggests that ligand binding to the orthosteric site results in changes in conformation which affect the transmembrane domain, resulting in channel opening.[6,18,19] More precisely, agonist binding results in the LBD clamshell closing, and the resulting force from this conformational change prompts opening of the pore through mechanical propagation. Following channel opening, receptor desensitization results when disruption of the interface between the two closed agonist-bound clamshells within a functional dimer, which become dissociated, resulting in loss of the force necessary to maintain the channel in the open state. Recently, pioneering studies from Gouaux and co-workers has given further insight into the specific conformational changes that occur.[16] Significant changes in the quaternary structure of the receptor take place, with the amino-terminal domain being directed toward the LBD, resulting in overall compression of the protein assembly. This results in the LBD becoming flattened, exerting a mechanical force on the transmembrane domains leading to the channel closing. Clearly, this level of understanding of the structural features which encode the function of the receptor can significantly facilitate the design and development of allosteric AMPA modulators located at the dimer interface, as based on the structural changes intimated above, compounds which are able bind and stabilize this protein–protein interaction (Figure 14.1C) have the potential to block both desensitization and deactivation, and thereby enhance AMPA-evoked currents.[12,13,20]

From consideration of the above, it is therefore apparent that control of the crucial properties of desensitisation and deactivation of the AMPA receptor can be achieved through compounds which act as positive allosteric modulators. Slowing or inhibition of the processes of deactivation will result in prolongation of glutamate-evoked currents arising from the AMPA receptor, and this is anticipated to relieve NMDA receptors from Mg^{2+}-associated block through membrane depolarisation. Many synapses contain only NMDA receptors, and blocking by Mg^{2+} renders these systems unable to contribute to post-synaptic transmission, hence the label silent synapses.[21] Recruitment of AMPA receptors to the post-synaptic membrane is believed to activate silent NMDA receptors, allowing them to participate as functional components of neural networks.[22] This aspect of AMPA receptor pharmacology is believed to represent one of the mechanisms underlying forms of synaptic plasticity such as long-term potentiation, which is defined as a use-dependent enhancement in synaptic function that is thought to encode both learning and memory formation.[23,24]

Accordingly, this approach may have utility in a therapeutic setting for a range of CNS disorders linked to attenuated glutamatergic signalling, and in particular, cognitive impairment associated with schizophrenia.

14.3 Chemical Classes of AMPA Receptor Positive Allosteric Modulators

In the past two decades, an emerging body of compounds covering a variety of chemotypes has emerged from both the chemical and patent literature, with that number expanding considerably in recent years, reflecting the heightened level of interest in the area. Reporting and associated interpretation of biological data associated with AMPA receptor modulators is frequently complicated through the use of varying assay formats in addition to different measures of efficacy. The primary aim of this section is to provide an overarching view of the known classes of AMPA receptor modulators. The principal focus will be on the four predominant chemical classes that have emerged, as well giving an overview of other notable compounds of interest from the point of pharmacology.

Among the first series of AMPA receptor positive modulators to be identified were the so-called benzamide derivatives (Figure 14.2). This compound class was discovered by researchers based at the University of California through modification of the known nootropic agents **1** (piracetam)[25] and **2** (aniracetam).[26] In initial studies, compound **1** had been shown to potentiate AMPA-induced $^{45}Ca^{2+}$ uptake into neuronal cultures when tested at 10 μM. Compound **2** was evaluated at rat AMPA receptors and demonstrated to enhance steady-state currents evoked by L-glutamate, with a relatively low (1 mM) potency. In addition, benzamide **2** also exhibited a degree of selectivity for the flop *vs.* flip isoform of the receptor.

Notable progenitors of the benzamide series include **3** (CX546)[27,28] and **4**.[27–29] Like compound **2**, these analogues were shown to be relatively low potency potentiators of glutamate-evoked steady-state currents, with measured concentration giving 50% of the maximum effect (EC_{50}) values generally in excess of 100 μM. Whole-system *ex vivo* slice electrophysiology has also been extensively described for compounds from the benzamide class, through the recording of excitatory post-synaptic currents (EPSCs), which gives an indication of the efficacy of a particular compound in the context of intact neuronal circuitry. Despite their comparatively low tonic potency, both compounds **3** and **4** are able to slow the deactivation time of the receptor 10-fold and 3.2-fold in native rat membranes, respectively. In addition to

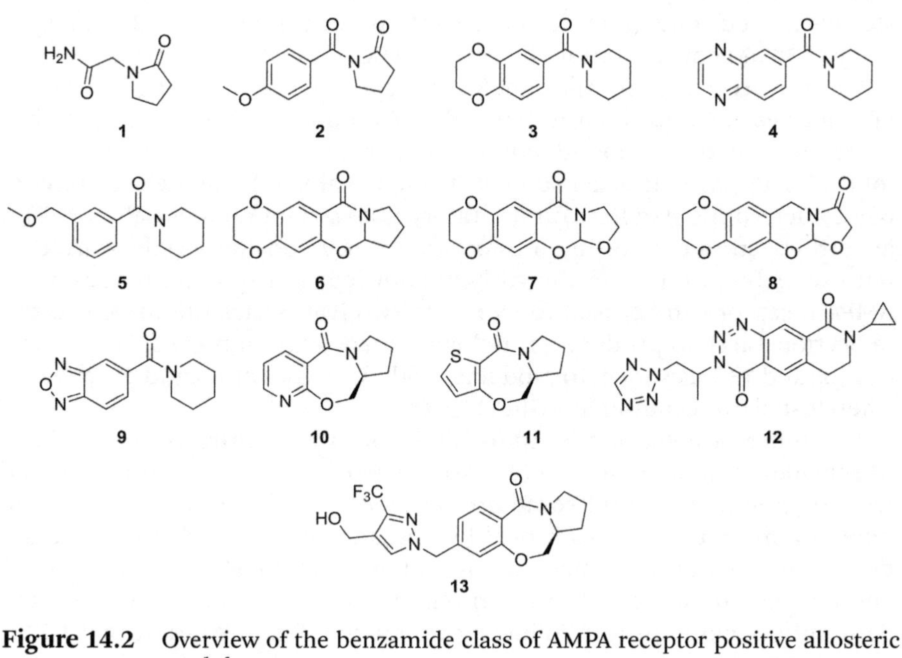

Figure 14.2 Overview of the benzamide class of AMPA receptor positive allosteric modulators.

its effects on deactivation, compound **4** also exhibits increases in EPSC half-width of 28% and amplitude of 68% when evaluated at a concentration of 2 μM. When evaluated using slice electrophysiology at a concentration of 30 μM, compound **3** is reported to increase the slice EPSC amplitude and half-width by approximately 40% and 80%, respectively, and to slow the deactivation time constant by a factor of 8.4.

Reduced-complexity analogues of compound **3** have also been claimed to have been developed by researchers at Cortex. For example, Compound **5** is claimed to increase the slice EPSC amplitude by 25% at a concentration of 300 μM.[30] In addition, applying conformational restraint to compound **3** furnished the comparatively more potent analogue **6**.[31,32] In a related fashion, constrained analogues derivatives **7** (CX614)[11,33] and **8**[34] are claimed to increase the slice EPSC amplitude by 25% when tested at concentrations of 30 μM and 300 μM, respectively. Compound **7** also slows both receptor deactivation and desensitisation when tested at 100 μM concentration and has been shown to be selective for flop rather than flip splice isoforms.

Further modifications of benzamide **4** resulted in researchers at Cortex producing the more potent analogue **9** (Org 24448, farampator).[29,35] This compound has been licensed to Organon (now Merck & Co.). Testing using patch clamp electrophysiology indicates that **9** is approximately an order of magnitude more potent than **4**. Additional pharmacological evaluation revealed that compound **9** can also increase the slice EPSC amplitude by 10% when tested at 3 μM concentration. The related benz- and pyrido/thienyl-oxazepinone series of compounds have been disclosed by workers at Organon (now Merck). Analogues **10** (Org 26576)[36] and **11**[37] are reported to enhance steady-state currents in rat hippocampal neurons by 20% and 32%, respectively at 30 μM concentration, however, no EC_{50} values were reported.

The more recent benzamide analogues reported include **12**, again from the laboratories of Cortex Pharmaceuticals.[38] Compound **12** has an EC_{2x} activity (concentration of compound required to increase steady-state current by two fold) of 0.46 μM and is active *in vivo* when evaluated using electrophysiology to determine its effect on excitatory post-synaptic potentials (EPSP) in anaesthetized rats following intraperitoneal (*i.p.*) dosing. Lastly, work from our own laboratories has shown how knowledge of binding modes at the AMPA receptor can be used to hybridize two lead series (discussed further in Section 14.4), to produce a novel chemotype with improved properties.[39] Compound **13** was shown to modulate both desensitization and deactivation when tested on recombinant GluA2 receptors.

A number of compounds from this class to have progressed to clinical use including **4**, **9**, **10** and CX717 (structure not disclosed). In initial studies, Compound **4** showed some promise with modest effects on cognition noted in phase I studies with healthy human volunteers.[40,41] However, later double-blind studies in schizophrenic patients did not show any therapeutic effect, when employed either as a monotherapy[42] or as an adjunct[43] to standard antipsychotic drugs. Following up on the demonstration that CX717 was able to enhance cognitive performance in sleep-deprived primates,

Cortex subsequently reported that the compound was able to surmount deficiency in alertness when trialed in sleep-deprived healthy young men, and showed some reversal of associated cognitive deficits. Despite this encouraging initial data, the positive effects on cognition could not be replicated in a double-blind placebo-controlled study in volunteers participating in simulated night shift work.[44] More promising clinical data around this asset have been reported by Cortex, including a phase IIa study in subjects with attention deficit–hyperactivity disorder (ADHD) where CX717 was demonstrated to have a statistically significant effect on the symptoms. Further phase IIa studies with CX717 have demonstrated that the compound is efficacious at protecting against respiratory depression induced by an opioid, without compromising the opioid's pain-relieving effects.[45] Following positive results on short-term memory in phase I studies,[29] Compound **9** (Org 24448) was progressed to clinical trials for schizophrenia and major depressive disorder,[46] however, most of these trials were either halted or the study terminated before commencement. Org 26576 successfully completed phase I studies in healthy volunteers,[47] and also proceeded to phase II clinical trials for major depressive disorder although no data has been reported for this study.

A second important class of AMPA positive allosteric modulator that has been evolved are the benzothiadiazine systems (Figure 14.3). The progeny of these compounds can generally be traced to **14** (cyclothiazide, CTZ), which was developed originally as a diuretic agent, however, has since been extensively profiled as an AMPA receptor positive allosteric modulator, with an EC_{2x} of 1.6 μM when measured in *Xenopus* oocytes transiently expressing GluA1.[48] The *in vivo* utility of this compound is, however, somewhat limited due to its lack of exposure in the CNS. Biostructural data detailing the binding mode of **14** to the AMPA ligand binding domain is discussed in Section 14.4. Perhaps a more useful member of this compound class is **15** (S-18986) which was developed by Servier. Compound **15** was shown to have an EC_{2x} of 60 μM[49] on the current induced by AMPA when added to *Xenopus* oocytes. The Servier group demonstrated that only the (*S*)-enantiomer was biologically active as a positive allosteric modulator of the AMPA receptor. Despite its comparatively lower potency compared with the progenitor **14**, compound **15** has measurable exposure in the CNS, thus facilitating a range of *in vivo* studies.[50] These include the novel object recognition test in rodents, which demonstrated that when dosed at 0.3 to 3 mg kg^{-1} orally compound **15** improved the retention of memory, and at the 0.3 mg kg^{-1} dose was able

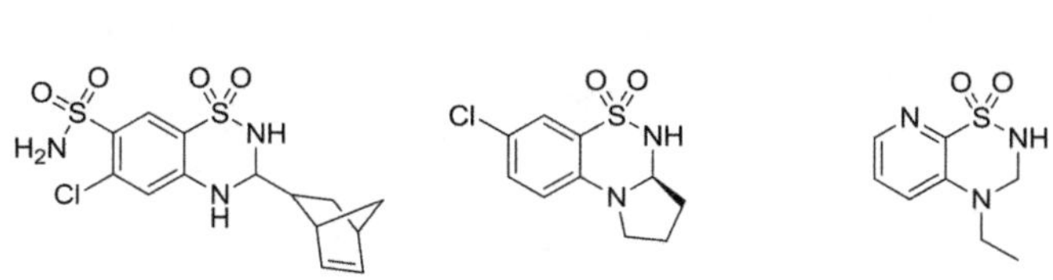

Figure 14.3 Representative exemplars of the benzothiadiazine series.

to surmount age-related deficits in memory.[51] Compound **15** has also been evaluated clinically,[46] although no data has been reported for this study. Subsequent investigations have shown that closely related analogues of **15** have been shown to have superior levels of potency (*e.g.* **16** EC_{2x} = 8.8 μM).[48] It has yet to be established if this enhanced potency results in improvements in efficacy *in vivo* using pharmacodynamic models of cognition.

A third significant class of AMPA receptor positive allosteric modulator that has been reported are the so-called phenethyl sulfonamide derivatives (Figure 14.4). The original examples in this class of AMPA positive modulators were first disclosed by Eli Lilly & Co. in 2000.[52] The progenitor compound in this alkyl-sulfonamide series (**17**) was identified from a high-throughput screening campaign using a recombinant assay in which compounds were assessed for their ability to potentiate the response mediated by 100 μM L-glutamate in a stable cell line expressing homomeric GluA4 flip receptors as determined by a Ca^{2+} fluorescence assay. In this screen, compound **17** exhibited an EC_{50} of 19.6 μM, which was only fivefold less than one of the

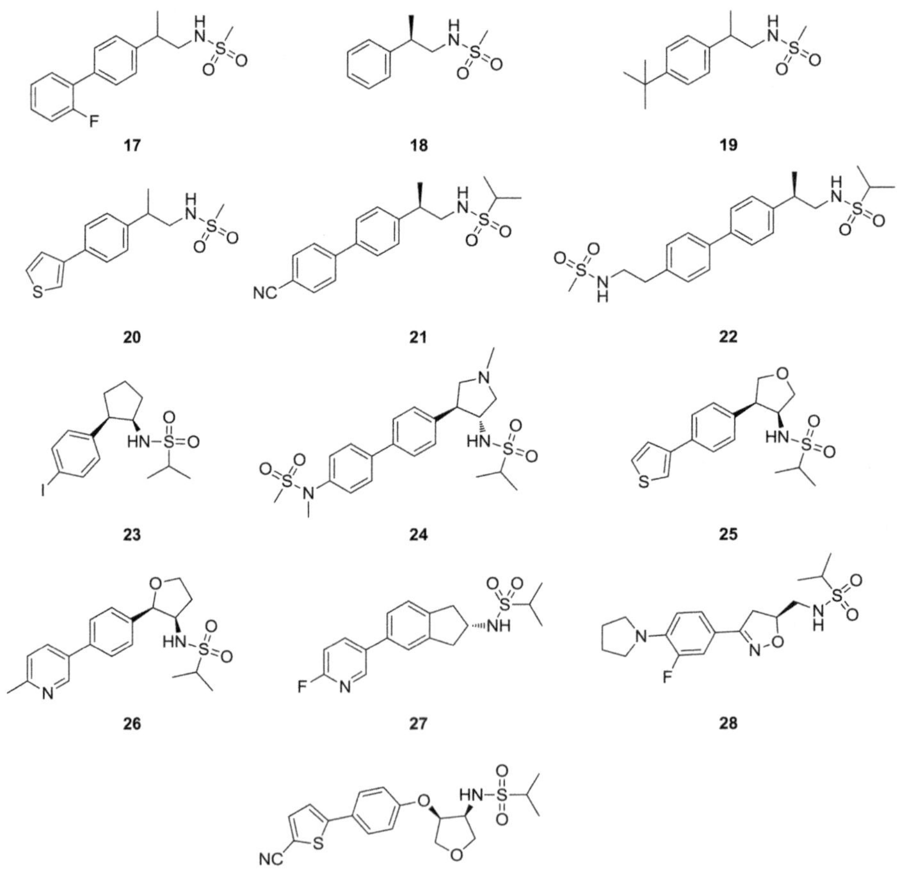

Figure 14.4 Members of the phenethyl sulfonamide chemotype.

most active benzothiadiazine compounds reported at the time (cyclothiazide, **14**), and by comparison, displayed significantly higher activity as an AMPA receptor positive allosteric modulator than that previously observed for benzamide compounds such as **4**, although there are differences in the precise assay format employed in each case. Preliminary work to enhance the potency of compound **17** indicated that the structure–activity relationship (SAR) was somewhat flat in the sulfonamide region of the molecule. This is evidenced by the fact that the ethyl- and iso-propyl- analogues displayed approximately a four-fold improvement in potency relative to the progenitor methyl sulfonamide **17**. In contrast, the more sterically demanding analogues bearing *n*-butyl and benzyl groups were found to be inactive when tested in the same assay (EC_{50} > 100 μM). Still focusing on the sulfonamide region, an incremental increase in the sterics of the benzylic substituent confirmed that a methyl group was preferred in this position. The precise substitution pattern around the biaryl moiety was also found to be important in controlling activity at the AMPA receptor, however, the results of SAR studies indicated that this region is somewhat more tolerant to structural changes. For example, although the deletion analogue **18** showed a significant loss of activity (EC_{50} = 1980 μM),[13] exchanging the *o*-fluorophenyl group in **17** with replacement groups such as a *tert*-butyl **19** and 3-thiophenyl **20** provided approximately a fivefold to tenfold increase in potency (EC_{50} = 1.2 μM and EC_{50} = 4.5 μM, respectively). In combining the effect of substitution around the biaryl ring system with the preferred iso-propyl sulfonamide functionality, the group at Lilly were able to generate a 100-fold enhancement in potency for analogues containing methyl, formyl, amino, or nitrile groups in the 4′-position of the biaryl system.

As one of the most active AMPA positive allosteric modulators reported in the literature at the time with an EC_{50} of 0.29 μM when tested at homomeric GluA4 receptors, the cyano derivative **21** was further examined using a raft of both *in vitro* and *in vivo* assays, and on the basis of the results of these studies, has become a useful tool compound in validation of the role of AMPA receptors in neurological disorders.[53] Compound **21** was shown to be essentially equipotent at GluA2 with an EC_{50} of 0.15 μM, however, interestingly it was shown to be relatively selective with respect to the remaining two isoforms, GluA1 (EC_{50} = 5.65 μM) and GluR3 (EC_{50} = 1.66 μM).[54] In addition to displaying some isoform selectivity, compound **21** was shown to have around tenfold selectivity for the flip splice variant compared with the flop form of the receptor (GluR2o EC_{50} = 1.44 μM). Wider selectivity testing against a panel of ion channels, including other ionotropic glutamate receptors (*e.g.* kanaite or NMDA), and voltage-gated potassium, sodium, or calcium channels indicated that the compound was selective as an AMPA positive allosteric modulator.[55] In relation to its specific mechanism of action in potentiating glutamate-evoked currents, whole-cell patch-clamp electrophysiology experiments using rat prefrontal cortex neurons studies suggested that **21** potentiates AMPA receptors by acting as a blocker of desensitization, consequently maintaining the channel in an agonist-bound open state.[53]

The high potency observed in the recombinant assay using homomeric GluA4 receptors also resulted in potentiation of AMPA receptors when using *ex vivo* preparations, and taken as a whole, these studies paved the way towards determining the efficacy of the compound in mechanistically relevant *in vivo* models. For instance, *ex vivo* studies using Purkinje neurons in patch-clamp electrophysiology experiments have indicated that compound **21** is over three orders of magnitude more potent than well characterized benzamide modulators such as **4**.[56] In the *in vivo* setting, again using electrophysiological techniques, compound **21** demonstrated an increase in the firing rate of rat hippocampal neurons, with a measured ED_{50} of 12 μg kg^{-1}.[57] Compound **21** has also been tested pre-clinically in a variety of pharmacodynamic models aimed at evaluating its effect on various aspects of cognition. For example, **21** showed positive effects in animal models of working memory such as the Morris water maze and passive avoidance tests.[53] The positive outcomes of these studies provide strong validation of the connection between hypofunction of glutamate receptors and cognitive impairments associated with schizophrenia. In other pre-clinical studies it has been shown that sub-chronic treatment with **21** resulted in an increase in expression of Brain Derived Neutrotrophic Factor (BDNF) in rats. BDNF is neural growth factor, and reduced levels of this protein have been correlated with depressive disorders, potentially indicating that compound **21** may have anti-depressant properties.[58]

Based on the highly encouraging *in vitro* and *in vivo* data, further optimization efforts around compound **21** were initiated and this culminated in the identification of a related bis-sulfonamide series, from which compound **22** (LY451395) has progressed into clinical studies.[59,60] When evaluated in pre-clinical animal models of cognition, compound **22** showed encouraging levels of activity in behavioral paradigms including the Morris water maze and retention in the radial arm maze, and through these studies was shown to have a very similar *in vivo* profile to that of compound **21**.[58] When evaluated clinically, **22** was shown to be safe and well tolerated in human volunteers at doses of up to 5 mg.[61] However, in a phase II clinical study on 180 patients with Alzheimer's disease, **22** did not have any demonstrable positive effect on cognition.[62] This report indicated that the lack of efficacy may be attributable to the low doses of the compound employed (0.2 mg twice daily for 28 days in the first instance, then 1.0 mg twice daily to a maximum period of 8 weeks). A second clinical trial has been reported, with higher doses (up to 3 mg twice daily, less if not tolerated) employed in patients with Alzheimer's. The primary end-point in this case was to establish if the test compound had a positive effect in alleviating the symptoms of aggression and agitation in the patient cohort. Although the study was completed in 2011, no data has been published to indicate the success or otherwise of this trial.

Following on from the identification of LY451395, interest has continued in further optimization around the phenethyl sulfonamide chemotype. Workers at Lilly developed a range of cyclopentyl-containing analogues, which were up to 10-fold more potent than the corresponding acyclic systems.[63] For example, the *iodo* analogue (*R*,*R*) **23** which had an EC_{50} of 23 nM was the most potent modulator among of the AMPA receptor from the 29 examples detailed in this report.

Further related series of sulfonamide analogues have also been reported by researchers at both GlaxoSmithKline (GSK) and Pfizer. The GSK efforts centered around incorporation saturated heterocyclic ring systems, potentially in an effort to optimize physicochemical properties such as Log*P* (the log of the partition coefficient). From these studies, AMPA allosteric modulators such as **24** (EC_{50} = 1.6 μM), **25** and **26** emerged.[64–66]

In a parallel lead identification approach, GSK adopted a different cyclisation strategy which compliments the original Eli Lilly hypothesis, culminating in the discovery of an indane-derived series, exemplified by the sulfonamide analogue **27** (EC_{50} = 2.5 μM).[67–69] From the SAR data disclosed by GSK, it appears that the indane-derived systems are generally of lower potency at the AMPA receptor in comparison with the cyclopentane-derived systems reported by Lilly. When tested using electrophysiology, compound **27** exhibited robust potentiation of the AMPA receptor evoked current using recombinant GluA2(i) cell lines. Further pharmacological and developability profiling of **27** (GSK729327) revealed that it was selective against five common P450 isoforms, as well as having low intrinsic clearance in both rat and human microsomal preparations. Extended selectivity profiling against a panel of ion channels (including kainate, NMDA, and human ether-à-go-go related gene (hERG)), enzymes and GPCRs showed that the compound was a selective modulator of the AMPA receptor. *In vivo* drug metabolism and pharmacokinetics (DMPK) studies in rat showed low (5.4 ml min^{-1} kg^{-1}) clearance, allied with an excellent half life ($T_{1/2}$) of 5 hours and high oral exposure with a bioavailability of 61%. Additionally, the compound was shown not to be a P-glycoprotein substrate and had good CNS exposure, with a brain : plasma ratio of 2.1. Given the promising developability of the compound, **27** was tested in rodent models of cognition, including the Novel Object Recognition test, where it was found to have an MED (minimum effective dose) of 0.3 mg kg^{-1}. Based on all of the above, GSK729327 was selected as a clinical candidate for the treatment of cognitive deficits associated with schizophrenia. In phase I studies, compound **27** was well tolerated using both single and rising doses in the range 0.25–6 mg. Clearance of the compound was found to be low, resulting in relatively long half-lives of 107–168 h on average. No further development activity has been reported for this compound.

As stated above, Pfizer have also indicated an interest in the design and development of AMPA receptor positive allosteric modulators. Two lead molecules have come to prominence through these efforts, the first being PF-04725379 (**28**) which was identified following high throughput screening (HTS) and subsequent hit-to-lead effort which was expedited using Structure Based Drug Design (SBDD), exploiting the AMPA S1S2J LBD as a workhorse construct.[70] Compound **28** was a relatively potent modulator of the AMPA receptor, with an EC_{50} of 123 nM and was found to be CNS-penetrant, with a robust therapeutic index of around 500-fold between pro-convulsant activity and efficacy in models of cognitive impairment relevant to schizophrenia.

The second notable compound from the Pfizer portfolio is **29** (PF-04958242) which was identified using a scaffold-hopping approach from the biarylsulfonamide template reported by Lilly.[71] Again, the optimisation efforts were guided significantly through use of SBDD to determine the specific interactions between the compound and the AMPA LBD. Compound **29** had an EC_{50} at the AMPA receptor of 310 nM, showed low intrinsic clearance in human liver microsomal preparations, good cell permeability (A–B $P_{app} = 19.3 \times 10^{-6}$ cm s^{-1} in RRCK cells where P_{app} is the apparent permeability) and no evidence of efflux from MDR1-MDCK cells. Wider selectivity profiling showed that **29** had no significant off-target effects, and evaluation in rat cortical neurons using electrophysiology confirmed the compound acted as a robust potentiator of AMPA-evoked currents, with an EC_{50} of 43 nM). From an *in vivo* perspective, **29** was shown to be readily brain-penetrant and could reverse deficits in cognition induced through application of NMDA blockers such as MK-801. Target engagement in the brain was confirmed using a biomarker approach; compound **29** showed a dose-dependent effect on 2-[^{18}F] fluoro-D-glucose (FDG) uptake in brain regions associated with pro-cognitive effects. DMPK studies indicated a relatively high clearance rate in rat (169 ml min^{-1} kg^{-1}), although this was found to be lower in dogs (11 ml min^{-1} kg^{-1}) with concomitant enhancement of $T_{1/2}$. Allometric calculations were used to project to human pharmacokinetics, which suggested low to moderate clearance with a predicted half-life of 2.3 h and a low efficacious dose of 0.3 mg (twice daily). Following additional toxicology studies, **29** was progressed to clinical trials, where it has been examined in a total of 11 studies[46] including in healthy volunteers, and in schizophrenic patients being treated with standard antipsychotic therapies. The current clinical focus for PF-04958242 is in demonstrating efficacy in treating cognitive impairments associated with schizophrenia.

The fourth major class of AMPA receptor positive allosteric modulator to emerge are compounds largely based on the 3-trifluormethyltetrahydroindazole motif (Figure 14.5). These derivatives were contemporaneously reported by both GlaxoSmithKline and our own laboratories, with the initial exemplars identified through HTS efforts and the subsequent hit-to-lead campaigns being significantly expedited through application of SBDD.

Ward and co-workers at GSK[72] have reported on the identification of compound **30** which was selected as a clinical candidate. Following HTS, the optimisation efforts focused on balancing metabolic stability and potency, which were two issues identified when profiling early members of the emerging lead series. Areas of the molecule examined included incorporation of heteroatoms in to the saturated ring of the tetrahydroindazole, variation of the spacer and nature of the linker attached to the phenyl ring system, as well as replacement of the pyrrolidine portion. As stated above, much of the work was guided through the availability of crystal structures which greatly facilitated identification of areas of the molecule which could be modified without compromising binding to the AMPA LBD. In this way, prioritisation of

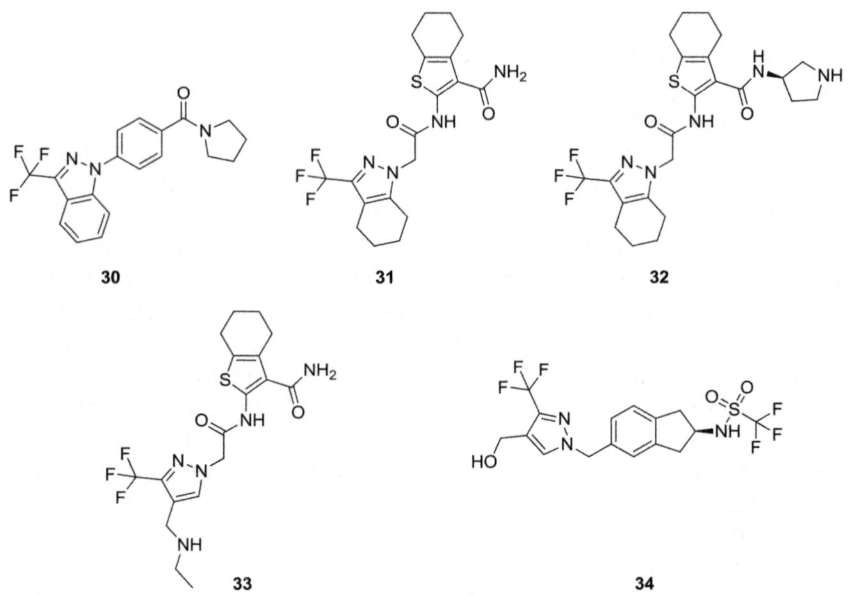

Figure 14.5 3-Trifluoromethyltetrahydropyran-derived systems.

compounds for synthesis could be effectively managed. Overall, compound **30** had the best balance of potency (EC_{50} = 10 μM), *in vitro* metabolic stability, and overall developability properties, and, accordingly, was selected for further evaluation. Electrophysiological studies showed robust potentiation of AMPA-evoked currents both in recombinant (GluA2) and native (rat cortical neuron) cells, with significant potentiation of steady-state currents noted at concentrations as low as 100 nM. From a rat DMPK perspective, compound **30** had moderate blood clearance (Cl_b = 40 ml min^{-1} kg^{-1}) and oral bioavailability (54%), with a low to moderate half-life (1.4 h, intravenous (*i.v.*); 2.3 h, oral). Results from subsequent allometric studies indicated that clearance in humans was likely to be low to moderate. Efficacy in a number of cognitive paradigms was demonstrated, including in the rat Novel Object Recognition model, where the compound was shown to have an MED of 0.3 mg kg^{-1}, accompanied with good levels of exposure in the brain. Results of selectivity profiling against a broad panel of ion channels, receptors and enzymes, including important P450 isoforms and hERG, suggested the compound was selective as an AMPA receptor modulator and further safety evaluation, including assessment of genotoxicity and 4-week rat toxicology studies, revealed no issues. Based on all of the above data, compound **30** was nominated as a clinical development candidate for the treatment of cognitive impairment associated with schizophrenia.

Work from our own group has also centred on a similar chemotype, with the identification and evolution of the chemical matter embracing a similar strategy, namely application of HTS followed by structure-guided hit-to-lead optimisation.[73] Starting from HTS hit **31** (EC_{50} = 200 nM), the substituted amide derivative **32** was identified; the initial hit optimisation efforts focused on improving aqueous solubility of **31** (measured as being <1 mg ml^{-1}) and to this end, the (*S*)-3-aminopyrollidine amide unit was installed. The design of this solubilising group was influenced by knowledge of ligand binding to the allosteric site which indicated that access to a hydrophilic pocket was feasible. Amide **32** had good levels of activity at the AMPA receptor (EC_{50} = 250 nM), improved aqueous solubility in comparison with **31** (measured as 20 mg ml^{-1}) allied with good *i.v.* pharmacokinetic data (Cl_b = 2.3 ml min^{-1} kg^{-1}; $T_{1/2}$ = 3.8 h). The measured oral bioavailability was low (3.8%), however, this was thought to be due to sub-optimal physicochemical properties which may limit absorption; the relatively high molecular weight (495 Da) and high Log*P* (4.5) are approaching the limit of what is generally regarded as being acceptable for an oral drug. Accordingly, we sought to identify a reduced-complexity analogue with the appropriate balance of potency and bioavailability.[74] A central tenet of our design strategy was to maintain solubility, and crystal structure information was invaluable in identifying alternative positions for installation of a solubilising group (discussed further in Section 14.4). From these studies, amine derivative **33** was identified. It possessed good levels of activity (EC_{50} = 790 nM), improved solubility (over 100 mg l^{-1}) and acceptable oral bioavailability (*F*% = 43). Additional *in vitro* testing confirmed that **33** was a selective AMPA positive modulator, with no activity at kainate, NMDA, or GABA. Extended selectivity measurements indicated no off-target activity against a panel of receptors and channels. Focusing on CNS exposure, measurement of brain to plasma ratio in Wistar BRL rats gave a value of 0.05, suggesting limited exposure in the central compartment. Having stated this, *in vivo* testing in a mechanistic model showed that amine **34** gave robust potentiation of AMPA-evoked single-unit activity as measured by electrophysiology in the hippocampus of anesthetized rats, with a measured MED of 0.3 mg kg^{-1} when dosed intravenously. This robust mechanistic data provided us with confidence that **34** could modulate AMPA activity *in vivo* despite its apparently limited CNS penetration.

Following on from the identification of amine **34**, we then sought to generate a back-up series of compounds which could be structurally differentiated from our existing assets.[75] To achieve this, we sought to leverage existing biostructural information around previously reported chemotypes. In particular, we focused on combining features from the phenethyl sulfonamide class (*e.g.* compound **21**) with our 3-trifluormethyl pyrazole series. Through understanding how the isopropylsulfonamide and 3-trifluoromethyl pyrazole motifs interacted with the AMPA allosteric site, we were able to expediently design and evaluate hybrid modulators which had promising overall properties. The lead exemplar from this effort is the pyrazole-sulfonamide derivative **35** which has acceptable potency at the AMPA receptor (EC_{50} = 2.7 μM) and excellent overall developability properties. Compound **35** has high solubility (89 mg ml^{-1}), displayed low clearance in rats (plasma clearance Cl_p = 1.9 ml min^{-1} kg^{-1}), accompanied with a long half-life of 7.4 h and excellent (100%)

oral bioavailability. In addition, the compound was found to be CNS-penetrant, with a maximum concentration (C_{max}) in rat cerebrospinal fluid of 0.12 μM. As stated above, a similar approach was used to hybridise with the benzamide series to furnish *e.g.* **13**, and taken as a whole, these studies therefore serve to illustrate the power of SBDD in enabling the rapid generation of new chemical matter for allosteric modulation of the AMPA receptor.

14.4 Impact of Biostructural Data

As intimated in the previous section, the availability of high resolution X-ray structures of the AMPA ligand binding domain has had a significant bearing on the design of new modulators within each of the chemical classes already described. In this final section, we aim to provide an overview of the currently available crystallographic information and link this to specific features of the known chemotypes discussed previously.

The overall topology of the ion channel was reviewed in Section 14.1 where it was noted that allosteric modulators occupied a distinct site remote from the orthosteric domain. In particular, these compounds were found to bind at the interface between two LBD clamshells, and serve to stabilise the protein–protein interaction between the two domains. This has the pharmacological effect of holding the channel in the open conformation (*i.e.* blocking desensitisation and deactivation), thereby potentiating the current produced.

The major classes of AMPA allosteric modulators have distinct yet overlapping binding sites and these are discussed below. The allosteric site itself is C_2 symmetric in nature and in two dimensions can be regarded as being an approximate 'gear-box' shape. In three dimensions, this becomes bent due to the presence of a hydrophobic saddle region formed by Pro494 and Ser497. On either side of this saddle exists a hydrophobic pocket and a relatively hydrophilic domain, and these symmetry-related sites are connected *via* a narrow channel formed by the saddle region. Figure 14.6 highlights these three main features.

Benzamide compounds such as aniracetam (**2**) constitute one of the three distinct binding modes that have been characterised. From consideration of the X-ray structure (Figure 14.7A), compound **2** does not make significant interactions with the receptor, occupying only the central pocket in the vicinity of the saddle region and makes hydrogen bonding interactions with a network of water molecules present in the hydrophobic pocket. In contrast, cyclothiazide (**14**) interacts more extensively with the receptor (Figure 14.7B). The norbornenyl system is buried in the hydrophobic pocket and the polar sulfonamide group extends towards the hydrophilic region, participating in hydrogen bonding interactions with the side chains of Ser494 and Ser497 and the backbone amide systems of Pro494 and Ser497. Additionally, two molecules of cyclothiazide are found to bind to the allosteric site; a second compound is found to interact with the symmetry-related domain.

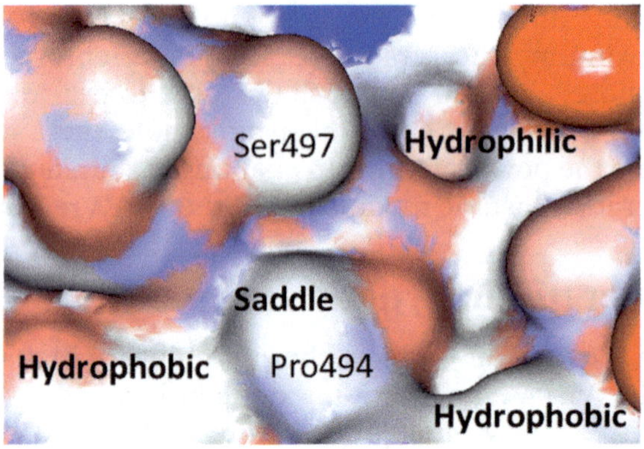

Figure 14.6 Annotated view of the AMPA allosteric binding site indicating key features. Several residues have been omitted from this and subsequent figures for clarity.

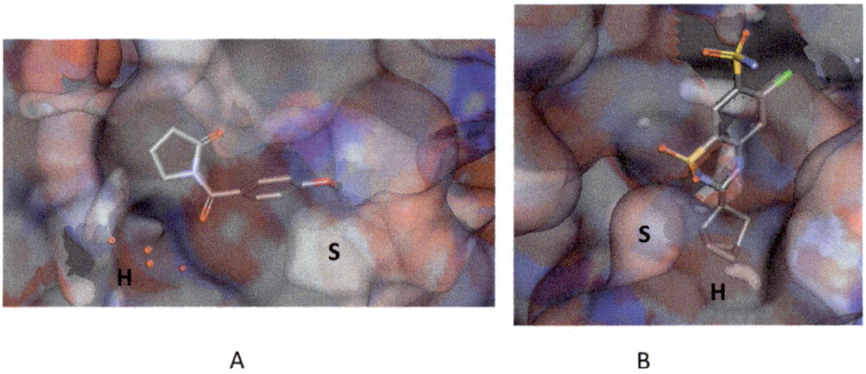

Figure 14.7 A. Aniracetam (**2**) bound to the GluA2 S1S2J LBD. B. Cyclothiazide (**14**) bound to the GluA2 S1S2J LBD. S, Saddle region; H, Hydrophobic pocket.

A third binding mode is evident when considering the structures of the sulfonamide systems such as **27** and **29**. The isopropyl sulfonamide group is found to interact with the hydrophobic pocket, and additional interactions are gained through hydrogen bonding with Pro494 (Figure 14.8). In the case of **27** the terminal pyridyl ring system is found to project across towards the symmetry-equivalent hydrophobic pocket.

Some degree of commonality exists when examining the binding mode of the 3-trifluoromethyltetrahydroindazole derivatives such as **34**, although subtle differences within this series have been noted (Figure 14.9). In the case of **34**, the trifluoromethyl group engages fully within the hydrophobic pocket, emulating the interactions made by the isopropyl group in compounds **27** and **29**. The tetrahydrobenzothiophene motif is then found to occupy the

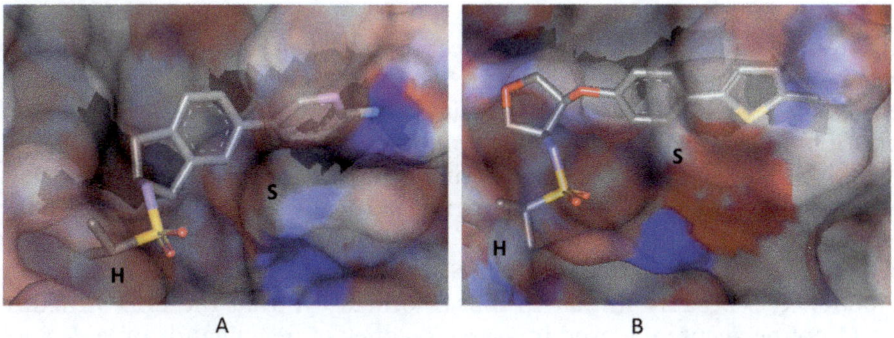

Figure 14.8 Sulfonamide derivatives bound to the GluA2 S1S2J LBD. A. GSK729327 (**27**) bound to the GluA2 S1S2J LBD. B. PF-04958242 (**29**). S, Saddle region; H, Hydrophobic pocket.

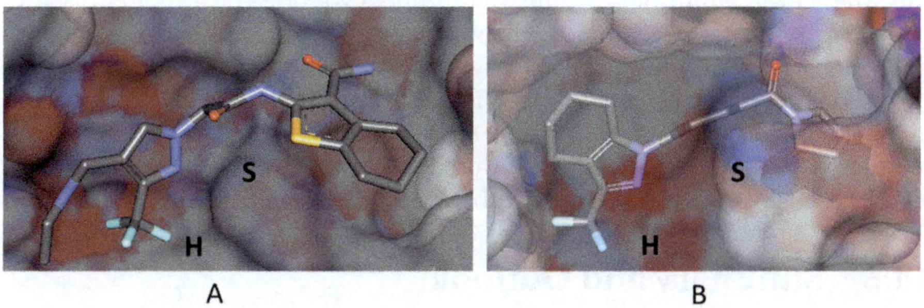

Figure 14.9 3-Trifluorotetrahydroindazole derived analogues bound to the GluA2 S1S2J LBD. A. Compound **34**. B. Compound **30**. S, Saddle region; H, Hydrophobic pocket.

symmetry-related site and the methylene amide spacer spans the saddle region. The secondary amine which was important in controlling the physicochemical properties of the compound is found to reside between Pro494 from one monomer and Ser729 from another, participating in a hydrogen bond with the backbone carbonyl of Ser729. The primary amide moiety lies primed to enable access to the hydrophilic pocket, which underlines the design hypothesis behind compound **33**.

Some differences can be noted when considering the binding mode of other amide derivatives such as **30**. The trifluoromethyl group still resides in the hydrophobic pocket, however, it is believed in this case to participate in a hydrogen bonding interaction between one of the fluorine atoms and the backbone amide of Gly731, contrasting with the earlier observations where the interaction was principally hydrophobic in nature. The amide portion of the molecule is able to make useful hydrogen bonding interactions with a network of water molecules in the other hydrophobic pocket in an analogous manner to aniracetam.

The impact that SBDD has had in recent years on the discovery of new allosteric modulators of the AMPA receptor has been significant. It has provided insight into both the channel biophysics and SAR observed within established chemotypes. Moreover, it has been an invaluable adjunct in prioritising the synthesis of new compounds to address specific design hypothesis as a detailed understanding of ligand–receptor interaction has allowed for ready identification of regions of a molecule that are potentially amenable to modification without significantly compromising primary activity. Having stated this, access to SBDD techniques is not a panacea. Our experience is broadly similar to that of others[76] in that the LBD structure could not be accurately used to assess potency of new ligands on a *de novo* basis. This is perhaps not unsurprising as the LBD construct used is only one component of a much larger macromolecular assembly present in the functional ion channel, therefore, subtleties in functional activity may not be discernible when considering ligand binding in isolation. In addition to this, there exist a number of transmembrane AMPA receptor regulatory proteins (TARPs) which can affect receptor trafficking as well as being able to potentiate receptor function.[77] Clearly, the interaction with protein modifiers cannot be taken into account when looking only at the LBD. Having stated all of this, access to high-resolution X-ray structures still affords significant opportunities both in the identification and optimisation of novel templates as positive allosteric modulators of the AMPA receptor.

14.5 Summary and Outlook

The past two decades have witnessed a significant and escalating level of interest in the development of positive allosteric modulators of the AMPA receptor. The therapeutic focus has shifted from initial emphasis on use as a monotherapy for the treatment of schizophrenia based on the disease association with glutamate hypofunction towards application as an adjunct therapy to address the cognitive domain of the disease. Clearly, a significant upside to development of this compound class is the potential to ameliorate cognitive decline in neurodegenerative disorders such as Alzheimer's disease.

A number of compound classes have emerged, a number of which are associated with promising developability properties and impressive levels of *in vivo* potency in both mechanistic studies focusing on potentiation of AMPA activity as well as animal models of cognitive impairment. In the last decade, two enabling technologies, namely HTS and SBDD have been established as a powerful combination for both hit identification and expedient optimisation of high-quality clinical candidates. This is particularly noteworthy as use of SBDD is still a relatively nascent strategy for membrane-bound targets compared with soluble enzymes such as protein kinases or proteases.

Despite these significant developments, no AMPA receptor modulator has yet demonstrated clinical proof-of-concept in phase II trials. There is no firm consensus behind the specific reasons for this, although a number of proposals can be put forward. Development of animal models of greater clinical relevance may assist late lead optimisation efforts in selecting the most appropriate compounds for development work as many of the existing models rely on a pharmacologically induced deficit in cognition. Similarly, embedding translatable biomarker strategies into the Lead Optimisation phase will again better inform candidate selection, and offers the potential to confirm CNS activity and target engagement in exploratory development, as well as supporting determination of appropriate clinical doses, an issue encountered in a number of clinical studies attempted thus far. Based on all of the above, application of appropriate biomarkers would be anticipated to enable better decision making regarding compound progression to the more expensive late-stage development activities.

In addition to tuning late-stage discovery efforts, further emphasis should be given to gaining a more holistic understanding of the pharmacology of AMPA receptor modulation. As stated in the introductory section, the functional receptor composition is heterogeneous at many levels; sub-unit composition and splice variants can significantly influence the current evoked by a positive allosteric modulator of the receptor. When the role of receptor-modifying proteins (*e.g.* TARPs) is considered, the overall picture becomes very complex, and accordingly, further studies in this area are warranted.

Despite these challenges, the general area of cognitive impairment or age-related decline remains an area of immense societal need. New classes of medicines are urgently required to address these needs, and based on the data presented in this chapter, it is evident that AMPA receptor positive allosteric modulators hold some considerable promise for this important therapeutic area.

References

1. J. N. C. Kew and J. A. Kemp, *Psychopharmacology*, 2005, **179**, 4–29.
2. M. L. Mayer and N. Armstrong, *Annu. Rev. Physiol.*, 2004, **66**, 161–181.
3. M. F. Green, *Am. J. Psychiatry*, 1996, **153**, 321–330.
4. M. F. Green, K. H. Nuechterlein, J. M. Gold, D. M. Barch, J. Cohen, S. Essock, W. S. Fenton, F. Frese, T. E. Goldberg, R. K. Heaton, R. S. E. Keefe, R. S. Kern, H. Kraemer, E. Stover, D. R. Weinberger, S. Zalcman and S. R. Marder, *Biol. Psychiatry*, 2004, **56**, 301–307.
5. C. L. Palmer, L. Cotton and J. M. Henley, *Pharmacol. Rev.*, 2005, **57**, 253–277.
6. A. I. Sobolevsky, M. P. Rosconi and E. Gouaux, *Nature*, 2009, **462**, 745–756.
7. Figures generated using Protein Data Bank entry 3KG2.
8. Y. Sun, R. Olson, M. Horning, N. Armstrong, M. Mayer and E. Gouaux, *Nature*, 2002, **417**, 245–253.
9. R. Jin, S. K. Singh, S. Gu, H. Furukawa, A. I. Sobolevsky, J. Zhou, Y. Jin and E. Gouaux, *EMBO J.*, 2009, **28**, 1812–1823.

10. A. Clayton, C. Siebold, R. J. C. Gilbert, G. C. Sutton, K. Harlos, R. A. J. McIlhinney, E. Y. Jones and A. R. Aricescu, *J. Mol. Biol.*, 2009, **392**, 1125–1132.
11. R. Jin, S. Clark, A. M. Weeks, J. T. Dudman, E. Gouaux and K. M. Partin, *J. Neurosci.*, 2005, **25**, 9027–9036.
12. N. Armstrong, Y. Sun, G. Q. Chen and E. Gouaux, *Nature*, 1998, **395**, 913–917.
13. B. H. Kaae, K. Harpsøe, J. S. Kastrup, A. C. Sanz, D. S. Pickering, B. Metzler, R. P. Clausen, M. Gajhede, P. Sauerberg, T. Liljefors and U. Madsen, *Chem. Biol.*, 2007, **14**, 1294–1303.
14. Y. Stern-Bach, B. Bettler, M. Hartley, P. O. Sheppard, P. J. O'Hara and S. F. Heinemann, *Neuron*, 1994, **13**, 1345–1357.
15. Y. Stern-Bach, S. Russo, M. Neuman and C. Rosenmund, *Neuron*, 1998, **21**, 907–918.
16. K. L. Durr, L. Chen, R. A. Stein, R. De Zori, I. M. Fola, T. Walz, H. S. McHaourab and E. Gouaux, *Cell*, 2014, **158**, 778–792.
17. M. D. Black, *Psychopharmacology*, 2005, **179**, 154–163.
18. A. Frandsen, D. S. Pickering, B. Vestergaard, C. Kasper, B. B. Nielsen, J. R. Greenwood, G. Campiani, C. Fattorusso, M. Gajhede, A. Schousboe and J. S. Kastrup, *Mol. Pharmacol.*, 2005, **67**, 703–713.
19. R. Jin, T. G. Banke, M. L. Mayer, S. F. Traynelis and E. Gouaux, *Nat. Neurosci.*, 2003, **6**, 803–810.
20. C. P. Ptak, A. H. Ahmed and R. E. Oswald, *Biochemistry*, 2009, **48**, 8594–8602.
21. J. T. Isaac, M. C. Crair, R. A. Nicoll and R. C. Malenka, *Neuron*, 1997, **18**, 269–280.
22. D. E. Feldman and E. I. Knudsen, *Neuron*, 1998, **20**, 1067–1071.
23. J. A. Esteban, *Mol. Interventions*, 2003, **3**, 375–385.
24. J. Lisman and S. Raghavachari, *Sci. STKE*, 2006, **2006**, re11.
25. T. H. Johansen, A. Chaudhary and T. A. Verdoorn, *Mol. Pharmacol.*, 1995, **48**, 946–955.
26. A. Copani, A. A. Genazzani, G. Aleppo, G. Casabona, P. L. Canonico, U. Scapagnini and F. Nicoletti, *J. Neurochem.*, 1992, **58**, 1199–1204.
27. A. C. Arai, Y.-F. Xia, G. Rogers, G. Lynch and M. Kessler, *J. Pharmacol. Exp. Ther.*, 2002, **303**, 1075–1085.
28. G. S. Lynch and G. A. Rogers, WO/1994/02475, 1994.
29. E. Wezenberg, R. J. Verkes, G. S. F. Ruigt, W. Hulstijn and B. G. C. Sabbe, *Neurophychopharm.*, 2007, **32**, 1272–1283.
30. G. A. Rogers and L. Nilssen, WO/1996/38414, 1996.
31. A. C. Arai, M. Kessler, G. Rogers and G. Lynch, *Mol. Pharmacol.*, 2000, **58**, 802–803.
32. G. A. Rogers and G. S. Lynch, WO/1997/36907, 1997.
33. G. A. Rogers and C. Marrs, WO/1999/33469, 1999.
34. G. A. Rogers and P. Johnstrom, WO/1999/51240, 1999.
35. G. A. Rogers and C. M. Marrs, WO/1998/35950, 1998.
36. S. J. A. Grove, M. Zhang and M. Shahid, WO/2002/100865, 2002.

37. S. J. A. Grove, J. Adam-Worrall, M. Zhang and R. Gilfillan, WO/2002/102808, 2002.
38. R. Mueller, S. Rachwal, M. A. Varney, S. A. Johnson, K. Alisala, S. Zhong, Y. X. Li, P. Haroldsen, T. Herbst and L. J. Street, *Bioorg. Med. Chem. Lett.*, 2011, **21**, 7455–7459.
39. N. Caldwell, J. E. Harms, K. M. Partin and C. Jamieson, *ACS Med. Chem. Lett.*, 2014, **6**, 392–396.
40. G. Lynch, R. Granger, J. Ambros-Ingerson, C. M. Davis, M. Kessler and R. Schehr, *Exp. Neurol.*, 1997, **145**, 89–92.
41. G. Lynch, M. Kessler, G. Rogers, J. Ambros-Ingerson, R. Granger and R. S. Schehr, *Int Clin Psychopharmacol.*, 1996, **11**, 13–19.
42. S. Marenco and D. R. Weinberger, *CNS Drugs*, 2006, **20**, 173–185.
43. D. C. Goff, J. S. Lamberti, A. C. Leon, M. F. Green, A. L. Miller, J. Patel, T. Manschreck, O. Freudenreich and S. A. Johnson, *Neuropsychopharmacol.*, 2008, **33**, 465–472.
44. N. J. Wesensten, R. M. Reichardt and T. J. Balkin, *Aviat., Space Environ. Med.*, 2007, **78**, 937–943.
45. B. G. Oertel, L. Felden, P. V. Tran, M. H. Bradshaw, M. S. Angst, H. Schmidt, S. Johnson, J. J. Greer, G. Geisslinger, M. A. Varney and J. Lötsch, *Clin. Pharmacol. Ther.*, 2010, **87**, 204–211.
46. Data from http://www.clinicaltrials.gov, accessed 10th Jan 2016.
47. K. R. Nations, R. Bursi, P. Dogterom, L. Ereshefsky, L. Gertski, T. Mant and J. Schipper, *Drugs R&D*, 2012, **12**, 127–139.
48. B. Pirotte, T. Podona, O. Diouf, P. de Tullio, P. Lebrun, L. Dupont, F. Somers, J. Delarge, P. Morain, P. Lestage, J. Lepagnol and M. Spedding, *J. Med. Chem.*, 1998, **41**, 2946–2959.
49. P. Desos, A. Cordi and P. Lestage, EP/1486503, 2004.
50. F. Bourasset, K. Bernard, C. Munoz, P. Genissel and J. M. Scherrmann, *Drug Metab. Dispos.*, 2005, **33**, 1137–1143.
51. P. Lestage, L. Danober, B. Lockhart, A. Roger, C. Lebrun, J.-L. Robin, P. Desos and A. Cordi, *Res. Pract. Alzheimer's Dis.*, 2002, **6**, 253–259.
52. P. L. Ornstein, D. M. Zimmerman, M. B. Arnold, T. J. Bleisch, B. Cantrel, R. Simon, H. Zarrinmayeh, S. R. Baker, M. Gates, J. P. Tizzano and D. Bleakman, *J. Med. Chem.*, 2000, **43**, 4354.
53. J. C. Quirk and E. S. Nisenbaum, *CNS Drug Rev.*, 2002, **8**, 255–282.
54. P. Miu, K. R. Jarvie, V. Radhakrishnan, M. R. Gates, A. Ogden, P. L. Ornstein, H. Zarrinmayeh, K. Ho, D. Peters, J. Grabell, A. Gupta, D. M. Zimmerman and D. Bleakman, *Neuropharmacol.*, 2001, **40**, 976–983.
55. P. J. Baumbarger, M. Muhlhauser, J. Zhai, C. R. Yang and E. S. Nisenbaum, *J. Pharmacol. Exp. Ther.*, 2001, **298**, 86–102.
56. M. Arundine and M. Tymianski, *Cell. Mol. Life Sci.*, 2004, **61**, 657–668.
57. J. Vandergriff, K. Huff, A. Bond and D. Lodge, *Neuropharmacol.*, 2001, **40**, 1003–1009.
58. M. J. O'Neill and J. M. Witkin, *Curr. Drug Targets*, 2007, **8**, 603–620.
59. M. B. Arnold, T. J. Bleisch, G. W. Cuff, P. L. Ornstein, and D. M. Zimmerman, WO/2001/090057, 2001.

60. W. D. Miller, A. H. Fray, J. T. Quatroche and C. D. Sturgill, Suppression of a Palladium-Mediated Homocoupling in a Suzuki Cross-Coupling Reaction, *Org. Process Res. Dev.*, 2007, **11**, 359–364.
61. S. S. Jhee, A. S. Chappell, V. Zarotsky, S. V. Moran, M. Rosenthal, E. Kim, S. Chalon, N. Toublanc, J. Brandt, D. E. Coutant and L. Ereshefsky, *J. Clin. Pharmacol.*, 2006, **46**, 424–432.
62. A. S. Chappell, C. Gonzales, J. Williams, M. M. Witte, R. C. Mohs and R. Sperling, *Neurology*, 2007, **68**, 1008–1012.
63. T. A. Shepherd, J. A. Aikins, D. Bleakman, B. E. Cantrell, J. P. Rearick, R. L. Simon, E. C. Smith, G. A. Stephenson, D. M. Zimmerman, A. Mandelzys, K. R. Jarvie, K. Ho, M. Deverill and R. K. Kamboj, *J. Med. Chem.*, 2002, **45**, 2101–2111.
64. D. M. Bradley, K. M. Thewlis and S. E. Ward, WO/2006/015828, 2006.
65. D. Andreotti and S. E. Ward, WO/2007/090841, 2007.
66. K. M. Thewlis and S. E. Ward, WO/2007/090840, 2007.
67. S. E. Ward and K. M. Thewlis, WO/2006/015829, 2007.
68. K. M. Thewlis and S. E. Ward, WO/2006/015827, 2006.
69. S. E. Ward, M. Harries, L. Aldegheri, D. Andreotti, S. Ballantine, B. D. Bax, A. J. Harries, A. J. Harker, J. Lund, R. Melarange, A. Mingardi, C. Mookherjee, J. Mosley, M. Neve, B. Oliosi, R. Profeta, K. J. Smith, P. W. Smith, S. Spada, K. M. Thewlis and S. P. Yusaf, *J. Med. Chem.*, 2010, **43**, 5801–5812.
70. N. C. Patel, J. Schwarz, X. J. Hou, D. J. Hoover, L. Xie, A. J. Fliri, R. J. Gallaschun, J. T. Lazzaro, D. K. Bryce, W. E. Hoffmann, A. N. Hanks, D. McGinnis, E. S. Marr, J. L. Gazard, M. Hajos, R. J. Scialis, R. S. Hurst, C. L. Shaffer, J. Pandit and C. J. O'Donnell, *J. Med. Chem.*, 2013, **56**, 9180–9191.
71. C. L. Shaffer, N. C. Patel, J. Schwarz, R. J. Scialis, Y. Wei, X. J. Hou, L. Xie, K. Karki, D. K. Bryce, S. M. Osgood, W. E. Hoffmann, J. T. Lazzaro, C. Chang, D. F. McGinnis, S. M. Lotarski, J. Liu, R. S. Obach, M. L. Weber, L. Chen, K. R. Zasadny, P. A. Seymour, C. J. Schmidt, M. Hajos, R. S. Hurst, J. Pandit and C. J. O'Donnell, *J. Med. Chem.*, 2015, **58**, 4291–4308.
72. S. E. Ward, M. Harries, L. Aldegheri, N. E. Austin, S. Ballantine, E. Ballini, D. M. Bradley, B. D. Bax, B. P. Clarke, A. J. Harris, S. A. Harrison, R. A. Melarange, C. Mookherjee, J. Mosley, G. Dal Negro, B. Oliosi, K. J. Smith, K. M. Thewlis, P. M. Woollard and S. P. Yusaf, *J. Med. Chem.*, 2011, **54**, 78–94.
73. C. Jamieson, S. Basten, R. A. Campbell, I. A. Cumming, K. J. Gillen, J. Gillespie, B. Kazemier, M. Kiczun, Y. Lamont, A. J. Lyons, J. K. F. Maclean, E. M. Moir, J. A. Morrow, M. Papakosta, Z. Rankovic and L. Smith, *Bioorg. Med. Chem. Lett.*, 2010, **19**, 5753–5756.
74. C. Jamieson, R. A. Campbell, I. A. Cumming, K. J. Gillen, J. Gillespie, B. Kazemier, M. Kiczun, Y. Lamont, A. J. Lyons, J. K. F. Maclean, F. Martin, E. M. Moir, J. A. Morrow, J. Pantling, Z. Rankovic and L. Smith, *Bioorg. Med. Chem. Lett.*, 2010, **19**, 6072–6075.

75. C. Jamieson, J. K. F. Maclean, C. I. Brown, R. A. Campbell, K. J. Gillen, J. Gillespie, B. Kazemier, M. Kiczun, Y. Lamont, A. J. Lyons, E. M. Moir, J. A. Morrow, J. Pantling, Z. Rankovic and L. Smith, *Bioorg. Med. Chem. Lett.*, 2011, **21**, 805–811.
76. S. E. Ward, *Ion Channel Drug Discovery*, ed. B. Cox and M. Gosling, Royal Society of Chemistry, Cambridge, 2015, RSC Drug Discovery Series, vol. 39, pp. 105–134.
77. E. Suzuki, M. Kessler and A. C. Arai, *Mol. Cell. Neurosci.*, 2008, **38**, 117–123.

CHAPTER 15

Allosteric Modulation of Neuronal Nicotinic Acetylcholine Receptors

MARK M. LEVANDOSKI*[a] AND SIVARAMAKRISHNA KOGANTI[a]

[a]Department of Chemistry and Programs in Biological Chemistry and Neuroscience, Grinnell College, Grinnell, IA 50112, USA
*E-mail: levandos@grinnell.edu

15.1 Introduction

Acetylcholine gates ionotropic receptors in muscle and nerve tissues across many species. Historically and ethnopharmacologically (for humans using tobacco) these ion channels were distinguished from the metabotropic (muscarinic) acetylcholine receptors by their response to nicotine. As drug targets, nicotinic receptors (nAChRs) have never gone out of fashion, but perhaps today are enjoying particular attention, especially where allosteric modulation is concerned.[1–3] Even focusing on molecular aspects of these receptors, the amount of current (and past) work on understanding their structures and mechanisms of action is prodigious, so we do not intend an exhaustive review of the literature in this chapter. Rather, we focus on some common themes of allosteric regulation of nAChRs, highlighting some key results, and we propose a conceptual framework for future investigations. We will briefly review the structure(s) of nAChRs, some main ideas from

RSC Drug Discovery Series No. 56
Allosterism in Drug Discovery
Edited by Dario Doller

Published by the Royal Society of Chemistry, www.rsc.org

allostery theory, and the current state of nicotinic ligands in clinical use. Our detailed analysis emphasizes the known or proposed binding sites of several allosteric modulators. Where appropriate, we refer the reader to numerous excellent reviews that are more comprehensive for particular topics; several of these appeared in two recent special journal issues dedicated to nicotinic receptors, namely *Biochemical Pharmacology*, volume 97, issue 4 and *Neuropharmacology*, volume 96, Part B. Overall, we will see that the field has made tremendous progress, yet some fundamental questions remain.

15.2 Nicotinic Receptors Display Broad Expression and Function

Nicotinic receptors have played an important role in the history of neuroscience as generations of scientists have elucidated the processes of neurotransmission, and the nAChRs have served as a model for all ligand-gated ion channels, particularly for the other members of the Cys-loop superfamily that includes gamma-aminobutyric acid A ($GABA_A$), serotonin, and glycine receptors.[4] However, as a family of proteins, they are important in their own right for their functions in both normal and pathological physiology. The fast synaptic transmission involving the muscle-type nAChR at the neuromuscular junction is a classic of neurobiology textbooks.[5,6] The nAChRs are also found throughout the autonomic nervous system, where they are responsible for the fast excitatory transmission in ganglia and display a diversity of subtypes.[7] These mediate the adverse autonomic side effects and toxicology associated with nicotinic agents.[8] Additionally, nAChRs are expressed in non-neuronal tissues such as immune cells, for which they have recently garnered therapeutic interest.[9] By and large, the interest in nAChRs currently as therapeutic targets rests on their implicated roles in central nervous system (CNS) function and dysfunction. Therefore, in this introduction, we survey the established and putative roles of neuronal nicotinic receptors in normal and pathological states.

Cellular and systems physiology of nAChR behavior bridges the molecular pharmacology and the roles of these receptors in disease. Just as molecular pharmacologists have labored to identify binding sites and mechanisms of action for drugs acting on individual nAChR molecules, physiologists have labored to identify the type of nAChRs present, sub-cellular locations thereof, and circuits involving nicotinic receptors. Diverse nAChR subtypes (see below) are distributed throughout the CNS where they function primarily as modulators of fast excitatory or inhibitory synapses;[10,11] for example, nAChRs mediate synaptic plasticity in the hippocampus, amygdala, and other areas.[12] Rodent models, not surprisingly given the use of knockout strains, have proven very powerful in elucidating correlations between nAChR expression and behaviors, including nicotine self-administration, conditioned place preference, and locomotion, as well as certain physiological functions such as nociception.[13] Along these lines, emerging optogenetic approaches should improve the clarity of our understanding of cholinergic micro-circuitry.[14] In humans, nAChRs function in processes such as working memory, attention, the reward system and emotion, and goal-directed behavior and decision making.[15–17] Although the mapping of nAChR subtypes at the level of specific nuclei and pathways is as yet incomplete, their expression in regions such as the hippocampus, amygdala, ventral tegmental area and nucleus accumbens, and prefrontal cortex correspond to the general localization of these functions.[18]

Consistent with their broad expression in the CNS and the variety of their neurophysiological roles, nicotinic receptors are implicated in several disorders whose brain localizations are not well-understood.[19] Attention-deficit–hyperactivity disorder (ADHD) affects both adolescents and adults, and is characterized as a general inability to focus, with decreased executive functions such as response inhibition, and variable speed of response.[20] These cognitive components are thought to be mediated by α4β2 receptors. In addition to the cognitive deficits, those suffering from ADHD often have associated emotional dysregulation, including irritability and mood lability.[20] Related to this, mounting evidence suggests that nAChRs are involved in depression and anxiety, as well as behaviors with an aspect of aggression.[21] Although much more work is needed to understand the mechanisms of these behaviors, limiting the throughput activity of α4β2 receptors generally appears to improve mood.[21] While we should consider pain a normal physiological function, recently the study of nAChRs in nociception has drawn interest for their therapeutic potential. Subtyping of the receptors involved in pain, primarily by pharmacology, shows the usual diversity in both ascending and descending tracts as well as in areas of the brainstem.[22]

We have greater mechanistic understanding of nicotinic signaling in several very severe neurological disorders. Schizophrenia is a multi-faceted, extreme mental illness marked by atypical perceptions, cognition, movement, and mood. Although schizophrenia is multigenic, the linkage between developing the disease and the gene for the α7 subunit in humans has motivated study of this nAChR subtype.[23] Low expression of α7 receptors on pyramidal cells and inhibitory interneurons projecting to the hippocampus is associated with an auditory sensory gating deficit; this perceptual system may be a component of control of attention.[23] Specific mutations (changes in amino acid sequence) of α4 and β2 subunits change the biophysical properties of α4β2-containing receptors in autosomal dominant nocturnal frontal lobe epilepsy;[24] however, how these "channelopathies" then give rise to seizures remains unclear. Alzheimer's disease (AD) is neurodegenerative and is the most prevalent form of dementia, showing a high correlation of cholinergic loss and cognitive deficit.[25] Common features of Alzheimer's include loss of episodic memory and loss of attention.[25] Like schizophrenia, Alzheimer's appears to involve a degradation of α7-mediated signaling in the neocortex and hippocampus.[23,25] Finally, numerous movement disorders involve nAChRs, the most prevalent of which is the neurodegenerative Parkinson's disease (PD).[26] The primary lesion in PD is loss of dopaminergic signaling in the basal ganglia, where nicotinic receptors occur on nigrostriatal dopamine neurons and modulate inputs from the cerebral cortex.[26,27] The role for nAChRs in dyskinesia induced by the most common PD treatment (levadopa (L-dopa)) is particularly well-established.[26] In addition, nicotinic receptor activity may be neuroprotective, and nicotinic therapy could address cognitive declines associated with PD.[28] Along these lines, difficulties in divided attention and sustained attention predict falls and freezing in PD patients, so α4β2 receptors of the prefrontal cortex in pathways governing these functions may be a distinct therapeutic target for ameliorating PD.[27] Although these diseases and disorders are complex, each with numerous functions affected, nAChRs definitively act in particular facets of them.

With some 15–20% of the human population using nicotine as a drug, nicotine addiction is the most significant health problem involving nAChRs, and the link between the receptors and addiction is direct. Considerable experimental effort has aimed at identifying the composition of receptors[18] and their pharmacological and physiological properties[11] that underlie behaviors related to addiction. The addictive aspect of nicotine use, typically measured as self-administration in animal models, arises from changes in nAChR number and perhaps composition in the ventral tegmental area, nucleus accumbens, and prefrontal cortex; these nAChRs modulate dopaminergic signaling in pathways associated with reward.[29] Dysregulation of aversion to nicotine and the effects of withdrawal from the drug involve the medial habenula and interpeduncular nucleus.[29] Withdrawal symptoms and their relation to appetite, anxiety, and depression are important for their impact on smoking cessation.[29,30] Being the molecular origin of addiction to nicotine, the nAChRs are most well-developed as therapeutic targets in this context.

Clearly nicotinic receptors figure importantly in numerous disorders, and nicotine addiction is central to what is considered the leading cause of preventable death worldwide.[30] The diversity of nAChR subtypes has presented a challenge to developing nicotinic therapeutics in regard to functional localization, but this diversity might also be viewed as an opportunity where drug specificity is concerned. Hundreds of nAChR ligands, both natural products and synthetic, are known; the majority of these are agonists or antagonists.[3,31,32] Despite the recognized structural homology of nAChRs and $GABA_A$ receptors, for which the allosteric modulator benzodiazepines are a well-developed and clinically used class of compounds,[33] identification and study of allosteric modulators of nAChRs was comparatively slow. Nonetheless, we recognize the therapeutic potential of nAChR modulators. Such compounds elicit little or no response from nAChRs alone, but when co-applied with agonist, the response is cooperatively enhanced or inhibited. Thus, allosteric modulators could act *in vivo* as coincidence detectors, working in concert with endogenous cholinergic signaling.[34,35] Although positive allosteric modulators have caused greater excitement in the field,[3,36–39] in principle either positive or negative modulation could be useful depending on whether the root pathology involved decreased or increased cholinergic throughput, respectively.

15.3 Nicotinic Receptors are Built for Diversity

Decades of experimental effort have given us a fairly complete understanding of the structure(s) of nicotinic acetylcholine receptors,[40–42] but we remain frustrated by a lack of high-resolution detail for all receptor subtypes. The full nAChR has resisted crystallization suitable for such high-resolution X-ray analysis because it is a membrane-embedded, multimeric glycoprotein of 290 kDa molecular weight. Additionally, the fact that many unique subtypes of the receptor are of pharmacological interest complicates the road to a high-resolution structure. We have available, for example: A 4 Å structure of a muscle-type receptor;[41] numerous 2–3 Å structures of molluscan homologues of the ligand binding domain, with and without ligands;[43–45] ~3 Å structures for bacterial Cys-loop homologues;[42] and a 2 Å structure of the mouse α1 extracellular domain.[46] In this context, we must keep well in mind that, despite the widespread use of these models in molecular studies of nAChR pharmacology, at present our atomic-level understanding of receptor–drug interactions is approximate at best, especially where human nAChRs are concerned.

The nAChR is a pentameric complex with the subunits arranged around a central, ion-conducting pore in a pseudo-symmetric manner. Each subunit thus contributes to three primary domains, namely the extracellular (N-terminal), transmembrane, and intracellular domains, as shown in Figure 15.1A. Each subunit crosses the membrane four times to form a bundle of α-helices to comprise the transmembrane domain, and the C-terminus is also extracellular. Subunit interfaces in the extracellular domain form the orthosteric ligand site, where acetylcholine and nicotine bind (Figure 15.1B, and see below). Many studies, employing the full spectrum of biochemical and biophysical techniques, have contributed to our understanding of the roles of these primary domains (and sub-domains thereof) in nAChR function; we refer the reader to several excellent reviews that describe these aspects of the system in more detail.[5,32,42]

Our purpose here is to relate key factors of nAChR structure necessary to understand ligand binding well. First, some 17 unique nAChR genes have been cloned from many species, ranging from *Caenorhabditis elegans*, through *Drosophila* and other insects, through fishes and birds, to mammals, including humans.[47] Although it is clear that in none of these species are all of the thousands of possible pentameric combinations of these subunits expressed,[48] many distinct subtypes are found throughout the central and peripheral nervous systems, in patterns that probably help define the functional specificity of these pathways, structures and regions.

The subunit composition of nAChRs defines their subtype, and these typically differ in both their pharmacology and biophysical properties.[49] Figure 15.1C shows a sampling of known subtypes, from the homomeric $(\alpha7)_5$ to the muscle-type $(\alpha1)_2\beta1\gamma\delta$ receptor. Again, not all possible combinations of subunit proteins are expressed, and enormous effort has gone into identifying the composition of native nAChRs.[18,48,50] The results of these molecular biology experiments to identify the native proteins are best compared with

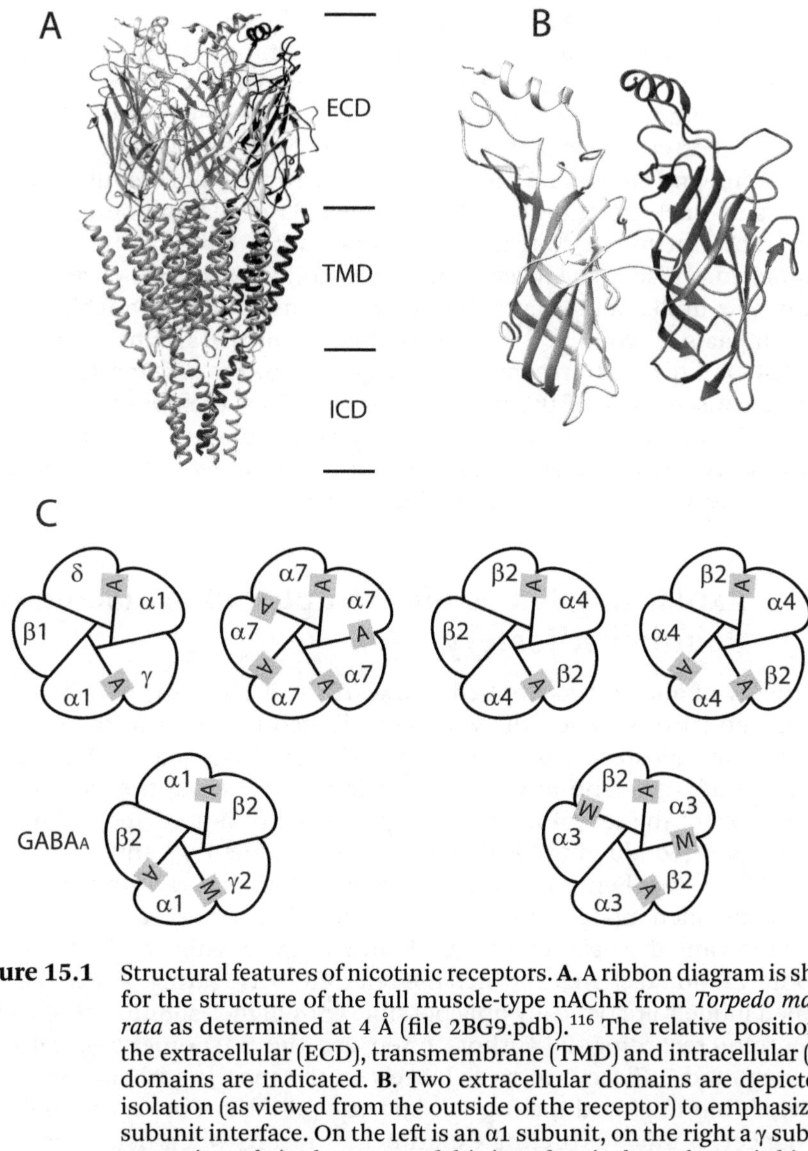

Figure 15.1 Structural features of nicotinic receptors. **A.** A ribbon diagram is shown for the structure of the full muscle-type nAChR from *Torpedo marmorata* as determined at 4 Å (file 2BG9.pdb).[116] The relative positions of the extracellular (ECD), transmembrane (TMD) and intracellular (ICD) domains are indicated. **B.** Two extracellular domains are depicted in isolation (as viewed from the outside of the receptor) to emphasize the subunit interface. On the left is an α1 subunit, on the right a γ subunit; approximately in the center of this interface is the orthosteric binding site with regions contributed by the principal (α) and complementary (non-α) faces. See also Figure 15.2. **C.** Schematics of several nAChR subtypes are shown, with their known or putative extracellular binding sites for agonists (A) and allosteric modulators (M). The view is from the synaptic space, and the subunits align head-to-tail (principal face to complementary) counter-clockwise. From left, top row, are the muscle-type $(\alpha1)_2\beta1\gamma\delta$, the homomeric $(\alpha7)_5$, and the two stoichiometries of α4β2 receptors. In the bottom row are a representative $GABA_A$ receptor, where M indicates the benzodiazepine site,[33] and a possible type of α3β2 receptor.[159] As elaborated in the text, growing evidence supports the idea that the structurally homologous pockets at non-orthosteric interfaces are viable ligand sites.

mapping studies by RNA transcript analysis and ligand-binding autoradiography. At this level, α7-and α4β2-subunit-containing nAChRs predominate in the CNS. However, many other specific subtypes have been localized to particular brain regions or pathways; a few examples include α4α5β2 receptors in the cortex, hippocampus, striatum and substantia nigra; α6β2β3 in the striatum and substantia nigra; α3β3β4 in the medial habenula and interpeduncular nucleus; and α4α5β2 and α4α6β2β3 on dopaminergic neurons in the ventral tegmental area.[18,29,48,50] Importantly, having identified which subunits comprise a given receptor is not sufficient to define the subtype, as it is now understood that subunit stoichiometry (and probably arrangement) impacts function;[51–53] we will see this in more detail when considering $(\alpha4)_2(\beta2)_3$ and $(\alpha4)_3(\beta2)_2$ receptors. As suggested by the schematics in Figure 15.1C, a consequence of the multiple combinations of subunits that define subtypes by virtue of their subunit identity, stoichiometry and arrangement, is *the diversity of subunit interfaces*, where ligand binding occurs. We give this aspect of these systems special consideration as a focus of this review.

15.4 Explaining Nicotinic Receptor Pharmacology Requires Allostery

The theory of allostery as applied to nAChRs is particularly well-developed, perhaps no surprise given the work contributed to this field by one of the pre-eminent allostery researchers, Jean-Pierre Changeux.[54] This theory seeks to explain quantitatively the relationships among the known functional states of the receptor – for example, receptor bound and unbound by ligand, open (conducting) and closed (non-conducting), and desensitized states.[34,38,42,55,56] When enough data are available, one can formulate intricate and detailed models of receptor function that specify rate constants for association and dissociation of ligands and for the opening and closing steps of conformational changes, which are referred to as "gating" steps; the most detailed models are typically only possible with high-resolution single-channel current recordings.[57] Authors often use the phenomenological terms "potency" and "efficacy" to refer loosely to the more rigorously defined or modeled "binding" and "gating" kinetic or equilibrium processes, but there is a danger in misinterpreting experimental results when using the former terms, especially for the very common methodology of mutation analysis.[55] Unfortunately, the neuronal forms of nAChRs are not as well-behaved as the muscle-type receptor in single-channel recordings, thus far precluding detailed modeling.[58–60] Nonetheless, some important characterizations, for example measuring unitary conductance, and open and closed time distributions, can be applied to neuronal nAChR systems,[59,61–63] including under conditions with and without allosteric modulators.[64–67]

A major conclusion from exhaustive studies of the muscle-type $[(\alpha1)_2\beta1\gamma\delta$ or $(\alpha1)_2\beta1\varepsilon\delta]$ nAChRs is the very high efficacy of this subtype – that is, when the receptor is bound by two agonist molecules, it has a very high probability of opening.[68,69] In this context, the distinction between homotropic (same-ligand) and heterotropic linkage is noteworthy.[70] Although apparent affinities for some ligands binding to the two orthosteric (agonist) sites can be distinguished,[6,71–74] these differences are often neglected in modeling.[5,55,75] Fundamentally, allosteric systems consist of multiple binding sites,[70] so we must consider how the various sites influence one another with regard to ligand binding and channel activity. Interestingly, Auerbach reported that agonist binding contributions to the two sites on muscle-type nAChRs are independent.[56] Presumably this is not the case for heterotropic allostery, activating a receptor by co-application of agonist and modulator, because many modulators have little to no effect on activation when applied alone, but increase the maximal response of the system (improve efficacy) upon co-application.[59,76–79] We consider in a later section the dogmatic idea that full activation of nAChRs requires the binding of two agonist molecules; this open question is of special relevance for mechanisms of allosteric modulation. Furthermore, being based on thermodynamic principles, allostery theory has no concern for molecular (structural) detail, but emerging evidence reveals that not only is the number of active sites important, whether for orthosteric (agonist) or allosteric ligands, but possibly also their spatial relationships.

Although allostery theory strives to explain the functional behavior of biomolecules, of course we desire to know the structure associated with each functional state. Here again the difficulties in determining high-resolution structures for the complete nAChR of any specific subtype have made for slow progress in this area. Unwin has described the differences in the open and closed states of the muscle-type receptor.[41] Further insight into these states has come from structures of higher resolution than the *Torpedo* nAChR, albeit of different sequence because they are from bacterial homologues of two different species.[42] The inter-conversion of these states has been explored in extreme detail at the level of functional contributions of individual amino acids, employing single-channel recordings, kinetic modeling, and thermodynamic cycle analysis of mutant receptors.[80,81] These studies have elucidated pathways through which agonist binding sites and the channel gate, separated by some 60 Å, communicate. This analysis has also been extended to the energetics of the system to evaluate how ligand binding energy is utilized to promote channel gating.[56,82] Despite the lower resolution imposed by the lack of single-channel data for neuronal nAChRs,[83] some have applied this thinking to studies of intra-molecular communication involving allosteric modulation.[84–86] For example, this intra-molecular communication depends on numerous contacts between the extracellular (ligand binding) and transmembrane (gate) domains.[68,69]

To date, nicotinic receptor research groups have done relatively little modeling of allosteric modulation. A few approaches have included a phenomenological description, typically of data approximated as equilibrium processes.[87–89] More progress has been made in this area for α7 nAChRs, possibly due to a greater number (comparatively) of known specific modulators and at least two fairly distinct pharmacological profiles relating to α7 desensitization properties.[38]

Categorizing allosteric drugs has its utility, for example in comparing the pharmacological profile of a novel compound with those of well-studied ligands, but we should not allow such categorization to take the place of mechanistic explanations of drug behavior. Common in the nAChR literature are the terms positive allosteric modulator (PAM) and negative allosteric modulator (NAM), with self-evident meanings for their actions.[38,39] More recently, authors have introduced the term "silent agonist" for compounds that do not modulate orthosteric agonists, but enhance currents of positive modulators.[90,91] This concept of a molecule that is active in conjunction with a second class of compounds yet is inactive or silent with a third has been extended to silent allosteric modulators.[92] Less descriptive are "type I" and "type II," terms introduced by Grønlien *et al.* to classify four positive allosteric modulators of α7 nAChRs the authors characterized side by side.[76] The primary distinction in these two classes is whether the modulator prevents desensitization (and can rescue channels from this state); importantly, both classes generally increased receptor efficacy, as determined by increased evoked currents for saturating agonist concentrations.[76] Leaving aside the possible proliferation of nomenclature, mixed effects of types I and II observed for some compounds suggests that these terms are not helpful.[38,39] We advocate that comparative pharmacology studies focus on (1) binding sites, especially attempting to identify them if unknown, and on (2) allosteric mechanism. It should be possible, for example, to formulate a single model of positive allosteric modulation that encompasses a given compound enhancing only efficacy and another compound decreasing the rate of entry into desensitized states. Ideally such models can be developed in quantitative detail. Whether continuing to use this nomenclature or not, we should continue to appeal to allosteric theory as a basis for understanding our experimental results and for framing new studies. The fact of the matter is that no modulator has a single, "pure" effect, and as we shall see, even activation of nAChRs by acetylcholine can be complicated. Therefore, interpretations of nAChR molecular pharmacology experiments require a foundation in allostery theory.

15.5 Nicotinic Receptors Offer Diverse Therapeutic Targets

Elucidating the allosteric nature of nicotinic receptor function is a huge challenge with intellectual rewards for their own sake. However, we recognize the practical applicability in understanding nAChR allostery as a foundation of drug development. Nicotinic receptors hold great promise as therapeutic targets, not least for being a diverse family of subtypes and being implicated in numerous pervasive disorders. Thus, in this section we survey the development and clinical use of some cholinergic agents, highlighting the current enthusiasm for allosteric nAChR ligands.[93] Again, this literature is extensive, so we hope to establish a sense of the scope of this area instead of providing a comprehensive review.

Sparingly few nicotinic ligands are in clinical use, and nicotine rather amazingly is one of them. Nicotine, delivered in a variety of ways, is marketed as a replacement therapy for smoking cessation.[30] The α4β2 receptor partial agonist varenicline, also prescribed for smoking cessation, has shown good efficacy despite some serious adverse effects for some people.[94] In this context, bupropion, originally identified as an antidepressant targeting monoamine systems, has also been quite effective;[30,94] this compound is a channel blocker of the predominant brain nAChRs.[95] Galantamine, rivastigmine, and donepezil are inhibitors of acetylcholinesterase, so function to prolong ACh residence in synapses; they are currently in use to treat cognitive deficits in Alzheimer's disease.[25] However, galantamine and related compounds, such as physostigmine, are also nAChR positive allosteric modulators,[11] and some evidence suggests that galantamine may function as a direct modulator of nAChRs in cholinergic signaling.[96,97] As drugs already approved for use in humans, nicotine, varenicline, and galantamine are also in clinical trials for treating other nAChR-mediated pathologies such as Parkinson's disease and other movement disorders,[26–28] schizophrenia,[98] ADHD,[20] mood disorders,[21] and pain.[22]

These same neurological disorders and others in which nAChRs probably have a role are targets of novel nAChR ligands. In this category are smoking cessation,[30] Alzheimer's[25] disease (AD), Parkinson's disease (PD),[26] schizophrenia,[23] ADHD,[20] anxiety and depression,[21] and pain.[22] The database www.ClinicalTrials.gov lists over 100 projects related to nicotinic receptor function, most of which address AD and PD, and hundreds more address smoking cessation. Some of the more promising cases, as judged largely by their advanced development include the following. The agonists ABT-418[99] and ABT-089[100] were originally developed for smoking cessation,[30] but have been used in clinical trials for ADHD.[20] In studies of non-human primates, ABT-089 also decreased the L-dopa-induced dyskinesias associated with PD.[26] The α7 agonist 3-(2,4-dimethoxybenzylidene)-anabaseine (DMXB-A; also known as GTS-21),[101] a derivative of the natural product anabaseine, improved aspects of attention and memory in clinical trials with schizophrenic patients,[23] and clinical studies of the compound in AD were initiated.[25] Similarly, cognitive enhancement seen in rodents with the α7 partial agonist EVP-6124[102] prompted clinical trials for schizophrenia and AD in humans.[23,25]

Perhaps reflecting the history of our molecular understanding of nAChRs, only a minority of nAChR drugs currently in development are allosteric modulators, despite the broad discussion of such compounds as a therapeutic approach.[3,36–39] One measure of the prevalence of allosteric modulators in development in academic and pharmaceutical labs is the number of patents filed for novel ligands. Gündisch and Eibl reported for the period 2006–2011 that only 30% (some 33 compounds) were identified by their inventors as modulators, although some others may have modulatory activity.[103] PNU-120596 is a positive allosteric modulator of α7 receptors and has served as an important standard in this class.[104] The compound has positive effects on learning and memory, and some physiological substrates thereof, in rodents[25,38] as well as anti-nociceptive properties.[22] Likewise, AVL-3288 (also called *N*-(4-chlorophenyl)-α-[[(4-chlorophenyl)amino]methylene]-3-methyl-5-isoxazoleacetamide (CCMI) and "Compound 6") is a positive allosteric modulator of α7 receptors and improves memory in rats;[105] it has entered clinical trials for cognitive enhancement in Alzheimer's and schizophrenia patients.[23,25]

Again, none of the aforementioned leads has become a marketed drug, but the high volume of work on these and many other nicotinic ligands, including clinical trials, is encouraging; the reviews cited above offer extensive detail on these areas. In sum, while developing novel nicotinic therapies holds promise for addressing many devastating ailments, the more multifarious the syndrome, the wider the gap between the molecular lesion and its higher-order effects.

15.6 How Can Ligand Site Identification Elucidate Allosteric Mechanisms?

Once a ligand with interesting pharmacological properties has been shown to act noncompetitively, for example by the failure of an agonist to compete away the effect at saturating concentrations,[106] the hard work has just begun: the ligand might bind *anywhere* else on the target macromolecule, and one must now find that site (or sites). Although the size of the nAChR is enough to make this a daunting task for novel nicotinic ligands, the receptor's recalcitrance to structural analysis makes this problem most difficult. Most of our knowledge of nAChR ligand binding sites comes from indirect evidence and has required multiple approaches to convincingly localize these sites.[32,39,107] That is, we have comparatively few high-resolution structural determinations,[40,42] and effectively none of these is of a "real" receptor.[46,108,109]

Instead of cataloguing the (now) many known nAChR allosteric modulators or the details of studies determining their binding sites, we develop here a theme of *the unity of these results*. We outline below several case studies that suggest the transmembrane domain and extracellular inter-subunit clefts are two important "hot spots" for allosteric modulator binding. We conclude this final section by proposing that recognition of these putative hot spots reveals an underlying principle of allostery mechanisms for nAChRs.

Before describing these two modulator binding regions, we note briefly that allosteric regulation of nAChRs *via* sites in the intracellular domain, such as by phosphorylation or association of the receptor with other proteins, has been documented.[110] In addition, modulation of nAChR activity by a variety of divalent cations through sites in or near subunit interfaces for many nicotinic receptor subtypes is well-established.[66,111–114] We focus on small-molecule modulators because these are more likely to see therapeutic development than simple inorganic cations or compounds that act intracellularly. Along these lines, most authors continue to distinguish the small-molecule (open) channel blockers as a distinct category of ligand;[107] these are not thought of as allosteric modulators *per se*, because their mechanism of inhibition differs fundamentally from those of other negative modulators. Likewise, agonists, competitive at the orthosteric site, and competitive antagonists binding to that same site, are not typically considered "allosteric." However, these two classes are simply homotropic ligands.[70] Therefore, in a generalized allosteric conceptualization of ligand binding, we could consider open channel blockers as antagonists competitive with the channel pore and/or gate. That is, we can view the orthosteric agonist binding site and channel gate as *functional locales* (like an "active site" of an enzyme) that are linked through the protein; so-called allosteric sites are then those that communicate with the orthosteric site, the gate, or both.

15.6.1 Transmembrane Domain Sites

The transmembrane domain has a structurally self-explanatory name and most importantly harbors the ion channel gate, the site in the protein that veritably defines open and closed states.[85,115,116] The domain consists of twenty α-helices, four per subunit, arranged in three concentric circles; the helices are roughly orthogonal to the plane of the membrane. The inner-most set is the second (TM2) membrane crossing for each subunit; these helices have a solvent-exposed face comprising the channel pore[32,115] and some determinants for open-channel-blocker binding (*e.g.*, QX-222,[117] or chlorpromazine).[107,118] The amino acid sequences of these transmembrane domain helices are quite homologous, and are highly conserved across all nAChR subunits and the Cys-loop family proteins generally.[119] The transmembrane domain also contains binding sites for numerous allosteric modulators.

Considerable evidence, obtained using a variety of techniques, supports the conclusion that nAChR transmembrane domains harbor modulator binding sites, and that they are strongly coupled to channel gating activity. Young *et al.* found that mutations in an intra-subunit cavity involving three of the four α-helices in the transmembrane domain of α7 nAChRs greatly reduced the potentiation of these receptors by PNU-120596 and LY-2087101;[120] the structurally similar positive modulator NS1738 also acts from this cavity.[121] Many of these mutations also diminished potentiation by ivermectin, and the set of perturbing mutations included all four α-helices in this case;[122] more recent evidence suggests that ivermectin and spinosad may bind between subunits in the transmembrane domain of some systems.[39] Interestingly, three of the α7 transmembrane domain mutations converted modulation by ivermectin from positive to negative. The Millar group has also demonstrated that a large series of cyclopenta[*c*]quinoline-8-sulfonamide compounds, subtly different plus or minus methyl moieties on the aryl substituent group, display the full range of behaviors on α7 receptors, from (allosteric) agonist, to positive, negative and silent allosteric modulation;[92] these compounds probably act from the intra-subunit cavity described above.[123] Olsen *et al.* found that the novel positive allosteric modulator NS206, specific for α4-containing nAChRs, binds in the transmembrane domain, although the authors did not identify specific residues comprising the site.[78]

Although nAChRs are not the primary effector molecule of general anesthetics, these diverse small molecules do act on nicotinic receptors.[124] The binding sites of many anesthetics, including halothane and short-chain alcohols, have been localized to the transmembrane domain of the nAChR, largely by photolabeling techniques. General anesthetics tend to be inhibitory (often by open channel blocking), but some compounds such as TDBzl-etomidate, which binds to an inter-subunit transmembrane site, potentiate agonist activity.[125] Possibly several of the transmembrane domain residues identified as being important for anesthetic activity correspond to those that interact with allosteric modulators,[39,124,126] but ambiguities in sequence alignments across species and subtypes and use of homology models for a structural interpretation of this idea suggest caution. It is not clear whether the extent of labeling for some anesthetics (10 or more residues) relative to their size is a function of the methodology or truly represents binding to, and action through, a population of sites within the transmembrane domain.

Whether binding to an inter- or intra-subunit site, if a modulator binds to a specific subunit, the total number of sites corresponds to the number of copies of that subunit (one to five) in the receptor. Thus, the number of sites occupied by a modulator in allosteric activation of the channel is an open question, despite the high homology of transmembrane domain regions in nAChRs. daCosta and Sine evaluated the stoichiometry of binding for PNU-120596 in potentiating α7 receptors by incorporating mutations of five amino acids in this pocket in conjunction with conductance reporter mutations. They concluded that modulator binding is highly cooperative and requires four or five molecules per receptor for potentiation.[127] This study highlights the importance of considering the number of sites, whether for agonist or modulator, whose occupation gives rise to allosteric modulatory activation.

We have long known that certain mutations in the pore-lining second transmembrane helix can alter the biophysical properties of nAChRs, making the channels hypersensitive to agonist.[128,129] Many mutations in this region can also increase channel efficacy[60,130,131] and although they are a subset of all nAChR mutations in these pathologies, naturally-occurring transmembrane domain mutations are prevalent in a particular rare gain-of-function epilepsy[24] and in some congenital myasthenias.[132] In short, transmembrane domain micro-locales couple strongly to gating, so it is no surprise that even small perturbations in the polarity of a moiety and/or in size on the order of a few atoms can affect channel function. Allosteric modulators binding to transmembrane domain sites can provide such perturbations.

15.6.2 Extracellular Inter-Subunit Cleft Sites

The agonist (orthosteric or canonical) binding sites in nAChRs occur at subunit interfaces, with closest-approach amino acids contributed by both subunits. Figure 15.2 shows these regions, the "loops" A–F.[133] We think of these as the canonical determinants of the binding site, but note that residues in other "loops" have also been shown to be in or near the agonist or competitive antagonist binding site.[31,134] Figure 15.2 also shows the five aromatic residues that comprise a "box" known to stabilize agonist binding through π-cation interactions.[6,135,136] For the human muscle-type nAChR, these residues are Tyr93 (loop A), Trp149 (loop B), Tyr190 and Tyr198 (loop C) from the α1 subunit in the principal, or "+" face, and Trp55/57 from the δ or γ/ε subunit in the complementary, or "−" face. We show the alignment of all human nAChRs for these binding site regions to emphasize not only their overall homology, but the high conservation of the aromatic box residues across *all* subunits. The conservation across species is of a similarly high degree.[47] We posit that (nearly) all extracellular domain interfaces are putative ligand binding sites,[137] especially for those compounds bearing amine-based positive charge.

The structural homology, yet radial asymmetry, of all nAChR subunits gives rise in the pentameric complex to one to five unique interfaces (see Figure 15.1). Considering the set of five aromatic residues as the primary determinants for an agonist site, we see that the homopentameric α7 nAChR has five potential agonist sites, whereas the muscle-type has only two (Figure 15.1C). Ample evidence beyond simple sequence comparison supports the hypothesis that the two orthosteric sites of the muscle-type receptor are not identical.[71,72,138] However, we note that the three non-orthosteric pockets of the muscle-type nAChR are particularly *non*-conserved for these aromatic residues, with none of the five present in any of the three interfaces, except for the sole semi-conserved β1Tyr55 of the β1(+)/α1(−) interface. Given that the muscle-type receptor came later than the neuronal forms in the evolution of these subunits and that β1, γ, δ, and ε are more closely related to one another than to α1,[119] this lack of conservation at these interfaces is especially interesting, suggesting a functional importance. This may relate to the "kinetic perfection" of the muscle-type receptor and its biological requirement of all-or-none response at the neuromuscular junction.[75]

As a homopentamer, of course the α7 subunit must contain all determinants of the agonist binding site. Modeling recovery from methyllycaconitine (MLA, a competitive antagonist) block, Palma *et al.* concluded that all five putative α7 sites were active.[139] Using an electrical fingerprinting approach (subunits with mutations that alter unitary conductance), Rayes *et al.* found that maximal activation of a chimeric α7-5HT_3 receptor, in which the extracellular domain is the α7 sequence and the remainder of the protein the serotonin receptor, occurs when three sites are occupied, only two of which are adjacent.[140] While activation was possible with two agonist molecules bound, open times were longer when occupancy was at non-adjacent

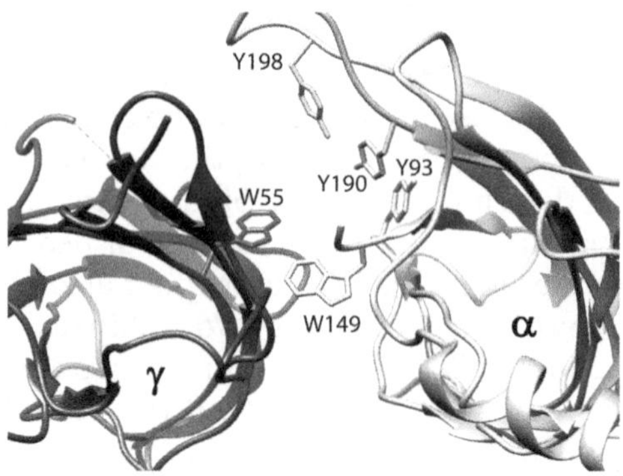


```
    D LOOP 55 + $                    A LOOP 93
α1  EVNQIVTTNVRLKQQW                 WRPDLVLYNNAD
α3  EVNQIMETNLWLKQIW                 WKPDIVLYNNAV
α6  EVNQIMETNLWLRHIW                 WKPDIVLYNNAV
α2  EKNQMMTTNVWLKQEW                 WIPDIVLYNNAD
α4  EKNQMMTTNVWVKQEW                 WRPDIVLYNNAD
α5  EKNQLMTTNVWLKQEW                 WTPDIVLFDNAD
β3  EKNQLMTTNVWLKQEW                 WLPDIVLFENAD
ε   EKEETLTTSVWIGIDW                 WLPEIVLENNID
γ   EREEALTTNVWIEMQW                 WRPDIVLENNVD
δ   EVEETLTTNVWIEHGW                 WLPEIVLENNND
β1  EKDEEMSTKVYLDLEW                 WLPDVVLLNNND
β2  EREQIMTTNVWLTQEW                 WLPDVVLYNNAD
β4  EREQIMTTNVWLKQEW                 WLPDIVLYNNAD
α10 ERNQVLTLYLWIRQEW                 WRPDIVLYNKAD
α9  ERNQILTAYLWIRQIW                 WRPDIVLYNKAD
α7  EKNQVLTTNIWLQMSW                 WKPDILLYNSAD
```

```
    E LOOP $      $                                  B LOOP
           + +    +++ $121               *            * 149
α1  FAIVKFTKVLLQYTGHITWTPP           CEIIVTHFPFDEQNCSMKLGTWTYDGSVVA
α3  FQVDDKTKALLKYTGEVTWIPP           CKIDVTYFPFDYQNCTMKFGSWSYDKAKID
α6  FQVEGKTKALLKYNGMITWTPP           CPMDITFFPFDHQNCSLKFGSWTYDKAEID
α2  FAVTHMTKAHLFSTGTVHWVPP           CSIDVTFFPFDQQNCKMKFGSWTYDKAKID
α4  FAVTHLTKAHLFHDGRVQWTPP           CSIDVTFFPFDQQNCTMKFGSWTYDKAKID
α5  FEGT-STKTVIRYNGTVTWTPP           CTIDVTFFPFDLQNCSMKFGSWTYDGSQVD
β3  FEGSLMTKVIVKSNGTVVWTPP           CTMDVTFFPFDRQNCSMKFGSWTYDGTMVD
ε   FGVAYDANVLVYEGGSVTWLPP           CAVEVTYFPFDWQNCSLIFRSQTYNAEEVE
γ   FEVALYCNVLVSPDGCIYWLPP           CSISVTYFPFDWQNCSLIFQSQTYSTNEID
δ   FQISYSCNVLVYHYGFVYWLPP           CPISVTYFPFDWQNCSLKFSSLKYTAKEIT
β1  FDVALDISVVVSSDGSVRWQPP           CSIQVTYFPFDWQNCTMVFSSYSYDSSEVS
β2  YEVSFYSNAVVSYDGSIFWLPP           CKIEVKHFPFDQQNCTMKFRSWTYDRTEID
β4  YEVSVYTNLIVRSNGSVLWLPP           CKIEVKYFPFDQQNCTLKFRSWTYDHTEID
α10 PPGSASTNVVLRHDGAVRWDAP           CRVDVAAFPFDAQHCGLTFGSWTHGGHQLD
α9  SSEPVNTNVVLRYDGLITWDAP           CVVDVTYFPFDNQQCNLTFGSWTYNGNQVD
α7  FDATFHTNVLVNSSGHCQYLPP           CYIDVRWFPFDVQHCKLKFGSWSYGGWSLD
```

```
            F LOOP                                       C LOOP
                   + +     172 +                   190 ** 198
α1  INPESDQ------------PDLSNFMESGEW        GWKHSVTYSCCPDTP--YLDITY
α3  LVLIGSS------------MNLKDYWESGEW        GYKHDIKYNCCEEI---YPDITY
α6  LLIIGSK------------VDMNDFWENSEW        GYKHDIKYNCCEEI---YTDITY
α2  LEQMEQT------------VDLKDYWESGEW        GTYNSKKYDCCAEI---YPDVTY
α4  LVNMHSR------------VDQLDFWESGEW        GTYNTRKYECCAEI---YPDITY
α5  IILEDQD------------VDKRDFFDNGEW        GSKGNRTDSCCW-----YPYVTY
β3  LILINEN------------VDRKDFFDNGEW        GMKGNRRDGVYS-----YPFITY
ε   FTFAVDN---DGKTINKIDIDTEAYTENGEW        GVIRRHHGGATDGP--GETDVIY
γ   LQLSQE----DGQTIEWIFIDPEAFTENGEW        AKMLLDPAAPAQEA--GHQKVVF
δ   LSLKQDAKENRTYPVEWIIIDPEGFTENGEW        ARVNVDPRAPLDSP--SRQDITF
β1  LQTGLGP---DGQGHQEIHIHEGTFIENGQW        SRLIQPPGDPRGGREGQRQEVIF
β2  LVLKSEV------------ASLDDFTPSGEW        GRRNENPDDS------TYVDITY
β4  MVLMTPT------------ASMDDFTPSGEW        GRRTVNPQDP------SYVDVTY
α10 VRPRGAAA------------SLADFVENVEW        ARRRVLTYGCCSEP---YPDVTF
α9  IFNALDSG------------DLSDFIEDVEW        AVKNVISYGCCSEP---YPDVTF
α7  L--QMQEA------------DISGYIPNGEW        GKRSERFYECCKEP---YPDVTF
```

Figure 15.2 Extracellular domain inter-subunit cleft binding site determinants. An alignment of the 16 human nAChR subunit amino acid sequences is shown, focusing on the extracellular domain "loop" regions first identified for the orthosteric site;[133] numbering is according to the mature sequence of the α1 subunit. Asterisks (*) indicate the two cysteines (C128 and C142) that define the Cys-loop superfamily, and the vicinal cysteines (C192 and C193) that are a hallmark of the nAChR α subunits. The positions of the five aromatic box residues are indicated by numbers and gray shading. The inset molecular structure shows sites.[140] Apparently for the native α7, open times are the same for occupancy of one to five sites, but the agonist sensitivity increases with increasing site occupancy.[141,142] Thus, the jury is still out on the question of agonist site occupancy requirements for full nAChR activation, which should be more complicated still when allosteric modulators are added to the mix (see Figure 15.1C); controlling the number and arrangement of sites with concatenated subunit constructs may help resolve this issue.[143–145]

Several groups recently showed that the homologous ligand pocket of the α4(+)/α4(−) interface in α4β2 receptors is an active agonist site.[73,144,145] In light of α7(+)/α7(−) sites constituting agonist binding mandatorily for the homopentamer, perhaps the α4 result is not surprising. Nonetheless, demonstrating this low-affinity activity in the context of the two orthosteric α4(+)/β2(−) sites was no mean feat. These studies exploited the observation, amply repeated,[51–53,146] that the two α4β2 receptor stoichiometries exhibit different potencies for ACh and several other agonists, often giving rise to biphasic concentration–response curves. These are the so-called high-sensitivity [$(\alpha4)_2(\beta2)_3$; HS] and low-sensitivity [$(\alpha4)_3(\beta2)_2$; LS] receptors (Figure 15.1C).[3,37]

Harpsøe *et al.* employed "free" (non-concatenated) subunits, expressed in varying ratios to form high- and low-sensitivity subtypes, and mutation analysis to demonstrate that the α4(+)/α4(−) interface constitutes a low-affinity agonist binding site.[144] Their mutations, a triad of complementary face residues (β2V136H/F144Q/L146T and α4H142V/Q150E/T152L), interconverted the relative sensitivity of the α4(+)/β2(−) and α4(+)/α4(−) sites. These experiments not only established the novel activity of this non-orthosteric site, but also identified residues important to the selectivity for ACh as agonist and for the novel agonists NS3573 and NS3920;[144,147] not surprisingly, these residues correspond to those found to be important as orthosteric complementary face specificity determinants in other contexts (see Figure 15.2). Using α4β2 receptors fully concatenated as pentamers, interface-specific mutations and chemical modification, the Bermudez group also showed that the α4(+)/α4(−) site in $(\alpha4)_3(\beta2)_2$ receptors constituted a lower-affinity ACh site[145] and contributed significantly to desensitization.[148] Substituting alanine for the principal face (loop B) tryptophan of the aromatic box decreased function from that site, and these effects segregated for mutation of the α4(+)/β2(−) *vs.* α4(+)/α4(−) sites. The demonstration of a role for the α4(+) face in this study,

---

the relative disposition of the aromatic box residues of the orthosteric α1(+)/γ(−) site as determined for the *Torpedo marmorata* muscle-type receptor (file 2BG9.pdb).[116] The view is from above in the synaptic space, with the channel vestibule toward the bottom of the page; a few stretches of ribbon in each subunit have been removed to reveal this pocket. A non-exhaustive sampling of other binding site determinants is shown: plus signs (+) indicate positions of complementary face residues that are important for agonist selectivity in orthosteric sites,[74,134,144,151] and dollar signs ($) indicate positions of residues contributing to modulator selectivity in non-orthosteric sites.[78,153,159] complicated due to its contribution to the orthosteric sites, fits well with the Danish group's results.

Eaton *et al.* used the highly selective and high-affinity agonist sazetidine-A and α4(+)/α4(−) site-specific mutations in concatemeric α4β2 receptors to deconvolute contributions to activation from the α4(+)/β2(−) and α4(+)/α4(−) sites.[73] They concluded that the third agonist site in $(\alpha4)_3(\beta2)_2$ receptors contributes the low sensitivity, but that these receptors natively produce about 20% high-sensitivity response. Furthermore, the authors demonstrated that the maximal current output, normalized to surface expression, was about five times greater for the low-sensitivity form, suggesting greater efficacy for these channels. These results are pleasingly logical, and show correspondence with the increased function for occupation of three *vs.* two agonist sites in α7-5HT$_3$ receptors.[140] The conclusion of three-site occupancy was confirmed by Mazzaferro *et al.*,[74] who also demonstrated that the α4(−) H142 position governs selectivity at the non-orthosteric site on the basis of ligand size. Ussing *et al.* synthesized new analogues of 3-(dimethylamino) butyl dimethylcarbamate which showed some selectivity for $(\alpha4)_3(\beta2)_2$ receptors over $(\alpha4)_2(\beta2)_3$, thus offering a proof-of-principle that we can target such non-orthosteric sites for drug development.[149]

Accumulating evidence points to several structurally diverse allosteric modulators binding to nAChR extracellular interface clefts. Henderson *et al.* used molecular modeling and mutagenesis to identify an α4(+)/β2(−) binding site, adjacent to the agonist site, for a family of novel negative allosteric modulators, represented by their lead compound KAB-18; this site includes residues of the canonical β2(−) orthosteric site.[150,151] Using mutations of principal-face residues in low- and high-sensitivity forms of α4β2 receptors, Weltzin and Schulte determined that the natural product desformylflustrabromine (dFBr), a positive modulator, interacts with the β2(+) face in the canonical loops A–C regions.[79] The authors suggest that dFBr may act at the β2(+)/β2(−) interface, but more experiments are needed to make this conclusive. The diversity of sites photolabeled by dFBr in the muscle-type receptor, which it inhibits, unfortunately sheds little light on this issue;[152] these authors did note some labeling in the non-orthosteric δ(+)/β1(−) interface. The novel positive modulator NS9283 acts from an α4(+)/α4(−) non-orthosteric site involving some canonical loop E residues, as determined by its lack of effect on $(\alpha4)_2(\beta2)_3$ receptors, and the finding that α4(−) loop E mutations eliminated its potentiation in $(\alpha4)_3(\beta2)_2$ receptors.[153]

The clinically used galantamine has been studied extensively enough to make the question of its nAChR binding site(s) more murky.[11] This natural product, a known acetylcholinesterase inhibitor, is either a noncompetitive agonist or positive allosteric modulator, depending on nAChR subtype and receptor preparation,[64,154] and some evidence supports a binding site near the principal-face (α) residues of the orthosteric site.[155] Additionally, interface residues in non-α subunits are implicated in binding galantamine, including some in the δ(+)/β1(−) interface of muscle-type receptor.[156,157]

Suffice it to say that, despite some ambiguity concerning the galantamine active sites, we think this case fits our general model of interface cleft binding. Many other compounds with well-characterized modulation behavior, and in some cases defined subtype specificity, now await binding site identification.[3,37,39,103,107]

In work from our laboratory, we have shown that the anthelmintic compounds morantel and oxantel act from sites at the β(+)/α(−) interfaces in α3β2 and α4β2 receptors;[158,159] these compounds, and the related pyrantel, positively modulate α3-containing receptors but inhibit α4 nAChRs. Both the β and α subunits are needed,[158] and the two conserved tryptophans (β2W149 and α3W53) in this site play a role in modulation.[159] We have also identified many of the determinants of specificity for morantel and oxantel, such as α3L107, α3E113, and α3I117,[159] some of which align with residues in the D and E loops of the complementary face important for agonist selectivity in orthosteric sites (see Figure 15.2). Morantel, pyrantel, and oxantel also directly activate α7 and muscle-type receptors (morantel and pyrantel are more efficacious agonists than ACh on α7), and the Bouzat group found that α7Q57 in the D loop of the complementary face was important for selectivity.[65,160] These results illustrate that, given a structurally homologous interface binding pocket, subtle differences in sequence can give rise to seemingly large differences in mode of activation as either agonist or modulator.

The extracellular and transmembrane domains are structurally nearly independent (per subunit). Combining the extracellular and transmembrane–intercellular domains from two Cys-loop family members gives functional channels[161] and the occurrence of extracellular domain homologues as free proteins suggest a certain modularity.[162] The coupling between the extracellular and transmembrane domains to effect the conformational transitions of agonist binding and channel gating depends on nearly all the contacts between the two domains.[85] In the light of these ideas, that molecules binding to extracellular interfaces or the transmembrane domain would activate these conformational changes is parsimoniously logical. Local movements,[41,42] supported by a network of residue-by-residue contacts,[163,164] constitute conformational changes in the nAChR. However, the extracellular–transmembrane domain interface, the subunit interfaces and the receptor gate (itself a locale of inter-subunit contact) offer the greatest degrees of freedom of movement at the quaternary level. Thus has nature exploited these three-dimensional pivots for the free energy donated by effector molecules.

15.7 Total Synthesis

In this review we have established that the subtype diversity of nicotinic acetylcholine receptors and their broad expression in the nervous system give rise to their involvement in many functions, from physiological to cognitive.

For this reason, the nAChRs are viable drug targets for treating a host of neurological problems. We argue that nAChR drug development should be grounded in allostery theory, because even the "basic" functions of these channels are complicated due to their allosteric properties. At the risk of ignoring important nuances of that behavior, we propose continued exploration of allosteric modulator binding sites and modes of action in the context of structural modules – folded domains and their interfaces.

The accepted notion that two molecules of agonist are required to fully activate nAChRs[56,75] has led to privileging four of the receptor subunits in nAChRs. Thus, we think of the importance of α1(+)/γ(−) and α1(+)/δ(−) pairs in the muscle-type receptor or the two pairs of α4(+)/β2(−) in the neuronal α4β2 receptor. Consequently, many authors refer to the remaining, fifth subunit as "structural" or "accessory".[2,13,18,48] This language connoting 'subordinate' or 'supplemental' is inappropriate because we expect, for example, all subunits to undergo global conformational changes in nAChR function. The α4(+)/α4(−) agonist site and its contributions to activation in $(\alpha4)_3(\beta2)_2$ receptors underscores the functional importance of the "fifth" subunit in that system,[73,74] and ample evidence supports the conclusion that all five subunits shape the biophysical and pharmacological profiles of nAChR subtypes.[140,165] Similarly, receptors of concatenated subunits, with defined arrangements of agonist sites, reproduce quite closely the properties of receptors composed of "free" subunits.[53,146,166] The possibility that optimal activation of pentameric channels occurs when three ligands are bound[33,73,74,140] suggests that interface-to-interface communication is important for function. In short, if our "pan-interface" ligand site model is correct, we must treat all subunits in nAChRs equally.

We recognize that our domain interface "hot spot hypothesis" may not account for all nAChR ligand binding sites,[167] but a goal-directed pursuit of the hypothesis has a major implication for drug specificity. The specificity of a binding site is dictated by the heterogeneity of its molecular determinants. Thus, we might expect sites in inter- and intra-subunit transmembrane domains, with their high degree of sequence homology, to be inherently less specific than extracellular domain interface sites. Leo Tolstoy wrote, "All happy families are alike; each unhappy family is unhappy in its own way".[168] Extrapolating, orthosteric sites are probably more similar to one another than allosteric sites, but this may be – happily, in fact – to our advantage. Restricting our attention to extracellular domain interface sites, we can draw on the vast knowledge of what makes an orthosteric site and the elements that are common for agonists,[134,144,145,169] and contrast this information with what makes non-orthosteric sites unique.[147,159] Although highly specific drugs may not be the sole therapeutic approach,[170] this notion still holds sway, and we may find non-orthosteric interface sites that uniquely define nAChR subtypes. Importantly, understanding not just the site specificity of nAChR ligands, but also the specificity with which these sites communicate – that is, the receptor's deeper allosteric nature – will provide for the long-term realization of the promise of nicotinic therapies.

Acknowledgements

We thank Monica St Angelo and Lizette of Big Sky for technical support and Dr Michael Leone for a critical reading of the manuscript. This work was supported by Academic Research Enhancement Award 1 R15 NS090368-01 from the National Institute of Neurological Disorders and Stroke.

References

1. J. A. Dani, D. Donnelly-Roberts and D. Bertrand, *Biochem. Pharmacol.*, 2013, **86**, 1041.
2. R. L. Papke, *Biochem. Pharmacol.*, 2014, **89**, 1–11.
3. M. Grupe, M. Grunnet, J. F. Bastlund and A. A. Jensen, *Basic Clin. Pharmacol. Toxicol.*, 2015, **116**, 187–200.
4. P. S. Miller and T. G. Smart, *Trends Pharmacol. Sci.*, 2010, **31**, 161–174.
5. S. M. Sine, *Physiol. Rev.*, 2012, **92**, 1189–1234.
6. A. Auerbach, *Biochem. Pharmacol.*, 2015, **97**, 601–608.
7. V. I. Skok, *Autonon. Neurosci.*, 2002, **97**, 1–11.
8. L. J. Schep, R. J. Slaughter and D. M. Beasley, *Clin. Toxicol.*, 2009, **47**, 771–781.
9. R. Zdanowski, M. Krzyzowska, D. Ujazdowska, A. Lewicka and S. Lewicki, *Cent. Eur. J. Immunol.*, 2015, **40**, 373–379.
10. H. D. Mansvelder, K. I. van Aerde, J. J. Couey and A. B. Brussaard, *Psychopharmacology (Berlin, Ger.)*, 2006, **184**, 292–305.
11. E. X. Albuquerque, E. F. Pereira, M. Alkondon and S. W. Rogers, *Physiol. Rev.*, 2009, **89**, 73–120.
12. M. Subramaniyan and J. A. Dani, *Ann. N. Y. Acad. Sci.*, 2015, **1349**, 46–63.
13. M. J. Marks, *Biochem. Pharmacol.*, 2013, **86**, 1105–1113.
14. L. Jiang, G. Y. Lopez-Hernandez, J. Lederman, D. A. Talmage and L. W. Role, *Rev. Neurosci.*, 2014, **25**, 755–771.
15. R. B. Poorthuis and H. D. Mansvelder, *Biochem. Pharmacol.*, 2013, **86**, 1089–1098.
16. F. Nees, *Neuropharmacology*, 2015, **96**(B), 289–301.
17. J. Naude, M. Dongelmans and P. Faure, *Neuropharmacology*, 2015, **96**(B), 244–254.
18. M. Zoli, F. Pistillo and C. Gotti, *Neuropharmacology*, 2015, **96**(B), 302–311.
19. A. S. Lewis and M. R. Picciotto, *Psychopharmacology (Berlin, Ger.)*, 2013, **229**, 477–485.
20. A. S. Potter, G. Schaubhut and M. Shipman, *CNS Drugs*, 2014, **28**, 1103–1113.
21. M. R. Picciotto, A. S. Lewis, G. I. van Schalkwyk and Y. S. Mineur, *Neuropharmacology*, 2015, **96**(B), 235–243.
22. I. C. Umana, C. A. Daniele and D. S. McGehee, *Biochem. Pharmacol.*, 2013, **86**, 1208–1214.
23. R. Freedman, *Annu. Rev. Med.*, 2014, **65**, 245–261.

24. O. K. Steinlein and D. Bertrand, *Pflugers Arch.*, 2010, **460**, 495–503.
25. A. S. Valles, M. V. Borroni and F. J. Barrantes, *CNS Drugs*, 2014, **28**, 975–987.
26. M. Quik, D. Zhang, X. A. Perez and T. Bordia, *Pharmacol. Ther.*, 2014, **144**, 50–59.
27. M. Sarter, R. L. Albin, A. Kucinski and C. Lustig, *Exp. Neurol.*, 2014, **257**, 120–129.
28. M. Quik, D. Zhang, M. McGregor and T. Bordia, *Biochem. Pharmacol.*, 2015, **97**, 399–407.
29. M. R. Picciotto and P. J. Kenny, *Cold Spring Harbor Perspect. Med.*, 2013, **3**, a012112.
30. T. S. Mohamed, S. S. Jayakar and A. K. Hamouda, *Front. Mol. Neurosci.*, 2015, **8**, 71.
31. M. N. Romanelli and F. Gualtieri, *Med. Res. Rev.*, 2003, **23**, 393–426.
32. A. A. Jensen, B. Frølund, T. Liljefors and P. Krogsgaard-Larsen, *J. Med. Chem.*, 2005, **48**, 4705–4745.
33. E. Sigel and B. P. Luscher, *Curr. Top. Med. Chem.*, 2011, **11**, 241–246.
34. S. J. Edelstein and J.-P. Changeux, *Adv. Protein Chem.*, 1998, **51**, 121–184.
35. A. Maelicke and E. X. Albuquerque, *Eur. J. Pharmacol.*, 2000, **393**, 165–170.
36. D. Bertrand and M. Gopalakrishnan, *Biochem. Pharmacol.*, 2007, **74**, 1155–1163.
37. A. Pandya and J. L. Yakel, *Biochem. Pharmacol.*, 2011, **82**, 952–958.
38. D. K. Williams, J. Wang and R. L. Papke, *Biochem. Pharmacol.*, 2011, **82**, 915–930.
39. A. Chatzidaki and N. S. Millar, *Biochem. Pharmacol.*, 2015, **97**, 408–417.
40. P. J. Corringer, M. Baaden, N. Bocquet, M. Delarue, V. Dufresne, H. Nury, M. Prevost and C. Van Renterghem, *J. Physiol.*, 2010, **588**(4), 565–572.
41. N. Unwin, *Q. Rev. Biophys.*, 2013, **46**, 283–322.
42. M. Cecchini and J.-P. Changeux, *Neuropharmacology*, 2015, **96**(B), 137–149.
43. K. Brejc, W. J. van Dijk, R. V. Klaassen, M. Schuurmans, J. van Der Oost, A. B. Smit and T. K. Sixma, *Nature*, 2001, **411**, 269–276.
44. P. Taylor, T. T. Talley, Z. Radic, S. B. Hansen, R. E. Hibbs and J. Shi, *Biochem. Pharmacol.*, 2007, **74**, 1164–1171.
45. P. Rucktooa, A. B. Smit and T. K. Sixma, *Biochem. Pharmacol.*, 2009, **78**, 777–787.
46. D. D. Dellisanti, Y. Yao, J. C. Stroud, Z. Z. Wang and L. Chen, *Nat. Neurosci.*, 2007, **10**, 953–962.
47. N. Le Novère and J.-P. Changeux, *Nucleic Acids Res.*, 1999, **27**, 340–342.
48. J. Wu and R. J. Lukas, *Biochem. Pharmacol.*, 2011, **82**, 800–807.
49. P. B. Sargent, *Annu. Rev. Neurosci.*, 1993, **16**, 403–443.
50. C. Gotti, F. Clementi, A. Fornari, A. Gaimarri, S. Guiducci, I. Manfredi, M. Moretti, P. Pedrazzi, L. Pucci and M. Zoli, *Biochem. Pharmacol.*, 2009, **78**, 703–711.
51. R. Zwart and H. P. M. Vijverberg, *Mol. Pharmacol.*, 1998, **54**, 1124–1131.

52. M. E. Nelson, A. Kuryatov, C. H. Choi, Y. Zhou and J. Lindstrom, *Mol. Pharmacol.*, 2003, **63**, 332–341.
53. A. L. Carbone, M. Moroni, P. J. Groot-Kormelink and I. Bermudez, *Br. J. Pharmacol.*, 2009, **156**, 970–981.
54. J.-P. Changeux, *J. Mol. Biol.*, 2013, **425**, 1396–1406.
55. D. Colquhoun, *Br. J. Pharmacol.*, 1998, **125**, 924–947.
56. A. Auerbach, *Neuropharmacology*, 2015, **96**(B), 150–156.
57. D. Colquhoun and B. Sakmann, *J. Physiol.*, 1985, **369**, 501–557.
58. A. Mathie, S. G. Cull-Candy and D. Colquhoun, *J. Physiol.*, 1991, **439**, 717–750.
59. T. Y. Wu, C. M. Smith, S. M. Sine and M. M. Levandoski, *Mol. Pharmacol.*, 2008, **74**, 466–475.
60. J. A. Bernal, J. Mulet, M. Castillo, M. Criado, S. Sala and F. Sala, *Biochim. Biophys. Acta*, 2009, **1788**, 410–416.
61. R. L. Papke, *Prog. Neurobiol.*, 1993, **41**, 509–531.
62. J. P. Boorman, M. Beato, P. J. Groot-Kormelink, S. D. Broadbent and L. G. Sivilotti, *J. Biol. Chem.*, 2003, **278**, 44033–44040.
63. P. Li, M. McCollum, J. Bracamontes, J. H. Steinbach and G. Akk, *Mol. Pharmacol.*, 2011, **80**, 818–827.
64. G. Akk and J. H. Steinbach, *J. Neurosci.*, 2005, **25**, 1992–2001.
65. M. Bartos, D. Rayes and C. Bouzat, *Mol. Pharmacol.*, 2006, **70**, 1307–1318.
66. B. Hsiao, K. B. Mihalak, K. L. Magleby and C. W. Luetje, *J. Neurophysiol.*, 2008, **99**, 999–1007.
67. C. Peng, M. R. Kimbrell, C. Tian, T. F. Pack, P. A. Crooks, E. K. Fifer and R. L. Papke, *Mol. Pharmacol.*, 2013, **84**, 459–475.
68. A. Jha, D. J. Cadugan, P. Purohit and A. Auerbach, *J. Gen. Physiol.*, 2007, **130**, 547–558.
69. W. Y. Lee, C. R. Free and S. M. Sine, *J. Neurosci.*, 2009, **29**, 3189–3199.
70. J. Wyman and S. J. Gill, *Binding and Linkage: Functional Chemistry of Biological Macromolecules*, University Science Books, Mills Valley, CA, 1990.
71. A. Figl, B. N. Cohen, M. W. Quick, N. Davidson and H. A. Lester, *FEBS Lett.*, 1992, **308**, 245–248.
72. R. J. Prince and S. M. Sine, *J. Biol. Chem.*, 1996, **271**, 25770–25777.
73. J. B. Eaton, L. M. Lucero, H. Stratton, Y. Chang, J. F. Cooper, J. M. Lindstrom, R. J. Lukas and P. Whiteaker, *J. Pharmacol. Exp. Ther.*, 2014, **348**, 46–58.
74. S. Mazzaferro, F. Gasparri, K. New, C. Alcaino, M. Faundez, P. Iturriaga Vasquez, R. Vijayan, P. C. Biggin and I. Bermudez, *J. Biol. Chem.*, 2014, **289**, 21795–21806.
75. M. B. Jackson, *Proc. Natl. Acad. Sci. U. S. A.*, 1989, **86**, 2199–2203.
76. J. H. Gronlien, M. Hakerud, H. Ween, K. Thorin-Hagene, C. A. Briggs, M. Gopalakrishnan and J. Malysz, *Mol. Pharmacol.*, 2007, **72**, 715–724.
77. J. K. Gill, P. Dhankher, T. D. Sheppard, E. Sher and N. S. Millar, *Mol. Pharmacol.*, 2012, **81**, 710–718.
78. J. A. Olsen, J. S. Kastrup, D. Peters, M. Gajhede, T. Balle and P. K. Ahring, *J. Biol. Chem.*, 2013, **288**, 35997–36006.

79. M. M. Weltzin and M. K. Schulte, *J. Pharmacol. Exp. Ther.*, 2015, **354**, 184–194.
80. C. Grosman, M. Zhou and A. Auerbach, *Nature*, 2000, **403**, 773–776.
81. W. Y. Lee and S. M. Sine, *Nature*, 2005, **438**, 243–247.
82. G. Maksay, *Prog. Biophys. Mol. Biol.*, 2011, **106**, 463–473.
83. K. R. Gleitsman, J. A. Shanata, S. J. Frazier, H. A. Lester and D. A. Dougherty, *Biophys. J.*, 2009, **96**, 3168–3178.
84. J. T. McLaughlin, J. Fu, A. D. Sproul and R. L. Rosenberg, *Mol. Pharmacol.*, 2006, **70**, 16–22.
85. C. Bouzat, *J. Physiol. (Paris)*, 2012, **106**, 23–33.
86. C. A. Short, A. T. Cao, M. A. Wingfield, M. Doers, E. M. Jobe, N. Wang and M. M. Levandoski, *Neuropharmacology*, 2015, **91**, 157–168.
87. R. Zwart and H. P. M. Vijverberg, *Mol. Pharmacol.*, 1997, **52**, 886–895.
88. R. Zwart and H. P. M. Vijverberg, *Eur. J. Pharmacol.*, 2000, **393**, 209–214.
89. B. W. Madsen and G. F. Yeo, *Arch. Biochem. Biophys.*, 2000, **373**, 429–434.
90. N. T. Burford, M. J. Clark, T. S. Wehrman, S. W. Gerritz, M. Banks, J. O'Connell, J. R. Traynor and A. Alt, *Proc. Natl. Acad. Sci. U. S. A.*, 2013, **110**, 10830–10835.
91. K. Chojnacka, R. L. Papke and N. A. Horenstein, *Bioorg. Med. Chem. Lett.*, 2013, **23**, 4145–4149.
92. J. K. Gill-Thind, P. Dhankher, J. M. D'Oyley, T. D. Sheppard and N. S. Millar, *J. Biol. Chem.*, 2015, **290**, 3552–3562.
93. J. A. Dani, D. Donnelly-Roberts and D. Bertrand, *Biochem. Pharmacol.*, 2015, **97**, 351.
94. E. J. Mills, P. Wu, I. Lockhart, K. Thorlund, M. Puhan and J. O. Ebbert, *Ann. Med.*, 2012, **44**, 588–597.
95. A. Pandhare, A. K. Hamouda, B. Staggs, S. Aggarwal, P. K. Duddempudi, J. R. Lever, D. J. Lapinsky, M. Jansen, J. B. Cohen and M. P. Blanton, *Biochemistry*, 2012, **51**, 2425–2435.
96. B. Schilstrom, V. B. Ivanov, C. Wiker and T. H. Svensson, *Neuropsychopharmacology*, 2007, **32**, 43–53.
97. Y. Noda, A. Mouri, Y. Ando, Y. Waki, S. N. Yamada, A. Yoshimi, K. Yamada, N. Ozaki, D. Wang and T. Nabeshima, *Int. J. Neuropsychopharmacol.*, 2010, **13**, 1343–1354.
98. D. L. Boggs, J. Carlson, J. Cortes-Briones, J. H. Krystal and D. C. D'Souza, *Curr. Pharm. Des.*, 2014, **20**, 5077–5092.
99. S. P. Arneric, J. P. Sullivan, C. A. Briggs, D. Donnelly-Roberts, D. J. Anderson, J. L. Raszkiewicz, M. L. Hughes, E. D. Cadman, P. Adams, D. S. Garvey, J. T. Wasicak and M. Williams, *J. Pharmacol. Exp. Ther.*, 1994, **270**, 310–318.
100. N.-H. Lin, D. E. Gunn, K. B. Ryther, D. S. Garvey, D. L. Donnelly-Roberts, M. W. Decker, J. D. Brioni, M. J. Buckley, A. D. Rodrigues, K. G. Marsh, D. J. Anderson, J. J. Buccafusco, M. A. Prendergast, J. P. Sullivan, M. Williams, S. P. Arneric and M. W. Holladay, *J. Med. Chem.*, 1997, **40**, 385–390.
101. D. S. Woodruff-Pak, Y.-T. Li and W. R. Kem, *Brain Res.*, 1994, **645**, 309–317.

102. J. Prickaerts, N. P. van Goethem, R. Chesworth, G. Shapiro, F. G. Boess, C. Methfessel, O. A. Reneerkens, D. G. Flood, D. Hilt, M. Gawryl, S. Bertrand, D. Bertrand and G. Konig, *Neuropharmacology*, 2012, **62**, 1099–1110.
103. D. Gündisch and C. Eibl, *Expert Opin. Ther. Pat.*, 2011, **21**, 1867–1896.
104. R. S. Hurst, M. Hajos, M. Raggenbass, T. M. Wall, N. R. Higdon, J. A. Lawson, K. L. Rutherford-Root, M. B. Berkenpas, W. E. Hoffmann, D. W. Piotrowski, V. E. Groppi, G. Allaman, R. Ogier, S. Bertrand, D. Bertrand and S. P. Arneric, *J. Neurosci.*, 2005, **25**, 4396–4405.
105. H. J. Ng, E. R. Whittemore, M. B. Tran, D. J. Hogenkamp, R. S. Broide, T. B. Johnstone, L. Zheng, K. E. Stevens and K. W. Gee, *Proc. Natl. Acad. Sci. U. S. A.*, 2007, **104**, 8059–8064.
106. R. R. Neubig, M. Spedding, T. Kenakin and A. Christopoulos, *Pharmacol. Rev.*, 2003, **55**, 597–606.
107. H. R. Arias, *Adv. Protein Chem. Struct. Biol.*, 2010, **80**, 153–203.
108. S. X. Li, S. Huang, N. Bren, K. Noridomi, C. D. Dellisanti, S. M. Sine and L. Chen, *Nat. Neurosci.*, 2011, **14**, 1253–1259.
109. V. Bondarenko, D. D. Mowrey, T. S. Tillman, E. Seyoum, Y. Xu and P. Tang, *Biochim. Biophys. Acta*, 2014, **1838**, 1389–1395.
110. J.-P. Changeux, *Drug Discovery Today: Technol.*, 2013, **10**, e223–228.
111. R. Zwart, R. G. D. M. van Kleef, J. M. Milikan, M. Oortgiesen and H. P. M. Vijverberg, *Eur. J. Pharmacol.*, 1995, **291**, 399–406.
112. J. L. Galzi, S. Bertrand, P. J. Corringer, J.-P. Changeux and D. Bertrand, *EMBO J.*, 1996, **15**, 5824–5832.
113. L. K. Lyford, A. D. Sproul, E. Eddins, J. T. McLaughlin and R. L. Rosenberg, *Mol. Pharmacol.*, 2008, **64**, 650–658.
114. M. Moroni, R. Zwart, E. Sher, B. K. Cassels and I. Bermudez, *Mol. Pharmacol.*, 2006, **70**, 755–768.
115. A. Karlin and M. H. Akabas, *Neuron*, 1995, **15**, 1231–1244.
116. N. Unwin, *J. Mol. Biol.*, 2005, **346**, 967–989.
117. P. Charnet, C. Labarca, R. J. Leonarad, N. J. Vogelaar, L. Czyzyk, A. Gouin, N. Davidso and H. A. Lester, *Neuron*, 1990, **2**, 87–95.
118. D. C. Chiara, A. K. Hamouda, M. R. Ziebell, L. A. Mejia, G. Garcia, 3rd and J. B. Cohen, *Biochemistry*, 2009, **48**, 10066–10077.
119. K. Tsunoyama and T. Gojobori, *Mol. Biol. Evol.*, 1998, **15**, 518–527.
120. G. T. Young, R. Zwart, A. S. Walker, E. Sher and N. S. Millar, *Proc. Natl. Acad. Sci. U. S. A.*, 2008, **105**, 14686–14691.
121. T. Collins, G. T. Young and N. S. Millar, *Neuropharmacology*, 2011, **61**, 1306–1313.
122. T. Collins and N. S. Millar, *Mol. Pharmacol.*, 2010, **78**, 198–204.
123. J. K. Gill, M. Savolainen, G. T. Young, R. Zwart, E. Sher and N. S. Millar, *Proc. Natl. Acad. Sci. U. S. A.*, 2011, **108**, 5867–5872.
124. S. A. Forman, D. C. Chiara and K. W. Miller, *Neuropharmacology*, 2015, **96**(B), 169–177.
125. S. Nirthanan, G. Garcia 3rd, D. C. Chiara, S. S. Husain and J. B. Cohen, *J. Biol. Chem.*, 2008, **283**, 22051–22062.

126. A. K. Hamouda, S. S. Jayakar, D. C. Chiara and J. B. Cohen, *J. Mol. Neurosci.*, 2014, **53**, 480–486.
127. C. J. daCosta and S. M. Sine, *Proc. Natl. Acad. Sci. U. S. A.*, 2013, **110**, 6595–6600.
128. F. Revah, D. Bertrand, J.-L. Galzi, A. Devillers-Thiery, C. Mulle, N. Hussy, S. Bertran, M. Ballivet and J.-P. Changeux, *Nature*, 1991, **353**, 846–849.
129. C. Labarca, J. Schwarz, P. Deshpande, S. Schwarz, M. W. Nowak, R. Nashmi, P. Kofuji, H. Dang, W. Shi, M. Fidan, B. S. Khakh, Z. Chen, B. J. Bowers, J. Boulter, J. M. Wehner and H. A. Lester, *Proc. Natl. Acad. Sci. U. S. A.*, 2001, **98**, 2786–2791.
130. C. Grosman and A. Auerbach, *J. Gen. Physiol.*, 2000, **115**, 637–651.
131. R. Rush, A. Kuryatov, M. E. Nelson and J. Lindstrom, *Mol. Pharmacol.*, 2002, **61**, 1416–1422.
132. A. G. Engel, X.-M. Shen, D. Selcen and S. M. Sine, *Lancet Neurol.*, 2015, **14**, 420–434.
133. J. L. Galzi and J.-P. Changeux, *Curr. Opin. Struct. Biol.*, 1994, **4**, 554–565.
134. H. R. Arias, *Neurochem. Int.*, 2000, **36**, 595–645.
135. S. M. Sine, P. Quiram, F. Papanikolaou, H.-J. Kreienkamp and P. Taylor, *J. Biol. Chem.*, 1994, **269**, 8808–8816.
136. W. Zhong, J. P. Gallivan, Y. Zhang, L. Li, H. A. Lester and D. A. Dougherty, *Proc. Natl. Acad. Sci. U. S. A.*, 1998, **95**, 12088–12093.
137. M. M. Levandoski, B. Piket and J. Chang, *Eur. J. Pharmacol.*, 2003, **471**, 9–20.
138. R. J. Prince and S. M. Sine, *J. Biol. Chem.*, 1998, **273**, 7843–7849.
139. E. Palma, D. Bertrand, T. Binzoni and D. Bertrand, *J. Physiol.*, 1996, **491.1**, 151–161.
140. D. Rayes, M. J. De Rosa, S. M. Sine and C. Bouzat, *J. Neurosci.*, 2009, **29**, 6022–6032.
141. N. Andersen, J. Corradi, S. M. Sine and C. Bouzat, *Proc. Natl. Acad. Sci. U. S. A.*, 2013, **110**, 20819–20824.
142. D. K. Williams, C. Stokes, N. A. Horenstein and R. L. Papke, *J. Gen. Physiol.*, 2011, **137**, 369–384.
143. J. H. Steinbach and G. Akk, *Anesthesiology*, 2011, **115**, 1328–1337.
144. K. Harpsøe, P. K. Ahring, J. K. Christensen, M. L. Jensen, D. Peters and T. Balle, *J. Neurosci.*, 2011, **31**, 10759–10766.
145. S. Mazzaferro, N. Benallegue, A. Carbone, F. Gasparri, R. Vijayan, P. C. Biggin, M. Moroni and I. Bermudez, *J. Biol. Chem.*, 2011, **286**, 31043–31054.
146. Y. Zhou, M. E. Nelson, A. Kuryatov, C. Choi, J. Cooper and J. Lindstrom, *J. Neurosci.*, 2003, **23**, 9004–90015.
147. A. Shahsavar, P. K. Ahring, J. A. Olsen, C. Krintel, J. S. Kastrup, T. Balle and M. Gajhede, *Mol. Pharmacol.*, 2015, **88**, 697–707.
148. N. Benallegue, S. Mazzaferro, C. Alcaino and I. Bermudez, *Br. J. Pharmacol.*, 2013, **170**, 304–316.
149. C. A. Ussing, C. P. Hansen, J. G. Petersen, A. A. Jensen, L. A. Rohde, P. K. Ahring, E. Ø. Nielsen, J. S. Kastrup, M. Gajhede, B. Frølund and T. Balle, *J. Med. Chem.*, 2013, **56**, 940–951.

150. B. J. Henderson, R. E. Pavlovicz, J. D. Allen, T. F. Gonzalez-Cestari, C. M. Orac, A. B. Bonnell, M. X. Zhu, R. T. Boyd, C. Li, S. C. Bergmeier and D. B. McKay, *J. Pharmacol. Exp. Ther.*, 2010, **334**, 761–774.
151. B. J. Henderson, T. F. Gonzalez-Cestari, B. Yi, R. E. Pavlovicz, R. T. Boyd, C. Li, S. C. Bergmeier and D. B. McKay, *ACS Chem. Neurosci.*, 2012, **3**, 682–692.
152. A. K. Hamouda, Z. J. Wang, D. S. Stewart, A. D. Jain, R. A. Glennon and J. B. Cohen, *Mol. Pharmacol.*, 2015, **88**, 1–11.
153. J. A. Olsen, P. K. Ahring, J. S. Kastrup, M. Gajhede and T. Balle, *J. Biol. Chem.*, 2014, **289**, 24911–24921.
154. M. Samochocki, A. Hoffle, A. Fehrenbacher, R. Jostock, J. Ludwig, C. Christner, M. Radina, M. Zerlin, C. Ullmer, E. F. Pereira, H. Lübbert, E. X. Albuquerque and A. Maelicke, *J. Pharmacol. Exp. Ther.*, 2003, **305**, 1024–1036.
155. J. Ludwig, A. Hoffle-Maas, M. Samochocki, E. Luttmann, E. X. Albuquerque, G. Fels and A. Maelicke, *J. Recept. Signal Transduction Res.*, 2010, **30**, 469–483.
156. S. B. Hansen and P. Taylor, *J. Mol. Biol.*, 2007, **369**, 895–901.
157. A. K. Hamouda, T. Kimm and J. B. Cohen, *J. Neurosci.*, 2013, **33**, 485–494.
158. S. Seo, J. T. Henry, A. H. Lewis, N. Wang and M. M. Levandoski, *J. Neurosci.*, 2009, **29**, 8734–8742.
159. L. C. Cesa, C. A. Higgins, S. R. Sando, D. W. Kuo and M. M. Levandoski, *Mol. Pharmacol.*, 2012, **81**, 239–249.
160. M. Bartos, K. L. Price, S. C. Lummis and C. Bouzat, *J. Biol. Chem.*, 2009, **284**, 21478–21487.
161. J.-L. Eisele, S. Bertrand, J.-L. Galzi, A. Devillers-Thiery, J.-P. Changeux and D. Bertrand, *Nature*, 1993, **366**, 479–483.
162. A. B. Smit, N. I. Syed, D. Schaap, J. van Minnen, J. Klumpermank, K. S. Kits, H. Lodder, R. C. van der Schors, R. van Elk, B. Sorgedrager, K. Brejc, T. K. Sixma and W. P. M. Geraerts, *Nature*, 2001, **401**, 261–268.
163. A. Auerbach, *Proc. Natl. Acad. Sci. U. S. A.*, 2005, **102**, 1408–1412.
164. Y. Chen, K. Reilly and Y. Chang, *J. Biol. Chem.*, 2006, **281**, 18184–18192.
165. C. Stokes and R. L. Papke, *Neuropharmacology*, 2012, **63**, 538–546.
166. A. A. George, L. M. Lucero, M. I. Damaj, R. J. Lukas, X. Chen and P. Whiteaker, *J. Biol. Chem.*, 2012, **287**, 25151–25162.
167. R. Spurny, S. Debaveye, A. Farinha, K. Veys, A. M. Vos, T. Gossas, J. Atack, S. Bertrand, D. Bertrand, U. H. Danielson, G. Tresadern and C. Ulens, *Proc. Natl. Acad. Sci. U. S. A.*, 2015, **112**, E2543–E2552.
168. L. Tolstoy, in *Anna Karenina. Trans.*, ed. R. Pevear and L. Volokhonsky, Penguin Books, New York, NY, 2002.
169. C. B. Marotta, C. N. Dilworth, H. A. Lester and D. A. Dougherty, *Neuropharmacology*, 2014, **77**, 342–349.
170. P. Iturriaga-Vasquez, J. Alzate-Morales, I. Bermudez, R. Varas and M. Reyes-Parada, *Pharmacol. Res.*, 2015, **101**, 9–17.

CHAPTER 16

Allosteric Binding in the Serotonin Transporter – Pharmacology, Structure, Function and Potential Use as a Novel Drug Target

CLAUS J. LOLAND[a], CONNIE SANCHEZ[b], PER PLENGE[a], KLAUS P. BØGESØ[c] AND BENNY BANG-ANDERSEN*[c]

[a]Department of Neuroscience and Pharmacology, Faculty of Health and Medical Sciences, University of Copenhagen, DK-2200 Copenhagen, Denmark; [b]Brintellix Science Team, Lundbeck USA, 215 College Road, Paramus, NJ 07652, USA; [c]H. Lundbeck A/S, Discovery Chemistry and DMPK, 9 Ottiliavej, DK-2500 Valby, Denmark
*E-mail: ban@lundbeck.com

16.1 Introduction

Serotonin (5-hydroxytryptamine, 5-HT) regulates through its at least 14 identified 5-HT receptor subtypes,[1] a variety of functions in the human psychology and physiology, such as mood, sleep, sexual drive and appetite.[2] Second after the glutamatergic neuronal network, the serotonergic is the largest neurochemical network in the central nervous system (CNS).

RSC Drug Discovery Series No. 56
Allosterism in Drug Discovery
Edited by Dario Doller

Published by the Royal Society of Chemistry, www.rsc.org

When an action potential arrives at the synaptic terminal of a serotonergic neuron, 5-HT is released to the synaptic cleft where it activates post and pre-synaptic 5-HT receptors. One of the main regulators of synaptic serotonergic transmission is the serotonin transporter (SERT). SERT is an integral membrane protein located in the vicinity of the presynaptic terminal, where it mediates rapid and efficient reuptake of released 5-HT. 5-HT is transferred into vesicles by the vesicular monoamine transporter and stored ready for reuse. Accordingly, SERT serves three main purposes: (1) it increases the speed at which the released 5-HT is cleared from the synaptic cleft, (2) it may prevent chemical crosstalk from diffusing 5-HT at adjacent synapses, and (3) it ensures reuse of transmitter with presumed savings in synthetic cost. Because of this, SERT is a central protein in controlling 5-HT signaling and an important pharmacological target in the treatment of diseases such as depression, anxiety, panic disorders, obsessive compulsive disorder, generalized anxiety disorder, post-traumatic stress disorder, bulimia, irritable bowel syndrome and premature ejaculation.[3–6]

SERT belongs to the family of Neurotransmitter:Sodium Symporters (NSS) which also include the dopamine transporter (DAT), the norepinephrine transporter (NET), gamma-aminobutyric acid (GABA) transporters (GAT1–4), glycine transporters (GlyT1–2) as well as several bacterial transporters.[7] They are all characterized by utilizing the sodium gradient across the cell membrane as driving force for the transport of its solute against its chemical gradient. Most NSS proteins have binding sites for two sodium ions (Na^+) even though in some of the transporters, only one Na^+ is stoichiometrically transported with substrate. For the mammalian NSS proteins, the binding of a chloride ion is also required for the transport process to occur.[6] In the case of SERT, a counter transport of K^+ has been reported.[8] It is generally believed that the transport process by NSS proteins follows the alternating access mechanism first proposed by Peter Mitchell[9] and refined by Oleg Jardetzky.[10] In this model, the transporter has the capacity to change conformation upon substrate binding rendering the binding site, at a given time, accessible only from the intracellular or the extracellular side of the membrane. At all times an impermeable barrier exists between the binding site and one side of the membrane, but the barrier can change from one side of the binding site to the other, giving the site alternate access to the two aqueous compartments that the membrane separates.[10]

Compelling evidence from X-ray crystallography suggests that the overall structure is conserved between all NSS family proteins from bacterial to mammalian members.[11–13] The most detailed structural information on NSS proteins comes from LeuT, a hydrophobic amino acid transporter from the thermophilic bacterium *Aquifex aeolicus*. The first structure was solved at 1.65 Å resolution[11] and showed a protein with 12 transmembrane segments (TMs) with TM1–5 having a similar folding as the following TM6–10 but inverted in the plane of the membrane (Figure 16.1). This TM5 + 5 repeat has since been shown to be a signature motif for several classes

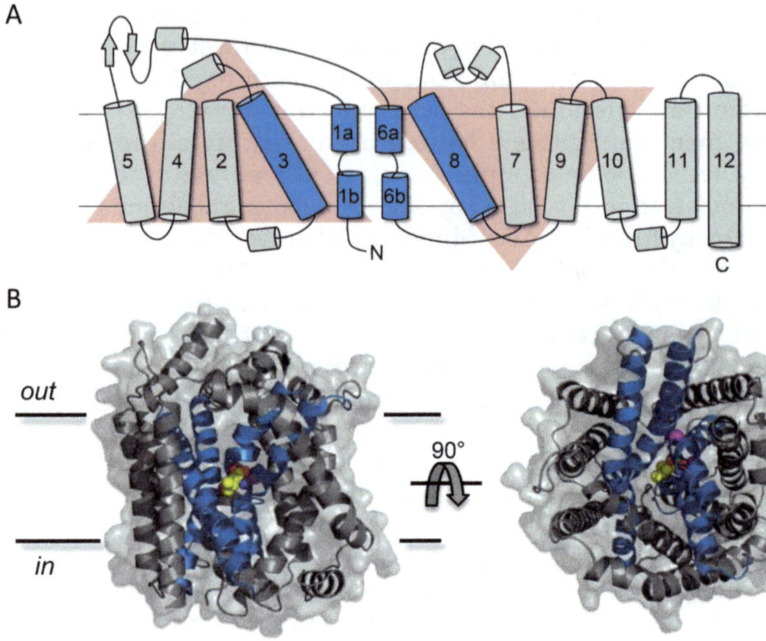

Figure 16.1 Topology and X-ray structure of LeuT, a bacterial NSS protein. (A) Topology of LeuT with the inner ring (TM1, TM3, TM6, and TM8) that forms the substrate binding site shown in blue. The red triangles highlight the 5 + 5 inverted repeats, formed by TM1–5 and TM6–10, which are related by an apparent two-fold symmetry. (B) Structure of LeuT in the substrate bound outward-facing occluded conformation [PDB 2A65]. The substrate binding site is located at the core of the transporter, and the substrate and two Na^+-ions are shown as van der Waals spheres in yellow and purple, respectively. The TM domains forming the inner ring are highlighted in blue seen from side-view (left; TM10 is omitted for clarity) and top-view (right; ECL1 and ECL4 are omitted for clarity). Figure modified from Kristensen *et al.*[6] with permission.

of secondary active transporters.[14] The core of LeuT was formed mainly by TM1, 3, 6, and 8 with participation from TM10 on the extracellular side and TM5 on the intracellular end of the protein. TM1 and TM6 possess unwound breaks in the helical structure (dividing both into an "a" and a "b" section) in the middle of the lipid bilayer exposing backbone carbonyl oxygen and nitrogen atoms that participate in the binding site for the substrate; in LeuT this has been found to be a leucine molecule.[11] This central substrate binding site is believed to have the same location in all NSS proteins, and is from here on termed the S1 site. The S1 site is completed by residues in TM3 and 8 as well as the coordination of one of the Na^+ binding to the Na1 site. The other bound Na^+ (Na2 site), has no direct interaction with the substrate bound to the S1 site. The central binding site was occluded from solution on both the extracellular and cytoplasmic sides; that is, both the extracellular and intracellular gates were closed. Even though LeuT only bears 20–25% amino acid homology to the mammalian NSS proteins, it is a general assumption that – at least in terms of the transmembrane domains – LeuT and mammalian NSS show high homology, making LeuT a suitable candidate as a model protein. This has also been confirmed by biochemical evidence,[15] but more importantly: X-ray structures of DAT[12] from *Drosophila melanogaster* (dDAT), and human SERT[13] (hSERT) both show an almost identical helical packing to LeuT.[12,13] The major difference between LeuT and the eukaryotic NSS are in the loop structures connecting the TM domains, which are considerably longer in DAT and SERT than in LeuT and include a disulfide bridge and glycosylation sites on the extracellular side. In the TMs the major differences lie in the coordination of a cholesterol molecule located in a groove between TM5 and TM7 and a kink in TM12 about halfway through the membrane.[12,13] The bound cholesterol is proposed to be poised to modulate the movement of TM1a that occurs during the transport cycle.[12] The TM12 kink causes the second half of the helix to turn away from the transporter.[12,13]

Due to the functional importance of SERT in controlling 5-HT homeostasis, it has been a central pharmacological target for drugs for decades.[3–6] Accordingly, SERT is the primary target for antidepressants, such as the tricyclic antidepressants (TCAs) (*e.g.*, imipramine, and clomipramine), the selective serotonin reuptake inhibitors (SSRIs) (*e.g.*, escitalopram (*S*-citalopram), fluoxetine, and paroxetine), as well as the serotonin and norepinephrine (NE) reuptake inhibitors (SNRIs) (*e.g.*, duloxetine, and venlafaxine).[6] Also the newly introduced multimodal antidepressants (*e.g.*, vilazodone, and vortioxetine)[16–18] possess a SERT-binding component in addition to other targets. Besides these prescribed drugs, also drugs of abuse such as cocaine, amphetamine, 3,4-methylenedioxymethamphetamine (MDMA or ecstasy), and ibogaine bind to SERT. They exhibit competitive binding to 5-HT, suggesting that their primary component is binding to the S1 site. This has been confirmed from the X-ray structure of hSERT, in which paroxetine and escitalopram were observed to coordinate to the S1 site.[13] Also mutagenesis studies[19–23] as well as molecular dynamic (MD) simulations in combination with Induced Fit Docking (IFD) protocols with the investigated ligands have confirmed this.[17,23–26] The only exception is ibogaine, which has been proposed to bind noncompetitively to 5-HT.[27] Results indicate that ibogaine induces an inward-facing structure of SERT at a binding site which is only accessible from the extracellular side of the membrane.[28]

In addition to the S1 site, compelling data indicates the existence of an additional binding site in SERT: the allosteric binding site. The following sections review the current knowledge about the allosteric site with respect to pharmacology, structure, and function. In addition, we will provide an overview of the chemical probes that are emerging for this site.

16.2 The Allosteric Binding Site in SERT

16.2.1 Early Findings of Allosteric Properties with SERT Ligands

It has been known for more than three decades that SERT possesses an allosteric binding site.[29] The first evidence was based on the observation that the dissociation of pre-bound [^{3}H]imipramine could be affected by the application of high-concentrations (50 μM) of 5-HT.[29] As can be derived from the law of mass action, the affinity of a ligand for its biological target (receptor, transporter, enzyme, *etc.*) is dependent on the on-rate and off-rate of the ligand. Whereas both reactions are equally dependent on temperature, the on-rate, or association constant, is also dependent on ligand concentration. In contrast, the off-rate, or dissociation constant, only depends on the interaction between ligand and receptor. Thus, a competitive (orthosteric) ligand cannot displace an already bound ligand, *i.e.* in the above case the competitive ligand cannot bind until [^{3}H]imipramine has dissociated from its binding site. Therefore, the inhibition of dissociation by the application of 5-HT to pre-bound [^{3}H]imipramine, can only be explained if the added ligand binds to another (allosteric) site. In other words, only in the case of an allosteric binding mode the added ligand is able to affect the dissociation of the pre-bound ligand. Indeed, the impeded dissociation of pre-bound [^{3}H]imipramine by 5-HT led the authors to conclude that 5-HT could bind to a site different from the imipramine binding site. The proposed allosteric site was further investigated for affinities towards other drugs. Accordingly, it was found that citalopram decreased the dissociation of pre-bound [^{3}H]imipramine and [^{3}H]paroxetine from their respective binding sites in both human platelets and rat neuronal membranes.[30] In contrast, paroxetine increased the dissociation rate of pre-bound [^{3}H]imipramine.[30] Imipramine did not affect the binding properties of any of the two radioligands. Citalopram dose-dependently inhibited the dissociation of [^{3}H]imipramine with a minimum effective concentration of 10 μM and with complete inhibition of dissociation at 200 mM. Importantly, the affinity of citalopram for the S1 site is in the low nanomolar range.[31] Accordingly, the high micromolar concentrations that were needed to produce these effects also suggested that citalopram binds to another site. Furthermore, paroxetine's facilitation of [^{3}H]imipramine dissociation also suggested a 'true' allosteric mechanism in the sense that binding to one site affects the binding of a drug to another site as opposed to a simple steric hindrance of the exit pathway for the orthosterically bound drug. The allosteric effects were not solely attributable to [^{3}H]imipramine and [^{3}H]paroxetine dissociation from its high-affinity site. Also, the dissociation of [^{3}H]citalopram was shown to be inhibited by high (200 μM) concentrations of SERT-uptake inhibitors such as citalopram and clomipramine both in human platelets and rat brain membranes.[32] Furthermore, the dissociation of racemic [^{3}H]citalopram from SERT was less affected by the *R*-enantiomer than the *S*-enantiomer.[33] Finally, [^{3}H]paroxetine dissociation could also be decreased by the application of 5-HT, whereas its effect on [^{3}H]citalopram was less pronounced.[32] Taken together, the data indicated that in addition to the high-affinity (nanomolar) binding site, the investigated tissue (either blood platelets or brain) possessed an additional low-affinity (micromolar) binding site for several of the same drugs, which allosterically modulated the binding properties to the high-affinity site.

The described experiments were performed on membranes prepared from brain or platelets (human or rat). Thus, after the tissue was isolated and homogenized it was suspended in a buffer solution. The suspension was aliquoted into test tubes and the radioligand added. After a suitable incubation time, a large volume of buffer containing the ligand to be tested for allosteric binding was added to the suspension. The reaction was stopped at different time points and the amount of remaining bound radioligand was assessed by scintillation counting. The advantage of this system was that SERT was in its native environment, suggesting that the allosteric effects were present in an *ex vivo* setting. However, a system like this could also raise questions since it contains many proteins which potentially could bind the investigated ligands. Thus, the possibility cannot be ruled out that a protein other than SERT was responsible for the observed allosteric effects. Also, only SERT wild type could be studied. To overcome the limitations in the system, the assay was instead applied to a heterologous expression system. Accordingly, SERT was cloned into an expression vector and transiently expressed into immortalized cell lines such as COS cells. COS cells are a fibroblast-like cell line derived from African green monkey kidney tissue expressing the Simian Virus 40 (SV40) large T antigen. This allows high-level expression of proteins in expression vectors carrying the SV40 promoter. The expression of SERT in COS cells was used to perform a comprehensive characterization of the allosteric effects of several drugs (*i.e.*, escitalopram, paroxetine, sertraline, fluoxetine, venlafaxine, *R*-citalopram, and 5-HT) on several high-affinity pre-bound radioligands (*i.e.*, [^{3}H]escitalopram, [^{3}H]paroxetine, [^{3}H]fluoxetine, [^{3}H]sertraline, [^{3}H]venlafaxine, [^{3}H]*N*,*N*-dimethyl-2-(2-amino-4-methylphenylthio)benzylamine (MADAM) and [^{125}I]RTI-55).[34] The conclusion was that the allosteric effect of a ligand depends on which compound was bound to the S1 site.[34] This agrees with the differential observation on allosteric effects of citalopram and paroxetine on [^{3}H]imipramine dissociation reported earlier.[30] It was also supported by the finding that when comparing the allosteric effects of escitalopram and *R*-citalopram, escitalopram was the most potent inhibitor of [^{3}H]escitalopram dissociation and *R*-citalopram was a more potent inhibitor of [^{3}H]imipramine dissociation.[35]

16.2.2 Location of the Allosteric Binding Site in SERT

The use of transient transfections of COS cells enables genetic modification of SERT by simple molecular biology and thereby investigation of the impact of specific residues in SERT on allosteric binding. With the application of this expression platform, it was possible to bring new perspectives and applications into the investigations of the allosteric binding site. First, COS cells expressing SERT did still possess allosteric properties, suggesting that the allosteric observations from previous investigations were indeed attributable to SERT.[36] Next, a series of mutations in SERT were introduced to delineate the allosteric site. The first investigations using mutagenesis were based on the observation that bovine SERT (bSERT) possessed a threefold lower affinity for escitalopram binding to the allosteric site compared with human SERT (hSERT). A systematic substitution of the residues which differ between hSERT and bSERT revealed three amino acid residues with effects on allosteric binding: Met180 (TM3), Tyr495 (TM10) and Ser513 (TM10).[36] The species-scanning approach was further explored with the observation that the chicken SERT (gSERT) possessed an allosteric site with even lower affinity than that observed for bSERT.[37] Two of the three residues found in bSERT are conserved between hSERT and gSERT. The latter (Ser513) was not determined to change the allosteric mechanism between the two species. Instead nine other residues were found to contribute to the difference in allosteric binding of escitalopram. The residues were distributed throughout the C-terminus in three distinct amino acid segments with three residues in TM10 (Ala505, Leu506, and Ile507), four residues in TM11 (Ile552, Ile553, Met558, and Ser559) and two residues in TM12 (Ser574, and Ile575). Together these residues almost completely revert allosteric binding of escitalopram in hSERT to the affinity found in gSERT. Elegantly, the hSERT residues could also be substituted into the corresponding gSERT positions, resulting in an allosteric affinity similar to the one observed for hSERT.[37] Assuming that the SERT structure resembles that of LeuT, the identified residues can be put into a structural context. Applying the residues to the cognate positions in LeuT reveals that the residues are not located in a confined area which could resemble an allosteric binding site for escitalopram. This was later confirmed from the SERT structure.[13] Instead the residues lie in an interface between LeuT monomers. The authors proposed that the mutations could affect communication between sites on monomers in an oligomeric protein: the binding of high-affinity ligands to the primary binding site at one subunit may perturb neighboring subunits, thus leaving the escitalopram binding site on the second subunit in a low-affinity state.[37] This statement was challenged by the finding that two specific mutations (Y95F and I172M) disrupted the high-affinity binding site for escitalopram by more than a 1000-fold,[19] but completely preserved allosteric binding of the drug,[35] suggesting that the high-affinity site and the allosteric site are two different entities.

Further information regarding the location of the allosteric binding site came from LeuT. Crystal structures of this protein were solved in complex with the antidepressants clomipramine,[38] imipramine,[38] desipramine,[38,39] fluoxetine,[40] and sertraline.[40] All drugs were found to coordinate to a site distinct from the S1 site located to the extracellular vestibule of LeuT forming part of the permeation pathway and located between the extracellular solution and the S1 site. The binding site was termed the S2 site. The bound antidepressants are cradled by the TM1b, TM6a, the top of TM10, and the approximate midpoint of TM3. The S2 site is demarcated from the top by the sharp turn of extracellular loop (ECL) 4. The S2-bound antidepressant is prevented from obtaining access further towards the S1-bound leucine by the conserved TM1–TM10 salt bridge between Arg30 and Asp404. The distance between the S1-bound leucine and S2-bound clomipramine is about 11 Å.[38] Binding of a ligand to the S2 site would probably have the effect of impeding the dissociation of a S1-bound ligand, if not in an allosteric manner then by a simple steric hindrance by occupying the exit pathway. All the antidepressants bound to LeuT with low affinity[38–40] but it was not possible to locate any antidepressant binding to the S1 site. Based on these observations and supported by the results of mutagenesis studies of the cognate position in human SERT and DAT, the authors proposed that the S2 site in LeuT corresponded to the high-affinity site for antidepressants in its mammalian counterparts.[39,40] Others were more reluctant in their interpretation.[38] Indeed the finding was controversial. Even though the helical packing might be similar between LeuT and mammalian NSS proteins, the residues involved in the binding sites are not particularly conserved. Most of the drugs had small planar tricyclic hydrophobic structures which might have low-affinity towards many different sites. The high-affinity binding of antidepressants in SERT might be at a completely different site. Indeed, results of both mutagenesis studies and molecular docking simulations did suggest that both TCAs,[23] escitalopram,[21] as well as cocaine[41] bound competitively to substrate in the S1 site of mammalian NSS proteins.

The further delineation of the allosteric binding site was performed with a combination of molecular docking simulations and experimental evidence.[42] The hypothesis was proposed that the identified binding site for antidepressants in LeuT, corresponded to the allosteric site in SERT. To embrace the simultaneous binding of two ligands, a complicated docking protocol was developed which combined IFD and MD simulations.[42] The approach allowed a SERT configuration with escitalopram bound to S1 and a subsequent clomipramine or escitalopram bound to the S2 site (Figure 16.2).

The IFD–MD protocol revealed an allosteric site in SERT similar to the S2 site in LeuT. The S2-bound escitalopram in SERT shares a few common positions with clomipramine in LeuT, but the site in SERT is positioned more extracellularly. The proposed allosteric binding site for escitalopram consists of side-chain interactions with residues in TM1 (Leu99, TRp103, and Arg104), TM3 (Ile179), TM10 (Ala486, Val489, and Lys490) and closed from the extracellular environment by ECL2 (Val236, and Leu237) and ECL4 (Gly402).[42] As in LeuT, the access towards the S1 site is blocked by the TM1–TM10 salt bridge (in SERT numbering: Arg104–Glu493). Systematic mutagenesis of the involved residues all decreased the potency of the allosterically bound molecule. In particular, the mutation of Gly402 (G402H) caused a complete elimination of the allosteric effect of escitalopram.[42] Further evidence regarding the location of the allosteric site was obtained by the insertion of two histidines into the S2 site of SERT. The histidine residues were in such a close proximity that it enabled the coordination of a zinc ion (Zn^{2+}). Zn^{2+} binding

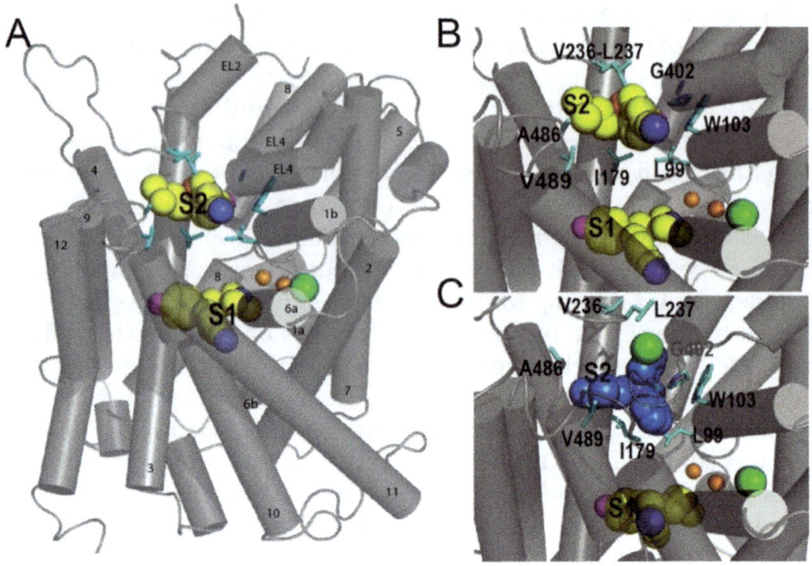

Figure 16.2 Binding modes of escitalopram and clomipramine in the S2 site of SERT. (A) The relative locations of the S1 and S2 sites in our SERT model viewed from an angle parallel to the membrane with escitalopram (yellow spheres) docked into both sites. The TMs and ECL2 and ECL4 are indicated (B + C) Top views of the escitalopram (yellow spheres) and clomipramine (blue spheres) in the S2 site, respectively, which is composed of residues from TM1, -3, and -10, and ECL2 and -4. S1 bound escitalopram (yellow) as well as the putative location of the two Na^+ (orange) and the Cl^- (green) are shown. Figure from Plenge *et al.*[42] with permission.

resulted in a significant decrease in the allosteric effect of escitalopram without affecting high-affinity S1 binding of [^{3}H]escitalopram.[42] Similar observations were seen for S2-bound clomipramine.

The SERT crystal structure confirms the presence of an allosteric binding site for escitalopram.[13] Here, two escitalopram molecules were found to be coordinated in the protein: one located to the S1 site and the other to the S2 site. The location of the S2-bound escitalopram is not identical between the IFD–MD model and the crystal structure: while escitalopram was located more towards TM3 and TM8 in the IFD–MD model, the molecule has twisted around TM10 and leans more towards TM6a and TM12 in the crystal structure. The Arg104–Glu493 salt bridge seems to play an important role in the formation of both binding sites. Accordingly, the experimental data showing decreased S2 affinity due to mutations in TM1 and TM10 could still be explained by the S2 site in the crystal structure. The Gly402 in ECL4 is more than 8 Å away from the ligand, but due to the flexible structure of the ECL4 that is supposed to protrude into the vestibule in the inward-facing conformations, it might still influence S2 binding in the dynamic protein.

Also, substituting the glycine for the histidine in that position will put the imidazole ring closer to the *N*,*N*-dimethyl propylamine group of escitalopram. Clearly, more experiments have to be performed to further elaborate the S2 binding site proposed from the crystal structure.

Support of low-affinity escitalopram binding at the S2 site has also come from performing Atomic Force Microscopy (AFM) on SERT-expressing cells with escitalopram covalently linked to the cantilever.[43] When the citalopram moiety on the cantilever bound to SERT, a pulling force developed during the upward movement of the tip. This force could be categorized into two populations which correspond to a S1 binding site (high-pulling-force) and a S2 site (low-pulling-force).[43] Introduction of the G402H mutant abolished the low-pulling-force population.[43] In contrast, pre-binding of escitalopram to the S1 site abolished the high-pulling-force population and, at the same time, increased the force necessary to pull the tip from the S2 site.

Taken together the data suggest that SERT contains two binding sites; a site (S1) which binds 5-HT, TCAs, and SSRIs with high affinity and an allosteric (S2) site which also binds 5-HT, TCAs (*e.g.* clomipramine), and SSRIs (*e.g.*, paroxetine, and escitalopram) but with a lower affinity. Binding to the S2 site can both impede and facilitate dissociation from the S1 site depending on the combination of bound ligands.

16.2.3 Is Allosteric SERT Modulation Therapeutically Relevant?

A series of animal studies of the citalopram enantiomers fueled the interest in allosteric modulation of SERT as a putatively relevant mechanism at therapeutic exposure levels. Shortly after the citalopram enantiomers were separated in 1988, results from *in vitro* studies in rat brain synaptosomes and *in vivo* behavioral studies in mice led to the conclusion that the 5-HT reuptake inhibition of racemic citalopram was mediated exclusively by the *S*-enantiomer.[44] The *R*-enantiomer was for about a decade thought to be pharmacologically inert. However, observations in a rat ultrasonic vocalization model of anxiety where escitalopram abolished the vocalization response, whereas citalopram only produced a partial reduction of the vocalization, triggered the hypothesis that *R*-citalopram counteracts the effect of escitalopram.[45] Shortly after, a series of microdialysis experiments demonstrated that *R*-citalopram counteracted the escitalopram-induced increase in extracellular 5-HT in freely moving rats both after systemic administration and after local application into the prefrontal cortex.[46] Subsequently, results from several other studies in animals confirmed the inhibitory actions of *R*-citalopram on escitalopram effects, both mechanistically and in models of depression and anxiety (reviewed by Sanchez *et al.*[47,48]).

After having addressed the possibility of a non-SERT-related mechanism or a pharmacokinetic interaction between the two enantiomers, the hypothesis was brought forward that the inhibitory effect of *R*-citalopram on escitalopram's regulation of 5-HT neurotransmission was due to a pharmacodynamic interaction between the enantiomers at SERT.[47] It was proposed that *R*-citalopram interferes with the action of escitalopram at the allosteric site and thereby prevents stabilization of escitalopram binding to the high-affinity binding site. However, as described in the previous sections, allosteric mechanisms of antidepressants had only been demonstrated *in vitro* using very high, micromolar, concentrations that far exceed the therapeutically active nanomolar concentrations of SERT inhibitors. Therefore, to further explore the hypothesis, activities were initiated to identify the location of the allosteric site relevant for the interaction between escitalopram and *R*-citalopram[36,37] and to enable generation of a transgenic mouse in which the allosteric site had been knocked out. Such a mouse would allow for studies of the physiological role of the allosteric mechanism as well as pharmacological studies.

Two independent projects were undertaken to generate lines of transgenic mice in which a proposed allosteric site had been removed.[49,50] In both cases mice were generated based on the mutations from the species-scanning mutagenesis project[37] carrying wild type hSERT or mutated hSERT with five or six amino acid substitutions, respectively, *i.e.* A505V, L506F, I507L, S574T, and I575T[49] or I552V, I553T, M558S, S559N, S574T, and I575T.[50] The comprehensive study of the five mutation line by Jacobsen *et al.*,[49] in which the *in vitro* allosteric mechanism at micromolar concentrations was confirmed in dissociation studies in the wild-type and abolished in the mutant mouse, failed to show any *in vivo* implications of the loss of the allosteric site with respect to escitalopram's and *R*-citalopram's pharmacology in microdialysis and behavioral assays of depression (marble burying and tail suspension tests).[49,50] Thus, in this study the escitalopram-induced increase of extracellular 5-HT after acute dosing, that fully occupies SERT, did not differ between wild-type and mutant mice.[49] Also *R*-citalopram did not affect binding of escitalopram. Similarly, studies in the six-mutation line failed to show a measurable difference from wild-type littermates with respect to changes in extracellular 5-HT after escitalopram treatment.[50] In contrast, mutant mice required a tenfold higher escitalopram dose than wild-type mice to achieve the same effects in the in marble burying assays, which could indicate a reduced sensitivity to drug treatment in the mutated line. A recent, as yet unpublished, study of SSRI effects on dorsal raphe nucleus firing in the five-mutation line revealed that escitalopram inhibited neuronal firing significantly more slowly in the absence of the allosteric site compared with the situation in mice expressing wild-type hSERT, and that the effects of paroxetine and sertraline were the same in wild-type and mutant mice (personal communication from Nasser Haddjeri). One possible explanation for the apparently contradictory results could be that serotonergic neuronal firing in the dorsal raphe nucleus is a more sensitive assay than microdialysis in projection regions to measure *in vivo* effects of allosteric SERT modulation, since the raphe nucleus has the highest density of SERT in the brain and the density in projection regions such as the cortex is relatively low.[51] Thus, the studies in transgenic mice generated from the domains identified in the cross-species study of hSERT *vs.* gSERT[37] have so far not provided a clear-cut answer to physiological and pharmacological implications of allosteric SERT modulation. It remains an empirical question whether a transgenic mouse derived from mutations of the allosteric site proposed by Plenge *et al.*[42] could provide a more clear-cut answer concerning the *in vivo* relevance of allosteric SERT modulation.

Taking all these findings together, it is therefore still a subject of debate whether the clinical efficacy of antidepressants that are S1 inhibitors as well as allosteric SERT modulators differ from that of antidepressants that only are orthosteric SERT inhibitors. The higher efficacy and faster onset of antidepressant action observed for escitalopram, as compared with racemic citalopram, has been suggested to depend on the allosteric interaction between escitalopram and *R*-citalopram.[52] However, the hypothesis relies at this point in time largely on *in vitro* studies and remains speculative from a clinical point of view. Availability of selective allosteric SERT modulators for clinical testing might hopefully permit this question to be addressed in the future.

16.3 Ligands that Bind to the S2 Site on SERT

More than 30 years ago, Plenge and Mellerup showed that antidepressant drugs could bind differently to SERT[30] and that different antidepressants could modulate this binding *via* an affinity-modulating site in an allosteric manner.[32] Later, an allosteric citalopram-binding site was proposed as already discussed, and it was shown that the dissociation of [^{3}H]escitalopram from SERT is most potently inhibited by escitalopram followed by *R*-citalopram, sertraline, 5-HT, and paroxetine.[36] The concentration giving half the maximal effect (EC_{50}) values for escitalopram and *R*-citalopram were in the micromolar range (5.1 μM and 25 μM, respectively[53]), whereas fluoxetine, venlafaxine, and duloxetine had no significant effect on the dissociation of [^{3}H]escitalopram.[36] However, escitalopram is much more potent on S1 (concentration giving 50% of maximum inhibition (IC_{50}) = 2.1 nM) than on S2, giving a S1 : S2 ratio of 0.004. Since then, several compounds with putative SERT S2 activity have been described and the structure–activity relationships (SARs) of these compounds will be discussed next. Key questions are: (1) how can the potency at SERT S2 be increased and (2) how can the selectivity between S1 and S2 binding be modified. The main goal is to identify a potent SERT S2 chemical probe that displays selectivity for SERT S2 *versus* S1 and other receptors, enzymes, and transporters. A high-quality chemical probe for SERT S2 will be instrumental for the future exploration of this target, as such a compound will make it possible to elucidate the role of SERT and specifically SERT S2 in healthy and diseased cells and tissues, leading to an increased understanding of the fundamental disease biology. Ideally, such a compound will also be suitable for *in vivo* administration and for probing the role of S2 in the living brain. Hopefully this will lead to new therapeutic opportunities for the discovery of medicines based on SERT.

The compounds that will be discussed in the following section have mainly been shown to have allosteric effects on the dissociation of [^{3}H]escitalopram from SERT. Some work has also been published on compounds inhibiting the dissociation of [^{3}H]imipramine from SERT, and it can be envisioned that other ligands binding to S1 might present their own SERT S2 SAR. Importantly, it has only been shown (indirectly) in a few cases that these compounds bind to SERT S2 (as defined by Plenge *et al.*[42]), but for ease of the following discussions this terminology will be used interchangeably for compounds from the citalopram series that have been shown to interact allosterically with SERT.

There are several reports on compounds based on citalopram or escitalopram as the lead structure, and which inhibit [^{3}H]escitalopram dissociation from SERT.[53–55] An overview of the structural changes that have been described in relation to the S2 SAR on the citalopram scaffold is shown in Figure 16.3.

The specific compounds from these reports are shown in Tables 16.1 and 16.2, which give a comprehensive SAR overview and ease pinpointing the preferred compound(s) across the literature. In the work of Banala *et al.*,[54] the aim was to study the SAR of citalopram at the basic nitrogen atom as well as at the 4, 5, and 4′ positions. It was proposed that the S2 site would appear larger and perhaps more promiscuous than the S1 site. Thus, the compounds that were prepared represented a diverse set of compounds from a structural point of view and it was concluded that "...analogs with sterically bulky and, in some cases, multifunctional substituents appended to the terminal amine of citalopram provide the basis for the design of a new series of compounds in which binding at S1 may be decreased, while the binding at S2 is improved".[54] In practice, the paper presented ideas for how to decrease SERT S1 binding but not for how to increase SERT S2 potency. The seven most potent S2 compounds had potencies similar to that of citalopram, in the micromolar range. The most interesting chemical probe disclosed in the paper is compound **19**, displaying a profile very similar to citalopram and escitalopram with an S2 activity of 4.9 μM and a S1 : S2 ratio of 0.006.

In the paper by Bøgesø *et al.*[53] a more focused SAR discussion was offered and especially on the replacement of the 5-fluorophenyl ring of citalopram

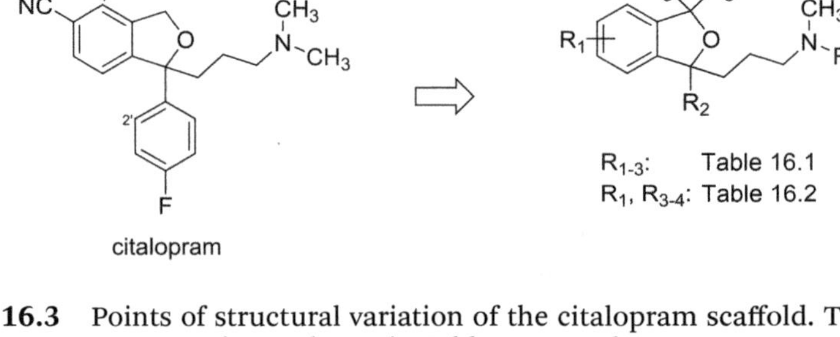

Figure 16.3 Points of structural variation of the citalopram scaffold. The specific compounds are shown in Tables 16.1 and 16.2.

Table 16.1 Allosteric SERT modulators based on the scaffold of citalopram–escitalopram (side chain kept constant).

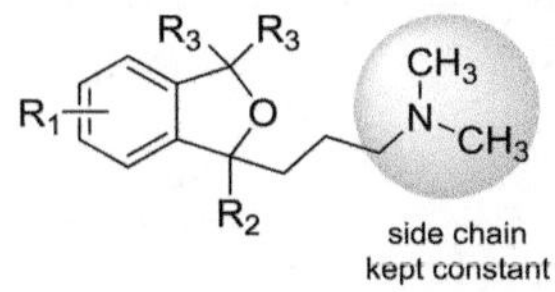

Compounds	R_1	R_2	R_3	IC_{50} (nM) SERT[a] S1	EC_{50} (nM) SERT S2
Citalopram[b]	5-CN	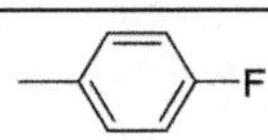	H	6.7	8700
Escitalopram[b]	5-CN	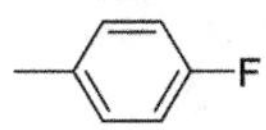	H	2.1	5100
R-Citalopram[b]	5-CN	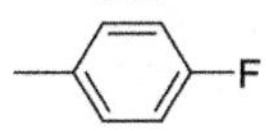	H	170	25 000
1[c]	5-CN	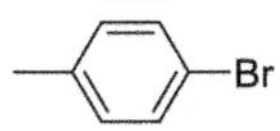	H	11.4[d]	4700
2[b]	5-CN	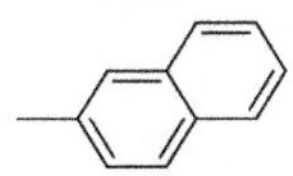	H	34	68 000
3[b]	5-CN	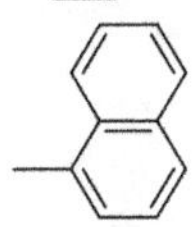	H	250	19 000
3a[b,e]	5-CN	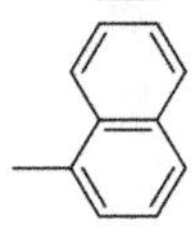	H	740	14 000
3b[b,e]	5-CN	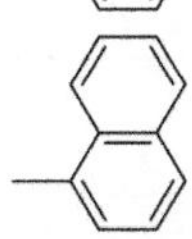	H	280	19 000
4[b]	5-CN	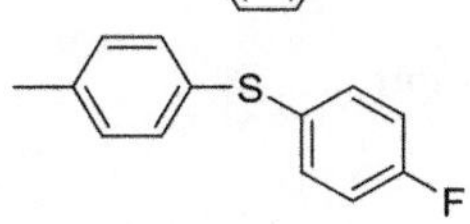	H	200	12 000
4a[b,e]	5-CN	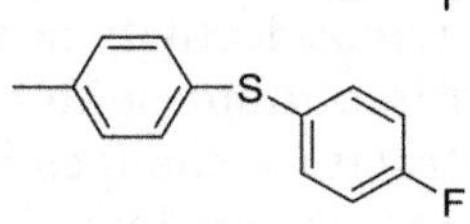	H	310	8300
4b[b,e]	5-CN	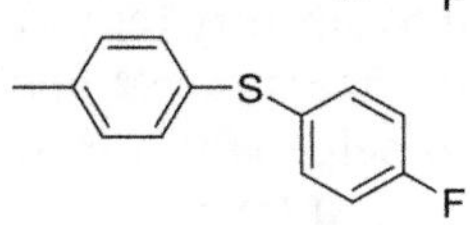	H	330	22 000
5[f]	5-CN	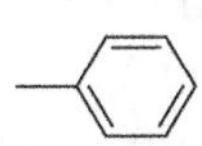	CH_3	10 000	12 000
6[f]	5-CN	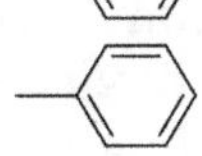	H	160	17 000

(continued)

Table 16.1 (*continued*)

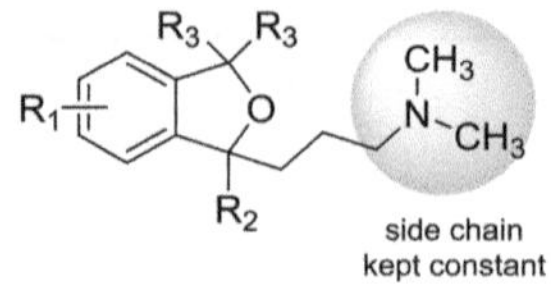

Compounds	R_1	R_2	R_3	IC_{50} (nM) SERT[a] S1	EC_{50} (nM) SERT S2
7[b]	5-Cl	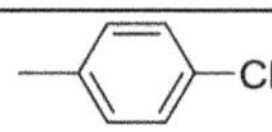	H	7.4	12 100
8[b]	4-Cl		H	9.2	6100
9[b]	H	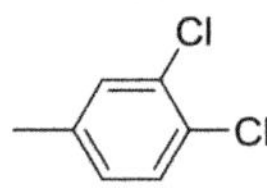	H	13	29 600
10[b]	H	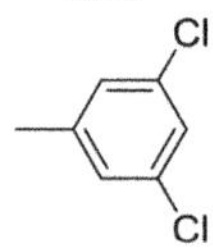	H	340	15 000
10a[b,e]	H	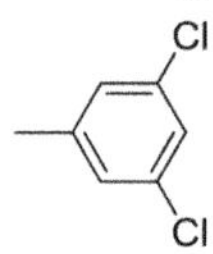	H	350	8800
10b[b,e]	H	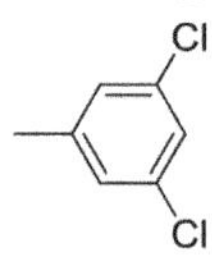	H	300	20 000

[a]Data from different but comparable assays.
[b]Bøgesø *et al.*[53]
[c]Banala *et al.*[54]
[d]Dissociation constant (K_i) value.
[e]a and b refer to different enantiomeric forms.
[f]Larsen *et al.*[55]

with other ring systems. The stereoselectivity at the S2 site for citalopram and key analogues was another important point of discussion in the paper. It was concluded that minor structural changes to the 5-fluorophenyl ring would lead to major changes in S1 affinity (compare compounds **2** and **3** as well as **7–9** and **10**) but limited effects on S2 activity. Again, this showed a path forward towards increased selectivity for S2 over S1 but no guidance on how to increase the activity at SERT S2. It was also shown that one of the enantiomers could be more potent at SERT S2 than the other enantiomer, but the eudismic ratio varied from 1 (compare compounds **3a** and **3b**) to maximally 2.7 (compare compounds **4a** and **4b** as well as **10a** and **10b**). In the case of escitalopram and *R*-citalopram, where the absolute configuration was determined, it was shown that the *S*-enantiomer was the more potent

Table 16.2 Allosteric SERT modulators based on the scaffold of citalopram–escitalopram (fluorophenyl ring kept constant).

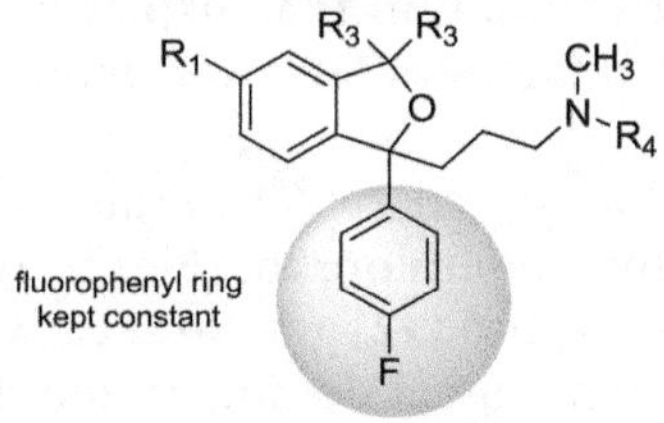

Compounds	R_1	R_3	R_4	IC_{50} (nM) SERT[a] S1	EC_{50} (nM) SERT S2
Citalopram[b]	CN	H	CH_3	6.7	8700
11[c]	CN	H	H	4.1	10 100
12[c]	CN	CH_3	CH_3	6400	3600
13[c]	CN	CH_3	H	500	3600
14[b]	Br	H	CH_3	13	30 200
15[d]	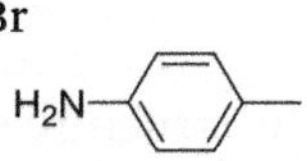	H	CH_3	29.2[e]	3600
S-(+)-**16**[d,f]	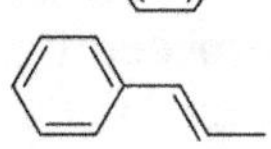	H	CH_3	10.6[e]	3400
17[b]	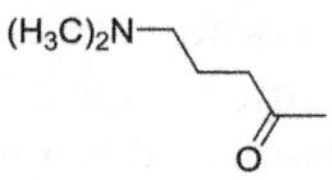	H	CH_3	1.6	714 000
18[d]	CN	H	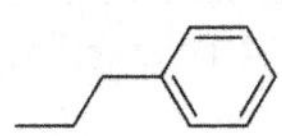	40.9[e]	5300
19[d]	CN	H	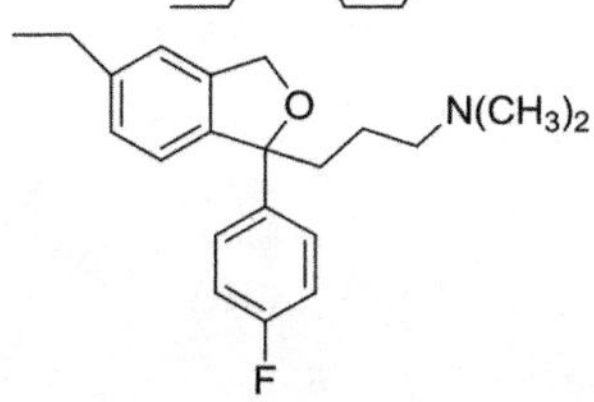	30.2[e]	4900
20[d]	CN	H	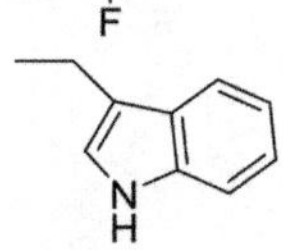	53.1[e]	3000
21[d]	CN	H	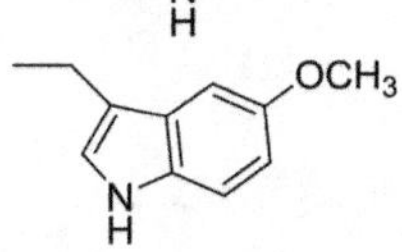	48.5[e]	5800

[a]Data from different but comparable assays.
[b]Bøgesø *et al.*[53]
[c]Larsen *et al.*[55]
[d]Banala *et al.*[54]
[e]K_i value.
[f]Zhang *et al.*[61]

enantiomer at SERT S2. In summary, the most potent compounds had potencies similar to that of citalopram, and the overall most interesting chemical probe from this paper was compound **10a**, displaying a similar EC_{50} value for S2 to citalopram, but with a factor of 50 lower potency at S1 giving an S1 : S2 ratio of 0.040.

Larsen *et al.*[55] provided a systematic SAR study where the four positions that distinguish citalopram from talopram (Figure 16.4) were varied. In this study the most potent compound (**12**) gave an EC_{50} value in the micromolar range (EC_{50} = 3.6 μM) and once again demonstrated the challenge of improving the potency at S2. However, as compound **12** was much weaker on S1 than citalopram by a factor of nearly a thousand, compound **12** is the first compound that has an S1 : S2 ratio of more than one (1.8).

Compound **22**[56] (Figure 16.4) has only been disclosed in the patent literature and a more comprehensive pharmacological characterization and SAR study of the series is warranted. Interestingly, compound **22** is structurally quite different from any of the citalopram analogs and it appears to be very potent on S2, allosterically inhibiting the dissociation of [^{3}H]escitalopram from SERT.

A few preliminary reports have been published on compounds affecting the dissociation of [^{3}H]imipramine from SERT and the structure of Lu AF60097 (Figure 16.4) was recently disclosed in a conference proceeding.[57] Lu AF60097, which is the *S*-enantiomer, was shown to be very potent on S2 (EC_{50} = 30 nM) and a factor of 10 weaker on S1 (IC_{50} = 300 nM), making this compound the first chemical probe on SERT S2 to show nanomolar activity at S2 and pronounced selectivity toward S1 (S1 : S2 ratio of 10). Lu AF60097 displays an S2 EC_{50} value of 6200 nM toward the dissociation of [^{3}H]escitalopram from

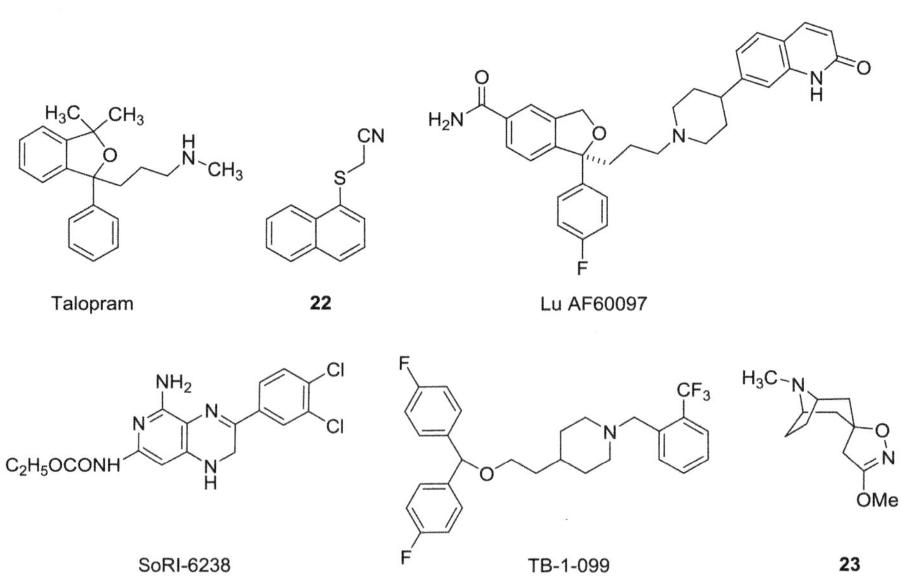

Figure 16.4 Other allosteric SERT modulators and talopram.

S1 (personal communication Per Plenge), substantiating that the allosteric effect of a ligand depends on which compound has been bound to the S1 site. It still needs to be understood what it means for the pharmacology when a compound is identified using dissociation of citalopram, imipramine, or maybe another ligand from the S1 site, especially if the compound is administered alone. Thus, it will be interesting to see further work being presented on Lu AF60097, in particular *in vivo* studies, as this compound is the first compound having nanomolar S2 activity.

A number of other compounds with allosteric activity at SERT have been described (SoRI-6238,[58] TB-1-099,[59] and the spirocyclic tropanyl-Δ^2-isoxazoline (**23**)[60]) but with weak activity and limited information on the pharmacology (Figure 16.4). It is not clear from the published descriptions whether these compounds also exert their allosteric effects *via* the same S2 site as discussed above or *via* another site on SERT. The reader is referred to the original papers for further details of these compounds.

16.4 Conclusions and Perspective

The majority of currently used antidepressants are potent SERT S1 inhibitors. In contrast, even though the existence of a SERT allosteric modulator site was realized more than 30 years ago, the research into this mechanism is still in its initial phase. For the last 10–15 years, there has been an increased awareness of allosteric SERT modulation and important progress has been made. The focus has been, to a large extent, on studies of citalopram and its enantiomers. Accordingly, *in vitro* strategies to determine the location and function of the allosteric site have progressed successfully, and have recently led to the hypothesis that the allosteric binding site in SERT is located in its extracellular vestibule, similar to the S2 site in LeuT. More work is clearly needed to substantiate this hypothesis and to understand the physiological roles of an allosteric site at that location. The possibility that allosteric modulator sites constitute a shared mechanism across the family of neurotransmitter–sodium symporters could eventually open up a whole new area of drug research.

The preliminary *in vivo* results from studies of transgenic mouse lines in which domains of relevance for the allosteric mechanism have been knocked out have so far been contradictory, even though a more thorough investigation using a broader variety of assays may help to clarify the situation. However, and more importantly, as *in vitro* studies advance our understanding of the location and function of the allosteric S2 site, they should be followed up by generation of more refined transgenic mouse lines where the allosteric site is directly affected. The availability of potent and selective S2 ligands is another very important prerequisite for advancing our understanding of the physiological role of the allosteric mechanism and assessing its potential as a valid therapeutic target. In that context it will be important to assess the therapeutic potential of a potent allosteric SERT modulator alone or in combination with an S1 inhibitor, first in animals but ultimately also in clinical studies in healthy subjects and in patients.

References

1. P. Celada, M. V. Puig and F. Artigas, *Front. Integr. Neurosci.*, 2013, **7**, 25.
2. L. F. Mohammad-Zadeh, L. Moses and S. M. Gwaltney-Brant, *J. Vet. Pharmacol. Ther.*, 2008, **31**(3), 187–199.
3. J. M. Gorman and J. M. Kent, *J. Clin. Psychiatry*, 1999, **60**(suppl. 4), 33–38.
4. G. F. Lasker, F. Halis and A. Gokce, *Curr. Drug Saf.*, 2014, **9**(2), 118–126.
5. P. Dharmshaktu, V. Tayal and B. S. Kalra, *J. Clin. Pharmacol.*, 2012, **52**(1), 6–17.
6. S. Kristensen, J. Andersen, T. N. Jørgensen, L. Sørensen, J. Eriksen, C. J. Loland, K. Strømgaard and U. Gether, *Pharmacol. Rev.*, 2011, **63**(3), 585–640.
7. M. H. Saier Jr., C. V. Tran and R. D. Barabote, *Nucleic Acids Res.*, 2006, **34**, D181–D186.
8. P. J. Nelson and G. Rudnick, *J. Biol. Chem.*, 1979, **254**(20), 10084–10089.
9. P. Mitchell, *Nature*, 1957, **180**(4577), 134–136.
10. O. Jardetzky, *Nature*, 1966, **211**(52), 969–970.
11. A. Yamashita, S. K. Singh, T. Kawate, Y. Jin and E. Gouaux, *Nature*, 2005, **437**(7056), 215–223.
12. A. Penmatsa, K. H. Wang and E. Gouaux, *Nat. Struct. Mol. Biol.*, 2015, **22**(6), 506–508.
13. J. A. Coleman, E. M. Green and E. Gouaux, *Nature*, 2016, **532**(7599), 334–339.
14. C. J. Loland, *Biochim. Biophys. Acta*, 2015, **1850**(3), 10.
15. J. Kniazeff, C. J. Loland, N. Goldberg, M. Quick, S. Das, H. H. Sitte, J. A. Javitch and U. Gether, *Neuropharmacology*, 2005, **49**(6), 715–723.
16. L. A. Dawson and J. M. Watson, *CNS Neurosci. Ther.*, 2009, **15**(2), 107–117.
17. J. Andersen, L. K. Ladefoged, D. Wang, T. N. Kristensen, B. Bang-Andersen, A. S. Kristensen, B. Schiøtt and K. Strømgaard, *ACS Chem. Neurosci.*, 2015, **6**(11), 1892–1900.
18. B. Bang-Andersen, T. Ruhland, M. Jørgensen, G. Smith, K. Frederiksen, K. G. Jensen, H. Zhong, S. M. Nielsen, S. Hogg, A. Mørk and T. B. Stensbøl, *J. Med. Chem.*, 2011, **54**(9), 3206–3221.
19. L. K. Henry, J. R. Field, E. M. Adkins, M. L. Parnas, R. A. Vaughan, M. F. Zou, A. H. Newman and R. D. Blakely, *J. Biol. Chem.*, 2006, **281**(4), 2012–2023.
20. L. K. Henry, J. R. Field and E. M. Adkins, *et al.*, Antidepressant recognition by serotonin transporters: TMs I and III interact to establish high affinity citalopram recognition, *Soc. Neurosci.*, 2004, 54.3.
21. J. Andersen, L. Olsen, K. B. Hansen, O. Taboureau, F. S. Jørgensen, A. M. Jørgensen and B. Bang-Andersen, *J. Biol. Chem.*, 2010, **285**(3), 2051–2063.
22. J. Andersen, O. Taboureau, K. B. Hansen, L. Olsen, J. Egebjerg, K. Strømgaard and A. S. Kristensen, *J. Biol. Chem.*, 2009, **284**(15), 10276–10284.

23. S. Sinning, M. Musgaard, M. Jensen, K. Severinsen, L. Celik, H. Koldsø, T. Meyer, M. Bols, H. H. Jensen, B. Schiøtt and O. Wiborg, *J. Biol. Chem.*, 2010, **285**(11), 8363–8374.
24. J. Grouleff, L. K. Ladefoged, H. Koldsø and B. Schiøtt, *Front. Pharmacol.*, 2015, **6**, 235.
25. H. Koldsø, K. Severinsen, T. T. Tran, L. Celik, H. H. Jensen, O. Wiborg, B. Schiøtt and S. Sinning, *J. Am. Chem. Soc.*, 2010, **132**(4), 1311–1322.
26. L. Celik, S. Sinning, K. Severinsen, C. G. Hansen, M. S. Møller, M. Bols, O. Wiborg and B. Schiøtt, *J. Am. Chem. Soc.*, 2008, **130**(12), 3853–3865.
27. M. T. Jacobs, Y. W. Zhang, S. D. Campbell and G. Rudnick, *J. Biol. Chem.*, 2007, **282**(40), 29441–29447.
28. S. Bulling, K. Schicker, Y. W. Zhang, T. Steinkellner, T. Stockner, C. W. Gruber, S. Boehm, M. Freissmuth, G. Rudnick, H. H. Sitte and W. Sandtner, *J. Biol. Chem.*, 2012, **287**(22), 18524–18534.
29. L. P. Wennogle and L. R. Meyerson, *Eur. J. Pharmacol.*, 1982, **86**(2), 303–307.
30. P. Plenge and E. T. Mellerup, *Eur. J. Pharmacol.*, 1985, **119**(1–2), 1–8.
31. E. L. Barker and R. D. Blakely, *Methods Enzymol.*, 1998, **296**, 475–498.
32. P. Plenge, E. T. Mellerup and H. Laursen, *Eur. J. Pharmacol.*, 1991, **206**(3), 243–250.
33. P. Plenge and E. T. Mellerup, *Pharmacol. Toxicol.*, 1997, **80**(4), 197–201.
34. F. Chen, M. B. Larsen, C. Sanchez and O. Wiborg, *Eur. Neuropsychopharmacol.*, 2005, **15**(2), 193–198.
35. P. Plenge, U. Gether and S. G. Rasmussen, *Eur. J. Pharmacol.*, 2007, **567**(1–2), 1–9.
36. F. Chen, M. B. Larsen, H. A. Neubauer, C. Sanchez, P. Plenge and O. Wiborg, *J. Neurochem.*, 2005, **92**(1), 21–28.
37. H. A. Neubauer, C. G. Hansen and O. Wiborg, *Mol. Pharmacol.*, 2006, **69**(4), 1242–1250.
38. S. K. Singh, A. Yamashita and E. Gouaux, *Nature*, 2007, **448**(7156), 952–956.
39. Z. Zhou, J. Zhen, N. K. Karpowich, R. M. Goetz, C. J. Law, M. E. Reith and D. N. Wang, *Science*, 2007, **317**(5843), 1390–1393.
40. Z. Zhou, J. Zhen, N. K. Karpowich, C. J. Law, M. E. Reith and D. N. Wang, *Nat. Struct. Mol. Biol.*, 2009, **16**(6), 652–657.
41. T. Beuming, J. Kniazeff, M. L. Bergmann, L. Shi, L. Gracia, K. Raniszewska, A. H. Newman, J. A. Javitch, H. Weinstein, U. Gether and C. J. Loland, *Nat. Neurosci.*, 2008, **11**(7), 780–789.
42. P. Plenge, L. Shi, T. Beuming, J. Te, A. H. Newman, H. Weinstein, U. Gether and C. J. Loland, *J. Biol. Chem.*, 2012, **287**(47), 39316–39326.
43. R. Zhu, D. Sinwel, P. S. Hasenhuetl, K. Saha, V. Kumar, P. Zhang, C. Rankl, M. Holy, S. Sucic, O. Kudlacek, A. Karner, W. Sandtner, T. Stockner, H. J. Gruber, M. Freissmuth, A. Hauck Newman, H. H. Sitte and P. Hinterdorfer, *Angew. Chem., Int. Ed. Engl.*, 2016, **55**(5), 1719–1722.
44. J. Hyttel, K. P. Bøgesø, J. Perregaard and C. Sanchez, *J. Neural Transm.: Gen. Sect.*, 1992, **88**(2), 157–160.

45. C. Sanchez, *Eur. J. Pharmacol.*, 2003, **464**(2–3), 155–158.
46. A. Mørk, M. Kreilgaard and C. Sanchez, *Neuropharmacology*, 2003, **45**(2), 167–173.
47. C. Sanchez, K. P. Bøgesø, B. Ebert, E. H. Reines and C. Braestrup, *Psychopharmacology*, 2004, **174**(2), 163–176.
48. C. Sanchez, *Basic Clin. Pharmacol. Toxicol.*, 2006, **99**(2), 91–95.
49. J. P. Jacobsen, P. Plenge, B. D. Sachs, A. L. Pehrson, M. Cajina, Y. Du, W. Roberts, M. L. Rudder, P. Dalvi, T. J. Robinson, S. P. O'Neill, K. S. Khoo, C. S. Morillo, X. Zhang and M. G. Caron, *Psychopharmacology*, 2014, **231**(23), 4527–4540.
50. K. E. Murray, K. J. Ressler and M. J. Owens, *Pharmacol., Biochem. Behav.*, 2016, **141**, 50–57.
51. P. Plenge, E. T. Mellerup and H. Laursen, *Prog. Neuro-Psychopharmacol. Biol. Psychiatry*, 1990, **14**(1), 61–72.
52. C. Sanchez, E. H. Reines and S. A. Montgomery, *Int. Clin. Psychopharmacol.*, 2014, **29**(4), 185–196.
53. K. P. Bøgesø and C. Sanchez, The discovery of citalopram and its refinement to escitalopram, in *Analogue-based drug discovery III*, ed. J. Fischer, C. R. Ganellin and D. P. Rotella, Wiley-VCH Verlag GmbH & Co. KGaA, Weinheim, Germany, 2013, pp. 269–294.
54. A. K. Banala, P. Zhang, P. Plenge, G. Cyriac, T. Kopajtic, J. L. Katz, C. J. Loland and A. H. Newman, *J. Med. Chem.*, 2013, **56**(23), 9709–9724.
55. M. A. Larsen, P. Plenge, J. Andersen, J. N. Eildal, A. S. Kristensen, K. P. Bøgesø, U. Gether, K. Strømgaard, B. Bang-Andersen and C. J. Loland, *Br. J. Pharmacol.*, 2016, **173**(5), 925–936.
56. H. A. Neubauer, O. K. Wiborg and H. H. Jensen, *Allosteric binding compounds*, 2010.
57. B. Bang-Andersen, P. Plenge and K. Larsen, *et al.*, Discovery and structure–activity relationship of ligands for the serotonin transporter S2 binding site based on the citalopram scaffold, in *paper presented at: EFMC-ISMC, XXIII International symposium on medicinal chemistry*, Lisbon, Portugal, 2014.
58. A. Nandi, C. M. Dersch, M. Kulshrestha, S. Ananthan and R. B. Rothman, *Synapse*, 2004, **53**(3), 176–183.
59. B. Nightingale, C. M. Dersch, T. L. Boos, E. Greiner, W. J. Calhoun, A. E. Jacobson, K. C. Rice and R. B. Rothman, *J. Pharmacol. Exp. Ther.*, 2005, **314**(2), 906–915.
60. C. Dallanoce, M. Canovi, C. Matera, T. Mennini, M. De Amici, M. Gobbi and C. De Micheli, *Bioorg. Med. Chem.*, 2012, **20**(21), 6344–6355.
61. P. Zhang, G. Cyriac, T. Kopajtic, Y. Zhao, J. A. Javitch, J. L. Katz and A. H. Newman, *J. Med. Chem.*, 2010, **53**(16), 6112–6121.

CHAPTER 17

Allosteric Inhibition of Abl Kinase

ANNA LUCIA FALLACARA[a], SILVIA SCHENONE[b] AND MAURIZIO BOTTA*[a,c]

[a]Dipartimento di Biotecnologie, Chimica e Farmacia, Università degli Studi di Siena, Via Aldo Moro 2, 53100 Siena, Italy; [b]Dipartimento di Farmacia, Università degli Studi di Genova, Viale Benedetto XV, 3, 16132 Genova, Italy; [c]Sbarro Institute for Cancer Research and Molecular Medicine, Center for Biotechnology, College of Science and Technology, Temple University, BioLife Science Building, Suite 333, 1900 N 12th Street, Philadelphia, Pennsylvania 19122, US

*E-mail: botta.maurizio@gmail.com

17.1 Introduction

Tyrosine kinases belong to a class of enzymes involved in a wide variety of key signaling processes in cells. The family of human protein kinases, of which there are about 500 members, represents the third largest and the most important enzyme class; they are estimated to be responsible for modifying one-third of the human proteome.[1,2] These enzymes catalyze the transfer of the γ-phosphate group from ATP to specific tyrosine hydroxyl groups on target protein substrates. It has been conclusively demonstrated that kinase alterations (especially hyperactivation, hyperproduction, or mutations),

RSC Drug Discovery Series No. 56
Allosterism in Drug Discovery
Edited by Dario Doller

Published by the Royal Society of Chemistry, www.rsc.org

leading to the disruption of cell signaling cascades, play important roles in several diseases, including cancer, inflammation, neurological disorders, and diabetes.[3]

In this chapter we will focus our attention on cellular Abelson murine-leukemia viral-oncogene homolog-1 (c-Abl), which represents one of the most ubiquitously conserved tyrosine kinases. c-Abl (encoded by the ABL1 gene) is a non-receptor tyrosine kinase expressed in most tissues. In cells, Abl is localized in both the nucleus and cytoplasm, and it can easily shuttle between the two compartments.[2,3]

The activation state of Abl is controlled by a number of molecular mechanisms[4,5] which, if dysregulated by mutations, translocations, deletions, or duplications, often result in an overactive oncogenic protein implicated in the development of cancer or other malignancies. The following paragraphs will describe in detail the structure of Abl kinase and the complex network of intramolecular interactions which occur between its domains and determine the activation state of the enzyme.

17.2 Structure of Abl Kinase

The Abl family of tyrosine kinases consists of two members, Abl and Arg (Abl-related gene), encoded by ABL1 and ABL2 genes, respectively.[6,7] Each family member has two isoforms: one that is myristoylated at the N-terminus (1b isoform or IV in mouse) and the other that is deficient in *N*-myristoylation site due to alternative splicing of the first exon (1a isoform or I in mouse). Abl kinases share the core catalytic unit with the majority of other cytoplasmic tyrosine and serine–threonine kinases and they all have a C-terminal tail that carries numerous protein–protein interaction sites. The N-terminus of the 1b variant of c-Abl has an N-terminal cap (N-cap) of about 80 amino acids myristoylated on the Gly2 residue (Figure 17.1), whereas the 1a variant is 19 amino acids shorter and does not undergoes this modification. Following the N-cap there are the SH3 (Src homology 3) and SH2 (Src homology 2) domains, which like in Src (sarcoma) family kinases, play a key role in substrate interaction and auto-regulation (Figure 17.1).

The SH3 and SH2 domains are among the most common regulatory domains found in human kinases.[8] A range of proteins are reported to interact directly or indirectly with the SH3 and SH2 domains of c-Abl, and are implicated in the proper regulation of the kinase activities of Abl family members in the cytoplasm.[9–11] The SH3 domain is composed of two short, antiparallel β-sheets packed against each other to form a barrel-shaped structure. The Abl SH3 domain binds to proline-rich peptides which are able to form poly-proline type II helix (PPII).[12] The Abl SH3 domain was the first SH3 domain to be crystallized bound by a ligand peptide, thus revealing how protein–protein SH3–substrate interactions are mediated.[13] In the down-regulated conformation of Abl, the SH3 domain binds a PPII motif located on the linker sequence which connects the SH2 domain to the catalytic domain.

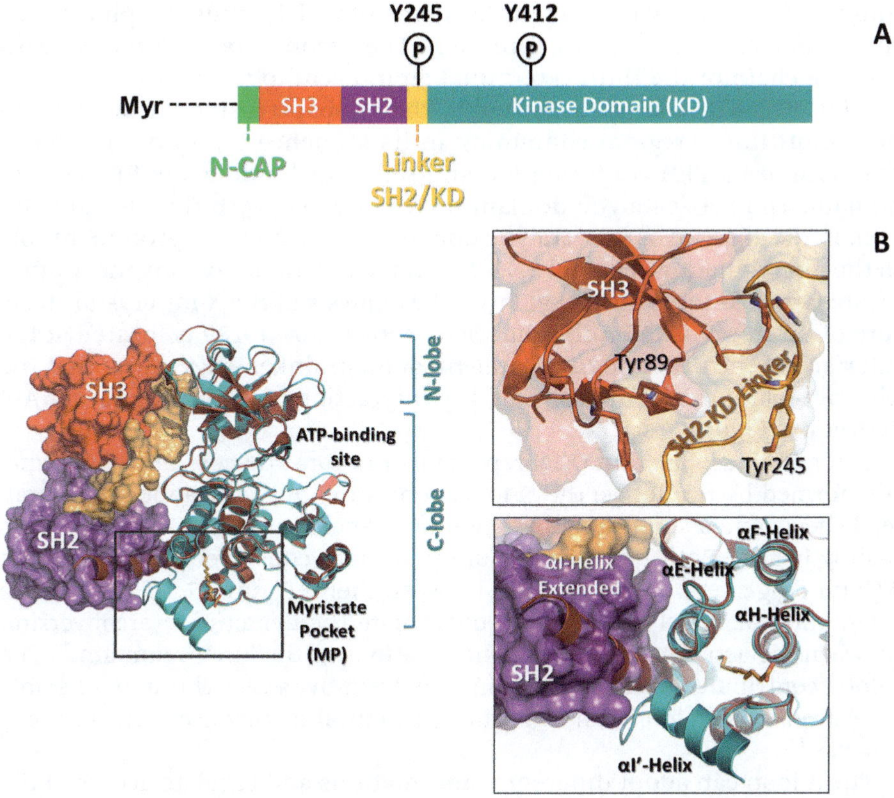

Figure 17.1 (A) Schematic representation of c-Abl kinase structure including domain organization; (B) *Left*. Overview of c-Abl structure in its downregulated conformation. SH3 (red), SH2 (violet), SH2–KD linker (orange) and KD (teal) have been highlighted. The Abl structure characterized by the αI-helix in an extended conformation (ruby, PDB code 3IK3) was overlapped. *Top Right*. Focus on the contacts between the SH3 domain (red) and the SH2–KD linker (orange). The key residues Tyr89 and Tyr245 have been displayed as sticks. *Bottom Right*. Focus on the MP. The comparison suggests a myristoyl switching mechanism: the binding of the myristoyl group into the pocket causes the bending of the αI-helix and the subsequent approach of SH2 domain to the kinase domain.

Deletion or mutation within the SH3 domain causes up-regulation of Abl kinase activity, supporting a negative regulatory role for this domain in enzyme modulation.[14]

The Abl SH2 domain mediates sequence-specific recognition of phosphotyrosine-containing sequences. This phosphotyrosine-binding function contributes to recognition and phosphorylation of some Abl substrate proteins. Structurally, the Abl SH2 domain consists of a central β-sheet flanked by α-helices on each side. The central β-sheet divides the domain into two functionally distinct pockets, one of which binds the phosphotyrosine side chain in the target protein. The second pocket interacts with the side chain of the third C-terminal amino acid to the phosphotyrosine residue. The SH2 domain is linked to the bilobal tyrosine kinase domain by a short linker region, containing in its sequence a poly-proline motif that assumes a PPII conformation and serves as binding site for the SH3 domain. The SH2-catalytic domain linker interacts with the SH3 domain even in the absence of the catalytic domain[15] confirming its prominent role in the Abl regulation. The Abl SH3 domain could undergo phosphorylation by Src family kinases on Tyr89. From the analysis of the X-ray crystal structure of down-regulated Abl, it has been shown that Tyr89 is located at the interface of the SH3 and SH2-catalytic domain linker, suggesting that the phosphorylation may disturb this key interaction contributing to full Abl activation (Figure 17.1).

The Abl kinase domain (KD) represents the catalytic core of the enzyme. It is formed by two lobes: the N-lobe, consisting of five-stranded antiparallel β-sheets and a single α-helix called the αC-helix, and the C-lobe, which is mainly helical. Both lobes contribute important conserved residues to the ATP binding site, a cleft located in between them (Figure 17.1). The relative orientation of the N- and C-lobes coordinate the dynamic interconversion between the active and inactive conformations of the kinase domain.[16] The C-lobe contributes to the regulation of protein dynamics through a flexible activation loop (A-loop), an important structural feature common to many kinases.[1]

The A loop can adopt different conformations and regulate access of the substrate to the catalytic cleft.[17] The A-loop is implicated in the regulation of the catalytic activity of Abl not only through the change of its conformation, but also by the phosphorylation of Tyr412, related to full kinase activation. The activation loop contains a highly conserved sequence, called the DFG (Asp–Phe–Gly) motif, adjacent to the ATP-binding site, which is very important for determining the activation state of the enzyme. This sequence is present within nearly every eukaryotic kinase domain, and the aspartate residue in the DFG triad is believed to be important for catalysis: in the active kinase conformation this motif points toward the ATP-binding site, where it can coordinate an ATP-bound magnesium ion, stabilizing the so-called "DFG-in" conformation (Figure 17.2A). Conversely, in the inactive conformation of the enzyme, the DFG-motif can adopt a dramatically different conformation, called "DFG-out", in which the side chains of aspartate (Asp400) and phenylalanine (Phe401) swap positions, leaving the aspartate pointing away from the ATP-binding site (Figure 17.2B). In the "DFG-out" conformation, these amino acids are not oriented in a manner that allows "phosphotransfer" and that blocks the approach of both the ATP and the substrate. The conformational change between "DFG-in" and "DFG-out" conformations is called "DFG-flip" and is a central feature in Abl activation.[18]

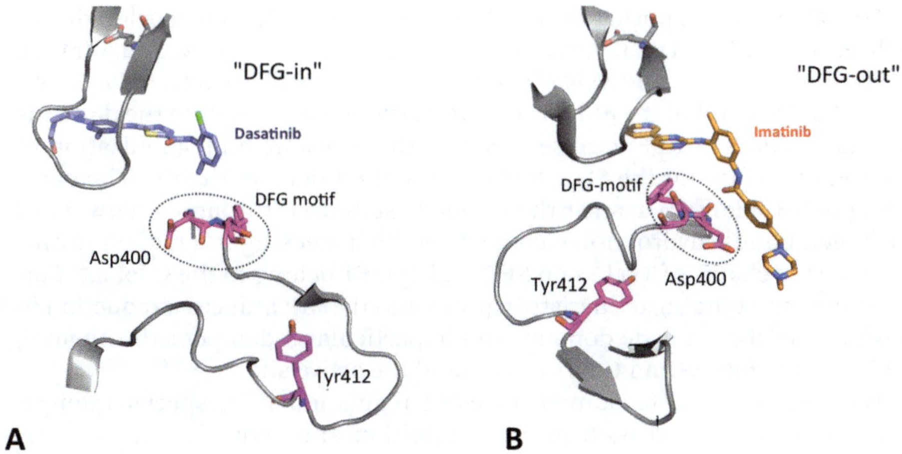

Figure 17.2 Details of "DFG-in" (A) and "DFG-out" (B) conformation of the DFG-motif in the activation loop of Abl.

17.3 Intramolecular Interactions Regulating Abl Activity

17.3.1 The Complex Mechanism of Abl Inactivation

The 1b splice form of Abl carries a myristate group, a saturated fatty acid with 14 carbon atoms, on Gly2 at its N-terminus. Protein *N*-myristoylation is a post-translational modification implicated in targeting proteins to membranes, but myristoylation alone is not sufficient for stable membrane binding. For membrane targeting, dual myristoylation and palmitoylation or additional polybasic amino acids that interact with negatively charged phospholipids at the inner leaflet of the plasma membrane are necessary.[19] In Abl 1b, neither of those two latter additional membrane targeting signals have been observed, and in line with this, only a minor fraction of Abl is localized in the proximity of the plasma membrane. Further evidence has confirmed that non-myristoylated Abl was not differentially localized from the myristoylated protein, demonstrating that myristoylation alone is not enough for membrane localization.[20] In contrast, Abl myristoylation was found to be involved in the regulation of kinase activity. Mutants of Abl 1b that lack the myristoylation site show strongly dysregulated cellular and *in vitro* tyrosine kinase activity.[21]

The myristoyl group penetrates deep into a hydrophobic pocket, called the myristate pocket (MP), at the base of the C-lobe on the KD. When the myristoyl group is absent the C-terminal helix, αI, extends from Ser504 to Ser522 with subsequent residues being disordered.[22] The binding of the myristate acid to the MP induces a break in the αI helix at Phe516, with residues Phe516–Ser519 forming a four-residues turn, which produces a 90° bend forming another helix, αI′, which spans residues Ser520 to Lys531 (Figure 17.1).

The MP is a small pocket formed by hydrophobic side chains belonging to helices αE (Ala356, Leu359, and Leu360), αF (Leu448, Ala452, and Tyr454), αH (Cys483, Pro484, and Val487), and the new αI′-helix (Ile521, Val525, and Leu529). The bend of the αI-helix has dramatic consequences in the domains rearrangements of Abl because it causes the exposure and stabilization of a docking surface for the SH2 domain onto the C-lobe of the KD. The packaging of SH2 domain against the C-lobe is stabilized through a network of hydrogen bonds, hydrophobic contacts and a π-stacking interaction involving the side chains of Tyr158 on SH2 and Tyr361 belong to the C-lobe.[23] Furthermore, the binding of myristoyl group into the MP induces a reduction in flexibility of the catalytic domain, and in particular it dampens the opening and closing of its N- and C-termini around the active site.[17,20,24,25]

The role of the SH2 domain in c-Abl regulation is of special interest, because it is involved both in auto-inhibition and activation of the catalytic domain. In the autoinhibited state, the SH3 and SH2 domains and the proline-rich sequence localized in the linker between the SH2 domain and the KD form a rigid clamp around the catalytic site.[26] This SH3–SH2 clamp, which packs tightly against the distal face of the KD, causes a further conformational change in the ATP-binding pocket.

The interaction of the myristoylated N-cap with the C-lobe is critical for the maintenance of the autoinhibition state (also called "closed" conformation, Figure 17.3), as mutation of the myristoylation signal sequence results in a highly active kinase.[21]

Interestingly, small molecules that bind to MP also modulate kinase activity, supporting a connection between this allosteric pocket and the kinase active site.[27] Opening the closed conformation by any means is predicted to induce catalytic activation. Accordingly, SH2 and SH3 binders can induce c-Abl activation, and disruption of these interactions by mutation leads to transforming activity.

17.3.2 Mechanisms of Abl Activation

Upon activation, Abl kinase undergoes extensive domain rearrangements. Results from recent studies show that the SH2 domain may undergo significant reorientation during kinase activation. In the highest activated state, evidence suggests that the SH2 domain moves from its negative regulatory position in the back of the C-lobe, to a new position on the top of the KD. Sometimes referred to as the "top-hat" conformation,[27,28] this orientation allows the SH2 domain to interact with the N-terminal lobe to stabilize the active form of the KD (Figure 17.3). Mutagenesis of amino acids involved in this SH2–N-lobe interaction, particularly Ile164, impairs the kinase activity, supporting a positive regulatory role for this interaction in kinase function.[29,30]

The positioning of the SH2 domain in the "top-hat" conformation exposes the residue Tyr245 located on the SH2–KD linker (Figure 17.3).[6] The SH2–KD linker in Abl, as described above, contains a PPII motif in which the second proline is replaced by Tyr245, and its phosphorylation was predicted to

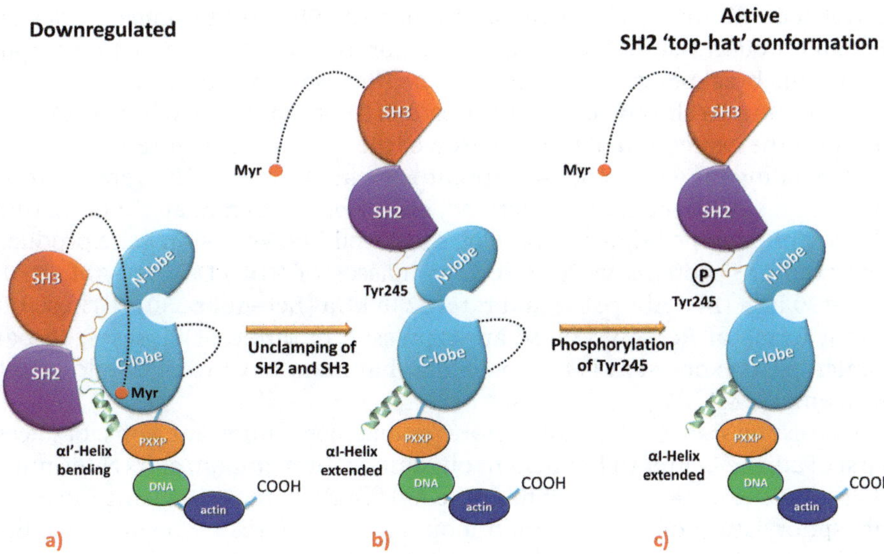

Figure 17.3 Model of Abl kinase switching between downregulated and fully activated conformation. (a) The assembled state is unlatched by release of the cap, anchored by the myristoyl acid to the MP of the Abl 1b variant. (b) SH2 and SH3 unclamping from the back of the KD expose Tyr245 to phosphorylation. (c) The kinase domain can then be fully activated by phosphorylation of Tyr245 and Tyr412 (not shown) in the activation loop.

disrupt the autoinhibited structure, by creating a steric hindrance which does not allow the SH3 domain to keep contact with the PPII motif, in agreement with the high levels of Abl activity observed upon phosphorylation of Tyr245.[18]

17.4 The Importance of Abl in Cancer Development

In humans, Abl kinases are involved in a number of chromosomal abnormalities in different cancers that lead to the expression of fusion proteins, but no point mutations in ABL1 and ABL2 genes have been yet identified. In all human Abl fusion proteins, regions upstream of the Abl kinase domain are replaced by other short protein sequences that drive dimerization or polymerization of the kinase.[31] In general, the fusion events lead to the a loss of kinase autoinhibition by removing the myristoylation site and, in some cases, the SH3 and SH2 domains.[26]

17.4.1 Bcr–Abl Kinase

The most well-known trigger protein deriving from the fusion of Abl gene with a short sequence from another protein is the Bcr–Abl kinase. The (9;22)(q34;q11) chromosomal translocation that results in the formation of the Philadelphia (Ph) chromosome has been one of the first chromosomal aberrations directly correlated with the development of cancer.[32] The Ph chromosome encodes the Bcr–Abl protein that is formed by the fusion of ABL1 gene (on chromosome 9) and the breakpoint cluster region gene (Bcr, on chromosome 22). The Bcr translocation on chromosome 9 upstream of the second exon of the ABL1 gene prevents the expression of the first exon of the ABL1 gene (Figure 17.4).

Depending on where the breakpoint is located, several different forms of Bcr–Abl fusion protein are generated, each leading to a related but distinct disease phenotype (Figure 17.4). The most well-known fusion gene products observed are proteins with molecular masses of 210 kDa (Bcr–Abl p210), 185–190 kDa (Bcr–Abl p185) and rarely 230 kDa (Bcr–Abl p230).[33] These different forms of Bcr–Abl kinase are expressed in different diseases. In particular, p210 expression is the molecular hallmark of Chronic Myelogenous Leukemia (CML).[34]

The expression of Bcr–Abl fusion protein has important consequences. First of all, the Bcr motif has a point oligomerization domain at its N-terminus that allow the juxtaposition of multiple Abl kinase domains causing the trans-phosphorylation of the activation loop and other tyrosine residues leading to kinase activation and to a deregulation of downstream pathways. Another important point is that Bcr–Abl, because of the translocation of Bcr upstream of the ABL1 gene, completely lacks the N-cap and the myristoylation site.[18,21] As a consequence, the negative regulatory function of the myristic acid and the interactions between the N-cap and SH2 and SH3 domains are lost.

17.4.2 The T315I Mutant

The Bcr–Abl inhibitors currently on the market, mostly ATP-competitive inhibitors, are not capable of inhibiting the form in which Thr is mutated to Ile at the highly conserved gatekeeper residue Thr315 (T315I). Gorre *et al.*

Figure 17.4 Schematic domain representation of the different isoforms of Bcr-Abl mutant genes.

were the first to document the development of Bcr–Abl mutations in 11 patients with advanced-phase CML or Ph$^+$ ALL who relapsed on imatinib (IM).[35] In six out of nine assessable patients, resistance was associated with the presence of the T315I mutation. T315 is referred to as the "gatekeeper" residue because it separates the ATP-binding site from an internal cavity constituted by a hydrophobic region of the Bcr–Abl active site, not contacted by ATP. The nature of the gatekeeper residue, that controls the access to this cavity in different kinases, is an important determinant of inhibitors' specificity.[36] The mutation at the "gatekeeper" residue is called pan-resistant[37] or the Achilles' heel of tyrosine kinase inhibitors (TKIs) and may have profound effects on small molecules binding in the context of different kinases. The biological importance of mutations in this position is also due to the fact that they do not inactivate the kinase functions.

In a more recent and extensive analysis, the T315I mutation was detected in 15% of 112 patients with CML who failed the first-line therapy resulting in complete insensitivity to IM[38] and the second-generation TKIs dasatinib,[39] nilotinib,[40,41] and bosutinib[42] (Figure 17.5) at clinically achievable concentrations. In addition, results of recent studies showed that bulkier hydrophobic residues in position 315 tend to stabilize a "hydrophobic spine", shifting the equilibrium towards the active conformation of Bcr–Abl.[36] Targeting Bcr–Abl with compounds able to avoid unfavorable clashes with I315 or compounds able to bind pockets different from the ATP-binding site (for example the MP) is becoming an effective strategy in the search for new drugs active on the highly resistant T315I mutant.[43]

17.5 Bcr–Abl Allosteric Modulation: From ATP Pocket Binders to Allosteric Inhibitors

The availability of different sites and different conformational states that could be targeted for the inhibition of protein kinases has definitely been conducive to the classification of the kinase inhibitors into five different types according to their potential mechanisms of action.[44,45] *Type 1* inhibitors targeting the active conformation of the ATP binding-pocket (DFG-in) and thus having a competitive mechanism of action with respect to the ATP substrate (dasatinib). *Type 2* inhibitors are those targeting the inactive conformation of the ATP binding-pocket (DFG-out) and locking the enzyme in the inactive form (imatinb and nilotinib). *Type 3* inhibitors are allosteric modulators which bind to sites distinct from the ATP binding site to regulate the kinase activity, providing the advantage of a greater selectivity and higher affinity (GNF-2 and GNF-5, described in the next paragraph). *Type 4* inhibitors are those binding covalently to the active site of a protein kinase, generally reacting with a nucleophilic cysteine present in the binding site. *Type 5* inhibitors are those interfering with the protein–protein interactions, blocking the access of the protein substrate to the kinase without affecting the conformation of the kinase.

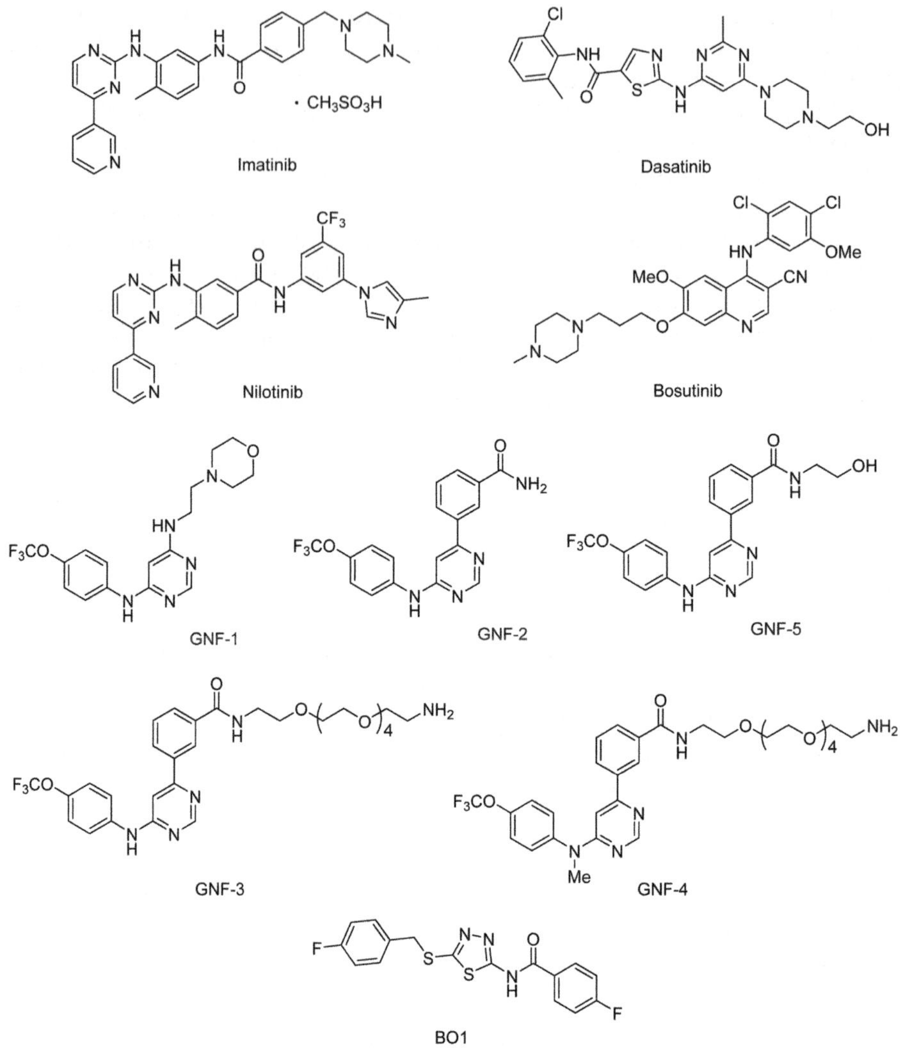

Figure 17.5 2D molecular structures of imatinib (IM), dasatinib, nilotinib, bosutinib, GNF derivatives, and BO1.

17.5.1 Myristate Pocket Binders: Abl Inhibitors

One of the major drawbacks of ATP-competitive TKIs is their low selectivity and, as a consequence, their possible side effects. In order to overcome this problem, expensive preliminary screenings to quantitatively measure the interactions between the tested compound and a wide panel of different kinases, including clinically relevant mutants, is usually planned in order to identify promising hits endowed with less pronounced off-target activities. Even if this approach turned out to be useful, it is too expensive for widespread application in a preliminary screening phase. However, while the ATP binding site is conserved among kinases, the evolutionary pressure on allosteric sites has been less severe, providing the unique opportunity to develop compounds with a higher degree of specificity. The discovery of new allosteric sites may allow better understanding of the basic biological processes but also provides the possibility of identifying novel drug-like inhibitors and activators of specific enzymes.[46] Targeting these sites, outside the ATP-binding pocket, may allow fine tuning of the kinase activity without affecting the basal activity of the enzymes, thus preserving some beneficial functions of the kinases required for the regulation of many cellular processes.

17.5.1.1 GNF-2

During the course of differential cytotoxicity screening aimed at identifying new molecules active against Bcr–Abl, a novel family of allosteric inhibitors was discovered.[47] A initial fast fluorescence-based proliferation assay, was performed to screen a library of around 50 000 heterocyclic molecules against 32D.p210 cells, selecting those compounds able to inhibit cell proliferation by 80% at 1 μM and at the same time showing no toxicity against parental cell lines. Next, evaluation of kinase inhibition and Bcr–Abl autophosphorylation allowed identification of the promising hit GNF-1 (Figure 17.5) endowed with a 4,6-disubstituted pyrimidine scaffold and possessing a concentration giving 50% of maximum inhibition (IC_{50}) of 400 nM for 32D.p210 Bcr–Abl-expressing cells. The subsequent hit-optimization process led to the identification of GNF-2 (Figure 17.5), characterized by an IC_{50} of 138 nM, apoptosis induction after 48 h treatment at 1 μM, specificity for cell lines expressing Bcr–Abl and no toxicity against non-transformed cells. Culture of Bcr–Abl-transformed Ba/F3 cells with increasing concentrations of GNF-2 resulted in the identification of a total of 306 mutants, 80% of which were located in the myristate-binding pocket: three of the most common mutations conferring complete resistance to GNF-2 were C464Y, P465S, and E505K. These results led scientists to speculate that GNF-2 was able to bind the MP of Bcr–Abl. However, complete resistance was also found for the T315I gatekeeper mutation, located far from the putative GNF-2 binding pocket.

In order to definitely identify the binding site of these GNF compounds two derivatives bearing a polyethylene linker were synthesized (GNF-3 and GNF-4) and supported on a Sepharose matrix. GNF-3 (Figure 17.5) represented the active analog of GNF-2 while GNF-4 acted as a negative control since *N*-methylation had been proved to generate inactive GNF-derivatives. Incubation of such solid-supported compounds with human Abl followed by displacement experiments with selected ligands (GNF-2, IM, ATP, and myristoylated peptide) allowed proposal of the MP as the interaction site of GNF-2. In order to further support the interaction of GNF-2 with the MP, two different Bcr–Abl mutants carrying the point mutations A337N and A344L within the MP were expressed in Ba/F3 cells and tested for inhibition. As expected, the latter transformed cells were resistant to GNF-2 but sensitive to

IM. Surprisingly, radioenzymatic assays did not show any *in vitro* inhibition of Abl kinase activity both on the isolated KD and on the full length wild type Bcr–Abl. The absence of any *in vitro* inhibition observed in the radioenzymatic assays has been subsequently explained by the presence of a non-ionic detergent "polyoxoethylene lauryl ether" (Brij 35) in the buffer used for this particular experiment. Removal of Brij 35 allowed measurement of an IC_{50} of 0.24 μM for the inhibition of recombinant Abl exerted by GNF-2.[48] Kinetic experiments were then conducted showing a decrease of maximum rate (V_{max}) and an unchanged Michaelis constant (K_m) for ATP (indicative of non ATP-competitive mechanisms for GNF-2) while an increase in K_m for the peptide was observed in a separate experiment. These observations suggested that GNF-2 binding to the myristate pocket may indeed reduce the affinity of Abl for the peptide substrate.

X-ray crystallography experiments, conducted by soaking crystals of Abl–IM–myristate with an excess of GNF-2, unambiguously allowed identification of the myristoyl pocket of Abl as the binding site of GNF-2 (Figure 17.6).[49] It should be mentioned that the presence of IM in the ATP-binding pocket was fundamental to solving the crystal structure with GNF-2 bound to the MP. As shown in the crystal structure, GNF-2 binds in an extended conformation with the CF_3 group deeply buried within the myristate-pocket, giving a weak polar interaction with the backbone of Leu340.

A network of water-mediated hydrogen bonds between GNF-2, Tyr454, and Glu481 was also observed. However, most of the interactions between the inhibitor and the myristate binding pocket were hydrophobic. In addition,

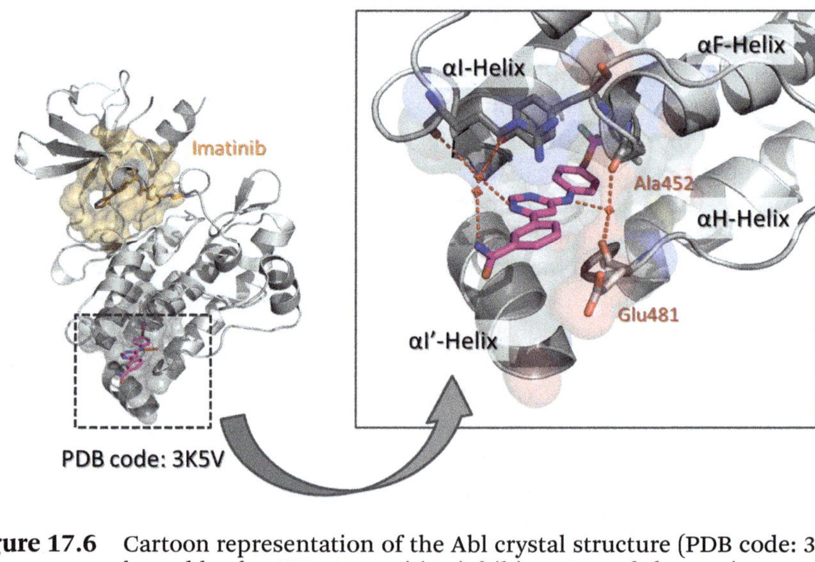

Figure 17.6 Cartoon representation of the Abl crystal structure (PDB code: 3K5V) bound by the ATP-competitive inhibitor IM and the myristate-pocket binder GNF-2. Details of the interactions between GNF-2 and the MP are shown in the right insert. For reasons of clarity only key residues have been labeled. Hydrogen bonds are represented as red dashes.

superimposition of the IM–Abl (PDB code: 1IEP),[22] IM–myristate–Abl (PDB code: 1OPJ)[18] and IM–GNF-2–Abl (PDB code: 3K5V)[49] complexes showed very little difference in the ATP binding site. Lead optimization studies resulted in the identification of compound GNF-5 (Figure 17.5), possessing a similar activity to GNF-2, but having more favorable pharmacokinetic properties. Compounds GNF-2 and GNF-5 combined with ATP-pocket-binders showed a synergistic effect in inhibiting the kinase activity of Abl carrying the SH3 and SH2 domains. It should be noted that no evidence for a significant difference in additivity between type I (dasatinib) and type II (IM and nilotinib) inhibitors was found. Similar additive effects were also shown in Ba/F3 cells expressing wild type p210-Bcr–Abl. Interestingly, a combination of GNF-5 and nilotinib inhibited T315I Bcr–Abl cell growth and, more importantly, led to *in vivo* efficacy, giving complete remission in a T315I Bcr–Abl mutant murine bone-marrow transplantation model.[49]

In order to understand the mechanisms of action of the GNF-derivatives, hydrogen-exchange mass spectrometry studies were conducted exposing unbound Abl and GNF-5 to deuterium oxide (D_2O) following a well-established experimental protocol.[50] The aim of the work was to study conformational changes in the protein structure after binding of the ligands and explain why a mutation in the ATP-binding site affects the efficacy of a MP binder. Changes in deuterium incorporation were observed not only in the myristate-pocket (indicative of GNF-5 binding) but also near the ATP-binding site, suggesting that the interaction of GNF-5 with the myristate-pocket can cause dynamic or conformational changes to the ATP-binding site with functional consequences for the catalytic machinery. All these data suggest a cross-talk between the myristate- and ATP-binding sites that could be somehow disrupted in the presence of the T315I gatekeeper mutation which proved to be unresponsive to treatment with myristate-pocket-binders only.

Recently, a combination of molecular dynamics (MD) studies and molecular mechanics–Generalized Born model augmented with the hydrophobic solvent accessible surface area (MM–GBSA) analysis have been performed on GNF-2 bound to MP of both Abl wild-type (WT) and T315I mutant to try to explain from a theoretical point of view the biological behavior of this compound.[51] What emerged, is that GNF-2 when bound to the MP of Abl WT is able to stabilize the closed (autoinhibited) conformation of the protein. The key residues driving the favorable interactions between KD and SH2–SH3 domains were found to be Tyr158, Arg189, Arg194, Tyr245, and Val247 from the SH2–SH3 domains and Lys282, Tyr283, Val357, Tyr361, Phe512, Glu518, and Glu526 from the kinase domain. Among them, Tyr245, Val247, Val357, Tyr361, Lys282, Tyr283, and Phe512 were mainly involved in hydrophobic interactions. Conversely, polar contacts were made by the other residues. Particularly, stable hydrogen bonds occur between Arg189 and both Glu526 and Asp523 as well as from Arg194 and Glu518. Interestingly, Tyr158 made π–π interaction with Tyr361 but also established permanent hydrogen bonds with His394 and Asn393. All these data are in agreement with previous experimental observations. MD movie analysis also highlighted a favorable interaction between the aniline NH on GNF-2 and a water molecule which was observed during most of the simulation (occupancy of 68%). This water molecule forms bridges between GNF-2 and the main chain carbonyls of Ala344 and Glu462 (Figure 17.7A). Remarkably, the water molecule selected from the dynamics is located exactly in the same position as a structural water found in the 3K5V[49] crystal structure (Figure 17.7).

To further investigate this aspect, molecular interaction fields (MIFs) were calculated for the binding site by means of the software Grid[52] using the water probe (OH2). MD results and Grid analysis were then combined together. Interestingly, a minimum grid point characterized by a significant favorable energy (-9.83 Kcal mol^{-1}) was found near the water molecule involved in the binding of GNF-2 in both the dynamics frame and 3K5V[49] X-ray structure. Overall, these results highlighted that a water molecule plays a crucial role in mediating the interactions of GNF-2 with Abl WT. As a consequence, the stability of the complex between GNF-2 and Abl WT was associated with a stabilization of the interactions between the SH2, SH3, and KD.

On the other hand, MD simulations and MM–GBSA analysis of GNF-2 bound to the MP of the Abl T315I mutant revealed that the efficiency of GNF-2 in stabilizing the interactions between SH2, SH3, and KD was strongly reduced when Thr315 was mutated into Ile within the ATP binding site. This single mutation seemed to be able to completely destabilize the binding mode of GNF-2 which comes partially out of the pocket by changing its pose during the simulation (Figure 17.7B). Furthermore, no water-mediated

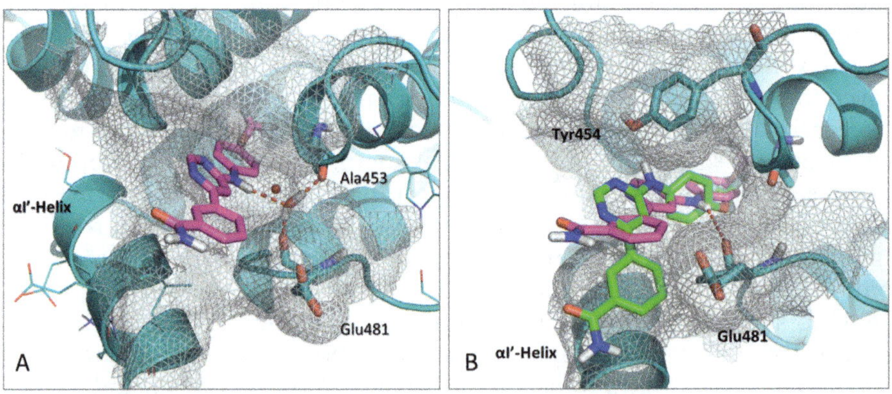

Figure 17.7 (A) Snapshot from MD simulation of GNF-2 bound to the MP of Abl WT. A water bridge involving GNF-2, Glu481, and Ala452 stabilized the complex. GRID analysis was performed and the most favourable interaction point determined for the probe OH2 (water, dark red sphere) was found to be located in the close proximity of a water molecule. (B) GNF-2 binding pose after 30 ns MD simulation in MP of Abl T315I mutant (green sticks). The binding pose of GNF-2 in the last frame of MD in Abl WT was superimposed (magenta sticks). In the T315I mutant the ligand partially comes out of the pocket, demonstrating a low stability of the complex.

hydrogen bridges were identified between the ligand and the protein. In line with experimental evidence, a synergistic effect of GNF-2 and IM in the stabilization of the closed conformation of the T315I mutant was also observed. The binding of IM to the ATP binding site strongly stabilizes the dynamics of the whole protein, allowing GNF-2 to maintain its binding pose within MP. This stabilization effect is translated into a further stabilization of the contacts at the interface of SH2–SH3 and KD.

In 2013, Skora *et al.* published an interesting work in which by the use of NMR in combination with small-angle X-ray scattering (SAXS) the authors were able to determine the domain motions in the autoinhibited conformation of Abl^{83-534} carrying the SH3, SH2, and KD domains.[53] The data provided the first structural information on the apo form of c-Abl in the absence of inhibitors, which is shown to adopt the "closed" conformation. It was demonstrated that the apo form of c-Abl is in a conformational equilibrium between the closed and open states, predominantly on the side of the closed state: the SH3 domain keeps contact with the polyproline helix in the SH2–KD region linker but the myristoyl group is not docked into the MP and the αI′ helix is not bent but is flexible. Unexpectedly, the addition of the catalytic site inhibitor IM induces a large structural rearrangement characterized by the detachment of the SH3–SH2 domains from the kinase domain and the formation of a dynamic "open" inactive state, which is inhibited in the ATP site. As a consequence of this opening, Tyr245 within the SH2-kinase linker becomes exposed. The relevance of this unexpected opening is corroborated by an observed increase in Tyr245 phosphorylation at moderate IM concentrations in cellular assays on full-length Bcr–Abl. In contrast to IM, addition of the myristoyl pocket inhibitor GNF-5[49] to apo c-Abl induces only local changes around the myristoyl-binding pocket and keeps the protein in the "closed" state. However, addition of GNF-5 to the "open" c-Abl–IM complex restores the "closed," inactive conformation.

In 2014, a multicenter Phase I clinical trial was initiated on GNF-5 (called ABL001).[54] The study was aimed at establishing the maximum tolerated dose (MTD) and/or recommended dose for expansion (RDE) of single agent ABL001 in CML or Ph-positive ALL patients, and in combination with nilotinib in Ph-positive CML patients.

The findings on the allosteric actions of the myristoyl pocket inhibitors have revealed molecular details of their synergism with ATP pocket binders to overcome drug resistance and opened the way to innovative strategies for new drug development.

17.5.1.2 BO1

In 2010, the search for novel scaffolds to be tested as Src and Bcr–Abl inhibitors lead to the identification of some interesting commercially available compounds showing moderate inhibitory activity towards both kinases.[55] Among these new molecular scaffolds, some thiadiazole derivatives were demonstrated to be able to inhibit both Src and Abl (dual inhibitors) or alternatively to only one of the two kinases, depending on the substitution pattern on the benzamido moiety. To fully understand the SAR of these new Src/Abl inhibitors, a combinatorial approach was applied to the synthesis of a small library of thiadiazole analogues bearing a wide range of substituents on both aromatic rings. These compounds were then screened, in a cell-free assay, against recombinant human c-Src and Abl, using IM as a reference compound: few derivatives were characterized by high affinity towards the isolated Abl with an inhibitory activity similar to that of IM. Among them, BO1 showed also a promising biological profile on IM-sensitive and Bcr–Abl-independent IM-resistant leukemia cells, being able to reduce the clonogenic activity of Bcr–Abl-expressing clones (median lethal dose (LD_{50}) = 2.2 μM). Surprisingly, compound BO1 showed in cell free assays only a modest loss of potency (four fold) towards the AblT315I mutant (inhibitory constant (K_i) =400 nM) when compared with IM (20-fold) or GNF-2 (40-fold), thus making it a very promising lead compound for the development of Abl T315I effective drugs.

To better understand the mechanism of action of this compound, a thorough kinetic study was conducted on Abl WT and Abl T315I.[56] In the case of Abl WT, BO1 was suggested to target both the free enzyme and the enzyme–peptide complex, preventing ATP binding. BO1 caused an increase in the apparent K_m (K_{mapp}) values for the ATP and peptide substrates, respectively, as well as a decrease in the corresponding apparent V_{max} (V_{maxapp}) values, resulting in a mixed-type inhibition. This behavior can be explained by the fact that the inhibitor binds with higher affinity to the free enzyme than to the enzyme–substrate complex. In the case of T315I Abl, compound BO1 acted as a purely non-competitive inhibitor with respect to both ATP and the peptide substrates: no changes were in fact observed in the K_{mapp} values for both substrates, whereas the inhibitor decreased the V_{maxapp} of the reaction in a manner dependent on both substrates. This evidence lead to the hypothesis that BO1 can overcome the structural barrier imposed by the drug-resistant Abl mutant T315I by virtue of its ability to 'adapt' its mechanism of action to the specific enzymatic form of Abl: BO1 proved in fact to be an ATP-competitive inhibitor of WT Abl while the same compound proved to act as a noncompetitive inhibitor with respect to both the ATP and peptide substrates, in the case of Abl T315I.

In order to suggest a possible binding site for compound BO1, a series of docking studies were conducted. Initially it was speculated that the T315I mutation could force BO1 to bind in the deep allosteric pocket on the back of the gatekeeper residue, which is partially exploited by the inhibitor AP24534 (PDB code: 3IK3).[57] Unfortunately, docking and dynamics simulations conducted on this site were unable to rationalize the activity of BO1 against the T315I mutant.

The release on the Protein Data Bank of the co-crystal structure of Bcr–Abl (PDB code: 3K5V)[49] bound by the ATP-competitive inhibitor IM and the myristate pocket binder GNF-2, opened the way to another hypothesis. Docking simulations of BO1 on the myristate binding site showed a profitable interaction pattern of the compound. The 4-fluorobenzylthio chain completely filled the hydrophobic tunnel defined by the residues Leu448, Phe512, Ala363, Ile451, Ala356, Leu360, and Val487, while a hydrogen bond with Ala452 and edge-to-face aromatic interactions with Tyr452 contribute to maintaining BO1 in a binding orientation similar to that of the original ligand GNF-2.

Molecular dynamics simulations followed by MM-GBSA analysis clarify[51] that BO1 is able to maintain profitable interactions with the MP of both Abl WT and T315I mutant thanks to stable hydrogen bonding found between the ligand and the backbone carbonyl group of Glu481 and a π–π interaction between the *p*-fluoro phenyl ring of BO1 and the guanidine group of Arg351 (Figure 17.8).

From molecular modeling studies, BO1 was shown to be able to stabilize, much more than GNF-2, the compact and autoinhibited conformation of T315I mutant.[51] Accordingly, the effect of the two compounds on protein dynamics was found to be very different. BO1 stabilizes the bending of the αI-helix, assisting the docking of SH2 to the KD and inducing the stabilization of many interactions. The following residues have been identified as hot spots: Arg189, which was involved in hydrogen bonds with Glu526 and Asp523, Tyr158, Tyr245, Tyr361, Val357, and Ser520. BO1 promoted the formation of hydrogen bonds between Asp96 and Arg282 as well as between Ser93 and Val247 which were not detected in the dynamics of GNF-2. The hydrogen bond between Glu117 and Lys313 was also stabilized. In agreement with biological data, MD simulations helped to elucidate the mechanism of allosteric inhibition of BO1 against the T315I mutant.

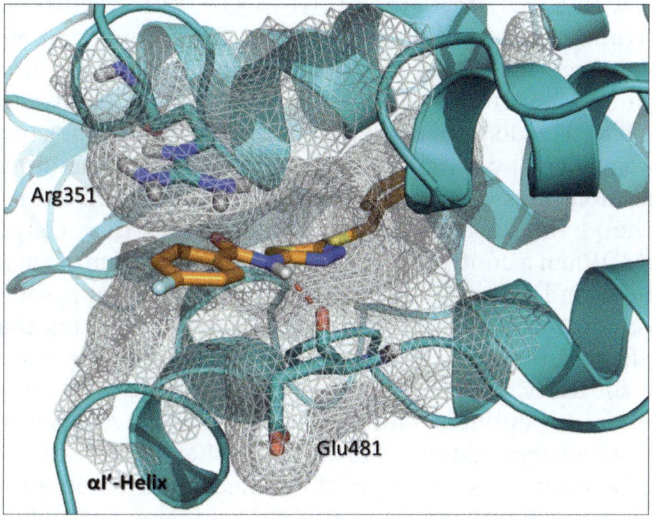

Figure 17.8 Snapshot from MD simulation of BO1 bound to MP of Abl T315I. The binding mode of the ligand is stabilized by a hydrogen bond (involving Glu481) and a π–π interaction with Arg351.

Beside kinetic enzymatic and molecular modeling studies, further experimental data are needed to confirm that compound BO1 binds to the myristate bonding pocket only in the presence of the T315I mutation.

17.5.2 Myristate Pocket Binders: Abl Activators

Although less effort has been devoted to the development of kinase activators compared with ATP-binding inhibitors, some examples can be found in the literature.[58] The idea is spreading that specific kinase activators can not only serve as important tools for enhancing our understanding of kinase regulation, but also have therapeutic applications in a growing number of diseases. It was suggested that c-Abl activation has a potential role in inhibiting mammary tumorigenesis, breast cancer cell mobility and invasiveness[59,60] and in treating neutropenia, since it has been suggested that c-Abl is critical in maintaining normal myelopoiesis.[61,62]

In 2011, scientists identified c-Abl 1b as a model system for a kinase activator screening effort.[63] Through a high-throughput screening (HTS) campaign on a compound library conducted by using two *in vitro* activation assays (an end-point activation assay IMAP and a continuous activation assay, Omnia), DPH (5-[3-(4-fluorophenyl)-1-phenyl-1*H*-pyrazol-4-yl]-2,4-imidazolidinedione or 5-(1,3-diaryl-1*H*-pyrazol-4-yl)hydantoin) was identified as the first Abl activator. To determine if DPH induces c-Abl activation by binding to the myristate-binding site displacing the myristoyl group, a fluorescence polarization (FP) competition binding assay was utilized. This assay used a N-terminal truncated kinase c-Abl$^{46-534}$ (D382N) and a carboxytetramethylrhodamine (TAMRA)-labeled peptide ligand that bears the same sequence as c-Abl 1b N terminus with a myristoyl group (Myr-GQQPGKVLGDQRRPSL(K-TAMRA) G-amide). Truncation of the N-terminus of c-Abl produces an empty myristoyl pocket for peptide ligand and small-molecule binding. Binding of the peptide ligand to the MP of c-Abl can compete with small molecules that also bind to this site. The displacement of the peptide ligand leads to a change in fluorescence anisotropy which can be used to calculate the binding affinity of the small molecules. The binding of the peptide ligand to c-Abl was confirmed in the FP binding experiment with a binding dissociation constant (K_d) of 38 nM. When a control peptide with an identical amino acid sequence to that of the ligand peptide, but lacking the myristoyl group, was used in the binding assay, no anisotropy change was observed, indicating that the binding is specific and requires the presence of the myristoyl group. When ATP was used in the assay, no competition with the peptide ligand was observed, indicating that the peptide ligand does not bind to the ATP pocket. FP binding data strongly suggested that DPH binds to the MP of c-Abl.

To gain a better understanding of the molecular interactions that govern the binding, a co-crystal structure of DPH with the c-Abl kinase domain spanning residues 248 to 531 (c-Abl$^{248-531}$) has been solved (PDB code: 3PYY).[63] The crystal structure showed that the more active enantiomer has the *R* configuration and thus is referred to as *R*-DPH. *R*-DPH binds to c-Abl in the myristoyl binding site and straightens the αI′ helix to elongate the αI helix (Figure 17.9A). The fluorophenyl ring of *R*-DPH occupies a hydrophobic pocket defined by Leu359, Leu360, Ala363, Val487, Leu448, and Phe512. The 2-NH of the hydantoin ring of *R*-DPH is hydrogen-bonded to the side chain of Tyr454 *via* a mediating water molecule, while the 4-NH of the hydantoin forms a direct hydrogen bond with the backbone carbonyl group of Glu481. The pyrazole core of *R*-DPH has van der Waals contacts with Pro484. The phenyl ring extends toward the elongated αI helix and sits in a lipophilic groove with Ala356 and Pro484 defining the respective walls of the groove. One edge of the phenyl ring is solvent-exposed. An overlay of the *R*-DPH structure with the GNF-2-bound c-Abl structure illustrates the significant conformational differences between the αI and αI′ helices (Figure 17.9B). The phenyl ring of *R*-DPH occupies the space where αI′ helix would reside in the inhibited form

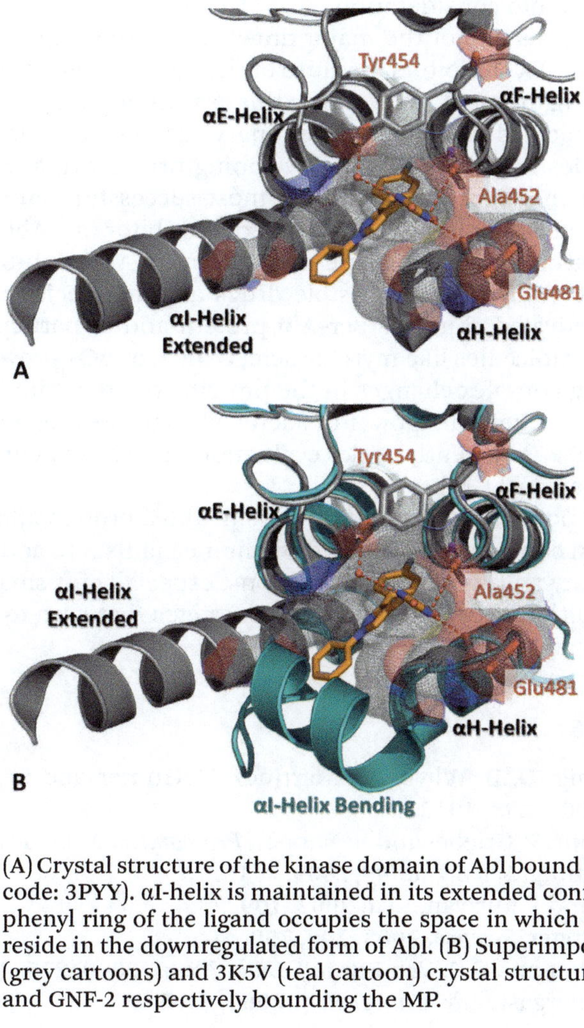

Figure 17.9 (A) Crystal structure of the kinase domain of Abl bound by *R*-DPH (PDB code: 3PYY). αI-helix is maintained in its extended conformation. The phenyl ring of the ligand occupies the space in which αI′-helix would reside in the downregulated form of Abl. (B) Superimposition of 3YYP (grey cartoons) and 3K5V (teal cartoon) crystal structures with *R*-DPH and GNF-2 respectively bounding the MP.

of c-Abl with GNF-2. These structural differences provide the basis for the opposing effects of GNF-2 and DPH on c-Abl activity.

17.6 Conclusions

Targeting tyrosine kinases is still one of the most promising approaches in the field of cancer therapy. In the last decades great strides have been made by academia and pharmaceutical companies, leading to the approval by the food and drug administration (FDA) of many small-molecule tyrosine kinases inhibitors for the treatment of different types of human cancers. TKs are mediators of many signaling pathways by which cancer cells promote their survival outside the standard cell controls, so inhibition of such enzymes has the potential to be a compelling anticancer strategy. Since kinases are also required for the function of normal cells, toxicity is an important issue that should be taken into consideration.

Low specificity is one of the major drawbacks associated with the use of ATP-competitive TK inhibitors, because of the high conservation of the catalytic sites among different kinases. For this reason targeting allosteric pockets, usually subjected to a lower evolutionary pressure, and unique for each enzyme provides the possibility of developing drugs with a high degree of specificity. For this purpose, one of the most successful examples is represented by the discovery of the first allosteric inhibitor of c-Abl kinase, GNF-2. The explanation of the allosteric mechanism of Abl's inhibition opened the way to the study of new possible drugs able to specifically target the constitutive activated chimeric Bcr–Abl protein and in particular its T315I mutant. Small molecules like myristic acid, GNF-2, or BO1 proved to be capable of inducing complex changes in the protein structure simply binding to a hydrophobic pocket. The power of such a unique event emerges when we think that GNF-2 is currently under evaluation in a Phase I clinical trial as a possible drug against CML.

Since it was postulated for the first time in 1965, protein allosteric modulation has been considered as a very difficult mechanism to address, because of its complexity and its many implications, especially for small molecules. Despite difficulties, the GNF-2 story demonstrates that even today the small David can defeat the great Goliath.

References

1. G. Manning, D. B. Whyte, R. Martinez, T. Hunter and S. Sudarsanam, *Science*, 2002, **298**, 1912.
2. R. Arsenault, P. Griebel and S. Napper, *Proteomics*, 2011, **11**, 4595.
3. B. D. Manning, *Sci. Signaling*, 2009, **2**, 15.
4. M. Huse and J. Kuriyan, *Cell*, 2002, **109**, 275.
5. J. Schlessinger, *Science*, 2003, **300**, 750.
6. B. B. Brasher and R. J. Van Etten, *J. Biol. Chem.*, 2000, **275**, 35631.
7. A. M. Pendergast, *Adv. Cancer Res.*, 2002, **85**, 51.

8. T. Pawson, *Nature*, 1995, **373**, 33.
9. S. T. Wen and R. A. Van Etten, *Genes Dev.*, 1997, **11**, 2456.
10. O. Hantschel and G. Superti-Furga, *Nat. Rev. Mol. Cell Biol.*, 2004, **5**, 33.
11. P. J. Woodring, T. Hunter and J. Y. Wang, *J. Cell Sci.*, 2003, **116**, 2613.
12. M. T. Pisbarro, L. Serrano and M. Wilmanns, *Mol. Cell*, 1998, **21**, 787.
13. A. Musacchio, M. Saraste and M. Wilmanns, *Nat. Struct. Biol.*, 1994, **1**, 546.
14. W. M. Franz, P. Berger and J. Y. Wang, *EMBO J.*, 1989, **8**, 137.
15. S. Chen, S. Brier, T. E. Smithgall and J. R. Engen, *Protein Sci.*, 2007, **16**, 572.
16. S. Cox, E. Radzio-Andzelm and S. S. Taylor, *Curr. Opin. Struct. Biol.*, 1994, **4**, 893.
17. S. S. Taylor, D. R. Knighton, J. Zheng, L. F. Ten Eyck and J. M. Sowadski, *Annu. Rev. Cell Biol.*, 1992, **8**, 429.
18. B. Nagar, O. Hantschel, M. A. Young, K. Scheffzek, D. Veach, W. Bornmann, B. Clarkson, G. Superti-Furga and J. Kuriyan, *Cell*, 2003, **112**, 859.
19. M. D. Resh, *Biochim. Biophys. Acta*, 1999, **1451**, 1.
20. O. Hantschel, S. Wiesner, T. Güttler, C. D. Mackereth, L. L. Rix, Z. Mikes, J. Dehne, D. Görlich, M. Sattler and G. Superti-Furga, *Mol. Cell*, 2005, **19**, 461.
21. O. Hantschel, B. Nagar, S. Guettler, J. Kretzschmar, K. Dorey, J. Kuriyan and G. Superti-Furga, *Cell*, 2003, **112**, 815.
22. B. Nagar, W. G. Bornmann, P. Pellicena, T. Schindler, D. R. Veach, W. T. Miller, B. Clarkson and J. Kuriyan, *Cancer Res.*, 2002, **62**, 4236.
23. B. Nagar, O. Hantschel, M. Seeliger, J. M. Davies, W. I. Weis, G. Superti-Furga and J. Kuriyan, *Mol. Cell*, 2006, **21**, 787.
24. M. A. Young, S. Gonfloni, G. Superti-Furga, B. Roux and J. Kuriyan, *Cell*, 2001, **105**, 115.
25. L. Skora, J. Mestan, D. Fabbro, W. Jahnke and S. Grzesiek, *Proc. Natl. Acad. Sci. U. S. A.*, 2013, **110**, E4437.
26. D. Barila and G. Superti-Furga, *Nat. Genet.*, 1998, **18**, 280.
27. R. E. Iacob, J. Zhang, N. S. Gray and J. R. Engen, *PLoS One*, 2011, **6**, e15929.
28. A. Dixit and G. M. Verkhivker, *PLoS Comput. Biol.*, 2009, **5**, e1000487.
29. P. Filippakopoulos, M. Kofler, O. Hantschel, G. D. Gish, F. Grebien, E. Salah, P. Neudecker, L. E. Kay, B. E. Turk, G. Superti-Furga, T. Pawson and S. Knapp, *Cell*, 2008, **134**, 793.
30. F. Grebien, O. Hantschel, J. Wojcik, I. Kaupe, B. Kovacic, A. M. Wyrzucki, G. D. Gish, S. Cerny-Reiterer, A. Koide, H. Beug, T. Pawson, P. Valent, S. Koide and G. Superti-Furga, *Cell*, 2011, **147**, 306.
31. J. M. Smith and R. A. Van Etten, *J. Biol. Chem.*, 2001, **276**, 24372.
32. J. D. Rowley, *Blood*, 2008, **112**, 2183.
33. S. Wong and O. n. Witte, *Annu. Rev. Immunol.*, 2004, **22**, 247.
34. M. W. Deininger, J. M. Goldman and J. V. Melo, *Blood*, 2000, **96**, 3343.
35. M. E. Gorre, M. Mohammed, K. Ellwood, N. Hsu, R. Paquette, P. N. Rao and C. L. Sawyers, *Science*, 2001, **293**, 876.
36. M. E. Noble, J. A. Endicott and N. L. Johnson, *Science*, 2004, **303**, 1800.

37. N. P. Shah, B. J. Skaggs, S. Branford, T. P. Hughes, J. M. Nicoll, R. L. Paquette and C. L. Sawyers, *J. Clin. Invest.*, 2007, **117**, 2562.
38. M. Deininger, E. Buchdunger and B. J. Druker, *Blood*, 2005, **105**, 2640.
39. M. R. Burgess, J. B. Skaggs, N. P. Shah, F. Y. Lee and C. L. Sawyers, *Proc. Natl. Acad. Sci. U. S. A.*, 2005, **102**, 3395.
40. T. O'Hare, D. K. Walters, M. W. Deininger and B. J. Druker, *Cancer Cell*, 2005, **7**, 117.
41. E. Weisberg, P. W. Manley, W. Breitenstein, J. Bruggen, S. W. Cowan-Jacob, A. Ray, B. Huntly, D. Fabbro, G. Fendrich, E. Hall-Meyers, A. L. Kung, J. Mestan, G. Q. Daley, L. Callahan, L. Catley, C. Cavazza, M. Azam, D. Neuberg, R. D. Wright, D. G. Gilliland and J. D. Griffin, *Cancer Cell*, 2005, **7**, 129.
42. J. M. Golas, K. Arndt, C. Etienne, J. Lucas, D. Nardin, J. Gibbons, P. Frost, F. Ye, D. H. Boschelli and F. Boschelli, *Cancer Res.*, 2003, **63**, 375.
43. M. Azam, M. A. Seeliger, N. S. Gray, J. Kuriyan and G. Q. Daley, *Nat. Struct. Mol. Biol.*, 2008, **15**, 1109.
44. J. Zhang, P. L. Yang and N. S. Gray, *Nat. Rev. Cancer*, 2009, **9**, 28.
45. R. M. Eglen and T. Reisine, *Expert Opin. Drug Discovery*, 2010, **5**, 277.
46. J. A. Hardy and J. A. Wells, *Curr. Opin. Struct. Biol.*, 2004, **14**, 1.
47. F. J. Adrián, Q. Ding, T. Sim, A. Velentza, C. Sloan, Y. Liu, G. Zhang, W. Hur, S. Ding, P. Manley, J. Mestan, D. Fabbro and N. S. Gray, *Nat. Chem. Biol.*, 2006, **2**, 95.
48. Y. Choi, M. A. Seeliger, S. B. Panjarian, H. Kim, X. Deng, T. Sim, B. Couch, A. J. Koleske, T. E. Smithgall and N. S. Gray, *J. Biol. Chem.*, 2009, **284**, 29005.
49. J. Zhang, F. J. Adrián, W. Jahnke, S. W. Cowan-Jacob, A. G. Li, R. E. Iacob, T. Sim, J. Powers, C. Dierks, F. Sun, G. R. Guo, Q. Ding, B. Okram, Y. Choi, A. Wojciechowski, X. Deng, G. Liu, G. Fendrich, A. Strauss, N. Vajpai, S. Grzesiek, T. Tuntland, Y. Liu, B. Bursulaya, M. Azam, P. W. Manley, J. R. Engen, G. Q. Daley, M. Warmuth and N. S. Gray, *Nature*, 2010, **463**, 501.
50. R. E. Iacob, T. Pene-Dumitrescu, J. Zhang, N. S. Gray, T. E. Smithgall and J. R. Engen, *Proc. Natl. Acad. Sci. U. S. A.*, 2009, **106**, 1386.
51. A. L. Fallacara, C. Tintori, M. Radi, S. Schenone and M. Botta, *J. Chem. Inf. Model.*, 2014, **54**, 1325.
52. *Grid, version 22*, Molecular Discovery Ltd, Pinner, Middlesex, UK.
53. L. Skora, J. Mestan, D. Fabbro, W. Jahnke and S. Grzesiek, *Proc. Natl. Acad. Sci. U. S. A.*, 2013, **110**, e4437.
54. http://www.cancer.gov/about-cancer/treatment/clinical-trials/search/view?cdrid=758878&version=HealthProfessional.
55. F. Manetti, G. A. Locatelli, G. Maga, S. Schenone, M. Modugno, S. Forli, F. Corelli and M. Botta, *J. Med. Chem.*, 2006, **49**, 3278.
56. E. Crespan, M. Radi, S. Zanoli, S. Schenone, M. Botta and G. Maga, *Bioorg. Med. Chem.*, 2010, **18**, 3999.
57. T. O'Hare, W. C. Shakespeare, X. Zhu, C. A. Eide, V. M. Rivera, F. Wang, L. T. Adrian, T. Zhou, W. S. Huang, Q. Xu, C. A. Metcalf III, C. J. W. Tyner, M. M. Loriaux, A. S. Corbin, S. Wardwell, Y. Ning, J. A. Keats, Y. Wang, R.

Sundaramoorthi, M. Thomas, D. Zhou, J. Snodgrass, L. Commodore, T. K. Sawyer, D. C. Dalgarno, M. W. Deininger, B. J. Druker and T. Clackson, *Cancer Cell*, 2009, **16**, 401.

58. G. L. Simpson, J. A. Hughes, Y. Washio and S. M. Bertrand, *Curr. Opin. Drug Discovery Dev.*, 2009, **12**, 585.
59. T. M. Allington, A. J. Galliher-Beckley and W. P. Schiemann, *FASEB J.*, 2009, **23**, 4231.
60. N. K. Noren, G. Foos, C. A. Hauser and E. B. Pasquale, *Nat. Cell Biol.*, 2006, **8**, 815.
61. D. Caracciolo, M. Valtieri, D. Venturelli, C. Peschle, A. M. Gewirtz and B. Calabretta, *Science*, 1989, **245**, 1107.
62. V. Rosti, G. Bergamaschi, C. Lucotti, M. Danova, C. Carlo-Stella, F. Locatelli, L. Tonon, G. Mazzini and M. Cazzola, *Blood*, 1995, **86**, 3387.
63. J. Yang, N. Campobasso, M. P. Biju, K. Fisher, X. Q. Pan, J. Cottom, S. Galbraith, T. Ho, H. Zhang, X. Hong, P. Ward, G. Hofmann, B. Siegfried, F. Zappacosta, Y. Washio, P. Cao, J. Qu, S. Bertrand, D. Y. Wang, M. S. Head, H. Li, S. Moores, Z. Lai, K. Johanson, G. Burton, C. Erickson-Miller, G. Simpson, P. Tummino, R. A. Copeland and A. Oliff, *Chem. Biol.*, 2011, **18**, 177.

CHAPTER 18

Allosteric Modulators of Heat Shock Protein 90 (HSP90)

YEN CHIN KOAY*[a] AND SHELLI McALPINE*[a]

[a]University of New South Wales, School of Chemistry, Kensington, NSW 2052 Australia
*E-mail: s.mcalpine@unsw.edu.au, koayyenchin37@gmail.com

18.1 Introduction: Molecular Chaperones

A typical mammalian cell expresses over 10 000 different kinds of proteins. These proteins are made from hundreds or thousands of amino acids connected together by the ribosome to produce linear polypeptides. In order to function effectively as proteins, these linear polypeptides must be folded into their native conformations.[1,2] Maintaining correctly folded proteins ensures that cells perform normal cellular functions. Proteins that are in misfolded states contribute to diseases associated with neurological disorders and cancer. Thus, the maintenance of protein homeostasis is critical to cellular processes and overall health of an organism. Imbalance in protein homeostasis gives rise to many human diseases, ranging from cystic fibrosis to cancer and neurodegeneration.[3] To avoid protein defects associated with misfolding and aggregation, cells express a network of several hundred proteins, notably molecular chaperones to assist in *de novo* folding or re-folding.[4]

Stress proteins or, as they are commonly referred to, Heat Shock Proteins (HSPs) are a family of molecular chaperones that are the most conserved proteins present in both prokaryotes and eukaryotes. HSPs range in size

RSC Drug Discovery Series No. 56
Allosterism in Drug Discovery
Edited by Dario Doller

Published by the Royal Society of Chemistry, www.rsc.org

between 20 and 100 kDa[5] and they facilitate numerous events in cellular physiology including: protein recycling, transportation, migration, post-translational modification, and the regulation of signalling pathways. The high levels of HSPs that are observed in cancer cells are the result of the adaptive response required to maintain homeostasis in a rapidly changing environment. These high levels of HSPs maintain homeostasis at the required accelerated rates associated with rapid division in tumor growth.[6,7]

18.2 Heat Shock Protein 90

The molecular chaperone heat shock protein 90 (HSP90) plays a particularly important role in maintaining homeostasis. As a highly conserved molecular chaperone, HSP90 is an abundant and ubiquitously expressed protein, constituting 1–2% of total cellular protein in normal cells. In order to protect mutated and over-expressed oncoproteins from degradation,[8] HSP90 increases to 3–6% of overall protein in cancerous cells. HSP90 is also a key regulator of the stress response,[9,10] in part because it folds, maintains, and stabilizes over 400 client proteins, many of which are involved in cancer cell growth.[11]

The number of HSP90 homologues in a cell varies from organism to organism.[12] Bacteria such as *Escherichia coli* possess one homologue, HtpG, while yeast has two homologues (HSP82 and HSC82), one localized to the cytosol and one localized to the mitochondria. Mammalian cells have four HSP90 isoforms: HSP90α (inducible form) and HSP90β (constitutive form) located in cytosol, Grp94 (94 kDa glucose-regulated protein), and TRAP-1 (tumor necrosis factor receptor associated protein 1) localized to the endoplasmic reticulum and mitochondria, respectively.[13] All isoforms function as dimers, and each monomer consists of three domains: an amino (N)-domain (25 kDa), a middle (M)-domain (35 kDa), and a carboxy (C)-domain (10 kDa). With the exception of TRAP-1, all isoforms contain a charged linker that connects the N-terminal and middle domains (Figure 18.1). The charged domains are residues: 217–275 for HSP90αβ, 283–329 for Grp94, 209–263 for HSP82 and HSC82, and 212–220 for bacterial Htpg.

The N-domain is responsible for ATP-binding and it recognizes and binds to key co-chaperones, including p23, Aha1, and Cdc37. The M-domain forms interactions with most of HSP90's client proteins, many of which contribute to the ten established hallmarks of cancer: cell growth, immortality, apoptosis, cell-cycle regulation, angiogenesis, metastasis, reprogramming of energy metabolism, avoiding immune destruction, tumor-promoting inflammation, genome instability, and inflammation[15,16] Examples of oncogenic clients associated with all ten hallmarks of cancer are shown in Table 18.1 and Figure 18.2.[17,18] These include transmembrane tyrosine kinases (HER-2, c-MET), metastable signalling proteins (AKT, Raf-1), mutated signalling proteins (p53), cell cycle regulators (CDK4, CDK6, WEE1, PLK, MYT1), TNF-mediated apoptosis (RIP), hypoxia-induced angiogenesis (HIF-1α), telomerase-associated protein (h-TERT).[19] Since these client proteins are

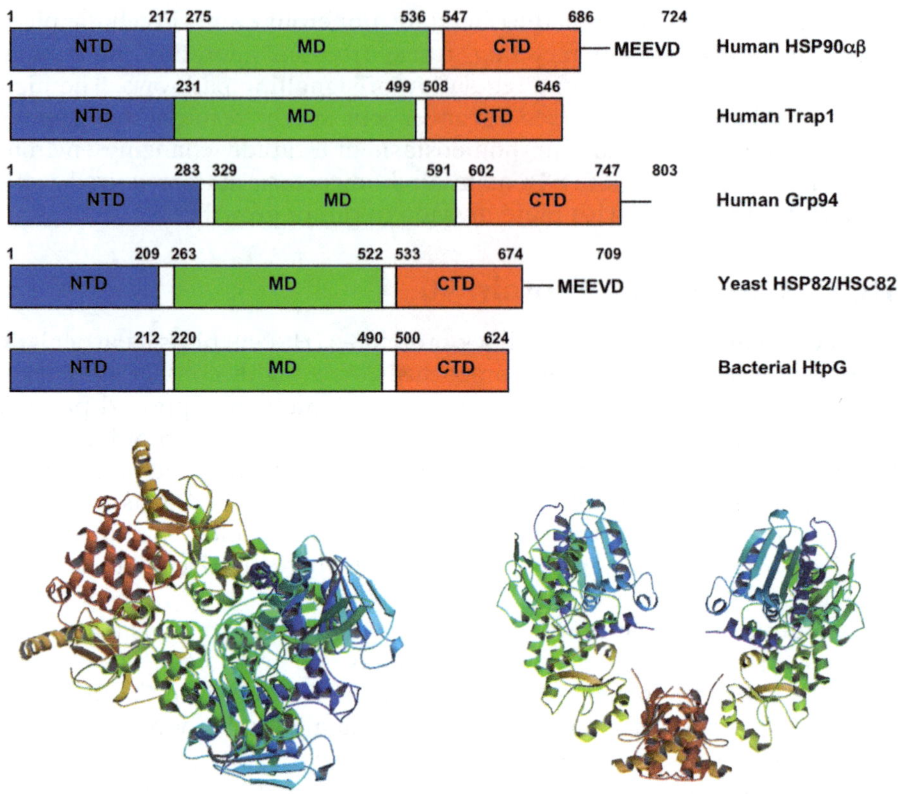

Figure 18.1 Top: Domain architecture for different HSP90 homologs. The domains consist of an amino-terminal domain (NTD) in blue, middle domain (MD) in green and the carboxy-terminal domain (CTD) in red. Reproduced from Quarterly Reviews of Biophysics, 44, K. A. Krukenberg; T. O. Street; L. A. Lavery; D. A. Agard, Conformational dynamics of the molecular chaperone HSP90, 229–255, Copyright 2011 with permission from Elsevier.[14] Bottom: Crystal structures of full-length HSP90 dimers from *Escherichia coli* (PDB code: 2IOQ). Colours indicate domains as in the top panel.

Table 18.1 Ten hallmarks of cancer and HSP90 related client proteins.

Hallmark of cancer	HSP90-dependent client protein (s)
1. Self-sufficiency in growth signals	RAF-1, AKT, HER-2, ERK
2. Insensitivity to anti-growth signals	PLK, WEEl, MYTl, CDK4, CDK6
3. Evading apoptosis	RIP, AKT, mutant p53, c-MET, IP6K2
4. Limitless replicative potential	Telomerase (h-TERT)
5. Sustained angiogenesis	AKT, HIF-1α, VEGFR, Src
6. Tissue-invasion and metastasis	c-MET
7. Reprogramming of energy metabolism	ARNT, ARRB1, HIF-1α, HMG1, SREBF1
8. Avoiding immune destruction	IRAK3
9. Tumor-promoting inflammation	IL-6, IL-8, IRAK1, IRAK2, IRAK3
10. Genome instability and inflammation	FANCA, MAFG, NEK8, NEK9, NEK11

mutated or overexpressed in cancers, HSP90 plays an important role in cancer progression as it facilitates the maturation and stabilization of these clients.[20]

The carboxy (C)-domain contains the site of dimerization of two 90-kDa monomers, where the dimer is considered the active form of HSP90. The C-terminal domain is also responsible for mediating interactions with co-chaperones through its MEEVD (M = methionine, E = glutamic acid, V = valine, D = aspartic acid) sequence, which is at the very end of the protein. The MEEVD is recognized by HSP90-associated co-chaperones that contain a tetratricopeptide repeat (TPR) domain. The TPR domains are comprised of 3–16 tandem helical repeats, each with 34 amino acids residues (Figure 18.3).

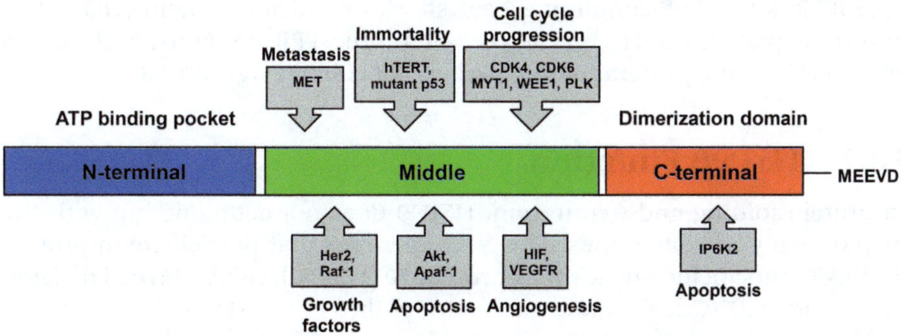

Figure 18.2 Schematic representation of some HSP90 client proteins.

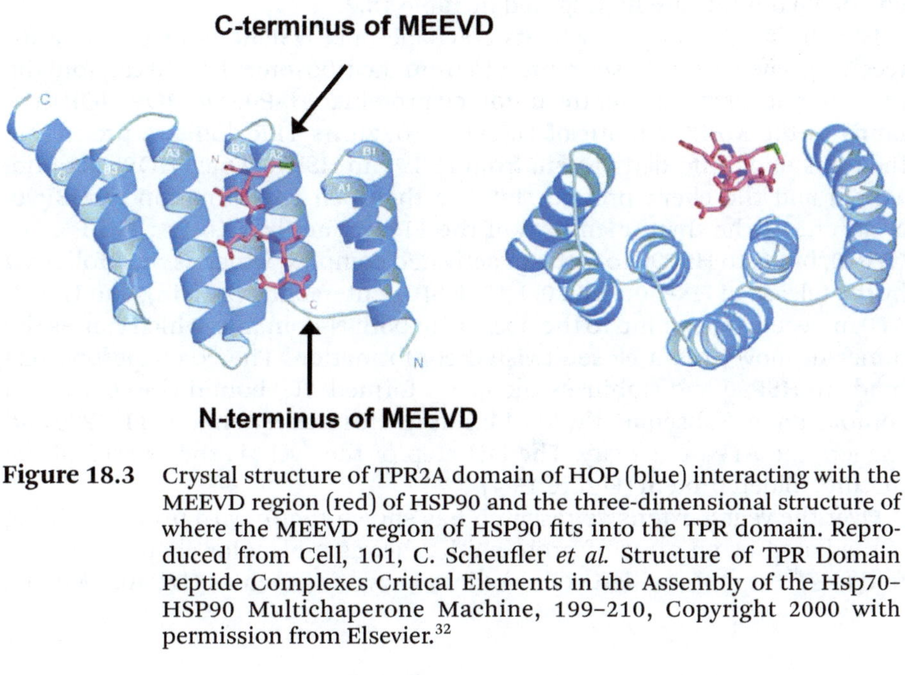

Figure 18.3 Crystal structure of TPR2A domain of HOP (blue) interacting with the MEEVD region (red) of HSP90 and the three-dimensional structure of where the MEEVD region of HSP90 fits into the TPR domain. Reproduced from Cell, 101, C. Scheufler *et al.* Structure of TPR Domain Peptide Complexes Critical Elements in the Assembly of the Hsp70–HSP90 Multichaperone Machine, 199–210, Copyright 2000 with permission from Elsevier.[32]

More than 20 co-chaperones have been identified as interacting with HSP90, and 16 of these co-chaperones contain TPR domains, indicating that they bind to the C-terminus of HSP90.[21,22] Among the group of 16 TPR-containing proteins, four are members of the immunophilin family with peptidylprolyl isomerase (PPIase) activity: cyclophilin 40 (CYP40; also known as peptidyl-prolyl isomerase D (PPID)),[23] FK506-binding protein 52 (FKBP52), FKBP51, and FKBP38.[24] Other TPR domain proteins that interact with HSP90 and are also important for protein homeostasis include: C-terminus of Hsc70 interacting protein (CHIP),[25] Ser–Thr protein phosphatase 5 (PP5),[26] WAF1–CIP1 stabilizing protein 39 (WISp39; also know as FK506-binding-protein-like (FKBPL)),[27] translocase of outer membrane 70 kDa (TOM70)[28] and the myosin folding factor (UNC-45).[29] Perhaps the most well-known co-chaperone of HSP90 is the TPR-containing heat shock organizing protein (HOP: also known as p60 and STI1).[30] HOP contains three TPR domains, and plays a central role in the protein folding pathway of HSP90 (Figure 18.3).[31]

18.3 HSP90 Function

In protein folding and maturation, HSP90 does not act alone but with the help of many co-chaperones. The co-chaperones that participate in protein folding events include heat shock protein 70 (HSP70), HOP, p23, cell division cycle protein 37 (Cdc37) and the activator of the HSP90 ATPase (Aha1). These co-chaperones are necessary to provide the conformational adjustments needed for the clients to prepare them for folding and to obtain the catalytically active state of HSP90. An overview of the co-chaperones and their effects on HSP90 binding are highlighted in Table 18.2.

HSP70, in conjunction with its co-chaperone, HSP40, is responsible for receiving a newly synthesized protein from the ribosome. HSP70 can fold the protein by itself or transfer the unfolded protein to HSP90 *via* HOP. HOP then binds to the MEEVD motif of HSP90 through its TPR domain, promoting the transfer of unfolded protein from HSP70 to HSP90. Once HOP is bound, HSP70 and the client protein stabilize the open conformation of HSP90. Next, one of the three members of the PPIase family, FKBP51, FKBP52, or CYP40, binds to HSP90 to form an activated complex, which is then followed by the release of HSP40, HSP70, HIP (HSP70-interacting protein), and HOP.[33] ATP molecules then bind to the pockets in both N-domains, which causes the dimer to move into a closed-twisted conformation. The co-chaperone p23 binds to HSP90 and stabilizes the newly formed ATP-bound closed–twisted conformation. Subsequently, Aha1 binds to the middle domain of HSP90 and triggers the ATPase activity. The last step of the cycle is the release of the mature, folded protein (Figure 18.4).[14]

Over the years, extensive research has shown that *in vivo* protein folding is mediated by an array of proteins that act as molecular chaperones and foldases.[46] In mammalian cells, foldase enzymes, such as CYP40, FKBP51,

Table 18.2 HSP90-associated co-chaperones that comprise the HSP90 protein folding machinery.

Protein	Description	HSP90 binding site	References
TPR co-chaperones			
CYP40	Peptidyl–prolyl isomerase involved in client protein maturation	C-terminus	23,35
FKBP51 and 52	Peptidyl–prolyl isomerase involved in client protein maturation	C-terminus	23,36
HOP	Facilitates the transfer of unfolded protein from HSP70 to HSP90	C-terminus	31,37
PP5	Protein phosphatase 5	C-terminus	38,39
TOM70	Mediates the import of mitochondrial preproteins	C-terminus	28,40
CHIP	Ubiquitin ligase, binds to HSP70 *via* TPR domains	None	25,41
HIP	Inhibits ATPase activity of HSP70	None	42
Non-TPR co-chaperones			
Cdc37	Activates protein kinase	N-terminus	43
p23	Stabilizes closed conformation of HSP90	N-terminus	44
Aha1	Activates ATPase activity	Middle domain	45

or FKBP52 from the PPIase family, are important components of the chaperone cycle. These foldases assist in the folding of newly-synthesized proteins and regulate the stability, localization, and activity of mature protein by catalysing the *cis–trans* isomerization of peptide bonds around a proline residue, thereby inducing conformational changes in target proteins.[47] Yet despite their similar activity as PPIases, these three proteins show different affinities for HSP90 and promote unique functions on HSP90. Each PPIase prompts different pathways to be activated when they form a complex with HSP90.[23]

18.4 HSP90 Inhibitors

Regulation of HSP90's ATPase activity, conformation, activation state of the co-chaperones, and client protein interactions are often affected by the crosstalk between the N-terminal and C-terminal domains of HSP90. Inhibiting the crosstalk between N- and the C-terminal binding sites of HSP90 can induce an allosteric regulation of the molecular chaperone machinery and alter the ability of HSP90 to complete its protein folding cycle.[48] High HSP90 expression in cancer cells has prompted the synthesis of small-molecule inhibitors that block HSP90's function, with the goal of inhibiting cancer cell growth.[49]

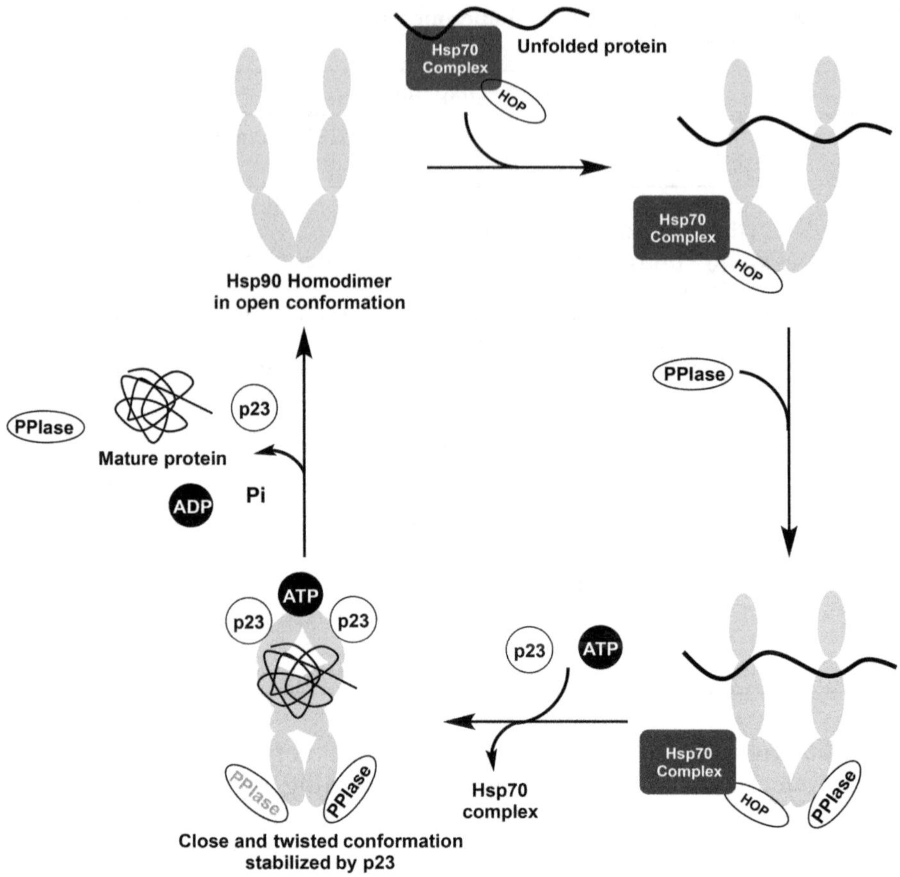

Figure 18.4 Model of HSP90 chaperone cycle. TPR-protein, HOP binds to HSP90 and to HSP70 to form a HSP90–HOP–HSPP70–HSP40 complex and facilitate unfolded protein delivery to HSP90. The other empty TPR acceptor site is then occupied by a PPIase, which leads to the formation of an asymmetric HSP90 complex. HSP90 converts to its closed conformation after ATP-binding, HOP is released from HSP90 and p23 binds. Another PPIase (grey) can potentially bind to form the final complex together with HSP90 and p23. After ATP hydrolysis, p23, PPIase, and the folded client protein are released from HSP90.[34]

HSP90 expression plays a crucial role in the maturation of many client proteins that contribute to the malignant phenotype, which includes an increase in proliferation and a decrease in apoptosis. HSP90 client proteins established to facilitate the ten hallmarks of cancer include: human epidermal growth factor receptor 2 (HER-2),[50,51] B-Raf,[52] AKT (also known as protein kinase B),[53] anaplastic lymphoma kinase (ALK), transcription factors, steroid hormone receptors GR,[54] androgen receptor (AR), estrogen receptor (ER), cell cycle proteins (CDK2, CDK4, or CDK6),[55,56] mutant p53,[57] B-cell lymphoma 2 (Bcl-2),[58] angiogenic factors HIF-1α,[59–62] and telomerase.[63]

18.4.1 HSP90 Inhibitors that Target the N-Terminus

Despite a wide diversity of structural motifs, all HSP90 inhibitors studied in clinical trials can be classified according to their similarity to three structural classes: geldanamycin (GA), radicicol (RD) or to a purine-scaffold (Figure 18.5).[64,65] These N-terminal inhibitors (termed "classical inhibitors") have been extensively studied and comprehensive reviews are available throughout the HSP90 literature.[65–67] These classical structures are considered allosteric inhibitors because they are reported to bind to the N-terminal binding site but affect clients that bind to the middle domain of HSP90.

To date, 16 structurally unique HSP90 inhibitors have been tested in clinical settings, all of which are proposed to target the N-terminal ATP-binding domain of HSP90, while influencing the binding events of clients and co-chaperones targeting alternative HSP90 domains. Currently six are still under clinical investigation as anticancer therapies (www.clinicaltrials.gov). NVP AUY-922 (AUY) is one of six clinical inhibitors that have been tested both as a single agent and in combination with other treatments (Figure 18.5). Phase II evaluation of AUY is currently being pursued as a treatment against two different types of cancer (non-small-cell lung cancer (NSCLC) and gastrointestinal stromal tumor (GST)) as a single agent. However, Novartis has recently removed its sponsorship of AUY, probably due to the drug's limited effectiveness during clinical trials.[68,69]

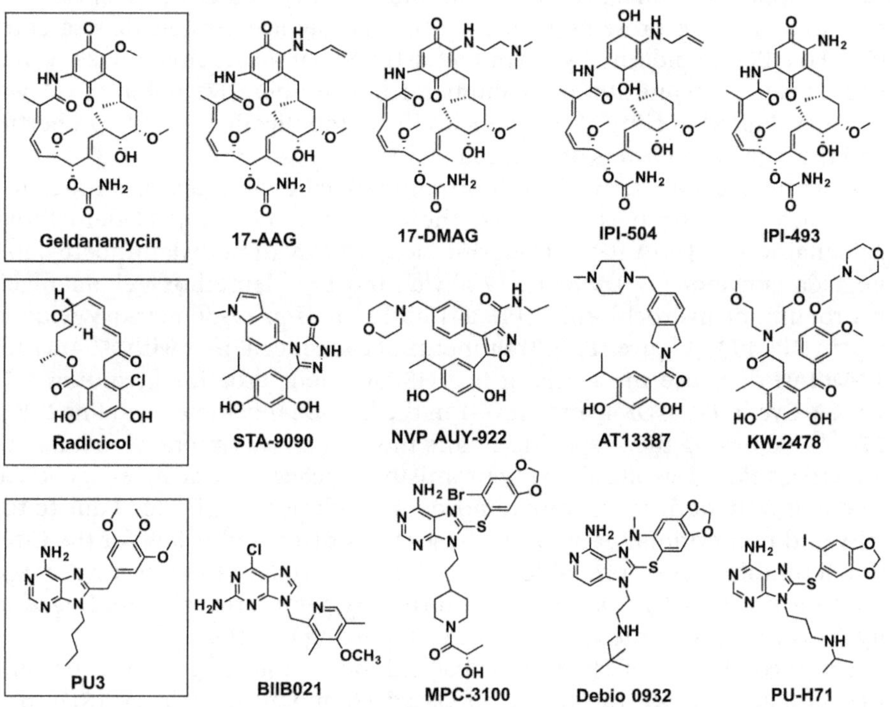

Figure 18.5 Structures of geldanamycin and derivatives, radicicol and derivatives, PU3 and purine-containing analogs.

Another most widely used HSP90 inhibitor, STA9090, is currently being clinically tested against more than ten different cancer types, with at least one more clinical study planned (Figure 18.5). However, STA9090 is not a successful "stand alone" drug for treating patients and only demonstrated clinical efficacy when used in combination with other standard cancer treatments such as paclitaxel, transtuzumab, pertuzumab, dexamethasone, sirolimus, fulvestrant, docetaxel, and doxorubicin.[70] Indeed, combinations of STA9090 with paclitaxel or docetaxel were found to synergistically enhance anti-tumor efficacies against triple-negative breast cancer (TNBC), NSCLC, and ovarian cancer models.[71–73] This combination has demonstrated promising clinical activity without the ocular toxicity observed when treating patients with 17-*N*,*N*-dimethylaminoethylaminogeldanamycin (17-DMAG) and AUY treatments. Another four HSP90 N-terminal inhibitors currently in clinical trials are SNX5422 and PU-H71, which are each used as a single agent, and XL888 and AT13387 which are used as single agents and combination with other drugs in cancer treatments. The patient drug responses with this class of drugs are still under investigation.[64,74–76]

Geldanamycin (GA) is a benzoquinone ansamycin antibiotic that was first isolated from *Streptomyces hygroscpicus* and was later shown to possess anti-cancer activity. It was proposed that this molecule acted *via* HSP90 inhibition.[77,78] GA was suspected to have inhibited multiple isoforms of mammalian HSP90 (HSP90α, HSP90β, GRP94, and TRAP-1). GA's apparent binding to HSP90 appears to affect the function and stability of steroid receptors and certain signalling proteins including the rapidly accelerated fibrosarcoma (Raf-1), HER-2, epidermal growth factor (EGF), viral sarcoma kinase (v-Src), and CDK4.[79] Despite its low nanomolar antiproliferative potency, GA was never evaluated in clinical trials due to its metabolically unstable properties and the significant levels of hepatotoxicity.[80]

Four ansamycin derivatives have entered clinical trials as anti-cancer therapies over the past 12 years, these include 17-*N*-allyl-17-demethoxygeldanamycin (17-AAG), 17-desmethoxy-17-*N*,*N*-dimethylaminoethylaminogeldanamycin (17-DMAG), 17-allylamino-17- demethoxygeldanamycin hydroquinone hydrochloride (IPI-504), and 17-amino-17-demethoxy geldanamycin (IPI-493) (Figure 18.5). Treatments of cancer cell lines with 17-AAG and 17-DMAG decrease the levels of five HSP90 client proteins including AKT, HER-2, Cyclin D1, CDK4, and cRAF-1 in the human melanoma cell line MEXF 276.[81] Decreasing these specific clients induces pro-apoptotic signaling and growth arrest.[81] Despite showing promising efficacy as anticancer agents, all geldanamycin derivatives were removed from clinical use by 2013 due to toxicity and non-optimal patient responses.[80] As discussed below for the C-terminal inhibitors, one probable reason for this is that these derivatives target multiple proteins. The nanomolar cytotoxicity, yet micromolar binding affinity for HSP90 is one piece of evidence supporting this theory.

The second class of classical HSP90 inhibitors that target the N-terminal ATP binding site are the resorcinol-based compounds: radicicol (RD), STA-9090, AUY, AT-13387, and KW-2478 (Figure 18.5). STA-9090, a resorcinol-based HSP90 inhibitor that is currently in clinical trials as a combination therapy against several cancer types.[82] *In vitro*, STA-9090 demonstrated 20-fold greater potency than 17-AAG, in the low nanomolar range against a panel of 57 transformed cell lines of both hematological and solid tumors, and has shown potent antitumor efficacy against human xenograft models.[83] Its anticancer activity is similar to that of AUY.[84] STA-9090 binds to the ATP-binding pocket at the N-terminus of HSP90 and induces the degradation of multiple oncogenic HSP90 client proteins that bind to the middle segment of HSP90, including HER-2, AKT, c-Kit, insulin-like growth factor-I receptor (IGF-1R), c-MET, and HIF-1∝.[82] STA-9090 demonstrates potent *in vitro* and *in vivo* activity in a range of solid and hematological tumor cells, which are dependent on Janus-associated kinase 2 (JAK2) activity for growth and survival. STA-9090 treatment results in sustained depletion of Janus-associated kinase 2 (JAK2) including the constitutively active $JAK2^{V617F}$ mutant, with subsequent loss of signal transducer and activator of transcription (STAT) activity and reduced STAT-target gene expression. It was reported that potent antitumor efficacy in the mouse models of xenograft JAK–STAT-driven leukemia was achieved due to its ability to interfere with cell growth and cell division signaling.[85]

The third class of classical Hsp90 inhibitors used in clinical trials includes purine-based compounds: PU3, CNF 2024/BIIB021, MPC-3100, Debio 0932/CUDC-305, and PU-H71 (Figure 18.5). PU3 was the initial lead purine scaffold HSP90 inhibitor, which exhibited low micromolar concentrations (≤50 μM) and anti-proliferative activity against genetically distinct breast cancer cells (*i.e.* MCF-7, ER+; SKBr3, HER2+; MDA-MB-468, ER− and HER2−). PU3 also interferes with the assembly of the mature progesterone complex in MCF-7 and SKBr3 breast cancer cells.[86] Other purine-scaffold HSP90 inhibitors such as BIIB021 and Debio0932 demonstrated low nanomolar cytotoxic properties against a broad range of cancer cell lines.[87,88] Most purine-based derivatives behave similarly to the geldanamycin class of compounds by inhibiting HSP90 client proteins from binding and inducing their degradation at 10–30 μM,[89] with the exception of BIIB021, Debio 0932, and PU-H71, which bind to HSP90 at ∼2–100 nM[87,88,90] All of these clients bind to HSP90's middle domain and are affected by molecules that target the N-terminal domain of HSP90, thereby making these allosteric inhibitors.

Classical HSP90 inhibitors are now primarily used in combination with other therapies because they have failed as single agents. There are a few possible explanations for the clinical failures of these molecules. First, they are highly toxic, and many produced liver toxicity in patients. Second, classical inhibitors all induce a large increase in heat shock proteins HSF1, HSP70, and HSP27. This response has been termed a heat shock response. However, a recent study showed that these classic inhibitors produced a phenotype that was distinct from a heat shock response.[91] Thus, although these molecules induce high levels of HSF1, HSP70, and HSP27, they also affect other cellular proteins and mRNA levels that are not observed when cells respond to high temperatures.[91] Third, comparing the growth inhibition values, (mid to low nM concentration giving 50% growth inhibition (GI_{50}) values in most cell lines), to the compound's binding affinity for HSP90 (typically low micromolar) indicates that inconsistencies exist within the reported mechanism of action, and that these compounds may target multiple proteins within the cell.[67]

Supporting evidence that the classical inhibitors may also bind to targets other than HSP90 includes the work of Villa *et al.*, who reported that GA and 17-AAG target telomerase.[92,93] In a widely cited paper explaining why 17-AAG had such a high affinity for HSP90 in cancer cells, but a low binding affinity for pure HSP90 or HSP90 in normal cells,[94] it was suggested that 17-AAG only bound to the active HSP90 conformer. However, there is still no conclusive evidence in support of these data and no follow-up data have been reported on this work. Indeed, contrary to this report, Petrucelli and co-workers showed that knocking down the HSP90α gene using small interfering RNA (siRNA) in a cell line led to a decrease in HSP70 protein levels, which is in contrast with the phenomenon observed when cells are treated with classic inhibitors.[95] Specifically, treating cells with 17-AAG produces a 4–6 fold increase in HSP70,[91,96,97] whereas the knockdown phenotype showed a decrease of approximately twofold in HSP70 protein levels. Together the data indicate that these classical inhibitors may not act primarily *via* HSP90, despite being considered one of the most well studied and accepted HSP90 inhibitors.

The purine and the radicicol molecules have also been extensively studied, and unlike the geldanamycin analogs, there is significant evidence that they bind to the HSP90 protein at nM ranges from isothermal calorimetry (ITC), crystallization techniques, and surface plasmon resonance (SPR). Indeed, the results of X-ray crystallography studies demonstrated that radicicol has a high binding affinity for HSP90 (dissociation constant (K_d) = 19 nM).[98] However, the affinity of PU3 for HSP90, which was measured by an immobilised geldanamycin displacement assay (*e.g.*, a binding assay) was found to be 15–20 μM.[99] Thus, the discrepancy in these data has not been resolved. Indeed, the purine and radicicol molecules also produce high levels of HSP70, HSF1, and HSP27 in cells, which is also in contrast to the HSP90 knockdown cell line.[95] Thus, it is likely that these two classes also act *via* an alternative mechanism in addition to inhibiting HSP90.

Another issue that needs to be considered when evaluating these compounds is that none of the binding techniques were used when other proteins were present. The binding affinities are all measured when the only protein available to the compound is HSP90, which may account for the high binding affinity in biochemical assays but the poor results in clinical trials. That is, it is possible that the binding affinity does not accurately represent what is going on in the cells. Indeed, there is no data showing that these compounds selectively target HSP90 in cell lysate.

18.4.2 C-Terminal Binders

There are a few classes of compounds reported to bind to HSP90's C-terminus. These are the coumarin-based antibiotics, which novobiocin (NB) (concentration giving 50% of maximum inhibition (IC_{50}) ~ 700 μM in SKBr3 cells) (Figure 18.6) is the most well-known HSP90 C-terminal inhibitor. Ratajczak and coworkers have shown that NB blocks the TPR immunophilins FKBP52, FKBP51, and CYP40 from binding to the C-terminus of HSP90. Through an allosteric interaction, NB also inhibits the binding of Cdc37 to the N-terminal domain. Blocking Cdc37 from binding to HSP90 inhibits HSP90's ability to fold proteins or mediate the maturation and stabilization of protein kinase clients.[100] NB also induces the degradation of HSP90 client proteins, including HER-2, v-Src, Raf-1, and mutated p53.[101]

KU-174 (Figure 18.6) has been identified as one of the most interesting novobiocin analogues. KU-174 exhibited a broad range of anti-proliferative activities against several cancer cell lines. It was cytostatic at the single dose of 10 μM in PC3 and DU145 prostate cancer cell lines, and has an IC_{50} of 128 nM against the androgen-dependent prostate cancer cell line, LNCap-LN3. As with NB, KU-174 binds to the C-terminus of HSP90 and induces degradation of many key HSP90 client proteins that bind to HSP90's M domain, including survivin, nestin, caspase-3, AKT, HER-2, and CXCR4. The K_d of KU-174 for HSP90 when analyzed by SPR is 94 μM, which is 12-fold more effective than that NB (K_d = 1.1 mM).[102,103] More importantly, this molecule does not induce a heat shock response.[102,103] However, evaluation of KU-174 in mice models was prevented because this compound was rapidly metabolized and cleared, as observed when pilot pharmacokinetic studies were performed.[102]

Another class of C-terminal HSP90 inhibitors related to the natural product silybin has only been characterized very recently (Figure 18.7). Silybin is a major component of the flavonolignan extract derived from the seed of milk thistle and has mild anticancer activity with a GI_{50} ~ 200 μM against various cancer cells.[104] Treatment with silybin induced cell-cycle arrest and caused the depletion of HSP90-dependent client kinases, such as cyclin-dependent kinase 2 (CDK2), cyclin-dependent kinase 4 (CDK4), cyclin E, and cyclin D1 proteins in colon cancer cells.[105] CDK2 and CDK4 are among the most prominent clients of HSP90–Cdc37 complexes that bind to the middle

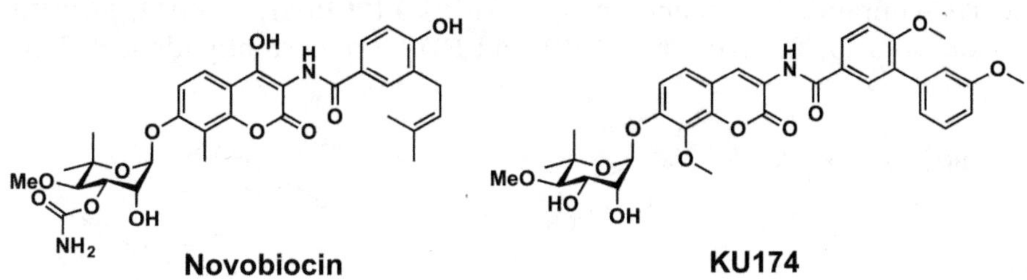

Figure 18.6 Structures of Novobiocin (NB) and KU-174.

domain of HSP90,[106] and CDKs-associated activities are associated with the regulation of the cell cycle and with tumor progression.[105] Silybin also induced degradation of other HSP90-dependent client proteins that bind to the middle domain of HSP90 including HER-2, Raf-1, and AKT without affecting the expression levels of HSP70 or HSP90.[107] Silybin is currently in a phase II clinical trial in combination with the epidermal growth factor receptor (EGFR) inhibitor, erlotinib (Figure 18.7), for the treatment of lung adenocarcinoma.Silybin also appears to potentially bind to the C-terminal domain of HSP90, although these results are not confirmed, and in doing so it releases mature glucocorticoid receptors from the HSP90 complex.[108]

18.4.3 C-Terminal Modulators

There are several significant issues associated with the first two types of inhibitors described above. The classical inhibitors, which target the N-terminal domain, induce high levels of HSF1, HSP70, and HSP27, which produce a cytoprotective cellular response. Accumulation of HSP70 and HSP27 rescues cancer cells because they stimulate proteins that cause drug resistance and metastasis.[109] These classical inhibitors also do not appear to selectively bind to HSP90 in cells, but rather appear to affect a number of proteins within the cell.

C-terminal HSP90 inhibitors have emerged as a promising alternative strategy for modulating HSP90 activity. These C-terminal inhibitors overcome the limitations associated with classical HSP90 inhibitors, specifically they do not induce high levels of HSF1, HSP70, and HSP27, and they selectively target HSP90 in cells.[67,91,97,110–113] This class of inhibitors can be broadly divided into two main categories: (1) C-terminal disruptors of HSP90–co-chaperones interactions and (2) direct C-terminal HSP90 inhibitors.

The first series, C-terminal disruptors, termed the "SM" compounds, was reported by McAlpine and co-workers.[67,91,97,110–113] These molecules allosterically modulate the C-terminus of HSP90 by binding to HSP90's N-middle domain and affecting binding with numerous C-terminal TPR-containing co-chaperones: FKBP51, FKBP52, FKBP38, CYP40, UNC-45, TOM70, PP5, CHIP, and HOP.

The C-terminus of HSP90 contains a MEEVD that selectively interacts with the TPR-containing co-chaperones, of which 9 (FKBP51, FKBP52, FKBP38, CYP40, UNC-45, TOM70, PP5, CHIP, and HOP) are currently identified. All

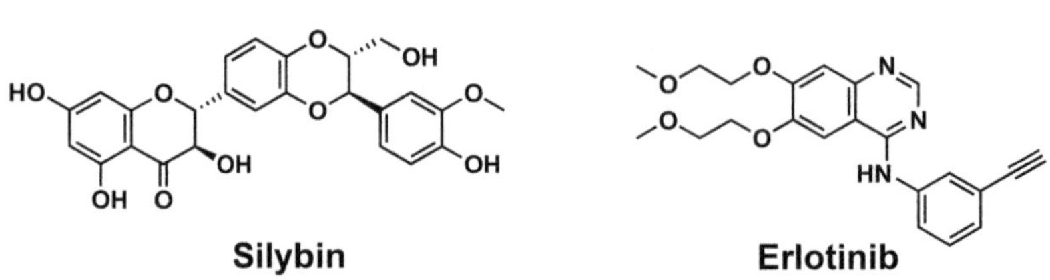

Figure 18.7 Structures of Silybin and Erlotinib.

three PPIases or immunophilins, (CYP40, FKBP51, and FKBP52), regulate cell growth by controlling hormone receptors.[114] Hormone receptors are important in cellular development, primarily at the gene transcription level, and over-expression of these receptors is associated with cancer development and progression.[115] Since the interaction of HSP90 with the immunophilins regulates hormone receptor function and maturation, disrupting the access of immunophilin's TPR domain to HSP90's MEEVD region affects hormone receptor protein levels.

Another TPR-containing protein that binds to HSP90's MEEVD region is FKBP38. This protein binds to HSP90 and induces a conformational change in HSP90, which inhibits Bcl-2, a pro-apoptotic protein, from binding to- and being stabilized by HSP90. Since Bcl-2 is no longer stabilized by HSP90 it is degraded, which produces an anti-apoptotic effect within the cell.[58] UNC-45 is another TPR-containing co-chaperone that binds to the C-terminus of HSP90. This protein facilitates the assembly and proper folding of myosin, which plays an important role in cell division and motility and thus affects cancer cell proliferation.[116,117] CHIP is an E3 ubiquitin ligase that also contains a TPR domain and binds to both HSP90 and HSP70. CHIP promotes the release of properly folded client proteins and also targets improperly folded proteins for degradation.[41] TOM70 functions as a chaperone-docking receptor for the import of pre-proteins to HSP90.[28,40] PP5 is a protein phosphatase that plays a key role in the regulation of both hormone- and stress-induced signaling networks that allow a cell to respond appropriately to genomic stress when it is activated by HSP90.[118] Importantly, the TPR-containing HOP is a key co-chaperone that regulates HSP90's function by bringing HSP70 and HSP90 together in a common complex.[119]

Making sure not to fall into the trap seen with the classical inhibitors, pulldown assays involving cell lysates and subsequently functional assays allowed McAlpine and co-workers to identify a cyclic peptide that bound to HSP90 (SM1) (Figure 18.8).[112] Pulldown assays using the N, middle, N–middle, middle–C, and C-terminal domains of HSP90 and SM1 tagged with biotin indicated that SM1 selectively bound to the N–middle domain of Hsp90. Results of testing SM1's ability to inhibit clients and co-chaperones that bind to Hsp90, showed that SM1 allosterically disrupted the binding

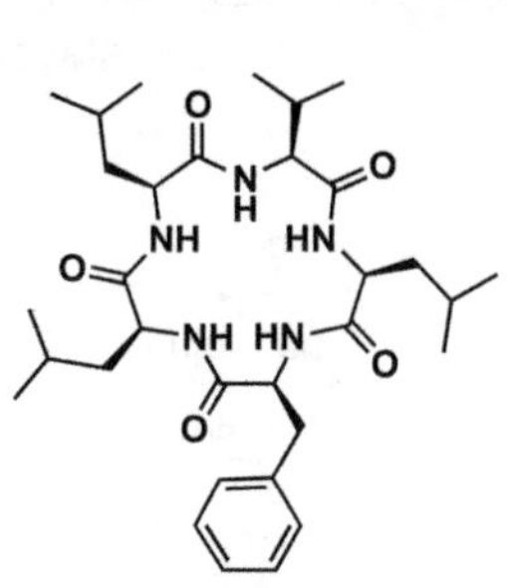

SM1

Figure 18.8 Structures of SM1.

between HSP90 and the C-terminal-binding client protein inositol hexakisphosphate kinase-2 (IP6K2).[112] SM1 also disrupted the C-terminal-binding co-chaperone FKBP52, but despite binding to the N–middle domain had no effect on HER-2's binding to HSP90.[112] Thus, this allosteric HSP90 inhibitor was affecting access to HSP90's C-terminus, but not the N-terminus function.

An extensive structure–activity relationship (SAR) study of SM1 produced two new structures: SM122 and SM145 (Figure 18.9). Both molecules showed ~8 μM GI50 against the colon cancer cell line HCT116. Evaluating the binding of SM122 and SM145 using cell lysates showed that both the compounds selectively pulled down HSP90 as the major target.[113,120] Subsequent pull-down assays with HSP90 domains proved that, like SM1, both molecules bound to the N-middle domain of HSP90.[113,120] Evaluating the effects of these compounds when HSP90 was knocked down *via* siRNA showed that the molecules were more effective when HSP90 was knocked down and less effective when HSP90 expression was increased.[113,120] Finally, assays testing these SM molecules for their ability to inhibit binding between HSP90 and all nine TPR-containing proteins (FKBP38, FKBP51, FKBP52, CYP40, TOM70, HOP, UNC-45, PP5, and CHIP) showed that both molecules effectively inhibited binding between all nine TPR-containing proteins and HSP90 at 500 nM to 1 μM concentrations.[91,110,113,121] Testing these compounds for their effects on HSF-1, HSP70, and HSP27 showed that both SM122 and SM145 decreased the levels of these proteins within the cell, making them consistent with the phenotype observed when the HSP90 alpha gene is knocked down.[91] Thus, these SM molecules target HSP90 in cell-based systems (according to pulldown data, siRNA, and over-expression studies), they are slightly more

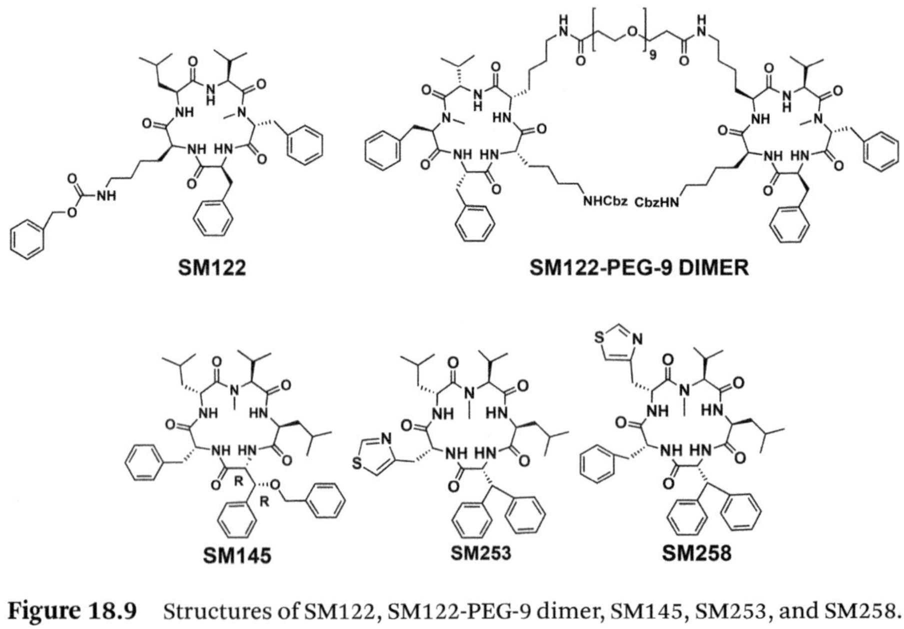

Figure 18.9 Structures of SM122, SM122-PEG-9 dimer, SM145, SM253, and SM258.

effective in biochemical binding assays than in cell-based assays, and they have a phenotype that matches the results of the knockdown studies.

Since SM145 effectively inhibited the binding events between all nine TPR-containing proteins and HSP90, it was anticipated that it would block HSP90 activity in cell lysate. Evaluating SM145's ability to inhibit protein-folding events regulated by HSP90 showed that SM145 was more effective at blocking HSP90's ability to fold the luciferase protein than 17-AAG.[110] SM145 was also extensively studied for its ability to affect hormone receptors.[110] Treating HeLa cancer cells with SM145 produced a dramatic decrease in the glucocorticoid receptor where receptor levels dropped to 10% compared with controls. Cells treated with SM145 also had decreases of FKBP52 and FKBP51 (60% and 20% respectively), indicating the degradation of these co-chaperones upon HSP90 inhibition.[110]

Dimerization of SM122 with a polyethylene glycol (9-PEG units) chain (Figure 18.9) blocked binding between HOP or FKBP52 and HSP90.[121] The dimerized molecule was more than twice as effective as two SM122 monomers at disrupting the binding interaction between HSP90 and TPR-containing proteins. Thus, SM122–PEG-9 dimer was not only an allosteric modulator of HSP90 but dimerizing the monomer had a synergistic effect. The synergistic effect on HSP90 was also seen in protein-folding assays, which involved monitoring HSP90's ability to refold luciferase in RRL lysates.[121] The monomer SM122 inhibited protein folding by 27% at 1 μM, whereas the SM122–PEG-9 dimer inhibited it by 47% at 0.5 μM. An increase of the compound concentrations to 5 μM SM122 and 2.5 μM SM122–PEG-9 dimer produced at 57% and 66% inhibition of protein folding. Thus, allosteric inhibitors can be dimerized in order to improve their biological effects.

New HSP90 inhibitors, SM253 and SM258, were reported in 2015 (Figure 18.9). These compounds share a structural similarity to SM145, and pull-down assays with these molecules using cell lysates from three different types of cancer cells (HCT116, colon; MiaPaca, pancreatic; and Hela, cervical) show that they also bind to selectively to HSP90.[111,119,122] Binding assays between TPR-containing co-chaperones and HSP90 show that both molecules block access to the C-terminal MEEVD region.[111,119,122] Both compounds are equally effective at inducing apoptosis and killing three different types of cancer cells (HCT116, MiaPaca, and Hela).[111,119,122] As observed with the other SM molecules, protein expression in HCT116 cells following treatment with both of these molecules reduced the expression level of HSF1, HSP70, and HSP27 in a dose-dependent manner. Indeed, treating HCT116 cells with SM258 completely depleted HSF1 levels and significantly (75%) decreased HSP70 and HSP27 levels. These data are in stark contrast with the corresponding protein levels when cells are treated with classical inhibitors 17-AAG or AUY922, yet they are consistent with the phenotype knockdown data.[91,111]

Allosteric modulation creates a challenging structure–activity relationship. This complex relationship is revealed in the SM series. Small modifications of the structure of SM253 and SM258 dramatically altered the biological activity and affected the function of HSP90 co-chaperones in cancer cells.[111,122,123] The SM series bind to the junction of HSP90's N–middle domain, which is a flexible region, often referred to as the charged region (residues 217–275 in human HSP90). However, it is this junction that allows the N-domain to rotate, change conformation and provide the chaperones' access to the C-terminal domain.[124] Placing the *N*-methyl moiety on an alternative residue instead of the L-valine residue in SM253 or SM258, markedly affects the biological activity, producing molecules with no biological activity. Removing the *N*-methyl in the case of SM258 creates a molecule with some activity, although less than that of SM258. Together, these data show that the binding pocket within the HSP90 N–middle domain is highly sensitive to subtle structural modifications of SM253 and SM258 despite being a flexible linker region.[111,123]

The unpredictable nature of producing active structures has made improving the binding affinity of this series highly challenging. NMR and mass spectrometry protein studies are ongoing to define the specific residues involved in the binding events, but the allosteric nature of these molecules' mechanism suggests that regulating HSP90's C-terminal exposure through a molecule that binds to the N–middle domain will remain a challenge. As such a series of inhibitors that target the C-terminal MEEVD region directly are being investigated, and these prove to have relatively simple structure–activity relationships.[125]

18.5 Concluding Remarks

Three classes of allosteric inhibitors have been described in this chapter. The first were molecules targeting the ATP binding site at the N-terminus of HSP90 (classical inhibitors), while influencing clients that bind to the middle domain. These molecules are the only ones that have reached clinical trials, but they had significant problems including induction of pro-survival responses and dose-limiting toxicities. Thus, these classical inhibitors are being used as part of dual inhibition regiments in current clinical trials. Data was presented in this chapter to suggest that the poor results are also due to the classical inhibitor's poor selectivity for HSP90 in a cellular environment. Indeed, knocking down HSP90 in cells produces a phenotype that is distinct from that reported with these classical inhibitors.

The second class of molecules described are those that bind to the C-terminus (*e.g.*, Novobiocin and KU-174), which have anti-cancer cellular activity without inducing high levels of HSF-1, HSP70, or HSP27. Although these compounds affected several clients that bound to the middle domain, efforts focused on improving the potency and pharmacokinetics of these molecules for *in vivo* efficacy. Silybin, also reported as a C-terminal HSP90 inhibitor, is currently in a combination study in Phase II clinical trials for the treatment of lung adenocarcinoma. However, strong evidence that silybin is definitively targeting HSP90 in a cellular environment has not been demonstrated. Indeed, the GI_{50} of silybin is 200 μM, which makes it an unusually poor molecule for clinical trials.

The third class of compounds described are the SM series; these molecules bind between the N- and middle domains of HSP90 but affect the C-terminus. The SM molecules bind to the flexible charged linker region and stop the N-terminus from rotating, thereby blocking access to the C-terminal MEEVD region. By inhibiting access to the MEEVD region, the SM series blocks all co-chaperones with a TRP domain from binding to HSP90. Downstream effects attributed to blocking these co-chaperones produce a reduction of immunophilins FKBP52, and FKBP51, and hormone receptor levels. These SM molecules are also the first to be proven to target HSP90 in the cell. Specifically, as determined using pulldown assays with tagged variants of the molecules in multiple cell lysates, the compounds selectively pull out HSP90. Evaluating their efficacy in cells treated with siRNA for HSP90 we see that the compounds are more effective when HSP90 is knocked down and less effective when HSP90 is over-expressed. Finally, cells treated with the SM compounds match the phenotype produced when the HSP90 alpha gene is knocked down. Thus, it is possible that there has not yet been a clinical evaluation of HSP90 inhibitors. However, as with all allosteric modulators, there are significant challenges associated with predicting active structures, and as such new direct C-terminal modulators are currently under evaluation.

References

1. C. M. Dobson, A. Šali and M. Karplus, *Angew. Chem., Int. Ed. Engl.*, 1998, **37**, 868–893.
2. A. I. Bartlett and S. E. Radford, *Nat. Struct. Mol. Biol.*, 2009, **16**, 582–588.
3. F. U. Hartl, A. Bracher and M. Hayer-Hartl, *Nature*, 2011, **475**, 324–332.
4. E. T. Powers, R. I. Morimoto, A. Dillin, J. W. Kelly and W. E. Balch, *Annu. Rev. Biochem.*, 2009, **78**, 959–991.
5. E. Schmitt, M. Gehrmann, M. Brunet, G. Multhoff and C. Garrido, *J. Leukocyte Biol.*, 2007, **81**, 15–27.
6. J. Anckar and L. Sistonen, *Annu. Rev. Biochem.*, 2011, **80**, 1089–1115.
7. S. O. Halacli, B. Halacli and K. Altundag, *Med. Oncol.*, 2013, **30**, 1–5.
8. G. Chiosis and L. Neckers, *ACS Chem. Biol.*, 2006, **1**, 279–284.
9. S. D. Westerheide and R. I. Morimoto, *J. Biol. Chem.*, 2005, **280**, 33097–33100.
10. F. Ritossa, *Cell Stress Chaperones*, 1996, **1**, 97.
11. M. Taipale, I. Krykbaeva, M. Koeva, C. Kayatekin, K. D. Westover, G. I. Karras and S. Lindquist, *Cell*, 2012, **150**, 987–1001.
12. J. L. Johnson and C. Brown, *Cell Stress Chaperon*, 2009, **14**, 83–94.
13. A. Zuehlke and J. L. Johnson, *Biopolymers*, 2010, **93**, 211–217.
14. K. A. Krukenberg, T. O. Street, L. A. Lavery and D. A. Agard, *Q. Rev. Biophys.*, 2011, **44**, 229–255.
15. D. Hanahan and R. A. Weinberg, *Cell*, 2000, **100**, 57–70.
16. D. Hanahan and R. A. Weinberg, *Cell*, 2011, **144**, 646–674.
17. E. Vartholomaiou, P. C. Echeverría and D. Picard, Chapter One–Unusual Suspects in the Twilight Zone Between the Hsp90 Interactome and

Carcinogenesis, in *Advances in Cancer Research*, ed. I. Jennifer and W. Luke, Academic Press, 2016, vol. 129, pp. 1–30.
18. B. S. Blagg and T. D. Kerr, *Med. Res. Rev.*, 2006, **26**, 310–338.
19. A. Maloney and P. Workman, *Expert Opin. Biol. Ther.*, 2002, **2**, 3–24.
20. L. Pearl, C. Prodromou and P. Workman, *Biochem. J.*, 2008, **410**, 439–453.
21. W. A. Houry, *The Molecular Chaperones Interaction Networks in Protein Folding and Degradation*, Springer, 2014, vol. 1.
22. J. Li, J. Soroka and J. Buchner, *Biochim. Biophys. Acta, Mol. Cell Res.*, 2012, **1823**, 624–635.
23. F. Pirkl and J. Buchner, *J. Mol. Biol.*, 2001, **308**, 795–806.
24. K. Dolinski, S. Muir, M. Cardenas and J. Heitman, *Proc. Natl Acad. Sci. U. S. A.*, 1997, **94**, 13093–13098.
25. P. Connell, C. A. Ballinger, J. Jiang, Y. Wu, L. J. Thompson, J. Hohfeld and C. Patterson, *Nat. Cell Biol.*, 2001, **3**, 93–96.
26. A. M. Silverstein, M. D. Galigniana, M.-S. Chen, J. K. Owens-Grillo, M. Chinkers and W. B. Pratt, *J. Biol. Chem.*, 1997, **272**, 16224–16230.
27. T. Jascur, H. Brickner, I. Salles-Passador, V. Barbier, A. El Khissiin, B. Smith, R. Fotedar and A. Fotedar, *Mol. Cell*, 2005, **17**, 237–249.
28. J. C. Young, N. J. Hoogenraad and F. U. Hartl, *Cell*, 2003, **112**, 41–50.
29. L. Gazda, W. Pokrzywa, D. Hellerschmied, T. Löwe, I. Forné, F. Mueller-Planitz, T. Hoppe, and T. Clausen, *Cell*, 2013, **152**, 183–195.
30. D. F. Smith, W. P. Sullivan, T. N. Marion, K. Zaitsu, B. Madden, D. J. McCormick and D. Toft, *Mol. Cell. Biol.*, 1993, **13**, 869–876.
31. S. Chen and D. F. Smith, *J. Biol. Chem.*, 1998, **273**, 35194–35200.
32. C. Scheufler, A. Brinker, G. Bourenkov, S. Pegoraro, L. Moroder, H. Bartunik, F. U. Hartl and I. Moarefi, *Cell*, 2000, **101**, 199–210.
33. S. Daniel, C. Söti, P. Csermely, G. Bradley and G. L. Blatch, Hop: An Hsp70/Hsp90 co-chaperone that functions within and beyond Hsp70/Hsp90 protein folding pathways, *Networking of chaperones by co-chaperones*, Springer, 2007, pp. 26–37.
34. J. Li, K. Richter and J. Buchner, *Nat. Struct. Mol. Biol.*, 2011, **18**, 61–66.
35. K. Hoffmann and R. E. Handschumacher, *Biochem. J.*, 1995, **307**, 5–8.
36. S. Chen, W. P. Sullivan, D. O. Toft and D. F. Smith, *Cell Stress Chaperon.*, 1998, **3**, 118.
37. B. D. Johnson, R. J. Schumacher, E. D. Ross and D. O. Toft, *J. Biol. Chem.*, 1998, **273**, 3679–3686.
38. M.-S. Chen, A. M. Silverstein, W. B. Pratt and M. Chinkers, *J. Biol. Chem.*, 1996, **271**, 32315–32320.
39. R. Conde, J. Xavier, C. McLoughlin, M. Chinkers and N. Ovsenek, *J. Biol. Chem.*, 2005, **280**, 28989–28996.
40. A. C. Fan, M. K. Bhangoo and J. C. Young, *J. Biol. Chem.*, 2006, **281**, 33313–33324.
41. S. Murata, Y. Minami, M. Minami, T. Chiba and K. Tanaka, *EMBO Rep.*, 2001, **2**, 1133–1138.
42. J. Höfeld, Y. Minami and F.-U. Hartl, *Cell*, 1995, **83**, 589–598.

43. C. K. Vaughan, M. Mollapour, J. R. Smith, A. Truman, B. Hu, V. M. Good, B. Panaretou, L. Neckers, P. A. Clarke and P. Workman, *Mol. Cell*, 2008, **31**, 886–895.
44. M. M. Ali, S. M. Roe, C. K. Vaughan, P. Meyer, B. Panaretou, P. W. Piper, C. Prodromou and L. H. Pearl, *Nature*, 2006, **440**, 1013–1017.
45. G. P. Lotz, H. Lin, A. Harst and W. M. Obermann, *J. Biol. Chem.*, 2003, **278**, 17228–17235.
46. N. Nagradova, *Curr. Protein Pept. Sci.*, 2007, **8**, 273–282.
47. G. Fischer, H. Bang and C. Mech, *Biomed. Biochim. Acta*, 1983, **43**, 1101–1111.
48. A. Dixit and G. M. Verkhivker, *PLoS One*, 2012, **7**, 1–19.
49. M. Taipale, D. F. Jarosz and S. Lindquist, *Nat. Rev. Mol. Cell Biol.*, 2010, **11**, 515–528.
50. P. Miller, R. C. Schnur, E. Barbacci, M. P. Moyer and J. D. Moyer, *Biochem. Biophys. Res. Commun.*, 1994, **201**, 1313–1319.
51. W. G. An, R. C. Schnur, L. Neckers and M. V. Blagosklonny, *Cancer Chemother. Pharmacol.*, 1997, **40**, 60–64.
52. O. Grbovic, A. Basso, A. Sawai, Q. Ye, P. Friedlander, D. Solit and N. Rosen, *Proc. Natl. Acad. Sci. U. S. A.*, 2006, **103**, 57–62.
53. S. Sato, N. Fujita and T. Tsuruo, *Proc. Natl. Acad. Sci. U. S. A.*, 2000, **97**, 10832–10837.
54. I. Grad and D. Picard, *Mol. Cell. Endocrinol.*, 2007, **275**, 2–12.
55. L. Lamphere, F. Fiore, X. Xu, L. Brizuela, S. Keezer, C. Sardet, G. F. Draetta and J. Gyuris, *Oncogene*, 1997, **14**, 1999–2004.
56. D. Mahony, D. A. Parry and E. Lees, *Oncogene*, 1998, **16**, 603–611.
57. D. Walerych, G. Kudla, M. Gutkowska, B. Wawrzynow, L. Muller, F. W. King, A. Helwak, J. Boros, A. Zylicz and M. Zylicz, *J. Biol. Chem.*, 2004, **279**, 48836–48845.
58. F. Edlich, F. Erdmann, F. Jarczowski, M.-C. Moutty, M. Weiwad and G. Fischer, *J. Biol. Chem.*, 2007, **282**, 15341–15348.
59. D. Picard, *Trends Endocrinol. Metab.*, 2006, **17**, 229–235.
60. S. D. Kuduk, F. F. Zheng, L. Sepp-Lorenzino, N. Rosen and S. J. Danishefsky, *Bioorg. Med. Chem. Lett.*, 1999, **9**, 1233–1238.
61. S. D. Kuduk, C. R. Harris, F. F. Zheng, L. Sepp-Lorenzino, Q. Ouerfelli, N. Rosen and S. J. Danishefsky, *Bioorg. Med. Chem. Lett.*, 2000, **10**, 1303–1306.
62. J. L. Johnson and D. O. Toft, *Mol. Endocrinol.*, 1995, **9**, 670–678.
63. A. Akalin, L. W. Elmore, H. L. Forsythe, B. A. Amaker, E. D. McCollum, P. S. Nelson, J. L. Ware and S. E. Holt, *Cancer Res.*, 2001, **61**, 4791–4796.
64. K. Jhaveri, T. Taldone, S. Modi and G. Chiosis, *Biochim. Biophys. Acta, Mol. Cell Res.*, 2012, **1823**, 742–755.
65. K. Sidera and E. Patsavoudi, *Recent Pat. Anti-Cancer Drug Discovery*, 2014, **9**, 1–20.
66. D. M. Ramsey, R. R. A. Kitson, J. I. Levin, C. J. Moody and S. R. McAlpine, *Macrocycles Drug Discovery*, 2014, **40**, 37.

67. Y. Wang and S. R. McAlpine, *Future Med. Chem.*, 2015, **7**, 87–90.
68. T. Doi, Y. Onozawa, N. Fuse, T. Yoshino, K. Yamazaki, J. Watanabe, M. Akimov, M. Robson, N. Boku and A. Ohtsu, *Cancer Chemother. Pharmacol.*, 2014, **74**, 629–636.
69. M. L. Johnson, A. Y. Helena, E. M. Hart, B. B. Weitner, A. W. Rademaker, J. D. Patel, M. G. Kris and G. J. Riely, *J. Clin. Oncol.*, 2015, **33**, 1666–1673.
70. K. Jhaveri and S. Modi, *OncoTargets Ther.*, 2015, **8**, 1849.
71. D. A. Proia, C. Zhang, M. Sequeira, J.-P. Jimenez, S. He, N. Spector, G. I. Shapiro, S. Tolaney, M. Nagai and J. Acquaviva, *Clin. Cancer Res.*, 2014, **20**, 413–424.
72. D. A. Proia, J. Sang, S. He, D. L. Smith, M. Sequeira, C. Zhang, Y. Liu, S. Ye, D. Zhou and R. K. Blackman, *Invest. New Drugs*, 2012, **30**, 2201–2209.
73. H. Liu, F. Xiao, I. G. Serebriiskii, S. W. O'Brien, M. A. Maglaty, I. Astsaturov, S. Litwin, L. P. Martin, D. A. Proia and E. A. Golemis, *Clin. Cancer Res.*, 2013, **19**, 5053–5067.
74. J. Bussenius, C. M. Blazey, N. Aay, N. K. Anand, A. Arcalas, T. Baik, O. J. Bowles, C. A. Buhr, S. Costanzo and J. K. Curtis, *Bioorg. Med. Chem. Lett.*, 2012, **22**, 5396–5404.
75. A. Rajan, R. J. Kelly, J. B. Trepel, Y. S. Kim, S. V. Alarcon, S. Kummar, M. Gutierrez, S. Crandon, W. M. Zein and L. Jain, *Clin. Cancer Res.*, 2011, **17**, 6831–6839.
76. G. Shapiro, E. Kwak, B. Dezube, D. Lawrence, J. Cleary, S. Lewis, M. Squires, V. Lock, J. Lyons and M. Yule, *Phase I pharmacokinetic and pharmacodynamic study of the heat shock protein 90 inhibitor AT13387 in patients with refractory solid tumors*, 2010, ASCO Annual Meeting, p. 3069.
77. C. DeBoer, P. Meulman, R. Wnuk and D. Peterson, *J. Antibiot.*, 1970, **23**, 442–447.
78. L. Whitesell, E. G. Mimnaugh, B. De Costa, C. E. Myers and L. M. Neckers, *Proc. Natl. Acad. Sci.*, 1994, **91**, 8324–8328.
79. L. Neckers and S. P. Ivy, *Curr. Opin. Oncol.*, 2003, **15**, 419–424.
80. Y. Miyata, H. Nakamoto and L. Neckers, *Curr. Pharm. Des.*, 2013, **19**, 347–365.
81. V. Smith, E. A. Sausville, R. F. Camalier, H.-H. Fiebig and A. M. Burger, *Cancer Chemother. Pharmacol.*, 2005, **56**, 126–137.
82. Y. Wang, J. B. Trepel, L. M. Neckers and G. Giaccone, *Curr. Opin. Invest. Drugs*, 2010, **11**, 1466–1476.
83. W. Ying, Z. Du, L. Sun, K. P. Foley, D. A. Proia, R. K. Blackman, D. Zhou, T. Inoue, N. Tatsuta and J. Sang, *Mol. Cancer Ther.*, 2012, **11**, 475–484.
84. S. A. Eccles, A. Massey, F. I. Raynaud, S. Y. Sharp, G. Box, M. Valenti, L. Patterson, A. de Haven Brandon, S. Gowan and F. Boxall, *Cancer Res.*, 2008, **68**, 2850–2860.
85. D. A. Proia, K. P. Foley, T. Korbut, J. Sang, D. Smith, R. C. Bates, Y. Liu, A. F. Rosenberg, D. Zhou and K. Koya, *PLoS One*, 2011, **6**, e18552.
86. T. Taldone and G. Chiosis, *Curr. Top. Med. Chem.*, 2009, **9**, 1436.

87. K. Lundgren, H. Zhang, J. Brekken, N. Huser, R. E. Powell, N. Timple, D. J. Busch, L. Neely, J. L. Sensintaffar, Y.-c. Yang, A. McKenzie, J. Friedman, R. Scannevin, A. Kamal, K. Hong, S. R. Kasibhatla, M. F. Boehm and F. J. Burrows, *Mol. Cancer Ther.*, 2009, **8**, 921–929.
88. R. Bao, C.-J. Lai, H. Qu, D. Wang, L. Yin, B. Zifcak, R. Atoyan, J. Wang, M. Samson and J. Forrester, *Clin. Cancer Res.*, 2009, **15**, 4046–4057.
89. P. Workman, F. Burrows, L. Neckers and N. Rosen, *Ann. N. Y. Acad. Sci.*, 2007, **1113**, 202–216.
90. A. Rodina, M. Vilenchik, K. Moulick, J. Aguirre, J. Kim, A. Chiang, J. Litz, C. C. Clement, Y. Kang and Y. She, *Nat. Chem. Biol.*, 2007, **3**, 498–507.
91. Y. Wang and S. McAlpine, *Chem. Commun.*, 2015, **51**, 1410–1413.
92. R. Villa, M. Folini, C. Della Porta, A. Valentini, M. Pennati, M. G. Daidone and N. Zaffaroni, *Carcinogenesis*, 2003, **24**, 851–859.
93. I. Hostein, D. Robertson, F. DiStefano, P. Workman and P. A. Clarke, *Cancer Res.*, 2001, **61**, 4003–4009.
94. A. Kamal, L. Thao, J. Sensintaffar, L. Zhang, M. F. Boehm, L. C. Fritz and F. J. Burrows, *Nature*, 2003, **425**, 407–410.
95. Y.-J. Zhang, T. F. Gendron, Y.-F. Xu, L.-W. Ko, S.-H. Yen and L. Petrucelli, *Mol. Neurodegener.*, 2010, **5**, 33.
96. Y. Wang and S. McAlpine, *Org. Biomol. Chem.*, 2015, **13**, 3691–3698.
97. Y. Wang and S. McAlpine, *Org. Biomol. Chem.*, 2015, **13**, 2108–2116.
98. S. M. Roe, C. Prodromou, R. O'Brien, J. E. Ladbury, P. W. Piper and L. H. Pearl, *J. Med. Chem.*, 1999, **42**, 260–266.
99. G. Chiosis, M. N. Timaul, B. Lucas, P. N. Munster, F. F. Zheng, L. Sepp-Lorenzino and N. Rosen, *Chem. Biol.*, 2001, **8**, 289–299.
100. K. Blacklock and G. M. Verkhivker, *PLoS One*, 2014, **9**, 1–21.
101. M. G. Marcu, T. W. Schulte and L. Neckers, *J. Natl. Cancer Inst.*, 2000, **92**, 242–248.
102. J. D. Eskew, T. Sadikot, P. Morales, A. Duren, I. Dunwiddie, M. Swink, X. Zhang, S. Hembruff, A. Donnelly and R. A. Rajewski, *BMC Cancer*, 2011, **11**, 468.
103. B. C. Shawna, A. V. George, A. M. Craig, K. C. Vamsee, S. Kristen, L. M. Robert, C. D. Alison, S. B. Brian and M. H. Jeffrey, *Prostate*, 2010, **70**, 27.
104. R. Gazak, D. Walterova and V. Kren, *Curr. Med. Chem.*, 2007, **14**, 315–338.
105. C. Agarwal, R. P. Singh, S. Dhanalakshmi, A. K. Tyagi, M. Tecklenburg, R. A. Sclafani and R. Agarwal, *Oncogene*, 2003, **22**, 8271–8282.
106. N. Lawless, K. Blacklock, E. Berrigan and G. Verkhivker, *Pharmaceuticals*, 2013, **6**, 1407–1428.
107. H. Zhao, G. E. Brandt, L. Galam, R. L. Matts and B. S. Blagg, *Bioorg. Med. Chem. Lett.*, 2011, **21**, 2659–2664.
108. M. Riebold, C. Kozany, L. Freiburger, M. Sattler, M. Buchfelder, F. Hausch, G. K. Stalla and M. Paez-Pereda, *Nat. Med.*, 2015, **21**, 276–280.
109. C. G. Evans, L. Chang and J. E. Gestwicki, *J. Med. Chem.*, 2010, **53**, 4585–4602.
110. J. R. McConnell, L. A. Alexander and S. R. McAlpine, *Bioorg. Med. Chem. Lett.*, 2014, **24**, 661–666.

111. Y. C. Koay, J. R. McConnell, Y. Wang and S. R. McAlpine, *RSC Adv.*, 2015, **5**, 59003–59013.
112. R. C. Vasko, R. A. Rodriguez, C. N. Cunningham, V. C. Ardi, D. A. Agard and S. R. McAlpine, *ACS Med. Chem. Lett.*, 2010, **1**, 4–8.
113. V. C. Ardi, L. D. Alexander, V. A. Johnson and S. R. McAlpine, *ACS Chem. Biol.*, 2011, **6**, 1357–1366.
114. M. Cox and J. Johnson, The Role of p23, Hop, Immunophilins, and Other Co-chaperones in Regulating Hsp90 Function, in *Molecular Chaperones*, ed. S. K. Calderwood and T. L. Prince, Humana Press, 2011, vol. 787, pp. 45–66.
115. S. K. Calderwood, *Scientifica*, 2013, 2013.
116. J. M. Barral, A. H. Hutagalung, A. Brinker, F. U. Hartl and H. F. Epstein, *Science*, 2002, **295**, 669–671.
117. Q. Yu and S. I. Bernstein, *Curr. Biol.*, 2003, **13**, R525–R527.
118. T. Golden, M. Swingle and R. E. Honkanen, *Cancer Metastasis Rev.*, 2008, **27**, 169–178.
119. P. E. Carrigan, L. A. Sikkink, D. F. Smith and M. Ramirez-Alvarado, *Protein Sci.*, 2006, **15**, 522–532.
120. J. B. Kunicki, M. N. Petersen, L. D. Alexander, V. C. Ardi, J. R. McConnell and S. R. McAlpine, *Bioorg. Med. Chem. Lett.*, 2011, **21**, 4716–4719.
121. H. Wahyudi, Y. Wang and S. R. McAlpine, *Org. Biomol. Chem.*, 2014, **12**, 765–773.
122. Y. C. Koay, J. R. McConnell, Y. Wang, S. J. Kim, L. K. Buckton, F. Mansour and S. R. McAlpine, *ACS Med. Chem. Lett.*, 2014, **5**, 771–776.
123. Y. C. Koay, N. L. Richardson, S. S. Zaiter, J. Kho, S. Y. Nguyen, D. H. Tran, K. W. Lee, L. K. Buckton and S. R. McAlpine, *ChemMedChem*, 2016, 881–892.
124. C. Graf, C.-T. Lee, L. E. Meier-Andrejszki, M. T. Nguyen and M. P. Mayer, *Front. Mol. Biosci.*, 2014, **1**, 1–15.
125. L. K. Buckton, H. Wahyudi and S. R. McAlpine, *Chem. Commun.*, 2016, **52**, 501–504.

Subject Index